T0327360

APPLIED UNIVARIATE, BIVARIATE, AND MULTIVARIATE STATISTICS

APPLIED UNIVARIATE, BIVARIATE, AND MULTIVARIATE STATISTICS: UNDERSTANDING STATISTICS FOR SOCIAL AND NATURAL SCIENTISTS, WITH APPLICATIONS IN SPSS AND R

Second Edition

DANIEL J. DENIS

WILEY

To Kaiser

CONTENTS

13 Principal Components Analysis   423

14 Factor Analysis   449

PREFACE

Technology is not progress. Empathy is. The dogs are watching us.

Now in its **second edition**, this book provides a general introduction and overview of univariate through to multivariate statistical modeling techniques typically used in the social, behavioral, and related sciences. Students reading this book will come from a variety of fields, including **psychology, sociology, education, political science, biology, medicine, economics, business, forestry, nursing, chemistry, law,** among others. The book should be of interest to anyone who desires a relatively compact and succinct survey and **overview of statistical techniques** useful for analyzing data in these fields, while also wanting to understand and appreciate some of the theory behind these tools. Spanning several statistical methods, the focus of the book is naturally one of **breadth** than of **depth** into any one particular technique, focusing on the unifying principles as well as what substantively (scientifically) can or cannot be concluded from a method when applied to real data. These are topics usually encountered by **upper division undergraduate** or beginning **graduate** students in the aforementioned fields.

The first edition has also been used widely as a reference resource for both students and researchers working on dissertations, manuscripts, and other publications. It is hoped to provide the student with a "big picture" overview of how applied statistical modeling works, while at the same time providing him or her the opportunity in many places to implement, to some extent at least, many of these models using SPSS and/or R software. References and recommendations for further reading are provided throughout the text for readers who wish to pursue these topics further. Each topic and software demonstration can literally be "unpacked" into a deeper discussion, and so long as the reader is aware of this, they will appreciate this book for what it is—**a bird's eye view of applied statistics**, and not the "one and only" source they should refer to when conducting analyses. The book does not pretend to be a complete compendium of each statistical method it discusses, but rather is a survey of each method in hopes of conveying how these methods generally "work," what technical elements unites virtually all of them, and the benefits and limitations of how they may be used in addressing scientific questions.

This second edition has been revised to make the book clearer and more accessible compared to the first edition. The book also contains a gentle introduction ("foot in the door") to a variety of new topics that did not appear in the first edition. All chapters have been edited to varying degrees to improve clarity of prose and in places provide more information or clarification of the concept under discussion. The following is a summary of updates and revisions in the second edition:

- Significant **revision** and **corrections of errata** appearing in the first edition. The second edition is a **stronger and better book** because it has been thoroughly re-read and edited in places where rewording was required. In this sense, the second edition has undergone very much "vetting" since the first edition. At the same time, some sections have been entirely deleted from the first edition due to their explanations being too brief to make them worthwhile. These are sections that did not seem to "work" in the first edition, so they were omitted in the second. This hopefully will help improve the "flow" of the book without the reader stumbling across sections that are insufficiently explained.

- **Bolded text** is used quite liberally to indicate **emphasis** and signal areas that are key for a good understanding of applied statistics. "Accentuate" bold text when reading the book. They are the **key words** and **themes** around which the book was built.

- The **images** in many chapters have been reproduced to make them clearer and more detailed than in the first edition. This is thanks to Wiley's team who has reconstructed many of the figures and diagrams.

- **Chapter** 2 now includes a brief survey of **psychometric validity** and **reliability**, along with a simple demonstration of computing **Cronbach's alpha** in SPSS.

- **Chapter** 3 features a bit more detail and better introduction on the nature of **nonparametric statistics** in the context of the analysis of variance.

- **Chapters** 7 and 8 on regression have been revised and edited in places to include expanded or new discussion, including a demonstration of power analysis using G*Power in addition to R. **Chapter** 8 now includes a more thorough and deeper discussion of model selection, and also features a new section that briefly introduces **ridge** and **lasso regression**, both penalized regression methods.

- **Chapter** 9 on interactions in regression now contains a brief software demonstration of the **analysis of covariance** (ANCOVA), conceptualized as a special case of the wider regression model. Some of the theory of the first edition has been removed as it did not seem to serve its intended goal. For readers who would like to delve into the subject of interactions in regression more deeply, additional sources and recommendations are provided.

- **Chapter** 11 now includes R and SPSS code for obtaining **Hotelling's** T^2. While readers can simply use a MANOVA program to evaluate mean vector differences on two groups, the inclusion of the relevant software code for Hotelling's T^2 is useful to make the MANOVA chapter a bit more complete.

- **Chapter** 14 on exploratory factor analysis now concludes with a brief introduction and overview of the technique of **multidimensional scaling** should readers wish to pursue this topic further. By relating the technique somewhat to previously learned techniques, the reader is encouraged to see the learning of new techniques as extending their current knowledge base. This is due to the book emphasizing **foundations** and **fundamental principles** of applied statistics, rather than a series of topics seemingly unrelated.

- **Chapter** 15 has been expanded slightly to include a basic demonstration of data analysis using **AMOS** software. Many users who perform SEM models use AMOS instead of R, and so it seemed appropriate to include a small sample of AMOS output in the context of building a simple path model. Additional references for learning and using AMOS are also provided for those who wish to venture further into structural equation models.

- The inclusion in select places brief discussions of, and references to, "**Big Data**," as well as **data science** and **machine learning**, and why understanding fundamentals and **classical statistics** is even more important today than ever before in light of these advancements. These fields are heavily computational, but for the most part, have technical origins in fundamental statistics and mathematics. We try our best to key the reader to where these topics "fit" in the wider data analytic landscape, so if they choose to embark on these topics in future study, or further their study of **computer science,** for example, they have a sense of how many of these techniques build on foundational elements.

- Select **chapter exercises** have been edited as to clarify what they are asking, while a few others have been deleted since they did not seem to work well in the first edition of the book. The majority of the exercises remain **conceptually-based** as to encourage a deep and far-reaching understanding of the material. Select data-analytic exercises have been either edited or substituted for better ones.

- Additional **references** and citations have been added to supplement the book which already features many classic references to pioneers in applied statistics.

- An on-line Appendix featuring a review of essential mathematics is available at www.datapsyc.com.

ACKNOWLEDGMENTS

I am indebted to all at Wiley who helped in the production of the book, both directly and indirectly. A sincere thank you to Mindy Okura-Marszycki, Editor at Wiley, who supported the writing of this second edition (the first edition was edited by Steve Quigley and Jon Gurstelle). Thank you as well to all other associates, both professional and unprofessional, who in one way or another influenced my own learning as it concerns statistics and research. Comments, criticism, corrections, and questions about the book are most welcome. Please e-mail your feedback to daniel.denis@umontana.edu or email@datapsyc.com. **Data sets** and **errata** are available at www.datapsyc.com.

DANIEL J. DENIS

ABOUT THE COMPANION WEBSITE

This book is accompanied by a companion website:

www.wiley.com/go/denis/appliedstatistics2e

The website contains appendix and preface of the first edition.

1

PRELIMINARY CONSIDERATIONS

> Still, social science is possible, and needs a strong empirical component. Even statistical technique may prove useful — from time to time.
>
> (Freedman, 1987, As Others See Us: A Case in Path Analysis, p. 125)

Before we delve into the complexities and details that is the field of applied statistics, we first lightly survey some germane philosophical issues that lay at the heart of where statistics fit in the bigger picture of science. Though this book is primarily about applied statistical modeling, the end-goal is to use statistical modeling in the context of scientific exploration and discovery. To have an appreciation for how statistics are used in science, one must first have a sense of some essential foundations so that one can situate where statistics finds itself within the larger frame of scientific investigation.

1.1 THE PHILOSOPHICAL BASES OF KNOWLEDGE: RATIONALISTIC VERSUS EMPIRICIST PURSUITS

All knowledge can be said to be based on fundamental philosophical assumptions, and hence empirical knowledge derived from the sciences is no different. There have, historically, been two means by which knowledge is thought to be attained. The **rationalist** derives knowledge primarily from mental, cognitive pursuits. In this sense, "real objects" are those originating from the mind via **reasoning** and the like, rather than obtained empirically. The **empiricist**, on the other hand, derives knowledge from **experience**, that is, one might crudely say, "objective" reality. To the empiricist, knowledge is in the form of tangible objects in the "real world."

Ideally, science should possess a healthy blend of both perspectives. On the one hand, science should, of course, be grounded in objective objects. The objects one studies should be independent

Applied Univariate, Bivariate, and Multivariate Statistics: Understanding Statistics for Social and Natural Scientists, With Applications in SPSS and R, Second Edition. Daniel J. Denis.
© 2021 John Wiley & Sons, Inc. Published 2021 by John Wiley & Sons, Inc.
Companion Website: www.wiley.com/go/denis/appliedstatistics2e

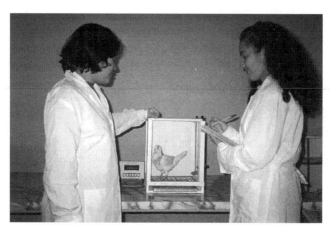

FIGURE 1.1   Observing the behavior of a pigeon in a Skinner box. Source: Dtarazona (1998). https://commons. wikimedia.org/wiki/File:UNMSM_PsiExperimental_1998_2.jpg. Public Domain.

of the psychical realm. A cup of coffee is a cup of coffee regardless of our belief or theory about the existence of the cup. On the other hand, void of **any** rationalist activity, science becomes the study of objects for which we are not allowed to assign meaning. For example, the behavior of a pigeon in a Skinner box[1] (see Figure 1.1) can be documented as to the number of times it presses on the lever for the reward of a food pellet. That the pigeon presses on the lever is empirical reality. **Why** the pigeon presses on the level is theoretical speculation, of which there could be many competing possibilities. Observing data is fine, but without theory, we have very little "guidance" to either explain current observations or predict new ones. B.F. Skinner's theory of **operant conditioning**, being such that the pigeon presses the lever because it is **reinforced** to do so, is a prime example of where a wedding of rationalism and empiricism takes place. The theory attempts to **explain** or **account** for the pigeon's behavior. It is a narrative for why the pigeon does what it does.

Of course, theorizing can go too far, much too far. One must be cautious to not "over-theorize" too emphatically without acknowledging the absence of empirical backing. Is there anything wrong with hypothesizing that cloudy days are associated with depressive moods? No, so long as you are prepared to state what evidence exists that may support or contradict your theory. If no evidence exists, you may still theorize, but you owe it to your audience to admit the lack of current empirical support for your hypothesis.

As an example of "heightened theorizing," recall the missing **Malaysia Airlines Flight 370** where a Being 777 aircraft vanished, apparently without a trace, originally destined from Kuala Lumpur to Beijing in March of 2014. Media were sometimes criticized for proposing numerous theories as to its disappearance, ranging from the plane being flown into a hidden location to it being hijacked or a result of pilot suicide. One theory even speculated that the plane was swallowed by a black hole! Speculation is fine and theorizing is a necessary scientific as well as human activity, so long as one is up front about existent available evidence to support the theory one is advancing. Indeed, one could assign probabilities to competing theories and revise such probabilities as new data become available. This is precisely what **Bayesian** philosophers and statisticians are wont to do. A theory should only be considered **credible** however when empirical reality and the theory coincide (see Figure 1.2). The fit may not be perfect, and seldom if ever is, but when the rational coincides well with the empirical, credibility of the

---

[1] B.F. Skinner was a psychologist known for his theory of operant conditioning within the behaviorist tradition in psychology. One of Skinner's primary investigatory tools was that of observing and recording the conditions that would lead a rat, pigeon, or other animal, to press a lever for a food pellet in a small chamber. This chamber came to be known as the **Skinner box**. For a read of Skinner, see Rutherford (2009) and Fancher and Rutherford (2011).

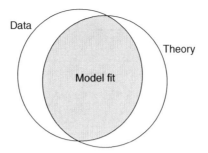

FIGURE 1.2 "Model fit" as an overlap of data with theory.

idea is at least tentatively assured, at least until potentially new evidence debunks it (e.g., the fall of Newtonian physics).

We must also ensure that our theories are not too **convenient of narratives** fit to data. If you have ever witnessed a sporting event where the deciding point occurred by the lucky bounce of a puck in hockey or the breezy push of a tennis ball in midair, only to hear post-match commentators laud the winning team or individual as suddenly so much better than the losing team, then you know what "convenient narratives" are all about. We must be careful not to exaggerate how well our given theory fits data simply because a few data points went "our way." George Box once said that **all models are wrong but some are useful**. In any scientific endeavor, guard against **falling in love with your theory** or otherwise exaggerating it far beyond what the data suggest. Otherwise, it no longer is a legitimate theory, but rather is simply **your brand** and more a product of subjective bias and "career-building" than anything scientific. After 20 years of advocating a theory, is the researcher you are speaking to really prepared to "accept" evidence that contradicts his or her theory? They have a lot of stakes in that theory, their whole career may have been built upon it, are they really willing to accept "defeat" of it? Indeed, one reason I believe why economic predictions, for instance, are often looked upon with suspicion, is because economists, like psychologists (and theoretical physicists, for that matter), are far too quick to advance theories as though they were near facts. "Sexy theories" sound great and may be marketable to uncritical consumers and media (make an outlandish claim on cable, you'll be a hero!), but to good scientists, theories are always only as good as the data that exist to support them. Science is exciting, to be sure, but should not be overly speculative. If you are looking for fireworks, then you are best to choose a field other than science.

1.2 WHAT IS A "MODEL"?

The word "model" is perhaps the most popular word featured in textbooks, tutorials, and lectures having anything to do with the application of quantitative methods. Attempting to define just what **is** a model in statistics can be a bit challenging. We discuss the concept by referring to Everitt's definition:

> A description of the assumed structure of a set of observations that can range from a fairly imprecise verbal account to, more usually, a formalized mathematical expression of the process assumed to have generated the observed data. The purpose of such a description is to aid in understanding the data.
>
> (Everitt, 2002, p. 247)

Models, are, essentially, and perhaps somewhat crudely, **equations**. They are equations fit to data that attempt to account for how the data came about or were **generated** in the first place. For example, if for every hour a student studied for an exam corresponded to exactly a 1-point increase in a student's grade, the model that would best explain how this data was generated would be a **linear model**. Even if

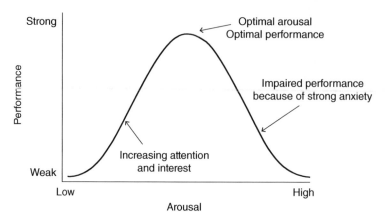

FIGURE 1.3   Hebbian Yerkes–Dodson performance–arousal curve. Source: Diamond et al. (2007). Licensed under CC by 3.0.

the relationship between hours studied and student grade was not perfect, a perfect line might still be the "best" summary. Models are often used to account for messy or imperfect data.

   Another example of a model is the classic Hebbian version of the **Yerkes–Dodson** curve expressing the relationship between performance and arousal, depicted in Figure 1.3.

   The curve is an inverted "U" shape (an approximate parabola) that provides a useful model relating these two attributes (i.e., performance and arousal). If one exhibits very low arousal, performance will be minimal. If one exhibits a very high degree of arousal, performance will likely also suffer. However, if one exhibits a moderate range of arousal, performance will likely be optimal. The model in this case, as in most cases, does not account for **all** the data one might collect. The extent to which it accounts for **most** of the data is the extent to which the model may be, in general, deemed "useful." The use of a model is also enhanced if it can make accurate predictions of future behavior.

   As another example of a model, consider the number of O-ring incidents on NASA's space shuttle (the fleet is officially, and sadly, retired now) as a function of temperature (Figure 1.4). At very low or high temperatures, the number of incidents appears to be elevated. A square function seems to adequately model the relationship. Does it account for all points? No. But nonetheless, it provides a fairly

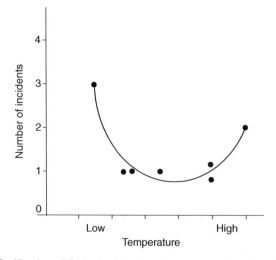

FIGURE 1.4   Number of O-ring incidents on boosters as a function of temperature.

good summary of the available data. Some have argued that had NASA had such a model (i.e., essentially the line joining the points) available before **Challenger** was launched on January 28, 1986, the launch may have been delayed and the shuttle and crew saved from disaster.[2] We feature this data in our chapter on logistic regression.

Why did George Box say that **all models are wrong, some are useful**? The reason is that even if we obtain a perfectly fitting model, there is nothing to say that this is the **only** model that will account for the observed data. Some, such as Fox (1997), even encourage divorcing statistical modeling as accounting for deterministic processes. In discussing the determinants of one's income, for instance, Fox remarks:

> I believe that a statistical model cannot, and is not literally meant to, capture the social process by which incomes are "determined" … No regression model, not even one including a residual, can reproduce this process … The unfortunate tendency to reify statistical models – to forget that they are descriptive summaries, not literal accounts of social processes – can only serve to discredit quantitative data analysis in the social sciences.
>
> (p. 5)

Indeed, psychological theory, for instance, has advanced numerous models of behavior just as biological theory has advanced numerous theories of human functioning. Two or more competing models may each explain observed data quite well. Sometimes, and unfortunately, the model we adopt may have more to do with our sociological (and even political) preferences than anything to do with whether one is more "correct" than the other. Science (and mathematics, for that matter) is a **human** activity, and often theories that are deemed valid or true have much to do with the **spirit of the times** (the so-called **Zeitgeist**) and what the scientific community will actually accept and tolerate as being **true**.[3] Of course, this is not true in all circumstances, but you should be aware of the factors that make theories popular, especially in fields such as social science where "hard evidence" can be difficult to come by. The reason the **experiment** is often considered the "gold standard" for evidence is because it often (but not always) helps us narrow down narratives to a few compelling possibilities. In strictly correlational research, isolating the correct narrative can be exceedingly difficult or nearly impossible, despite which narrative **we wish upon our data** the most. Good science requires a very critical eye. Whether the theory is that of the Big Bang, the determinants of cancer, or theories of bystander intervention, all of these are narratives to help account for observed data.

1.3 SOCIAL SCIENCES VERSUS HARD SCIENCES

There is often stated a distinction between the so-called "soft" sciences and the "hard" sciences (Meehl, 1967). The distinction, as is true in many cases of so many things, is **fuzzy** and **blurry** and requires deeper analysis to fully understand the issue. The difference between what is "soft" and what is "hard" science has usually only to do with the **object** of study, and not with the method of analytical inquiry.

For example, consider what distinguishes the scientist who studies temperature of a human organism compared to a scientist who studies the self-esteem of adolescents. Their analytical approaches, at their core, will be remarkably similar. They will both measure, collect data, and subject that data to curve-fitting or probabilistic analysis (i.e., statistical modeling). Their **objects**, however, are quite different. Indeed, some may even doubt the **measurability** of something called "self-esteem" in the first

---

[2] See Friendly (2000, pp. 208–211) for an analysis of the O-ring data. See Vaughan (1996) for an account of the social, political, and managerial influences at NASA that were also purportedly responsible for the disaster.

[3] The reader is strongly encouraged to consult Kuhn's excellent book **The Structure of Scientific Revolutions** in which an eminent philosopher of science argues for what makes some theories more longstanding than others and why some theories drop out of fashion. So-called **paradigm shifts** are present in virtually all sciences. An awareness of such shifts can help one better put "theories of the day" into their proper context.

place. Is self-esteem **real**? Does it actually exist? At the heart of the distinction, really, is that of **measurement**. Once measurement of an object is agreed upon, the debate between the hard and soft sciences usually vanishes. Both scientists, natural and social, are generally aiming to do the same thing, and that is to understand, document phenomena, and to identify **relations** among these phenomena. As Hays (1994) put it so well, the overreaching goal of science, at its core, is to determine **what goes with what**. Virtually every scientific investigation you read about has this underlying goal but may operationalize and express it in a variety of different ways.

Social science is a courageous attempt. Hard sciences are, in many respects, much easier than the softer social sciences, not necessarily in their subject matter (organic chemistry **is** difficult), but rather in what they attempt to accomplish. Studying beats-per-minute in an organism is relatively easy. It is not that difficult to measure. Studying something called **intelligence** is much, much harder. Why? Because even arriving at a suitable and agreeable operational definition of what constitutes intelligence is difficult. Most more or less agree on what "heart rate" means. Fewer people agree on what intelligence really means, even if everyone can agree that some people have more of the mysterious quality than do others. But the study of an object of science should imply that we can actually measure it. Intelligence, unlike heart rate, is not easily measured largely because it is a **construct** open to much scientific criticism and debate. Even if we acknowledge its existence, it is a difficult thing to "tap into."

Given the difficulty in measuring social constructs, should this then mean the social scientist give up and not study the objects of his or her craft? Of course not. But what it does mean is that she must be extremely **cautious**, **conservative**, and **tentative** regarding conclusions drawn from empirical observations. The social scientist must be up front about the weaknesses of her research and must be very careful not to **overstate** conclusions. For instance, we can measure the extent to which melatonin, a popular sleep aid, reduces the time to sleep onset (i.e., the time it takes to fall asleep). We can perform experimental trials where we give some subjects melatonin and others none and record who falls asleep faster. If we keep getting the same results time and time again across a variety of experimental settings, we begin to draw the conclusion that melatonin has a role in decreasing sleep onset. We may not know **why** this is occurring (maybe we do, but I am pretending for the moment we do not), but we can be reasonably sure the phenomenon exists, that "something" is happening.

Now, contrast the melatonin example to the following question—**Do people of greater intelligence, on average, earn more money than those of lesser intelligence**? We could correlate a measure of intelligence to income, and in this way, we are proceeding in a similar **empirical** (even if not **experimental**, in this case) fashion as would the natural scientist. However, there is a problem. There is a **big** problem. Since few consistently agree on what intelligence **is** or how to actually measure it, or even whether it "exists" in the first place, we are unsure of where to even **begin**. Once we agree on what IQ is, how it is measured, and how we will identify it and name it, the correlation between IQ and income is as reputable and respectable as the correlation between such variables as height and weight. It is getting to the very **measurement** of IQ that is the initial hard, and skeptics would argue, impossible part. But we know this already from experience. Convincing a parent that her son has an elevated heart rate is much easier than convincing her that her son has a deficit in IQ points. One phenomenon is measurable. The other, perhaps so, but not nearly as easily, or at minimum, **agreeably**.

Our point is that once we agree on the existence, meaning, and measurement of objects, soft science is just as "hard" as the hard sciences. If measurement is not on solid ground, no analytical method of its data will save it. All students of the social (and natural, to some extent) sciences should be exposed to in-depth coursework on the theory, philosophy, and importance of **measurement** to their field before advancing to statistical applications on these objects, since it is in the realm of measurement where the true controversies of scientific "reputability" usually lay. For general readable introductions to measurement in psychology and the social sciences, the reader is encouraged to consult Cohen, Swerdlik, and Sturman (2013), Furr and Bacharach (2013), and Raykov and Marcoulides (2011). For a deeper and philosophical treatment, which includes measurement in the physical sciences as well, consult Kyburg (2009). McDonald (1999) also provides a relatively technical treatment.

1.4 IS COMPLEXITY A GOOD DEPICTION OF REALITY?
ARE MULTIVARIATE METHODS USEFUL?

One of the most prominent advances in social statistics is that of **structural equation modeling**. With SEM, as we will survey in Chapter 15, one can model complex networks of variables, both **measurable** and **unmeasurable**. Structural equation modeling is indeed one of the most complex of statistical methods in the toolkit of the social scientist. However, it is a perfectly fair and reasonable question to ask whether structural equation modeling has helped advance the **cause** of social science. Has it increased our knowledge of social phenomena? Advanced as the tool may be **statistically**, has the tool helped social science build a bigger and better house for itself?

Such a question is open to debate, one that we will not have here. What needs to be acknowledged from the outset, however, is that statistical complexity has little, if anything, to do with **scientific complexity** or the guarantee of scientific advance. Indeed, the two may even rarely correlate. A classic scenario is that of the graduate student running an independent-samples t-test on well operationally defined experimental variables, yet feeling somewhat "embarrassed" that he used such a "simple" statistical technique. In the lab next door, another graduate student is using a complex structural equation model, struggling to make the model identifiable through fixing and freeing parameters at will, yet feeling as though she is more "sophisticated" scientifically as a result of her use of a complex statistical methodology. Not the case. True, the SEM user may be more sophisticated **statistically** (i.e., SEM **is** harder to understand and implement than t-tests), but whether her empirical project is advancing our state of knowledge more than the experimental design of the student using a t-test cannot even begin to be evaluated based on the statistical methodology used. It must instead be based on **scientific merit** and the overall strength of the scientific claim. Which scientific contribution is more noteworthy? That is the essential question, not the statistical technique used. The statistics used rarely have anything to do with whether good science versus bad science was performed. **Good science is good science**, which at times may require statistical analysis as a tool for communicating its findings.

In fact, much of the most rigorous science often requires the most simple and elementary of statistical tools. Students of research can often become dismayed and temporarily disillusioned when they learn that complex statistical methodology, aesthetic and pleasurable on its own that it may be (i.e., SEM models can be fun to work with), still does not solve their problems. Research wise, their problems are usually those of **design**, **controls**, **and coming up with good experiments, arguments, and ingenious studies**. Their problems are usually not statistical at all, and in this sense, an overemphasis on statistical complexity could actually delay their progress to conjuring up innovative, groundbreaking **scientific** ideas.

The cold hard facts then are that if you have poor design, weak research ideas, and messy measurement of questionable phenomena, your statistical model will provide you with anticlimactic findings, and will be nothing more than an exercise in the old adage **garbage in, garbage out**. Quantitative modeling, sophisticated as it has become, has not replaced the need for strict, rigorous experimental controls and good experimental design. Quantitative modeling has not made correlational research somehow more "on par" with the gold standard of experimental studies. Even with the advent of **latent variable modeling** strategies and methodologies such as **confirmatory factor analysis** and **structural equation modeling**, statistics does not purport to "discover," for real, hidden variables. Modeling is simply concerned with the partitioning of variability and the estimation of parameters. Beyond that, the remainder of the job of the scientist is to know his or her craft and to design experiments and studies that enlighten and advance our knowledge of a given field. When applied to sound design and thoughtful investigatory practices, statistical modeling does partake in this enlightenment, but it does nothing to save the scientist from his or her poorly planned or executed research design. Statistical modeling, complex and enjoyable as it may be on its own, guarantees nothing.

1.5 CAUSALITY

One might say that the ultimate goal of any science is still to establish **causal relations**, even if classical "Laplacian" determinism has been somewhat jettisoned by theoretical physicists, which would imply that there may actually not be "true causes" to events (despite our continued attempts to assign them). Our search for them may be entirely misguided. Still, and a bit more down to earth, nothing suggests a stronger understanding of a scientific field than to be able to speak of causation about the phenomena it studies. However, more difficult than establishing causation in a given research paradigm is that of understanding what causation **means** in the first place. There exist several definitions of causality. Most definitions have at their core that causation is a relation between two events in which the second event is assumed to be a **consequence**, in some sense, of the first event.

For example, if I slip on a banana peel and fall, we might hypothesize that the banana peel **caused** my fall. However, was it the banana peel that caused my fall, or was it the worn out soles on my shoes that I was wearing that day that caused the fall? Had I been wearing mountain climbers instead of worn-out running shoes, I might not have fallen. Who am I to say the innocent banana peel caused my fall? Causality is **hard**. Even if it seems that **A caused B**, there are usually many variables associated with the problem such that if adjusted or tweaked may threaten the causal claim. Some would say this is simply a trivial philosophical problem of specifying causality and it is "obvious" from the situation that the banana peel caused the fall. Nonetheless, it is clear from even such a simple example that causation is in no way an **easy** conclusion to draw. Perhaps this is also why it is extremely difficult to pinpoint true causes of virtually any behavior, natural or social. Hindsight is 20/20, but attributing causal attributes with any kind of methodological certainty in violent crimes, for instance, usually turns out to be speculative at best. True, we may accumulate evidence for **prediction**, but equating that with causation is under most circumstances the **wish**, not the **reality**, of a social theory.

In our brief discussion here we will not attempt to define causality. Books, dissertations, and treatises have been written exclusively on the topic. At most, what we can do in the amount of space we have is to simply heed the following advice to the reader—**If you are going to speak of causation with regard to your research, be prepared to back up your theory of causation to your audience**. It is simply not enough to say A causes B without subjecting yourself to at least some of the philosophical issues that accompany such a statement. Otherwise, it is strongly advised that you avoid words such as "cause" in hypothesizing or explaining results and findings. **Relations** and **predictions** are much epistemologically "safer" words to use, less prone to critique ending in quicksand. For a brief, but enlightening discussion of causality in the social sciences, see Fox (1997, pp. 3–14). For a more thorough treatment of the subject as it relates to structural equation models, see Mulaik (2009, pp. 63–117). Even a brief study of the **philosophy of science** goes a long way to understanding the complexities involved in using "causal" statements in research. These issues are not nearly as simple as they may at first appear.

1.6 THE NATURE OF MATHEMATICS: MATHEMATICS AS A REPRESENTATION OF CONCEPTS

Ian Stewart (1995) said it best when he wrote that the mathematician is not a juggler of numbers, he is a juggler of **concepts**. The greatest ambivalence to learning statistical modeling experienced by students outside (and even inside, I suppose) the mathematical sciences is that of the presumed mathematical complexity involved in such pursuits. Who wants to learn a mathematically-based subject such as statistics when one has "never been good at math?"

The first step in this pursuit is to critically examine assumptions and prior learned beliefs that have become implicit. One way to help "demystify" mathematics and statistics is to challenge your perception of what mathematics and statistics actually **are** in the first place. It is of great curiosity that so many students claim to dislike mathematics and statistics, yet at the same time cannot verbalize just what mathematics and statistics actually **are**, and then even worse, proceed to engage in real-life activities that utilize very much the same analytical cognitive capacities as would be demanded from doing mathematics and statistics!

More than likely, the "dislike" of these subjects has more to do with the **perceptions** one has learned to associate with these subjects than with an inherent ontological disdain for them. Human beings are creatures of **psychological association**. Any dislike of anything without knowing what that thing is in the first place is almost akin to disliking a restaurant dish you have never tried. You cannot dislike something until you at least know something about it and open your mind to new possibilities of what **it might be** that you are forming opinions **about**. Not to sound overly "Jamesian," (the analogy isn't perfect, but it's close) but perhaps you are afraid of mathematics because of your **fear** of it rather than the mathematics itself. That is, you run, not because of the mathematics, but because of the fear. If you accept that you are yet unsure of what mathematics is, and will not judge it until you are knowledgeable of it, it may delay derogatory opinion about it. **It is only when we assume we know something (to some extent, at least) that we usually feel free to judge and evaluate it**. Keep your perceptions open to revision, and what you may find is that what was disliked yesterday curiously becomes likable today, simply because you have now learned more about what that something actually **is**. But to learn more about it, you need to first drop, or at minimum suspend, previously held beliefs about it. Have an open mind from the outset, and refresh that mindset each time you open a book or attend a lecture in a course.

The first point is that **statistics is not mathematics**. Statistics is a discipline unto itself that **uses** mathematics, the way physics uses mathematics, and the way that virtually all of the natural and social sciences use mathematics. Mathematics is the tool statisticians use to express their statistical ideas, and statistics is the tool that scientists use to help make sense of their research findings. The field of **theoretical** or **mathematical statistics** is heavily steeped in theorem-building and proofs. **Applied statistics**, of the kind featured in this book, is definitely not. Thus, any fear of **real mathematics** can be laid to rest, because you will find no such mathematics in this book. Upon browsing this book, if you are of the opinion that it contains "lots of math," then quite simply, you do not know what "lots of math" looks like. Rest assured, the mathematics in this book is simply used as a vehicle for understanding statistics.

Mathematics and statistics are not things "mysterious" that can only be grasped by those with higher mental faculties. A useful working definition might be that mathematics is a set of well-defined and ever-expanding **rules** and consequences about symbolic abstract objects based on fundamental assumptions called **axioms**. The axioms of mathematics are typically assumed to be true without needing to be proved. Theorems and other results built on such axioms usually require proof. What is a proof? It is an analytical argument for why a proposition should be considered true. Any given proof usually relies on other theorems that have already been proven to be true. Make no mistake, mathematics is a very deep field of intellectual endeavor and activity. However, expecting something to be deeper than it is can also lead you to just as well not understand it. Sometimes, if you are not understanding something, it may very well be that you are looking far beyond what there is to be understood. If you retreat in your expectations slightly of what there is to see, it sometimes begins to make more sense. Thinking "too deep" where such depth is not required or encouraged, is a peril. Many "bright" students have this "gift" of **critical analysis**, and to understand a concept, need to actually retreat somewhat in their depth of inquiry (at least for the moment).

For a general overview of the nature of mathematics, the reader is encouraged to consult Courant, Robbins, and Stewart (1996), and for an excellent introduction to basic mathematical analysis, Labarre (1961). Hamming (1985) is another good introduction to the field of mathematics, as well as Aleksandrov, Kolmogorov, and Lavrent'ev (1999). For more philosophical treatments, the reader should consult Dunham (1994) and Stewart (1995). For an in-depth and very readable history of mathematics, consult Boyer and Merzbach (1991).

1.7   AS A SCIENTIST, HOW MUCH MATHEMATICS DO YOU NEED TO KNOW?

The answer to this question is, of course, **as much as possible**, for working through problems of any kind can only serve to hone your analytical and deductive abilities. Even working through problems completely unrelated to statistics can help your statistical abilities, because they give you practice in "figuring things out." However, that answer is, of course, a naïve if not idealistic one, since there is only so much time available for study and the study of statistics must be balanced by your own study of your chosen field.

For example, if the biology student became immersed in mathematics and statistics full-time, then that student would no longer be a student of biology. It can be exceedingly difficult to apply a statistical technique and interpret the results of such a technique in a field for which you are not familiar. If you are unaware of the substantive objects you are working with, that is, the "stuff" on which the statistics are being applied, then regardless of your quantitative expertise, you will often have difficulty interpreting the scientific result. Likewise, if spending too much time computing higher-order derivatives, the student of animal learning, for instance, will have little time remaining to study the learning patterns of the rats he is conditioning, or to speculate on theoretical advancements in his field. Hence, a "happy medium" is required that will balance your study of your substantive area along with the technical quantitative demands of your field of study. Indeed, even for those who specialize exclusively in statistics, the **American Statistical Association** strongly advises aspiring statisticians to choose a field of application. As a researcher, you will be expected to apply modeling techniques that are quite advanced (entire courses are devoted to the statistical technique you may be applying), and so you will face the opposite problem, that of choosing to specialize in statistics (to some extent) so that you may better understand the phenomena of your own science. Hence, regardless of whether one is coming from a mathematics or science background, one should aspire for a healthy mix of scientific and statistical expertise. **Computing** experience (e.g., R, SPSS, SAS, etc.) should also be part of your "repertoire" of skills. As an applied scientist, you should probably find yourself in the **data science** or **traditional research** intersections in the following Venn diagram.

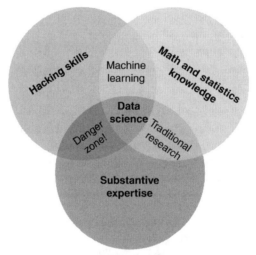

Source: From Drew Conway, THE DATA SCIENCE VENN DIAGRAM, Sep 30, 2010. Reproduced with permission from Drew Conway.

1.8   STATISTICS AND RELATIVITY

Statistical thinking is all about **relativity**. Statistics are not about numbers, they are about **distributions** of numbers (Green, 2000, personal communication). Rarely in statistics, or science for that matter, do we evaluate things in a vacuum.

Consider a very easy example. You board an airplane destined to your favorite vacation spot. How talented is the pilot who is flying your airplane? Is he a "good" pilot or a "bad" pilot? One would hope he is "good enough" to fulfill his duties and ensure your and other passengers' safety. However, when you start thinking like a statistician, you may ponder the thought of how good of a pilot he is **relative to other pilots**. Where on the curve does your pilot fall? In terms of his or her skill, the pilot of an airplane can be **absolutely** good, but still **relatively** poor. Perhaps that pilot falls on the lower end of the talent curve for pilots. The pilot is still very capable of flying the plane, they have passed an **absolute standard**, but he or she just isn't quite as good as most other pilots (see Figure 1.5).

We can come up with a lot of other examples to illustrate the **absolute** versus **relative** distinction. If someone asked you whether you are intelligent, ego aside, as a statistician, you may respond "relative to who?" Indeed, with a construct like IQ, **relativity is all we really have**. What does **absolute intelligence** look like? Should our species discover aliens on another planet one day, we may need to revise our definition of intelligence if such are much more (or much less) advanced than we are. Though of course, this would assume we have the intelligence to comprehend that their capacities are more than ours, a fact not guaranteed and hence another example of the **trap of relativity**.

Relativity is a benchmark used to evaluate much phenomena, from intelligence to scholastic achievement to prevalence of depression, and indeed much of human and nonhuman behavior. Understanding that events witnessed could be theorized to have come from known distributions (like the talent distribution of pilots) is a first step to thinking statistically. Most phenomena have distributions, either known or unknown. Statistics, in large part, is a study of such distributions.

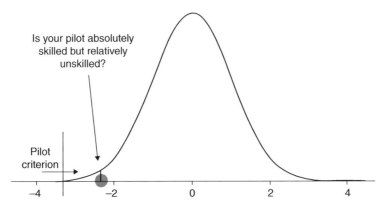

FIGURE 1.5   The "pilot criterion" must be met for *any* pilot to be permitted to fly your plane. However, of those skilled enough to fly, your pilot may still lay at the lower end of the curve. That is, your pilot may be ***absolutely*** good, but *relatively* poor in terms of skill.

1.9   EXPERIMENTAL VERSUS STATISTICAL CONTROL

Perhaps most pervasive in the social science literature is the implicit belief held by many that methods such as **regression** and **analysis of covariance** allow one to "control" variables that would otherwise not be controllable in the nonexperimental design. As is emphasized throughout this book, statistical methods, whatever the kind, do not provide methods of controlling variables, or "holding variables constant" as it were. Not in the **real way**. To get these kinds of effects, you usually need a strong and rigorous bullet-proof experimental design.

It is true, however, that statistical methods do afford a method, in some sense, for presuming (or guessing) **what might have been** had controls been put into place. For instance, if we analyze the correlation between weight and height, it may make sense to hold a factor such as age "constant." That is, we may wish to **partial out** age. However, partialling out the variability due to age in the bivariate correlation is not equivalent to actually **controlling** for age. The truth of the matter is that our statistical control is telling us nothing about what would actually be the case had we been able to truly control age, or any other factor. As will be elaborated on in Chapter 8 on multiple regression, statistical control is not a sufficient "proxy" whatsoever for experimental control. Students and researchers must keep this distinction in mind before they throw variables into a statistical model and employ words like "control" (or other **power** and **action** words) when interpreting effects. If you want to truly control variables, to actually hold them **constant**, you usually have to do **experiments**. Estimating parameters in a statistical model, confident that you have "controlled" for covariates, is simply not enough.

1.10   STATISTICAL VERSUS PHYSICAL EFFECTS

In the establishment of evidence, either experimental or nonexperimental, it is helpful to consider the distinction between **statistical** versus **physical** effects. To illustrate, consider a medical scientist who wishes to test the hypothesis that the more medication applied to a wound, the faster the wound heals. The statistical question of interest is—**Does amount of medication predict the rate at which a wound heals?** A useful statistical model might be a linear regression where amount of medication is the predictor and rate of healing is the response. Of course, one does not "need" a regression analysis to "know" whether something is occurring. The investigator can simply observe whether the wound heals or not, and whether applying more or less medication speeds up or slows down the healing process. The statistical tool in this case is simply used to **model** the relationship, not **determine** whether or not it **exists**. The variable in question is a physical, biological, "real" phenomenon. It exists independent of the statistical model, simply because we can see it. **The estimation of a statistical model is not necessarily the same as the hypothesized underlying physical process it is seeking to represent**.

In some areas of social science, however, the very observance of an effect cannot be realized without recourse to the statistics used to model the relationship. For instance, if I correlate self-esteem to intelligence, am I modeling a relationship that I know exists separate from the statistical model, or, is the statistical model the only recourse I have to say that the relationship exists in the first place? Because of **mediating** and **moderating** relationships in social statistics, an additional variable or two could drastically modify existing coefficients in a model to the point where predictors that had an effect before such inclusion no longer do after. As we will emphasize in our chapters on regression:

> **When you change the model, you change parameter estimates, you change effects. You are never, ever, testing individual effects in the model. You are always testing the model, and hence the interpretation of parameter estimates must be within the context of the model.**

This is one of the general problems of purely correlational research with nonphysical or "nonorganic" variables. It may be more an exercise in **variance partitioning** than it is in analyzing "true" substantive effects, since the effects in question may be simply statistical artifacts. They may have little other **bases**. Granted, even working with physical or biological variables this can be a problem, but it does not rear its head nearly as much. To reiterate, when we model a physical relationship, we have recourse to that physical relationship independent of the statistical model, because we have evidence that the physical relationship exists independent of the model. If we lost our modeling software, we could still "see" the phenomenon. In many models of social phenomena, however, the addition of one or two covariates in the model can make the relationship of most interest "disappear" and because of the nature of measured variables, we may no longer have physical recourse to justify the original relationship at all, **external to the statistical model**. This is why social models can be very "neurotic," frustrating, and context-dependent. Self-esteem may predict achievement in one model, but in another, it does not. Many areas of psychological, political, and economic research, for instance, implicitly operate on such grounds. The existence of phenomena is literally "built" on the existence of the statistical model and often does not necessarily exist separate from it, or at least not in an easily observed manner such as the healing of a wound. Social scientists working in such areas, if nothing else, must be aware of this. **Estimating a statistical model may or may not correspond to actual physical effects it is seeking to account for**.

1.11 UNDERSTANDING WHAT "APPLIED STATISTICS" MEANS

In this day and age of extraordinary computing power, the likes of which will probably seem laughable in even a decade from the date of publication of this book, with a few clicks of the mouse and a software manual, one can obtain a **principal components analysis**, **factor analysis**, **discriminant analysis**, **multiple regression**, and a host of other relatively theoretically advanced statistical techniques in a matter of seconds. The advance of computers and especially easy-to-use software programs has made performing statistical analyses seemingly quite easy because even a novice can obtain output from a statistical procedure relatively quickly. One consequence of this however is that there seems to have arisen a misunderstanding in some circles that "applied statistics" somehow equates with the idea of "statistics without mathematics" or even worse, "statistics via software."

The word "applied" in **applied statistics** should not be understood to necessarily imply the use of computers. What "applied" should mean is that the focus on the writing is on how to **use** statistics in the context of scientific investigation, oftentimes with demonstrations with real or hypothetical data. Whether that data is analyzed "by hand" or through the use of software does not make one approach more applied than the other. If analyzed via computer, what it does make it is more **computational** compared to the by-hand approach. Indeed, there is a whole field of study known as **computational statistics** that features a variety of software approaches to data analysis. For examples, see Dalgaard (2008), Venables and Ripley (2002), and Friendly (1991, 2000) for an emphasis on data visualization. Fox (2002) also provides good coverage of functions in S-Plus and R. And of course, **computer science** and the **machine-learning** movement have contributed greatly to software development and our ability to analyze data quickly and efficiently via algorithms, and implement new and classic procedures that would be impossible otherwise.

On the opposite end of the spectrum, if a course in statistics is advertised as **not** being applied, then most often what this implies is that the course is more theoretical or mathematical in nature with a focus on proof and the justification of results. In essence, what this really means is that the course is usually more **abstract** than what would be expected in an applied course. In such theoretical courses, very seldom will one see applications to real data, and instead the course will feature **proofs** of essential **statistical theorems** and the justification of analytical propositions. Hence, this is the true distinction between **applied versus theoretical** courses. The computer has really nothing to do with the distinction other than **facilitating** computation in either field.

REVIEW EXERCISES

1.1. Distinguish between **rationalism** versus **empiricism** in accounting for different types of knowledge, and why being a rationalist or empiricist exclusively is usually quite unreasonable and unrealistic.

1.2. Briefly discuss what is meant by a **model** in scientific research.

1.3. Compare and contrast the **social** versus so-called "**hard**" sciences. How are they similar? Different? In this context, discuss the statement "Social science is a courageous attempt."

1.4. Compare and contrast a **physical** quantity such as weight to a **psychological** one such as intelligence. How is one more "real" than the other? Can they be considered to be equally real? Why or why not?

1.5. Why would some people say that an attribute such as **intelligence** is not measurable?

1.6. Discuss George Box's infamous statement "**All models are wrong, some are useful.**" What are the implications of this for your own research?

1.7. Consider an example from your own area of research in which two competing explanations, one simple, and one complex, may equally well account for observed data. Then, discuss why the **simpler explanation** may be preferable to the more complex. Are there instances where the more complex explanation may be preferable to the simpler? Discuss.

1.8. Briefly discuss why using statistical methods to make **causal statements** about phenomena may be unrealistic and in most cases unattainable. Should the word "cause" be used at all in reference to nonexperimental social research?

1.9. Discuss why it is important to suspend one's beliefs about a subject such as **applied statistics** or **mathematics** in order to potentially learn more about it.

1.10. Statistical thinking is about **relativity**. Discuss what this statement means with reference to the pilot example, then by making up an example of your own.

1.11. Distinguish between **experimental** versus **statistical control**, and why understanding the distinction between them is important when interpreting a statistical model.

1.12. Distinguish between **statistical** versus **physical effects** and how the effect of a medication treating a wound might be considered different in nature from the correlation between intelligence and self-esteem.

1.13. Distinguish between the domains of **applied** versus **theoretical statistics**.

Further Discussion and Activities

1.14. William of Ockham (c. 1287–1347) is known for his infamous principle **Ockham's razor**, which essentially states that all things equal, given competing theories accounting for the same data, the simpler theory is the better theory. In other words, complex explanations for phenomena that could be explained by simpler means are not encouraged. Read Kelly (2007), and evaluate the utility of Ockham's razor as it applies to statistical modeling. Do you agree that the simpler statistical model is usually preferred over the more complex when it comes to modeling social phenomena? Why or why not?

1.15. Read Kuhn (2012). Discuss what Kuhn means by **normal science** and the essence of what constitute **paradigm shifts** in science.

1.16. As briefly discussed in this chapter, **statistical** control is not the same thing as **experimental** control or that of a **control group**. Read Dehue (2005), and provide a brief commentary regarding what constitutes a **real** control group versus the concept of statistical controls.

1.17. It was briefly discussed in the chapter potential problems with using the word **cause** or speaking of **causality** at all when describing findings in the social and (often) natural sciences. The topic of causality is a philosopher's career and a scientist's methodological nightmare. Epidemiology, the study of diseases in human and other populations, has, like so many other disciplines, had to grapple with the issue of causation. For example, if one is to make the statement **smoking causes cancer**, one must be able to defend one's philosophical position in advancing such a claim. Not everyone who smokes gets cancer. Further, some who smoke the most never get the disease, whereas some who smoke the least do. Tobacco companies have historically relied on the fact that **not everyone who smokes gets cancer** as a means for challenging the smoking-cancer "link." As an introduction to these issues, as well as a brief history of causal interpretations, read Morabia (2005). Summarize the historical interpretations of causality, as well as how epidemiology has generally dealt with the problem of causation.

1.18. Models are used across the sciences to help account for empirical observations. How to best relate mathematical models to reality is not at all straightforward. Read Hennig (2009), and discuss Hennig's account of the relation between reality and mathematical models. Do you agree with this account? What might be some problems with it?

2

INTRODUCTORY STATISTICS

> In spite of the immense amount of fruitful labour which has been expended in its practical applications, the basic principles of this organ of science are still in a state of obscurity, and it cannot be denied that, during the recent rapid development of practical methods, fundamental problems have been ignored and fundamental paradoxes left unresolved.
>
> (Fisher, 1922a, p. 310)

Our statistics review includes topics that would customarily be seen in a first course in statistics at the undergraduate level, but depending on the given course and what was emphasized by the instructor, our treatment here may be at a slightly deeper level. We review these principles with demonstrations in R and SPSS where appropriate. Should any of the following material come across as entirely "new," then a review of any introductory statistics text is recommended. For instance, Kirk (2008), Moore, McCabe, and Craig (2014), Box, Hunter, and Hunter (1978) are relatively nontechnical sources, whereas Degroot and Schervish (2002), Wackerly, Mendenhall III, and Scheaffer (2002) along with Evans and Rosenthal (2010) are much deeper and technically dense. Casella and Berger (2002), Hogg and Craig (1995) along with Shao (2003) are much higher-level theoretically oriented texts targeted mainly at mathematical and theoretical statisticians. Other sources include Panik (2005), Berry and Lindgren (1996), and Rice (2006). For a lighter narrative on the role of statistics in social science, consult Abelson (1995).

Because of its importance in the interpretation of evidence, we close the chapter with an easy but powerful demonstration of what makes a p-value small or large in the context of statistical significance testing and the testing of null hypotheses. It is imperative that as a research scientist, you are knowledgeable of this material before you attempt to evaluate **any** research findings that employ statistical inference.

2.1 DENSITIES AND DISTRIBUTIONS

When we speak of **density** as it relates to distributions in statistics, we are referring generally to theoretical distributions having **area under their curves**. There are numerous probability distributions or density functions. Empirical distributions, on the other hand, rarely go by the name of densities. They are in contrast "real" distributions of real empirical data. In some contexts, the identifier **normal distribution** may be given without reference as to whether one is referring to a density or to an empirical distribution. It is usually evident by the context of the situation which we are referring to. We survey only a few of the more popular densities and distributions in our discussion that follows.

The univariate normal density is given by:

$$f\left(x_i, \mu, \sigma^2\right) = \frac{1}{\sqrt{2\pi\sigma^2}} e^{-(x_i - \mu)^2/2\sigma^2}$$

where,

- μ is the population mean for the given density,
- σ^2 is the population variance,
- π is a constant equal to approximately 3.14,
- e is a constant equal to approximately 2.71,
- x_i is a given value of the independent variable, assumed to be a real number.

When μ is 0 and σ^2 is 1, which implies that the standard deviation σ is also equal to 1 (i.e., $\sqrt{\sigma^2} = \sigma = 1$), the normal distribution is given a special name. It is called the **standard normal distribution** and can be written more compactly as:

$$f\left(x_i, \mu, \sigma^2\right) = \frac{1}{\sqrt{2\pi\sigma^2}} e^{-(x_i - \mu)^2/2\sigma^2} = \frac{1}{\sqrt{2\pi(1)}} e^{-(x_i - 0)^2/2(1)} = \frac{1}{\sqrt{2\pi}} e^{-x_i^2/2} = \frac{e^{-x_i^2/2}}{\sqrt{2\pi}} = \frac{e^{-\frac{1}{2}x_i^2}}{\sqrt{2\pi}} \quad (2.1)$$

Notice that in (2.1), $e^{-(x_i - \mu)^2/2\sigma^2} = e^{-(x_i - 0)^2/2(1)}$ where μ is now 0 and σ^2 is now 1. Note as well that the density depends only on the **absolute** value of x_i, because both x_i and $-x_i$ give the same value x_i^2; the greater is x_i in absolute value, the smaller the density at that point, because the constant e is raised to the **negative** power $-x_i^2/2$.

The **standard normal distribution** is the classic z-distribution whose areas under the curve are given in the appendices of most statistics texts, and are more conveniently computed by software. An example of the standard normal is featured in Figure 2.1.

Scores in research often come in their own units, with distributions having means and variances different from 0 and 1. We can transform a score coming from a given distribution with mean μ and standard deviation σ by the familiar z-score:

$$z = \frac{x_i - \mu}{\sigma}$$

A z-score is expressed in units of the standard normal distribution. For example, a z-score of $+1$ denotes that the given raw score lay one standard deviation above the mean. A z-score of -1 means that the given raw score lay one standard deviation below the mean. In some settings (such as school psychology), t-scores are also useful, having a mean of 50 and standard deviation of 10. In most contexts, however, z-scores dominate.

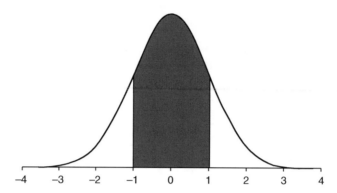

FIGURE 2.1   Standard normal distribution with shaded area from −1 to +1 standard deviations from the mean.

A classic example of the utility of z-scores typically goes like this. Suppose two sections of a statistics course are being taught. John is a student in section A and Mary is a student in section B. On the final exam for the course, John receives a raw score of 80 out of 100 (i.e., 80%). Mary, on the other hand, earns a score of 70 out of 100 (i.e., 70%). At first glance, it may appear that John was more successful on his final exam. However, scores, considered **absolutely**, do not allow us a comparison of each student's score relative to their **class distributions**. For instance, if the mean in John's class was equal to 85% with a standard deviation of 2, this means that John's z-score is:

$$z = \frac{x_i - \mu}{\sigma} = \frac{80 - 85}{2} = -2.5$$

Suppose that in Mary's class, the mean was equal to 65% also with a standard deviation of 2. Mary's z-score is thus:

$$z = \frac{x_i - \mu}{\sigma} = \frac{70 - 65}{2} = 2.5$$

As we can see, **relative to their particular distributions**, Mary greatly outperformed John. Assuming each distribution is approximately normal, the density under the curve for a normal distribution with mean 0 and standard deviation of 1 at a score of 2.5 is:

```
> dnorm(2.5, 0, 1)
[1] 0.017528
```

where dnorm is the density under the curve at 2.5. This is the value of $f(x)$ at the score of 2.5. What then is the probability of scoring 2.5 or greater? To get the cumulative density up to 2.5, we compute:

```
> pnorm(2.5, 0, 1)
[1] 0.9937903
```

The given area is represented in Figure 2.2. The area we are interested in is that at or above 2.5 (the area where the arrow is pointing). Since we know the area under the normal density is equal to 1, we can subtract pnorm(2.5, 0, 1) from 1:

```
> 1-pnorm(2.5, 0, 1)
[1] 0.006209665
```

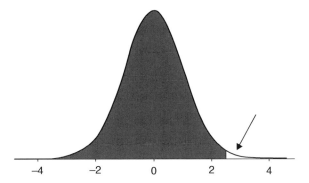

FIGURE 2.2 Shaded area under the standard normal distribution at a z-score of up to 2.5 standard deviations.

We can see then the percentage of students scoring higher than Mary is in the margin of approximately 0.6% (i.e., multiply the proportion by 100). What proportion of students scored better than John in his class? Recall that his z-score was equal to −2.5. Because we know the normal distribution is symmetric, we already know the area lying below −2.5 is the same as that lying above 2.5. This means that approximately 99.38% of students scored higher than John. Hence, we see that Mary drastically outperformed her colleague when we consider their scores **relative to their classes**. Be careful to note that in drawing these conclusions, we had to assume each score (that of John's and Mary's) came from a normal distribution. The mere fact that we transformed their raw scores to z-scores in no way normalizes their raw distributions. Standardization **standardizes**, but it does not **normalize**.

One can also easily verify that approximately 68% of cases in a normal distribution lie within −1 and +1 standard deviations, while approximately 95% of cases lie within −2 and +2 standard deviations.

2.1.1 Plotting Normal Distributions

We can plot normal densities in R by simply requesting the lower and upper limit on the abscissa:

```
> x <- seq(from = -3, to = +3, length.out = 100)
> plot(x, dnorm(x))
```

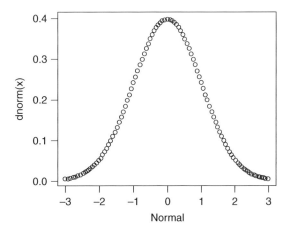

Distributions (and densities) of a single variable typically go by the name of **univariate** distributions to distinguish them from distributions of two (**bivariate**) or more variables (**multivariate**).

For example, we consider some of Galton's data on parent and child heights (the height of the children were measured when they were adults, not actual toddlers). Some of Galton's data appears below, retrieved from the `HistData` package (Friendly, 2014) in R:

```
> install.packages("HistData")
> library(HistData)
> attach(Galton)

> Galton
     parent child
1     70.5  61.7
2     68.5  61.7
3     65.5  61.7
4     64.5  61.7
5     64.0  61.7
6     67.5  62.2
7     67.5  62.2
8     67.5  62.2
9     66.5  62.2
10    66.5  62.2
```

We first install the package using the `install.packages` function. The `library` statement loads the package `HistData` into R's search path. From there, we `attach` the Galton data to insert the object (dataframe) into the search list. We generate a histogram of parent height:

```
> hist(parent, main = "Histogram of Parent Height")
```

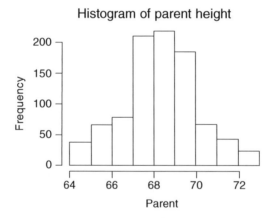

One can overlay a normal density over an empirical plot to show how closely observed data match that of a theoretical normal distribution, as was done by Fisher in 1925 displaying a distribution of the heights of 1375 women (see Figure 2.3, taken from **Classics in the History of Psychology**[1]). R.A. Fisher is usually regarded as the father of modern statistics and among his greatest contributions was the publication of **Statistical Methods for Research Workers** in 1925 in which he discussed such topics as tests of significance, correlation coefficients, and the analysis of variance.

---

[1] **Classics in the History of Psychology** is an on-line educational resource hosted by Christopher D. Green of York University in Toronto, Canada. It contains a huge selection of milestone papers and articles in the history of psychology. It can be accessed via http://psychclassics.yorku.ca/.

We can see that the normal density serves as a close, and very convenient, **approximation** to empirical data. Indeed, the normal density has figured prominent in the history of statistics largely because it serves as a useful **model** for many phenomena, and also because it provides a very convenient starting point for much work in theoretical statistics. Oftentimes the **assumption of normality** will be invoked in a derivation because it makes the problem simpler and easier to solve.

2.1.2 Binomial Distributions

The binomial distribution is given by:

$$p(r) = \binom{n}{r} p^r (1-p)^{n-r}$$
$$= \left(\frac{n!}{r!(n-r)!}\right) p^r (1-p)^{n-r}$$

where,
- $p(r)$ is the probability of observing r occurrences out of n possible occurrences,[2]
- p is the probability of a "success" on any given trial, and
- $1-p$ is the probability of a failure on any given trial, often simply referred to by "q" (i.e., $q = 1-p$).

The binomial setting provides an ideal context to demonstrate the essentials of hypothesis-testing logic, as we will soon see. In a binomial setting, the following conditions must hold:

- The variable under study must be **binary** in nature. That is, the outcome of the experiment can result in only one category or another. That is, the outcome categories are **mutually exclusive**. For instance, the flipping of a coin has this characteristic, because the coin can either come up "head" or "tail" and nothing else (yes, we are ruling out the possibility that it lands on its side, and I think it is safe to do so).
- The probability of a "success" on each trial remains constant (or **stationary**) from trial to trial. For example, if the probability of head is equal to 0.5 on our first flip, we assume it is also equal to 0.5 on the second, third, fourth flips, and so on.
- Each trial is **independent** of each other trial. That is, the fact that we get a head on our first flip of the coin in no way changes the probability of getting a head or tail on the next flip, and so on for the other flips (i.e., no outcome is ever "due" to occur, as the gambler sometimes believes).

We can easily demonstrate hypothesis testing in a binomial setting using R. For instance, let us return to the coin-flipping experiment. Suppose you would like to know the probability of obtaining two heads on five flips of a fair coin, where each flip is assumed to have a probability of heads equal to 0.5. In R, we can compute this as follows:

```
> dbinom(2, size = 5, prob = 0.5)
[1] 0.3125
```

---

[2] We can also extend the binomial distribution to one in which instead of n trials giving rise to r occurrences, we have n trials giving rise to outcomes in k categories:

$$p(\mathbf{x}) = p(x_1, x_2, \ldots, x_k) = \frac{n!}{x_1! x_2! \cdots x_k!} p_1^{x_1} p_2^{x_2} \cdots p_k^{x_k}$$

where $\mathbf{x}$ is now a vector of random variables $\mathbf{x} = [x_1, x_2, \ldots, x_k]'$.

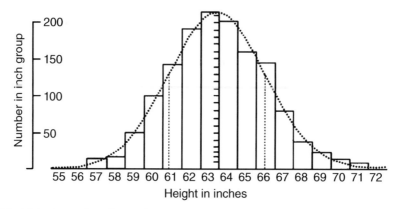

FIGURE 2.3 Fisher's overlay of normal density on empirical observations. Source: Fisher (1925, 1934).

where dbinom calls the "density for the binomial," "2" is the number of successes we are specifying, "size = 5" represents the number of trials we are taking, and "prob = 0.5" is the probability of success on any given trial, which recall is assumed constant from trial to trial.

Suppose instead of two heads, we were interested in the probability of obtaining five heads:

```
> dbinom(5, size = 5, prob = 0.5)
[1] 0.03125
```

Notice that the probability of obtaining five heads out of five flips on a fair coin is quite a bit less than that of obtaining two heads. We can continue to obtain the remaining probabilities and obtain the complete binomial distribution for this experiment:

Heads	0	1	2	3	4	5	
Prob	0.03125	0.15625	0.3125	0.3125	0.15625	0.03125	$\sum 1.0$

A plot of this binomial distribution is given in Figure 2.4.

Suppose that instead of wanting to know the probability of getting two heads out of five flips, we wanted to know the probability of getting two **or more** heads out of five flips. Because

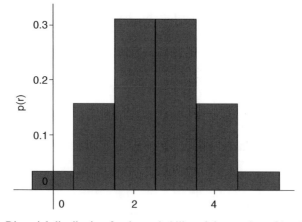

FIGURE 2.4 Binomial distribution for the probability of the number of heads on a fair coin.

the events 2 heads, 3 heads, 4 heads, and 5 heads are **mutually exclusive events**, we can add their probabilities by the probability rule that says $p\left(\cup_{i=1}^{\infty} A_i\right) = \sum_{i=1}^{\infty} p(A_i)$: $0.3125 + 0.3125 + 0.15625 + 0.03125 = 0.8125$. Hence, the probability of obtaining two or more heads on a fair coin on five flips is equal to 0.8125.

Binomial distributions are useful in a great variety of contexts in modeling a wide number of phenomena. But again, remember that the outcome of the variable must be **binary**, meaning it must have only **two** possibilities. If it has more than two possibilities or is continuous in nature, then the binomial distribution is not suitable. Binomial data will be featured further in our discussion of logistic regression in Chapter 10.

One can also appreciate the general logic of hypothesis testing through the binomial. If our null hypothesis is that the coin is fair, and we obtain five heads out of five flips, this result has only a 0.03125 probability of occurring. Hence, because the probability of this data is so low under the model that the coin is fair, we typically decide to reject the null hypothesis and infer the **statistical alternative** hypothesis that $p(H) \neq 0.5$. Substantively, we might infer that the coin is not fair, though this **substantive alternative** also assumes it is the coin that is to "blame" for it coming up five times heads. If the flipper was responsible for biasing the coin, for instance, or a breeze suddenly came along that helped the result occur in this particular fashion, then inferring the substantive alternative hypothesis of "unfairness" may not be correct. Perhaps the **nature** of the coin is such that it **is** fair. Maybe the flipper or other factors (e.g., breeze) are what are ultimately responsible for the rejection of the null. This is one reason why rejecting null hypotheses is quite easy, but inferring the **correct** substantive alternative hypothesis (i.e., the hypothesis that explains **why** the null was rejected) is much more challenging (see Denis, 2001). As concluded by Denis, "**Anyone can reject a null, to be sure. The real skill of the scientist is arriving at the true alternative**."

The binomial distribution is also well-suited for comparing proportions. For details on how to run this simple test in R, see Crawley (2013, p. 365). One can also use `binom.test` in R to test simple binomial hypotheses, or the `prop.test` for testing null hypotheses about proportions. A useful test that employs binomial distributions is the **sign test** (see Siegel and Castellan, 1988, pp. 80–87 for details). For a demonstration of the sign test in R, see Denis (2020).

2.1.3   Normal Approximation

Many distributions in statistics can be regarded as **limiting forms** of other distributions. What this statement means can be best demonstrated through an example of how the binomial and normal distributions are related. When the number of discrete categories along the x-axis grows larger and larger, the areas under the binomial distribution more and more resemble the probabilities computed under the normal curve. It is in this sense that for a large number of trials on the binomial, it begins to more closely **approximate** the normal distribution.

As an example, consider once again the binomial distribution for $n = 5$, $p = 0.5$, but this time with a normal density overlaying the binomial (Figure 2.5).

We can see that the normal curve "approximates" the binomial distribution, though perhaps not tremendously well for only five trials. If we increase the number of trials, however, to say, 20, the approximation is much improved. And when we increase the number of trials to 100, the binomial distribution looks virtually like a normal density. That is, we say that **the normal distribution is the limiting form of the binomial distribution**.

We can express this idea more formally. If the number of trials n in a binomial experiment is made large, the distribution of the number of successes x will tend to resemble a normal distribution. That is, the normal distribution is the limiting form of a binomial distribution as $n \rightarrow \infty$ for a fixed p (and where $q = 1 - p$), where $E(x_i)$ is the expectation of the random variable x_i (the meaning of "random variable" will be discussed shortly):

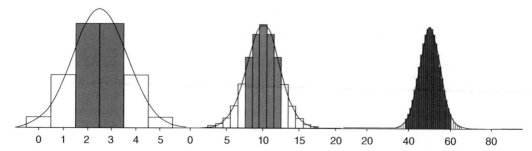

FIGURE 2.5 Binomial distributions approximated by normal densities for 5 (far left), 20 (middle), and 100 trials (far right).

$$z = \frac{x_i - \mu}{\sigma} \leftrightarrow z_m = \frac{x_i - E(x_i)}{\sigma} \leftrightarrow z_m = \frac{x_i - np}{\sqrt{npq}}$$

Notice that in a z-score calculation using the population mean μ, in the numerator, we are actually calculating the difference between the obtained score and the **expectation**, $E(x_i)$. We can change this to a binomial function by replacing the expectation μ with the expectation from a binomial distribution, that is, np, where np is the mean of a binomial distribution. Similarly, we replace the standard deviation from a normal distribution with the standard deviation from the binomial distribution, $\sqrt{npq}$. As n grows infinitely large, the normal and the binomial probabilities become identical for any standardized interval.[3]

2.1.4   Joint Probability Densities: Bivariate and Multivariate Distributions

A univariate density expresses the probability of a single random variable within a specified interval of values along the abscissa. A joint probability density, analogous to a joint probability, expresses the probability of simultaneously observing **two** random variables over a given interval of values. The **bivariate normal density** is given by:

$$f(x_1, x_2) = \frac{1}{2\pi\sigma_1\sigma_2\sqrt{1-\rho^2}} \exp\left\{ -\frac{1}{2(1-\rho^2)} \left[\frac{(x_1-\mu_1)^2}{\sigma_1^2} - 2\rho\frac{(x_1-\mu_1)(x_2-\mu_2)}{\sigma_1\sigma_2} + \frac{(x_2-\mu_2)^2}{\sigma_2^2} \right] \right\}$$

where ρ^2 is the squared Pearson correlation coefficient between x_1 and x_2.

When plotted, the bivariate density resembles a pile of raked leaves in the Autumn. A plot generated in R is given in Figure 2.6.

Empirical bivariate distributions (as opposed to bivariate **densities**) are those showing the joint occurrence on two variables. For instance, again using Galton's data, we plot parent height by child height, in which we also fit both regression lines (see Chapter 7) using `lm`:

```
> plot(parent, child, main = "Bivariate Plot of Parent and Child Height")
> abline(lm(parent~child))
> abline(lm(child~parent))
```

---

[3] For a more technical demonstration of how and why this convergence occurs, see Proschan (2008).

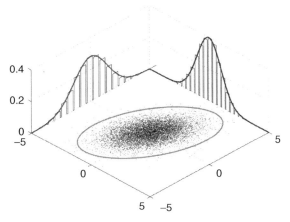

FIGURE 2.6 Bivariate density. Source: Data from Plotting bivariate normal distributions, Mon Sep 1 2003.

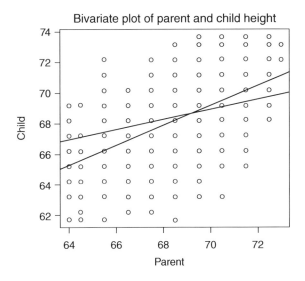

Note the relation between parent height and child height. Recall that a mathematical relation is a subset of the Cartesian product. The Cartesian product in the plot consists of **all** theoretically possible parent–child pairings. The fact that shorter than average parents tend to have shorter than average children and taller than average parents tend to have taller than average children reveals the linear form of the mathematical relation. In the plot are regression lines for child height as a function of parent height and parent height as a function of child height. Computing both the mean of child and of parent, we obtain:

```
> mean(child)
[1] 68.08847
> mean(parent)
[1] 68.30819
```

Notice that both regression lines, as they are required to do whatever the empirical data, pass through the means of each variable. The reason for this will become clearer in Chapter 7.

Turning now to multivariate distributions, the **multivariate density** is given by:

$$g(x_i) = \frac{1}{\left(\sqrt{2\pi}\right)^{p} |\Sigma|^{1/2}} e^{-(\mathbf{x}-\boldsymbol{\mu})'\Sigma^{-1}(\mathbf{x}-\boldsymbol{\mu})/2}$$

where p is the number of variables and $|\Sigma|$ is the determinant of the population covariance matrix, which can be taken as a measure of **generalized variance** since it incorporates both variances **and** covariances. Refer to the Appendix for examples of computing covariance and correlation matrices. Multivariate distributions represent the joint occurrence of three or more variables, and thus are quite difficult to visualize. One way, however, of representing a density in three dimensions is attempted in Figure 2.7.

Most multivariate procedures make some assumption regarding the **multivariate normality** of sampling distributions. Evaluating such an assumption is intrinsically difficult due to the high dimensionality of the data. The best researchers can usually do is attempt to verify univariate and bivariate normality through such devices as histograms and scatterplots. Fortunately, as is the case for methods assuming univariate normality, multivariate procedures are relatively robust, in most cases, to modest violations. Though **Mardia's test** (Mardia, 1970) is favored by some (e.g., Romeu and Ozturk, 1993), no single method for evaluating multivariate normality appears to be fully adequate. Visual inspections of Q–Q plots (to be discussed) are usually sufficient for applied purposes.

In cases where rather severe departures of normality exist, one may also choose to perform data transformations on the "offending" variables to better approximate normal distributions. However, it should be kept in mind that sometimes a severely nonnormal distribution can be evidence more of a **scientific** problem than symptomatic of a statistical issue. For example, if we asked individuals in a sample how many car accidents they got into this month, the vast majority of our responses would

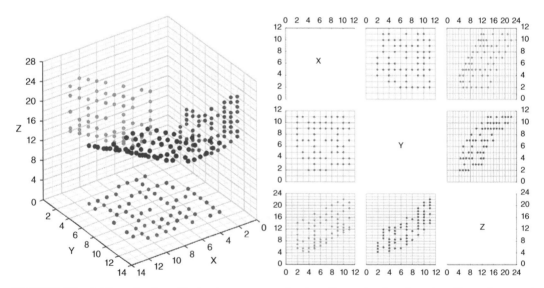

FIGURE 2.7   A 3D scatterplot with density contour and points (Image is taken from http://www.jmp.com/support/help/Scatterplot_3D_Platform_Options.shtml). Source: Figure taken from JMP 12 Essential Graphing, Copyright © 2015, SAS Institute Inc., USA. All Rights Reserved. Reproduced with permission of SAS Institute Inc, Cary, NC.

indicate a count of "0." Is the distribution skewed? Yes, but this is not a statistical problem alone, it is first and foremost a **substantive** one. We likely would not even have sufficient variability in our measurement responses to conduct any meaningful analyses since probably close to 100% of our sample will likely respond with "0." If virtually everyone in your sample responds with a constant, then one might say the very process of **measurement** may have been problematic, or at minimum, not very meaningful for scientific purposes. The difficulties presented in subjecting that data to statistical analyses should be an afterthought, second in priority to the more pressing scientific issue.

2.2 CHI-SQUARE DISTRIBUTIONS AND GOODNESS-OF-FIT TEST

The chi-square distribution is given by:

$$f(x) = \frac{1}{2^{v/2}\Gamma(v/2)} x^{[(v/2)-1]} e^{-x/2}$$

for $x > 0$, where v are degrees of freedom and Γ is the gamma function.[4] The chi-square distribution of a random variable is also equal to the sum of squares of n independent and normally distributed z-scores (Fisher, 1922b). That is,

$$\chi_n^2 = \sum_{i=1}^{n} z_i^2 = \sum_{i=1}^{n} \frac{(x_i - \mu)^2}{\sigma^2}$$

The chi-square distribution plays an important role in mathematical statistics and is associated with a number of tests on model coefficients in a variety of statistical methods. The multivariate analog to the chi-square distribution is that of the **Wishart distribution** (see Rencher, 1998, p. 53, for details).

The chi-square **goodness-of-fit test** is one such statistical method that utilizes the chi-square test statistic to evaluate the tenability of a null hypothesis. Recall that such a test is suitable for categorical data in which counts (i.e., instead of means, medians, etc.) are computed within each cell of the design. The goodness-of-fit test is given by

$$\chi^2 = \sum_{i=1}^{r} \sum_{j=1}^{c} (O_i - E_i)^2 / E_i$$

---

[4] For details on the gamma function, see Degroot and Schervish (2002, p. 295). A plot of the gamma function appears as follows (see Crawley, 2013, p. 264, for the R code):

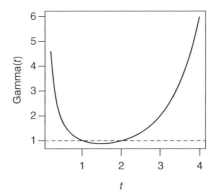

TABLE 2.1   Contingency Table for 2 × 2 Design

	Condition Present (1)	Condition Absent (0)	Total
Exposure yes (1)	20	10	30
Exposure no (2)	5	15	20
Total	25	25	50

where O_i and E_i represent **observed** and **expected** frequencies, respectively, summed across r rows and c columns.

As a simple example, consider the hypothetical data (Table 2.1), where the frequencies of those exposed to something adverse are related to whether a condition is present or absent. If you are a clinical psychologist, then you might define **exposure** as, perhaps, a variable such as combat exposure, and **condition** as posttraumatic stress disorder (if you are not a psychologist, see if you can come up with another example).

The null hypothesis is that the 50 counts making up the entire table are more or less randomly distributed across each of the cells. That is, there is no association between condition and exposure. We can easily test this hypothesis in SPSS by weighting the relevant frequencies by cell total:

exposure	condition	freq
1.00	0.00	10.00
1.00	1.00	20.00
2.00	0.00	15.00
2.00	1.00	5.00

```
WEIGHT BY freq.
CROSSTABS
  /TABLES=condition BY exposure
  /FORMAT=AVALUE TABLES
  /STATISTICS=CHISQ
  /CELLS=COUNT
  /COUNT ROUND CELL.
```

The output follows in which it is first confirmed that we set up our data file correctly:

Exposure ∗ Condition Crosstabulation

		Count		
		Condition		Total
		1.00	0.00	
Exposure	1.00	20	10	30
	2.00	5	15	20
Total		25	25	50

We focus on the Pearson chi-square test value of 8.3 on a single degree of freedom. It is statistically significant ($p = 0.004$), and hence we can reject the null hypothesis of no association between condition and exposure group.

Chi-square Tests					
	Value	df	Asymp. Sig. (two-sided)	Exact Sig. (two-sided)	Exact Sig. (one-sided)
Pearson chi-square	8.333[a]	1	0.004		
Continuity correction[b]	6.750	1	0.009		
Likelihood ratio	8.630	1	0.003		
Fisher's exact test				0.009	0.004
Linear-by-linear association	8.167	1	0.004		
No. of valid cases	50				

[a]0 cells (0.0%) have expected count less than 5. The minimum expected count is 10.00.
[b]Computed only for a 2 × 2 table.

In R, we can easily perform the chi-square test on this data. We first build the matrix of cell counts, calling it `diag.table`:

```
> diag.table <- matrix(c(20, 5, 10, 15), nrow = 2)
> diag.table
     [,1] [,2]
[1,]   20   10
[2,]    5   15

> chisq.test(diag.table, correct = F)

        Pearson's Chi-squared test

data:  diag.table
X-squared = 8.3333, df = 1, p-value = 0.003892
```

We see that the result in R agrees with what we obtained in SPSS. Note that specifying `correct = F` (correction = false) negated what is known as **Yates' correction for continuity**, which involves subtracting 0.5 from positive differences in $O - E$ and adding 0.5 to negative differences in $O - E$ in an attempt to better make the chi-square distribution approximate that of a multinomial distribution (i.e., in a crude sense, to help make discrete probabilities more continuous). To adjust for Yates, we can either specify `correct = T` or simply `chisq.test(diag.table)`, which will incorporate the correction. With the correction implemented, our p-value increases from 0.003 to 0.009 (not shown). We notice that this adjustment parallels that made in SPSS by adjusting for continuity. When expected counts per cell are relatively small (a working rule is that they should be at least five in each cell), one can also request **Fisher's exact test** (see Fisher, 1922a), which we note also mirrors the output generated by SPSS:

```
> fisher.test(diag.table)

        Fisher's Exact Test for Count Data

data:  diag.table
p-value = 0.008579
alternative hypothesis: true odds ratio is not equal to 1
95 percent confidence interval:
  1.466377 26.597383
sample estimates:
odds ratio
  5.764989
```

Other useful statistics for contingency tables include the **phi coefficient** and **Cramer's V**. Phi, ϕ, is a measure of association for 2×2 contingency tables, computed as

$$\phi = \sqrt{\frac{\chi^2}{n}}$$

where χ^2 is the chi-square statistic calculated on the 2×2 table, and n is the total sample size. The maximum ϕ can attain is 1.0, indicating maximal association. ϕ can be computed in SPSS by /statistics = phi and is available in R in the psych package (Revelle, 2015). Cramer's ϕ_c extends on ϕ in that it allows for contingency tables of greater than 2×2. It is included in the /statistics = phi command and also available in R's psych package. It is given by:

$$\phi_c = \sqrt{\frac{\chi^2}{n(k-1)}}$$

where k is the minimum of the number of rows or columns. The relationship between ϕ_c and ϕ is easily shown for $k = 2$:

$$\phi_c = \sqrt{\frac{\chi^2}{n(2-1)}} = \sqrt{\frac{\chi^2}{n}} = \phi$$

2.2.1 Power for Chi-Square Test of Independence

We can estimate power[5] and required sample size for the chi-square test of independence using the package pwr in R:

```
> library(pwr)
> pwr.chisq.test (w = , N = , df = , sig.level = , power = )
```

where w is the anticipated or required effect size, estimated as:

$$w = \sqrt{\sum_{i=1}^{m} \frac{(p0_i - p1_i)^2}{p0_i}}$$

and $p0_i$ and $p1_i$ are the probabilities in a given cell i under the null and alternative hypotheses, respectively. We demonstrate by estimating power for $w = 0.2$:

```
> pwr.chisq.test(w = 0.2, N =, df = 5, sig.level = .05, power = 0.90)

        Chi squared power calculation

              w = 0.2
              N = 411.7366
             df = 5
      sig.level = 0.05
          power = 0.9
NOTE: N is the number of observations
```

---

[5] Power will be discussed later in this chapter.

TABLE 2.2 Contingency Table for 2 × 2 × 2 Design

	Exposure	Condition Absent (0)	Condition Present (1)	Total
Males	Yes	10	20	30
	No	15	5	20
Females	Yes	13	17	30
	No	12	8	20
	Total	50	50	100

R estimates that a total of approximately 411 subjects are required to achieve power set at 0.90. Such a large sample is required because $w = 0.2$ constitutes a relatively small effect size (see Cohen (1988) for details).

The reader may ask at this point how one might go about analyzing data for higher-dimensional frequency tables. The example for the chi-square test of the data in Table 2.1 is only for that of a 2×2 layout. Suppose we added a third factor to our analysis, such as gender, making our contingency table appear as in Table 2.2.

For data such as that in Table 2.2 featuring higher-dimensional frequency data, **log-linear models** are a possibility (Agresti, 2002). Log-linear models are an option in the wider class of **generalized linear models**, to be discussed further in Chapter 10, where we discuss in some detail a special case of the generalized linear model called the **logistic regression model**.

2.3 SENSITIVITY AND SPECIFICITY

Sensitivity and specificity are measures historically used in diagnostic situations but can be applied to other contexts as well. We can easily adapt the data in Table 2.1 to suit a brief discussion of these measures. We keep the same cell frequencies, but modify variable names so the data become a bit more applicable to a discussion of sensitivity and specificity (see Table 2.3).

The **sensitivity** of the diagnostic instrument is the probability that the test is positive given that the individual has the disease. In the margins, we see that 30 people have the disease, of which 20 were diagnosed with it. Thus, the sensitivity of the test is 20/30 = 0.66. The **specificity** of the diagnostic instrument is the probability that the test is negative, given that the individual does not have the disease. In the margins, we see that 20 people do not have the disease, of which 15 were diagnosed with not having the disease. Hence, the specificity of the test is 15/20 = 0.75. The overall **prevalence** of the disease is equal to 30/50 (i.e., 30 people have the disease out of 50). One can also compute what are known as positive and negative **predictive values** from such tables. For these and other measures useful for diagnostic situations, see Dawson and Trapp (2004).

2.4 SCALES OF MEASUREMENT: NOMINAL, ORDINAL, INTERVAL, RATIO

Recall that in our discussion of the so-called "soft" versus "hard" sciences in Chapter 1, we concluded that a key principal difference between the two is not necessarily one of different statistical or analytical methods used in drawing conclusions, but rather in the actual **material** that is subjected to measurement. Though this book is not about measurement per se, we nonetheless wish to review the scales of measurement as first proposed by S.S. Stevens in 1946 (Stevens, 1946).

TABLE 2.3 Contingency Table for 2 × 2 Diagnostic Design

	Diagnosis Yes	Diagnosis No	Total
Disease Yes	20	10	30
Disease No	5	15	20
Total	25	25	50

Before we discuss these scales, it would do well to remind ourselves just what **is** measurement in the first place. We propose the following workable definition:

Measurement is the systematic assignment of numbers to observations according to a well-defined set of rules.

The job of the "rules" is to make good sense of the measurement process. For instance, if we simply assigned numbers to observations without a set of rules to govern the assignment, then even if I weigh more than you, I could be assigned 150 lbs and you 180 lbs. The requirement of having rules of measurement avoids such meaningless and contradictory assignments. If I weigh more than you, rules of measurement imply that my weight measurement will be larger than yours within the margin of measurement error.

2.4.1   Nominal Scale

Measurement at the **nominal** scale is hardly considered **real** measurement, because it is simply the process of grouping objects or subjects into **classes**. Each class is usually represented by a number, letter, name, etc. Other than naming these categories, no other properties are assumed or inferred, such as distance between objects or magnitude.

A classic example of measurement at the nominal level is that of hockey jersey numbers. That the number "99" is greater than the number "22" on the shirts of two hockey players does not imply anything about magnitude (though Wayne Gretzky did in this case wear "99" and was perhaps the best hockey player ever). The numbers 99 and 22 are simply "classes," they are symbols used to identify (or **name**) one class as different or distinct from the other. The fact that we use a rational system such as the real numbers to identify these different classes of "99" versus "22" does not imply anything about order or magnitude **at the level of substantive measurement**. Yes, to the mathematician, 99 is indeed numerically greater than 22. That is, an order property is implied in the numbers. However, to the scientist, nothing of order or magnitude needs to be implied when working with a nominal scale.

To briefly elaborate on this point, the concept of using numbers to represent classes makes for an ideal example of the distinction between mathematical measurement versus scientific measurement. In the mathematical measurement of the distance on the real line (e.g., the "length" between two real numbers), order is a necessary implication and differentiates any two numbers on the line. In scientific measurement, though we may still use the "objects" (i.e., the numbers) of pure mathematics, whether there exist order and magnitude in our **empirical** objects of study is for us to decide as scientists with the aid of our measurement tools. It is not solely a mathematical or "abstract" consideration.

As an example, consider the following objects:

$$* \quad \$ \quad \# \quad \%$$

Though we can say, at minimum, that nominal level measurement has been achieved (the objects have different symbols, that is, different names), we cannot say anything more about either the distance or magnitude between the objects, **unless we decide to impose an order relation** on the above objects. For instance, if we decide, based on our rules of measurement, that $ is greater than *, then not only have we measurement at the nominal level, we also have measurement at the **ordinal** level.

2.4.2   Ordinal Scale

In addition to categorizing objects into classes, measurement at the **ordinal** level imposes an **order relation** between objects. For instance, if $ is greater than * for some characteristic that these symbols represent, then we have measurement at the ordinal level. The imposition of an order relation is fundamental to any sort of true measurement. Consider that if your measurement system does not even allow you to say one thing has **more** of a characteristic than another, what could be the purpose of even measuring?

Ordinal measurement, however, does not say anything about the **precise** amount of magnitude between objects. For example, first place, second place, and third place in a race constitute measurement at the ordinal level, but that you finished second does not immediately tell us the **distance** between first and second, or the **distance** between second and third. To speak of distances between objects, we require measurement at the **interval** level.

2.4.3 Interval Scale

Measurement at the **interval** level possesses all the features of measurement of both nominal and ordinal scales, but with the extra requirement that **distances between measured objects are quantifiable, and that distances between successive measuring points on the scale are equal in magnitude**. For instance, consider the measurement of temperature in degrees Fahrenheit. The change in temperature from 10 degrees to 20 degrees essentially contains the same "amount" of temperature change as that from 20 to 30 degrees. That is, the **intervals** between measurement points are meaningful and represent an equal distance in the "thing" (i.e., temperature, in this case) we are measuring.

Is intelligence measurable on an interval scale? What would it mean for it to be measurable at the interval level? Well, supposing we base our measurements on a reputable standardized test, for IQ to be measurable at the interval scale would imply that the **distance** in the thing called "IQ" is equivalent from say, 90 to 100 as it is from 100 to 110. At first glance, this might appear an easy condition to satisfy, after all, the real number distance in each interval is equal to 10. However, recall that **that is a distance of real numbers, not necessarily of IQ.** As William James put it, we must not confuse the phenomena we study with the abstractions we use to study them. The real numbers are the abstraction. The IQ is the phenomenon. That we used a real line to measure these distances does not necessarily imply that the actual **true** distances in terms of "IQ substance" corresponds one-to-one (or even at all) to our measurement tool. It is entirely possible that 90 to 100 represents a greater increase in IQ than does 100 to 110, making the relation between our measurement of IQ versus "true IQ," **nonlinear**. Our measurement of IQ is simply not that precise to make such statements. **Numerical length** in this case may not translate to the **substantive length** of the difference under study.

2.4.4 Ratio Scale

The most sophisticated scale of measurement is that of the **ratio scale**. It is the most sophisticated because it is the only scale for which we can speak meaningfully about **ratios** between competing measurement intervals. By "ratio," we simply mean we have the power to make such statements as "object a is twice as large as object b." Up to now, no other scale has allowed us to make such statements. For instance, in the interval scale, concluding that a is any factor greater than b made no sense. We did not have a starting point to base such conclusions. An IQ of zero did not necessarily mean the **absence** of intelligence. Rather, it was simply an arbitrary point on the IQ scale presumably denoting a particular quantity of IQ (even if, in all probability, very small).

What gives us license to make statements of ratios? The element of the ratio scale that permits us to make such statements is the fact that the ratio scale has at its origin a **true zero point**. When something is deemed measurable at the ratio scale, a measurement of zero **actually means zero of the thing that is being measured**. Was this fact true of the interval scale? No, because zero degrees Fahrenheit did not equate to there being zero temperature. "Zero" was simply an arbitrary value on the scale. However, the fact that I have zero coins in my pocket actually means that I have **zero** coins. "Zero" is said to be, in this case, "absolute," meaning that there is truly nothing there.

Physical quantities such as weight, distance, velocity, motion, are all measurable at the ratio level. Variables such as reaction time in sensation experiments are also measurable at the ratio level. Phenomena such as intelligence, anxiety, attitude, are generally not. More often we deem them measurable at the interval level or less, and when we really get critical, it is even a stretch at times to consider the ordinal level of measurement as being satisfied for such variables. Then again, if we decided to operationally define anxiety by **beats per minute** of one's heart, then theoretically at least, one could conclude that an individual has zero anxiety if that individual has zero beats per minute (though of course, this could make for an awkward definition for the absence of anxiety!).

2.5 MATHEMATICAL VARIABLES VERSUS RANDOM VARIABLES

When we speak of a **mathematical variable** (or simply, **variable**), we mean a symbol that at any point could be replaced by values contained in a specified set. For instance, consider the mathematical variable y_i. By the subscript i is indicated the fact that y_i stands for a **set** of values, not all equal to the same number (otherwise y would be a constant) such that at any point in time any of these values in the set could serve as a temporary "replacement" for the symbol.

Of course, social and natural sciences are all about variables. Here are some examples:

- **Height** of persons in the world is a variable because persons of the world have different heights. However, height would be considered a constant if 10 people in a room were of the exact same height (and those were the only people we were considering).
- **Blood pressure** is a variable because persons, animals, and other living creatures have different blood pressure measurements.
- **Intelligence** (IQ) of human beings (difficult to measure to be sure, though psychology has developed instruments in an attempt to assess such things) is a variable because presumably people have differing intellectual capacities.
- **Earned run average** (ERA) of baseball players is a variable because players do not all have the name ERA.

A **random variable** is a mathematical variable that is associated with a probability distribution. That is, as soon as we assign probabilities to **values** of the variable, we have a random variable. More formally, we can say that a random variable is a **function from a sample space into the real numbers** (Casella and Berger, 2002), which essentially means that elements in the set (i.e., sample space) have probabilities associated with them (Dowdy, Wearden, and Chilko, 2004).

Consider a simple comparison between a mathematical variable and a discrete random variable in Table 2.4.

Notice that for the mathematical variable, probability does not enter the picture, it is not of any consideration. For the discrete random variable, each value of the variable has a probability associated with it. Note as well that the probabilities must sum to 1.0 for it to be a legitimate probability distribution (i.e., $0.20 + 0.50 + 0.30 = 1.0$). How the given probabilities are assigned is a matter to be governed by the specific context of the problem. Recall as well that variables can be classified as discrete or continuous (see Appendix for a review). This same distinction can be applied to random variables as to ordinary mathematical variables. In Table 2.4 features a discrete random variable. For continuous

TABLE 2.4   Mathematical versus Discrete Random Variable

Mathematical Variable y_i	Random Variable y_i
$y_1 = 1$	$y_1 = 1$ ($p = 0.20$)
$y_2 = 3$	$y_2 = 3$ ($p = 0.50$)
$y_3 = 5$	$y_3 = 5$ ($p = 0.30$)

random variables, since the probability of any particular value in a continuous distribution is theoretically zero, instead of associating probabilities with particular values, probabilities are associated with areas under the curve computed by way of integration in calculus.

The distinction between mathematical and random variables is important when we discuss such things as means, variances, and covariances. A reader first learning about random variables, having already mastered the concept of sample or population variance (to be discussed shortly), can be somewhat taken aback when encountering the variance of a random variable, given as

$$\sigma^2 = E(y_i - \mu)^2$$

and then attempting to compare it to the more familiar variance of a population:

$$\sigma^2 = \frac{\sum_{i=1}^{n}(y_i - \mu)^2}{n}$$

Realize, however, that both expressions are essentially similar, they both account for squared deviations from the mean. However, the variance of a random variable is stated in terms of its **expectation**, E. Throughout this book, we will see the operator E at work. What is an expectation? The expectation E of a random variable is the **mean** of that random variable, which amounts to it being a probability-weighted average (Gill, 2006). The operator E occurs several times throughout this book because in theoretical statistics, **long-run averages** of a statistic are of especial interest. As noted by Feller (1968, p. 221), should an experiment be repeated n times under identical conditions, the average of such trials should be **close to** expectation. Perhaps less formally, the operator E then tells us what we might expect to see in the **long run** for large n. Theoretical statisticians love taking expectations, because the short run of a variable is seldom of interest at a theoretical level. It is the long (probability) run that is often of most theoretical interest. As a crude analogy, on a personal level, you may be "up" or "down" now, but if your expectation E pointed to a favorable long-run endpoint, then perhaps that is enough to convince you that though "on the way" is a rough tumbly road, in the end, as the spiritual would say, we "arrive" at our expectation (which perhaps some would denote as an afterlife of sorts).

The key point is that when we are working with expectations, we are working with **probabilities**. Thus, instead of summing squared deviations of the kind $\sum_{i=1}^{n}(y_i - \mu)^2$ as one does in the sample or population variance for which there is specified n, one must rather assign to these squared deviations **probabilities**, which is what is essentially being communicated by the notation "$E(y_i - \mu)^2$." We can "unpack" this expression to read

$$\sum p(y_i)(y_i - \mu)^2$$

where $p(y_i)$ is the probability of the given deviation, $(y_i - \mu)$, for in this case, a discrete random variable.

2.6   MOMENTS AND EXPECTATIONS

When we speak of **moments** of a distribution or of a random variable, we are referring to such things as the mean, variance, skewness, and kurtosis.

The first moment of a distribution is its mean. For a discrete random variable y_i, the expectation is given by:

$$E(y_i) = \sum_{i=1}^{n} y_i p(y_i)$$

where y_i is the given value of the variable, and $p(y_i)$ is its associated probability. When y_i is a continuous random variable, the expectation is given by:

$$E(y_i) = \int_{-\infty}^{\infty} y_i p(y_i) dy$$

Notice again that in both cases, whether the variable is discrete or continuous, we are simply **summing products** of values of the variable with its probability, or **density** if the variable is continuous. In the case of the discrete variable, the products are "explicit" in that our notation tells us to take each value of y (i.e., y_i) and multiply by the probability of that given value, $p(y_i)$. In the case of a continuous variable, the products are a bit more **implicit** one might say, since the "probability" of any **particular** value in a continuous density is equal to 0. Hence, the product $y_i p(y_i)$ is equal to the given value of y_i multiplied by its corresponding density.

The arithmetic mean is a point such that $\sum_{i=1}^{n} (y_i - \bar{y}) = 0$. That is, the sum of deviations around the mean is always equal to 0 for any data set we may consider. In this sense, we say that the arithmetic mean is the **center of gravity** of a distribution, it is the point that "balances" the distribution (see Figure 2.8).

2.6.1   Sample and Population Mean Vectors

We often wish to analyze data simultaneously on several response variables. For this, we require vector and matrix notation to express our responses. The matrix operations presented here are surveyed more comprehensively in the Appendix and in any book on elementary matrix algebra.

FIGURE 2.8   Because the sum of deviations about the arithmetic mean is always zero, it can be conceptualized as a balance point on a scale.

Consider the following vector:

$$\mathbf{y} = \begin{pmatrix} y_1 \\ y_2 \\ . \\ . \\ . \\ y_n \end{pmatrix}$$

where y_1 is observation 1 up to observation y_n.

We can write the sample mean vector $\bar{\mathbf{y}}$ for several variables y_1 through y_p as

$$\bar{\mathbf{y}} = \frac{1}{n}\sum_{i=1}^{n}\mathbf{y}_i = \begin{pmatrix} \bar{y}_1 \\ \bar{y}_2 \\ . \\ . \\ . \\ \bar{y}_p \end{pmatrix}$$

where $\bar{y}_p$ is the mean of the pth variable.

The expectation of individual observations within each vector is equal to the population mean μ, of which the expectation of the sample vector $\mathbf{y}$ is equal to the population vector, μ. This is simply an extension of scalar algebra to that of matrices:

$$E(\mathbf{y}) = E\begin{pmatrix} y_1 \\ y_2 \\ . \\ . \\ . \\ y_n \end{pmatrix} = \begin{pmatrix} E(y_1) \\ E(y_2) \\ . \\ . \\ . \\ E(y_n) \end{pmatrix} = \begin{pmatrix} \mu_1 \\ \mu_2 \\ . \\ . \\ . \\ \mu_n \end{pmatrix} = \mu$$

Likewise, the expectations of individual sample means $\bar{y}_1, \bar{y}_2, \dots \bar{y}_p$ are equal to their population counterparts, $\mu_1, \mu_2, \dots \mu_p$. The expectation of the sample mean vector $\bar{\mathbf{y}}$ is equal to the population mean vector, μ:

$$E(\bar{\mathbf{y}}) = E\begin{pmatrix} \bar{y}_1 \\ \bar{y}_2 \\ . \\ . \\ . \\ \bar{y}_p \end{pmatrix} = \begin{pmatrix} E(\bar{y}_1) \\ E(\bar{y}_2) \\ . \\ . \\ . \\ E(\bar{y}_p) \end{pmatrix} = \begin{pmatrix} \mu_1 \\ \mu_2 \\ . \\ . \\ . \\ \mu_p \end{pmatrix} = \mu$$

We note also that $\bar{\mathbf{y}}$ is an **unbiased** estimator of μ since $E(\bar{\mathbf{y}}) = \mu$.

Recall that we said that the mean is the first moment of a distribution. We discuss the second moment of a distribution, that of the **variance**, shortly. Before we do so, a brief discussion of estimation is required.

2.7   ESTIMATION AND ESTIMATORS

The goal of statistical inference is, in general, to estimate parameters of a population. We distinguish between point estimators and interval estimators. A **point estimator** is a function of a sample and is used to estimate a parameter in the population. Because estimates generated by estimators will vary from sample to sample, and thus have a probability distribution associated with them, estimators are also often **random variables**. For example, the sample mean $\bar{y}$ is an estimator of the population mean μ. However, if we sample a bunch of $\bar{y}$ from a population for which μ is the actual population mean, we know, both from experience and statistical theory, that $\bar{y}$ will vary from sample to sample. This is why the estimator $\bar{y}$ is often a random variable, because its values will each have associated with them a given probability (density) of occurrence. When we use the estimator to obtain a particular number, that number is known as an **estimate**. An **interval estimator** provides a range of values within which the true parameter is hypothesized to exist within some probability. A popular interval estimator is that of the **confidence interval**, a topic we discuss later in this chapter.

More generally, if T is some statistic, then we can use T as an estimator of a population parameter θ. Whether the estimator T is any **good** depends on several criteria, which we survey now.

On average, in the long run, the statistic T is considered to be an **unbiased estimator** of θ if

$$E(T) = \theta$$

That is, an estimator is considered **unbiased if its expected value is equal to that of the parameter it is seeking to estimate**. The **bias** of an estimator is measured by how much $E(T)$ deviates from θ. When an estimator is biased, then $E(T) \neq \theta$, or, we can say $E(T) - \theta \neq 0$. Since the bias will be a positive number, we can express this last statement as $E(T) - \theta > 0$.

Good estimators are, in general, unbiased. The most popular example of an unbiased estimator is that of the arithmetic sample mean since it can be shown that:

$$E(\bar{y}) = \mu$$

An example of an estimator that is biased is the uncorrected sample variance, as we will soon discuss, since it can be shown that

$$E(S^2) \neq \sigma^2$$

However, S^2 is not **asymptotically** biased. As sample size increases without bound, $E(S^2)$ converges to σ^2. Once the sample variance is corrected via the following, it leads to an unbiased estimator, even for smaller samples:

$$E(s^2) = \sigma^2$$

where now,

$$s^2 = \frac{\sum\limits_{i=1}^{n} (y_i - \bar{y})^2}{n-1}$$

Consistency[6] of an estimator means that as sample size increases indefinitely, the variance of the estimator approaches zero. That is, $\sigma_T^2 \to 0$ as $n \to \infty$. We could also write this using a limit concept:

$$\lim_{n \to \infty} \sigma_T^2 = 0$$

---

[6] Though in this text we define consistency of an estimator quite simply, further distinctions exist between weak and strong consistency. See Shao (2003, pp. 132–133).

which reads "the variance of the estimator T as sample size n goes to infinity (grows without bound) is equal to 0." Fisher called this the **criterion of consistency**, informally defining it as "when applied to the whole population the derived statistic should be equal to the parameter" (Fisher, 1922a, p. 316). The key to Fisher's definition is **whole population**, which means, theoretically at least, an infinitely large sample, or analogously, $n \to \infty$. More pragmatically, $\sigma_T^2 \to 0$ when we have the entire population.

An estimator is regarded as **efficient** the lower is its mean squared error. Estimators with lower variance are more efficient than estimators with higher variance. Fisher called this the **criterion of efficiency**, writing "when the distributions of the statistics tend to normality, that statistic is to be chosen which has the least probable error" (Fisher, 1922a, p. 316). Efficient estimators are generally preferred over less efficient ones.

An estimator is regarded as **sufficient** for a given parameter if the statistic "captures" everything we need to know about the parameter and our knowledge of the parameter could not be improved if we considered additional information (such as a secondary statistic) over and above the sufficient estimator. As Fisher (1922a, p. 316) described it, "the statistic chosen should summarize the whole of the relevant information supplied by the sample." More specifically, Fisher went on to say:

> If θ be the parameter to be estimated, θ_1 a statistic which contains the whole of the information as to the value of θ, which the sample supplies, and θ_2 any other statistic, then the surface of distribution of pairs of values of θ_1 and θ_2, for a given value of θ, is such that for a given value of θ_1, the distribution of θ_2 does not involve θ. In other words, when θ_1 is known, knowledge of the value of θ_2 throws no further light upon the value of θ. (Fisher, 1922a, pp. 316–317)

2.8 VARIANCE

Returning to our discussion of moments, the **variance** is the second moment of a distribution. For the discrete case, variance is defined as:

$$\sigma^2 = \sum_{i=1}^{n} [y_i - E(y_i)]^2 p(y_i)$$

while for the continuous case,

$$\sigma^2 = \int_{-\infty}^{\infty} [y_i - E(y_i)]^2 p(y_i) dy$$

Since $E(y_i) = \mu$, it stands that we may also write $E(y_i)$ as μ. We can also express σ^2 as $E(y_i^2) - \mu^2$ since, when we distribute expectations, we obtain:

$$
\begin{aligned}
\sigma^2 &= E(y_i - \mu)^2 \\
&= E(y_i - \mu)(y_i - \mu) \\
&= E(y_i^2 - y_i\mu - y_i\mu + \mu^2) \\
&= E(y_i^2) - E(y_i)\mu - E(y_i)\mu + \mu^2 \\
&= E(y_i^2) - \mu\mu - \mu\mu + \mu^2 \\
&= E(y_i^2) - \mu^2 - \mu^2 + \mu^2 \\
&= E(y_i^2) - \mu^2
\end{aligned}
$$

Recall that the uncorrected and **biased** sample variance is given by:

$$S^2 = \frac{\sum\limits_{i=1}^{n}(y_i - \bar{y})^2}{n}$$

As earlier noted, taking the expectation of S^2, we find that $E(S^2) \neq \sigma^2$. The actual expectation of S^2 is equal to:

$$E(S^2) = [(n-1)/n]\sigma^2$$

which implies the degree to which S^2 is biased is equal to:

$$\frac{-\sigma^2}{n}$$

We have said that S^2 is biased, but you may have noticed that as n increases, $(n-1)/n$ approaches 1, and so $E(S^2)$ will equal σ^2 as n increases without bound. This was our basis for earlier writing $\lim\limits_{n \to \infty} E(S^2) = \sigma^2$. That is, we say that the estimator S^2, though biased for small samples, is **asymptotically unbiased** because its expectation is equal to σ^2 as $n \to \infty$.

When we lose a degree of freedom in the denominator and rename S^2 to s^2, we get

$$s^2 = \frac{\sum\limits_{i=1}^{n}(y_i - \bar{y})^2}{n-1}$$

Recall that when we take the expectation of s^2, we find that $E(s^2) = \sigma^2$ (see Wackerly, Mendenhall, and Scheaffer (2002, pp. 372–373) for a proof).

The population **standard deviation** is given by the positive square root of σ^2, that is, $\sqrt{\sigma^2} = \sigma$. Analogously, the sample standard deviation is given by $\sqrt{s^2} = s$.

Recall the interpretation of a standard deviation. It tells us **on average how much scores deviate from the mean**. In computing a measure of dispersion, we initially squared deviations so as to avoid our measure of dispersion always equaling zero for any given set of observations, since the sum of deviations about the mean is always equal to 0. Taking the average of this sum of squares gave us the variance, but since this is in squared units, we wish to return them to "unsquared" units. This is how the standard deviation comes about. Studying the analysis of variance, the topic of the following chapter, will help in "cementing" some of these ideas of variance and the squaring of deviations, since ANOVA is all about generating different sums of squares and their averages, which go by the name of **mean squares**.

The variance and standard deviation are easily obtained in R. We compute for parent in Galton's data:

```
> var(parent)
[1] 3.194561

> sd(parent)
[1] 1.787333
```

One may also wish to compute what is known as the **coefficient of variation**, which is a ratio of the standard deviation to the mean. We can estimate this coefficient for `parent` and `child` respectively in Galton's data:

```
> cv.parent <- sd(parent)/mean(parent)
> cv.parent
```

```
[1] 0.02616573
```

```
> cv.child <- sd(child)/mean(child)
> cv.child
```

```
[1] 0.03698044
```

Computing the coefficient of variation is a way of comparing the variability of competing distributions relative to each distribution's mean. We can see that the dispersion of child relative to its mean (0.037) is slightly larger than that of the dispersion of parent relative to its mean (0.026).

2.9   DEGREES OF FREEDOM

In our discussion of variance, we saw that if we wanted to use the sample variance as an estimator of the population variance, we needed to subtract 1 from the denominator. That is, S^2 was "corrected" into s^2:

$$s^2 = \frac{\sum_{i=1}^{n} (y_i - \bar{y})^2}{n-1}$$

We say we **lost a degree of freedom** in the denominator of the statistic. But what are degrees of freedom? They are **the number of independent units of information in a sample that are relevant to the estimation of some parameter** (Everitt, 2002). In the case of the sample variance, s^2, one degree of freedom is lost since we are interested in using s^2 as an estimator of σ^2. We are losing the degree of freedom because the numerator, $\sum_{i=1}^{n} (y_i - \bar{y})^2$, is not based on n independent pieces of information since μ had to be estimated by $\bar{y}$. Hence, a degree of freedom is lost. Why? Because values of y_i are not independent of what $\bar{y}$ is, since $\bar{y}$ is fixed in terms of the given sample data. In general, when we estimate a parameter, it "costs" a degree of freedom. Had we μ, such that $\sum_{i=1}^{n} (y_i - \mu)^2$, we would have not lost a degree of freedom, since μ is a known (not estimated) parameter.

A conceptual demonstration may prove useful in understanding the concept of degrees of freedom. Imagine you were asked to build a triangle such that there was to be no overlap of lines on either side of the triangle. In other words, the lengths of the sides had to join neatly at the vertices. We shall call this the "**Beautiful Triangle**" as depicted in Figure 2.9. You are now asked to draw the first side of the triangle. Why did you draw this first side the length that you did? You concede that the length of the first side is arbitrary, you were **free** to draw it whatever length you wished. In drawing the second length, you acknowledge you were also **free** to draw it whatever length you wished. Neither of the first two lengths in any way violated the construction of a beautiful triangle with perfectly adjoining vertices.

However, in drawing the third length, what length did you choose? Notice that to complete the triangle, you were not **free** to determine this length arbitrarily. Rather, the length was **fixed** given the constraint that the triangle was to be a beautiful one. In summary then, in building the beautiful triangle, you lost 1 degree of freedom, in that two of the lengths were of your free choosing, but the third was fixed. Analogously, in using s^2 as an estimator of σ^2, a single degree of freedom is lost. If $\bar{y}$ is equal to 10,

FIGURE 2.9   The "Beautiful Triangle" as a way to understanding degrees of freedom.

for instance, and the sample is based on five observations, then y_1, y_2, y_3, y_4 are freely chosen, but the fifth data point, y_5 is not freely chosen so long as the mean must equal 10. The fifth data point is fixed. We lost a single degree of freedom.

Degrees of freedom occur throughout statistics in a variety of statistical tests. If you understand this basic example, then while working out degrees of freedom for more advanced designs and tests may still pose a challenge, you will nonetheless have a conceptual base from which to build your comprehension.

2.10   SKEWNESS AND KURTOSIS

The third moment of a distribution is its **skewness**. Skewness of a random variable generally refers to the extent to which a **distribution lacks symmetry**. Skewness is defined as:

$$\gamma = \frac{E\left[(y_i - \mu)^3\right]}{\left(E\left[(y_i - \mu)^2\right]\right)^{3/2}}$$

- Skewness for a normal distribution is equal to 0, just as skewness for a rectangular distribution is also equal to 0 (one does not necessarily require a **bell-shaped** curve for skewness to equal 0)
- Skewness for a **positively** skewed distribution is greater than 0; these distributions have tails that stretch out into values on the abscissa of greatest value
- Skewness for a **negatively** skewed distribution is less than 0; these distributions have tails that stretch out to values on the abscissa of least value

An example of a positively skewed distribution is that of the typical F density, given in Figure 2.10.

The fourth moment of a distribution is its **kurtosis**, generally referring to the peakness of a distribution (Upton and Cook, 2002), but also having much to do with a distribution's tails (DeCarlo, 1997):

$$k = \frac{E\left[(y_i - \mu)^4\right]}{\left(E\left[(y_i - \mu)^2\right]\right)^2}$$

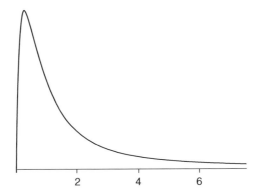

FIGURE 2.10   *F* distribution on 2 and 5 degrees of freedom. It is positively skewed since the tail stretches out to numbers of greater value.

With regard to kurtosis, distributions are defined:

- **mesokurtic** if the distribution exhibits kurtosis typical of a bell-shaped normal curve
- **platykurtic** if the distribution exhibits lighter tails and is flatter toward the center than a normal distribution
- **leptokurtic** if the distribution exhibits heavier tails and is generally more narrow in the center than a normal distribution, revealing that it is somewhat "peaked"

We can easily compute moments of empirical distributions in R or SPSS. Several packages in R are available for this purpose. We could compute skewness for parent on Galton's data by:

```
> library(psych)
> skew(parent)
[1] -0.03503614
```

The `psych` package (Revelle, 2015) also provides a range of descriptive statistics:

```
> library(psych)
> describe(Galton)
```

	vars	n	mean	sd	median	trimmed	mad	min	max	range	skew	kurtosis
parent	1	928	68.31	1.79	68.5	68.32	1.48	64.0	73.0	9	-0.04	0.05
child	2	928	68.09	2.52	68.2	68.12	2.97	61.7	73.7	12	-0.09	-0.35

	se
parent	0.06
child	0.08

The skew for child has a value of −0.09, indicating a **slight** negative skew. This is confirmed by visualizing the distribution (and by a relatively close inspection in order to spot the skewness):

```
> hist(child)
```

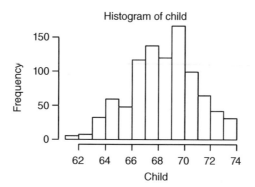

2.11 SAMPLING DISTRIBUTIONS

Sampling distributions are at the cornerstone of statistical inference. The sampling distribution of a statistic is a **theoretical probability distribution of that statistic**. As defined by Degroot and Schervish (2002), "the sampling distribution of a statistic tells us what values a statistic is likely to assume and how likely it is to assume those values prior to observing our data" (p. 391).

As an example, we will generate a theoretical sampling distribution of the mean for a given population with mean μ and variance, σ^2. The distribution we will create is entirely **idealized** in that it does not exist in nature anywhere. It is simply a statistical **theory** of how the distribution of means might look if we were able to take an **infinite number of samples** of a given size from a given population, and on each of these samples, calculate the sample mean statistic.

When we derive sampling distributions for a statistic, we are asking the following question:

If we were to draw an infinite number of samples of size n from this population and calculate the sample mean on each sample, what would the distribution of sample means look like?

If we can specify this distribution, then we can evaluate obtained sample means **relative** to it. That is, we will be able to compare our obtained means (i.e., the ones we obtain in real empirical research) to the theoretical sampling distribution of means, and answer the question:

If my obtained sample mean really did come from this population, what is the probability of obtaining a mean such as this?

If the probability is low, you might then decide to reject the assumption that the sample mean you obtained arose from the population in question. It could have, to be sure, but it **probably** did not. For continuous measures, our interpretation above is slightly informal, since the probability of any **particular value** of the sample mean in a continuous distribution is essentially equal to 0 (i.e., in the limit, the probability equals 0). Hence, the question is usually posed such that we seek to know the probability of obtaining a mean such as the one we obtained **or more extreme**.

2.11.1 Sampling Distribution of the Mean

Since we regularly calculate and analyze sample means in our data, we are often interested in the sampling distribution of the mean. If we regularly computed medians, we would be equally as interested in the sampling distribution of the median.

Recall that when we consider any distribution, whether theoretical or empirical, we are usually especially interested in knowing two things about that distribution: a measure of **central tendency** and a measure of **dispersion** or variability. Why do we want to know such things? We want to know these two things because they help summarize our observations, so that instead of looking at each individual data point to get an adequate description of the objects under study, we can simply request the mean and standard deviation as telling the story (albeit an incomplete one) of the obtained observations. Similarly, when we derive a sampling distribution, we are interested in the mean and standard deviation of that theoretical distribution of a statistic.

We already know how to calculate means and standard deviations for real empirical distributions. However, we do not know how to calculate means and standard deviations for sampling distributions. It seems reasonable that the mean and standard deviation of a sampling distribution should depend in some way on the given population from which we are sampling. For instance, if we are sampling from a population that has a mean $\mu = 20.0$ and population standard deviation $\sigma = 5$, it seems plausible that the sampling distribution of the mean should look different than if we were sampling from a population with $\mu = 10.0$ and $\sigma = 2$. It makes sense that **different populations should give rise to different theoretical sampling distributions**.

What we need then is a way to specify the sampling distribution of the mean for a given population. That is, if we draw sample means from this population, what does the sampling distribution of the mean look like for this population? To answer this question, we need both the expectation of the sampling distribution (i.e., its mean) as well as the standard deviation of the sampling distribution (i.e., its standard error (SE)). We know that the expectation of the sample mean $\bar{y}$ is equal to the population mean μ. That is, $E(\bar{y}) = \mu$. For example, for a sample mean $\bar{y} = 20.0$, the expected value of the sample mean is equal to the population mean μ of 20.0.

To understand why $E(\bar{y}) = \mu$ should be true, consider first how the sample mean is defined:

$$\bar{y} = \frac{(y_1 + y_2 + \cdots + y_n)}{n}$$

Incorporating this into the expectation for $\bar{y}$, we have:

$$E(\bar{y}) = E\left(\frac{(y_1 + y_2 + \cdots + y_n)}{n}\right)$$

There is a rule of expectations that says that **the expectation of the sum of random variables is equal to the sum of individual expectations**. This being the case, we can write the expectation of the sample mean $\bar{y}$ as:

$$E(\bar{y}) = \frac{E(y_1 + y_2 + \cdots + y_n)}{n}$$
$$= \frac{[E(y_1) + E(y_2) + \cdots + E(y_n)]}{n}$$

Since the expectation of each y_1 through y_n is $E(y_1) = \mu$, $E(y_2) = \mu$, ... $E(y_n) = \mu$, we can write

$$E(\bar{y}) = \frac{[\mu + \mu + \cdots + \mu]}{n}$$
$$E(\bar{y}) = \frac{n\mu}{n}$$

We note that the n values in numerator and denominator cancel, and so we end up with

$$E(\bar{y}) = \mu$$

Using the fact that $E(y_i) = \mu$, we can also say that the expected value of a sampling distribution of the mean is equal to the mean of the population from which we did the theoretical sampling. That is, $\mu_{\bar{y}} = \mu$ is true, since given $E(\bar{y}) = \mu$, it stands that if we have say, five sample means $\bar{y}_1, \bar{y}_2, \bar{y}_3, \bar{y}_4, \bar{y}_5$, the expectation of each of these means should be equal to μ, from which we can easily deduce $\mu_{\bar{y}} = \mu$. That is, **the mean of all the samples we could draw is equal to the population mean**.

We now need a measure of the **dispersion** of a sampling distribution of the mean. At first glance, it may seem reasonable to assume that the variance of the sampling distribution of means should equal the variance of the population from which the sample means were drawn. However, this is not the case. What is true is that the variance of the sampling distribution of means will be equal to only a **fraction** of the population variance. It will be equal to $\frac{1}{n}$ of it, where n is equal to the size of samples we are collecting for each sample mean. Hence, the variance of means of the sampling distribution is equal to

$$\frac{1}{n}\left(\sigma^2\right)$$

or simply,

$$\frac{\sigma^2}{n}$$

The mathematical proof of this statistical fact is in most mathematical statistics texts. A version of the proof can also be found in Hays (1994). The idea, however, can be easily and perhaps even more intuitively understood by recourse to what happens as n changes. We consider first the most trivial and unrealistic of examples to strongly demonstrate the point. Suppose that we calculate the sample mean from a sample size of $n = 1$, sampled from a population with $\mu = 10.0$ and $\sigma^2 = 2.0$. Suppose the sample mean we obtain is equal to 4.0. Therefore, the sampling variance of the corresponding sampling distribution is equal to:

$$\frac{\sigma^2}{n} = \frac{2}{1} = 2$$

That is, the variance in means that you can expect to see if you sampled an infinite number of means based on samples of size $n = 1$ repeatedly from this population is equal to 2. Notice that 2 is exactly equal to the original population variance. In this case, the variance in means is based on only a single data point.

Consider now the case where $n > 1$. Suppose we now sampled a mean from the population based on sample size $n = 2$, yielding

$$\frac{\sigma^2}{n} = \frac{2}{2} = 1$$

What has happened? What has happened is that the variance in sample means has decreased by 1/2 of the original population variance (i.e., 1/2 of 2 is 1). Why is this decrease reasonable? It makes sense, because we already know from the **law of large numbers** that as the sample size grows larger, one gets closer and closer to the true probability in estimating a parameter. That is, for a consistent estimator, our estimate of the true population mean (i.e., the expectation) should get better and better as sample size increases. This is exactly what happens as we increase n, our **precision** of that which is being estimated increases. In other words, the sampling variance of the estimator **decreases**. It's less variable, it doesn't "bounce around as much" on average from sample to sample.

Analogous to how we defined the standard deviation as the square root of the variance, it is also useful to take the square root of the variance of means:

$$\sqrt{\frac{\sigma^2}{n}} = \frac{\sigma}{\sqrt{n}}$$

which we call the **standard error of the mean**, σ_M. The standard error of the mean is the **standard deviation of the sampling distribution of the mean**. Lastly, it is important to recognize that $\frac{\sigma}{\sqrt{n}}$ is not "the" standard error. It is merely the standard error **of the mean**. Other statistics will have different SEs.

2.12 CENTRAL LIMIT THEOREM

It is not an exaggeration to say that the **central limit theorem**, in one form or another, is probably the most important and relevant theorem in theoretical statistics, which translates to it being quite relevant to applied statistics as well.

We borrow our definition of the central limit theorem from Everitt (2002):

> If a random variable y has a population mean μ and population variance σ^2, then the sample mean, $\bar{y}$, based on n observations, has an approximate normal distribution with mean μ and variance $\dfrac{\sigma^2}{n}$, for sufficiently large n. (p. 64)

Asymptotically, the distribution of a normal random variable **converges** to that of a normal distribution as $n \to \infty$. A multivariate version of the theorem can also be given (e.g., see Rencher, 1998, p. 53).[7]

The relevance and importance of the central limit theorem cannot be overstated: it allows one to know, at least on a theoretical level, what the distribution of a statistic (e.g., sample mean) will look like for increasing sample size. This is especially important if one is drawing samples from a population for which the shape is not known or is known a priori to be nonnormal. **Normality of the sampling distribution, for adequate sample size, is still assured even if samples are drawn from nonnormal populations**. Why is this relevant? It is relevant because if we know what the distribution of means will look like for increasing sample size, then we know we can compare our obtained statistic to a normal distribution in order to estimate its probability of occurrence. Normality assumptions are also typically required for assuming independence between $\bar{y}$ and s^2 in univariate contexts (Lukacs, 1942), and $\bar{y}$ (mean vector) and $\mathbf{S}$ (covariance matrix) in multivariate ones. When such estimators can be assumed to arise from normal or multivariate normal distributions (i.e., in the case of $\bar{y}$ and $\mathbf{S}$) we can generally be assured one is independent of the other.

2.13 CONFIDENCE INTERVALS

Recall that a goal of statistical inference is to estimate functions of parameters, whether a single parameter, a difference of parameters (for instance, in the case of population differences), or some other function of parameters. Though the sample mean $\bar{y}$ is an unbiased estimator of μ, the probability that $\bar{y}$ is

---

[7] We can also distinguish between weaker vs. stronger forms of the theorem. For details, see Casella & Berger (2002, pp. 236–238).

equal to μ in any given sample, for a continuous measure, converges to **zero** (Hays, 1994). For this reason, and to build some flexibility in estimation overall, the idea of **interval estimation** in the form of **confidence intervals** was developed. Confidence intervals provide a range of values for which we can be relatively certain lay the true parameter we are seeking to estimate. In what follows, we provide a brief review of 95 and 99% confidence intervals.

We can say that over all samples of a given size n, the probability is 0.95 for the following event to occur:

$$-1.96\sigma_M < \bar{y} - \mu < 1.96\sigma_M \tag{2.2}$$

How was (2.2) obtained? Recall the calculation of a z-score for a mean:

$$z = \frac{\bar{y} - \mu}{\sigma_M}$$

Suppose now that we want to have a 0.025 area on either side of the normal distribution. This value corresponds to a z-score of 1.96, since the probability of a z-score of ± 1.96 is $2(1 - 0.9750021) = 0.0499958$, which is approximately 5% of the total curve. So, from the z-score, we have

$$z = \frac{\bar{y} - \mu}{\sigma_M}$$

$$\pm 1.96 = \frac{\bar{y} - \mu}{\sigma_M}$$

We can modify the equality slightly to get the following:

$$\bar{y} - 1.96\sigma_M < \mu < \bar{y} + 1.96\sigma_M \tag{2.3}$$

We interpret (2.3) as follows:

> **Over all possible samples, the probability is 0.95 that the range between $\bar{y} - 1.96\sigma_M$ and $\bar{y} + 1.96\sigma_M$ will include the true mean, μ.**

Very important to note regarding the above statement is that μ is **not** the random variable. The part that is random is the sample on which is computed the interval. That is, the probability statement is not about μ but rather is about **samples**. The population mean μ is assumed to be **fixed**. The 95% confidence interval tells us that **if we continued to sample repeatedly, and on each sample computed a confidence interval, then 95% of these intervals would include the true parameter**.

The 99% confidence interval for the mean is likewise given by:

$$\bar{y} - 2.58\sigma_M < \mu < \bar{y} + 2.58\sigma_M \tag{2.4}$$

Notice that the only difference between (2.3) and (2.4) is the choice of different critical values on either side of μ (i.e., 1.96 for the 95% interval and 2.58 for the 99% interval).

Though of course not very useful, a 100% confidence interval, if constructed, would be defined as:

$$\bar{y} - \infty\,\sigma_M < \mu < \bar{y} + \infty\,\sigma_M$$

If you think about it carefully, the 100% confidence interval should make perfect sense. If you would like to be 100% "sure" that the interval will cover the true population mean, then you have to extend your limits to negative and positive infinity, otherwise, you could not be **fully** confident. Likewise, on the other extreme, a 0% interval would simply have $\bar{y}$ as the upper and lower limits:

$$\bar{y} < \mu < \bar{y}$$

That is, if you want to have **zero confidence** in guessing the location of the population mean, μ, then guess the sample mean $\bar{y}$. Though the sample mean is an unbiased estimator of the population mean, the probability that the sample mean covers the population mean exactly, as mentioned, essentially converges to 0 for a truly continuous distribution (Hays, 1994). As an analogy, imagine coming home and hugging your spouse. If your arms are open infinitely wide (full "bear hug"), you are 100% confident to entrap him or her in your hug because your arms (limits of the interval) extend to positive and negative infinity. If you bring your arms in a little, then it becomes possible to miss him or her with the hug (e.g., 95% interval). However, the precision of the hug is a bit more refined (because your arms are closing inward a bit instead of extending infinitely on both sides). If you approach your spouse with hands together (i.e., point estimate), you are sure to miss him or her, and would have 0% confidence of your interval (hug) entrapping your spouse. An inexact analogy to be sure, but useful in visualizing the concept of confidence intervals.

2.14   MAXIMUM LIKELIHOOD

When we speak of **likelihood**, we mean the probability of some sample data or set of observations conditional on some hypothesized parameter or set of parameters (Everitt, 2002). Conditional probability statements such as $p(D/H_0)$ can very generally be considered simple examples of likelihoods, where typically the set of parameters, in this case, may be simply μ and σ^2. A **likelihood function** is the likelihood of a parameter given data (see Fox, 2016).

When we speak of **maximum-likelihood** estimation, we mean the process of maximizing a likelihood subject to certain parameter conditions. As a simple example, suppose we obtain 8 heads on 10 flips of a presumably fair coin. Our null hypothesis was that the coin is fair, meaning that the probability of heads is $p(H) = 0.5$. However, our actual obtained result of 8 heads on 10 flips would suggest the true probability of heads to be closer to $p(H) = 0.8$. Thus, we ask the question:

Which value of θ makes the observed result most likely?

If we only had two choices of θ to select from, 0.5 and 0.8, our answer would have to be 0.8, since this value of the parameter θ makes the sample result of 8 heads out of 10 flips most **likely**. That is the essence of how maximum-likelihood estimation works (see Hays, 1994, for a similar example). ML is the most common method of estimating parameters in many models, including factor analysis, path analysis, and structural equation models to be discussed later in the book. There are very good reasons why mathematical statisticians generally approve of maximum likelihood. We summarize some of their most favorable properties.

Firstly, ML estimators are **asymptotically unbiased**, which means that bias essentially vanishes as sample size increases without bound (Bollen, 1989). Secondly, ML estimators are **consistent** and asymptotically **efficient**, the latter meaning that the estimator has a small asymptotic variance relative to many other estimators. Thirdly, ML estimators are asymptotically normally distributed, meaning that as sample size grows, the estimator takes on a normal distribution. Finally, ML estimators possess the **invariance** property (see Casella and Berger, 2002, for details).

2.15   AKAIKE'S INFORMATION CRITERIA

A measure of model fit commonly used in comparing models that uses the log-likelihood is **Akaike's information criteria**, or **AIC** (Sakamoto, Ishiguro, and Kitagawa, 1986). This is one statistic of the kind generally referred to as **penalized likelihood** statistics (another is the **Bayesian information criterion**, or **BIC**). AIC is defined as:

$$-2L_m + 2m$$

where L_m is the maximized log-likelihood and m is the number of parameters in the given model. Lower values of AIC indicate a better-fitting model than do larger values. Recall that the more parameters fit to a model, in general, the better will be the fit of that model. For example, a model that has a unique parameter for **each** data point would fit perfectly. This is the so-called **saturated** model. AIC jointly considers both the goodness of fit as well as the number of parameters required to obtain the given fit, essentially "penalizing" for increasing the number of parameters unless they contribute to model fit. Adding one or more parameters to a model may cause $-2L_m$ to decrease (which is a good thing substantively), but if the parameters are not worthwhile, this will be offset by an increase in $2m$.

The **Bayesian information criterion**, or **BIC** (Schwarz, 1978) is defined as $-2L_m + m \log(N)$, where m, as before, is the number of parameters in the model and N the total number of observations used to fit the model. Lower values of BIC are also desirable when comparing models. BIC typically penalizes model complexity more heavily than AIC. For a comparison of AIC and BIC, see Burnham and Anderson (2011).

2.16   COVARIANCE AND CORRELATION

The covariance of a random variable is given by:

$$\text{cov}(x_i, y_i) = \sigma_{xy} = E\left[(x_i - \mu_x)(y_i - \mu_y)\right]$$

where $E[(x_i - \mu_x)(y_i - \mu_y)]$ is equal to $E(x_i y_i) - \mu_x \mu_y$ since

$$
\begin{aligned}
\sigma_{xy} &= E\left[(x_i - \mu_x)(y_i - \mu_y)\right] \\
&= E\left(x_i y_i - x_i \mu_y - y_i \mu_x + \mu_x \mu_y\right) \\
&= E(x_i y_i) - E(x_i)\mu_y - E(y_i)\mu_x + \mu_x \mu_y \\
&= E(x_i y_i) - \mu_x \mu_y - \mu_y \mu_x + \mu_x \mu_y \\
&= E(x_i y_i) - \mu_x \mu_y
\end{aligned}
$$

The concept of covariance is at the heart of virtually all statistical methods. Whether one is running analysis of variance, regression, principal component analysis, etc. covariance concepts are central to all of these methodologies and even more broadly to science in general.

The sample covariance is a measure of relationship between two variables and is defined as:

$$\text{cov} = \frac{\sum_{i=1}^{n}(x_i - \bar{x})(y_i - \bar{y})}{n} \tag{2.5}$$

The numerator of the covariance, $\sum_{i=1}^{n}(x_i - \bar{x})(y_i - \bar{y})$, is the sum of products of respective deviations of observations from their respective means. If there is no **linear** relationship between two variables in a sample, covariance will equal 0. If there is a **negative** linear relationship, covariance will be a negative number, and if there is a **positive** linear relationship covariance will be positive. Notice that to measure covariance between two variables requires there to be **variability** on each variable. If there is no variability in x_i, then $(x_i - \bar{x})$ will equal 0 for all observations. Likewise, if there is no variability in y_i, then $(y_i - \bar{y})$ will equal 0 for all observations on y_i. This is to emphasize the essential fact that when measuring the extent of relationship between two variables, one requires variability on each variable to motivate a measure of relationship in the first place.

The covariance of (2.5) is a perfectly reasonable one to calculate for a sample if there is no intention of using that covariance as an estimator of the population covariance. However, if one wishes to use it as an unbiased estimator, similar to how we needed to subtract 1 from the denominator of the variance, we lose 1 degree of freedom when computing the covariance:

$$\text{cov} = \frac{\sum_{i=1}^{n}(x_i - \bar{x})(y_i - \bar{y})}{n - 1}$$

It is easy to understand more of what the covariance actually measures if we consider the trivial case of computing the covariance of a variable with itself. In such a case for variable x_i, we would have

$$\text{cov} = \frac{\sum_{i=1}^{n}(x_i - \bar{x})(x_i - \bar{x})}{n - 1}$$

But what is this covariance? If we rewrite the numerator as $(x_i - \bar{x})^2$ instead of $(x_i - \bar{x})(x_i - \bar{x})$, it becomes clear that the covariance of a variable with itself is nothing more than the usual **variance** for that variable. Hence, to better understand the covariance, it is helpful to start with the variance, and then realize that instead of computing the cross-product of a variable with **itself**, the covariance computes the cross-product of a variable with a **second** variable.

We compute the covariance between parent height and child height in Galton's data:

```
> attach(Galton)
> cov(parent, child)
[1] 2.064614
```

We have mentioned that the covariance is a measure of linear relationship. However, sample covariances from data set to data set are not comparable unless one knows more of what went into each specific computation. There are actually three things that can be said to be the "ingredients" of the covariance. The first thing it contains is a measure of the cross-product, which represents the degree to which variables are linearly related. This is the part in our computation of the covariance that we are especially interested in. However, other than concluding a negative, zero, or positive relationship, the size of the covariance does not by itself tell us the **degree** to which two variables are linearly related.

The reason for this is that the size of covariance will also be impacted by the degree to which there is variability in x_i and the degree to which there is variability in y_i. If either or both variables contain sizeable deviations of the sort $(x_i - \bar{x})$ or $(y_i - \bar{y})$, then the corresponding cross-products $(x_i - \bar{x})(y_i - \bar{y})$ will also be quite sizeable, along with their sum, $\sum_{i=1}^{n}(x_i - \bar{x})(y_i - \bar{y})$. However, we do not want our measure of relationship to be small or large as a consequence of variability on x_i or variability on y_i. We want our measure of relationship to be small or large as an exclusive result of **covariability**, that is, the extent to which there is actually a **relationship** between x_i and y_i. To incorporate the influences of variability in x_i and y_i (one may think of it as "purifying"), we divide the average cross-product (i.e., the covariance) by the product of standard deviations of each variable. The **standardized sample covariance** is thus:

$$r = \frac{\dfrac{\sum_{i=1}^{n}(x_i - \bar{x})(y_i - \bar{y})}{n-1}}{\sqrt{s_{x_i}^2 \cdot s_{y_i}^2}} = \frac{\text{cov}}{\sqrt{s_{x_i}^2 \cdot s_{y_i}^2}}$$

The standardized covariance is known as the **Pearson product-moment correlation coefficient**, or simply r, which is a **biased** estimator of its population counterpart, ρ_{xy}, except when ρ_{xy} is exactly equal to 0. The bias of the estimator r can be minimized by computing an adjustment found in Rencher (1998, p. 6), originally proposed by Olkin and Pratt (1958):

$$r^* = r\left[1 + \frac{1-r^2}{2(n-3)}\right]$$

Because the correlation coefficient is standardized, we can place lower and upper bounds on it. The minimum the correlation can be for any set of data is −1.0, representing a perfect negative relationship. The maximum the correlation can be is +1.0, representing a perfect positive relationship. A correlation of 0 represents the absence of a **linear** relationship. For further discussion on how the Pearson correlation can be a biased estimate under conditions of nonnormality (and potential solutions), see Bishara and Hittner (2015).

One can gain an appreciation for the upper and lower bound of r by considering the fact that the numerator, which is an average cross-product, is being divided by another product, that of the standard deviations of each variable. The denominator thus can be conceptualized to represent the total amount of cross-product variation **possible**, that is, the "base," whereas the numerator represents the total amount of cross-product variation actually existing between the variables **because of a linear relationship**. The extent to which cov_{xy} accounts for all of the possible "cross-variation" in $\sqrt{s_{x_i}^2 \cdot s_{y_i}^2}$ is the extent to which r will approximate a value of |1| (either positive or negative, depending on the direction of the relationship). It thus stands that cov_{xy} cannot be greater than the "base" to which it is being compared (i.e., $\sqrt{s_{x_i}^2 \cdot s_{y_i}^2}$). In the language of sets, cov_{xy} must be a **subset** of the larger set represented by $\sqrt{s_{x_i}^2 \cdot s_{y_i}^2}$.

It is important to emphasize that a correlation of 0 does not necessarily represent the **absence** of a relationship. What it does represent is the absence of a **linear** one. Neither the covariance or Pearson's r capture nonlinear relationships, and so it is possible to have very strong relations in a sample or population yet still obtain very low values (even zero) for the covariance or Pearson r. **Always plot your data to see what is going on before drawing any conclusions. Correlation coefficients should never be presented without an accompanying plot to characterize the form of the relationship.**

We compute the Pearson correlation coefficient on Galton's data between `child` and `parent`:

```
> cor(child, parent)
[1] 0.4587624
```

We can test it for statistical significance by using the `cor.test` function:

```
> cor.test(child, parent)

        Pearson's product-moment correlation

data:  child and parent
t = 15.7111, df = 926, p-value < 2.2e-16
alternative hypothesis: true correlation is not equal to 0
95 percent confidence interval:
 0.4064067 0.5081153
sample estimates:
      cor
0.4587624
```

We can see that observed t is statistically significant with a computed 95% confidence interval having limits 0.41 to 0.51, indicating that we can be 95% confident that the true parameter lies approximately between the limits of 0.41 and 0.51. Using the package `ggplot2` (Wickham, 2009), we plot the relationship between parent and child (with a smoother):

```
> library(ggplot2)
> qplot(child, parent, data = Galton, geom = c("point", "smooth"))
```

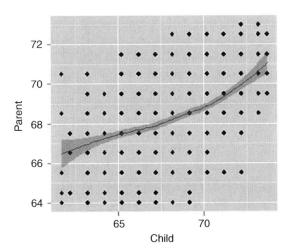

One drawback of such a simple plot is that the frequency of data points in the bivariate space cannot be known by inspection of the plot alone. **Jittering** is a technique that allows one to visualize the density of points at each parent–child pairing. By jittering, we can see where most of the data fall in the parent–child scatterplot (i.e., points are concentrated toward the center of the plot):

```
> qplot(child, parent, geom = "jitter")
```

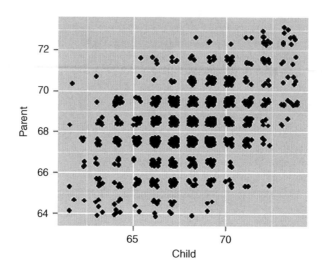

2.17   PSYCHOMETRIC VALIDITY, RELIABILITY:
A COMMON USE OF CORRELATION COEFFICIENTS

Correlation coefficients, specifically the Pearson correlation, are employed in virtually all fields of study, and without the invention or discovery of correlation, most modern-day statistics would simply not exist. This is especially true for the field of **psychometrics**, which is the science that deals with the measurement of psychological qualities such as intelligence, self-esteem, motivation, among others. Psychometrics features the development of psychometric tests purported to measure the construct of interest. For an excellent general introduction to psychometrics, consult McDonald (1999).

When developing psychometric instruments, two statistical characteristics of these tests are especially important: (1) **validity**, and (2) **reliability**. Validity of a test takes many forms, including **face validity**, **criterion validity**, and most notably, **construct validity**. Construct validity attempts to assess whether a purported psychometric test actually **measures what it was designed to measure**, and one way of evaluating construct validity is to correlate the newly developed measure with that of an existing measure that is already known to successfully measure the construct.

For example, in the area of depression assessment, the **Beck Depression Inventory (BDI)** is a popular self-report measure often used in evaluating one's level or symptoms of depression. Now, if we were to develop a new test, in order to learn whether that new test measures something called "depression," we may wish to compute a Pearson correlation of that measure with the BDI. To the extent that the correlation is relatively high, we might tentatively conclude that the new measure is assessing the same (or at least a **similar**) construct as that of the BDI. Not surprisingly, these correlations in this context often go by the name of **validities** in the psychometric literature. If a test lacks construct validity, then there is little guarantee that it is measuring the construct under investigation. Fields such as psychology depend on such construct validation to gain some sense of certainty that their measures are tapping into what they are most interested in. **Clinical psychology**, especially, depends on the strength of such things as construct validity to secure a sense of sureness that their diagnostic tests are measuring what they are thought to measure. Without psychometrics, clinical testing in this way would be no more advanced than folk or "pop" psychology tests we often find on the internet, which are usually wholly unscientific.

The second area of concern, that of **reliability**, is just as important. Two popular and commonly used forms of reliability in psychometrics are those of test–retest and internal consistency reliability. **Test–retest** reliability evaluates the consistency of test scores across one or more measurement time points. For example, if I measured your IQ today, and the test was worth its salt, I should expect that a measurement of your IQ a month from now should, within a reasonable margin of error, generate a similar score, assuming it was administered under standardized conditions both times. If not, we might doubt the test's reliability. The Pearson correlation coefficient is commonly used to evaluate test–retest reliability, where a higher-than-not coefficient between testings is desirable. In addition to test–retest, we often would like a measure of what is known as the **internal consistency** of a measure, which, though having potentially several competing meanings (e.g., see Tang et al., 2014), can be considered to assess how well items on a scale "hang together," which is informal language for whether or not items on a test are **interrelated** (Schmitt, 1996). For this assessment, we can compute **Cronbach's alpha**, which we will now briefly demonstrate in SPSS.

As a very small-scale example, suppose we have a test having only five items (items 1 through 5 in the SPSS data view), and would like to assess the internal consistency of the measure using Cronbach's alpha. Suppose the scores on the items are as follows:

	Item_1	Item_2	Item_3	Item_4	Item_5
1	10.00	12.00	15.00	11.00	12.00
2	12.00	18.00	12.00	12.00	1.00
3	8.00	16.00	14.00	14.00	4.00
4	6.00	8.00	16.00	8.00	6.00
5	4.00	7.00	8.00	7.00	5.00
6	6.00	6.00	3.00	7.00	3.00
7	3.00	4.00	6.00	5.00	8.00
8	7.00	3.00	7.00	9.00	9.00
9	8.00	9.00	4.00	10.00	10.00
10	9.00	5.00	6.00	11.00	12.00

To compute a Cronbach's alpha, and obtain a handful of statistics useful for conducting an **item analysis**, we code in SPSS:

```
RELIABILITY
  /VARIABLES=Item_1 Item_2 Item_3 Item_4 Item_5
  /SCALE('ALL VARIABLES') ALL
  /MODEL=ALPHA
  /STATISTICS=DESCRIPTIVE SCALE CORR
  /SUMMARY=TOTAL.
```

The MODEL = ALPHA statement requests SPSS to compute a Cronbach's alpha. Select output now follows:

Reliability Statistics		
Cronbach's Alpha	Cronbach's Alpha Based on Standardized Items	No of Items
0.633	0.691	5

Item Statistics			
	Mean	Std. Deviation	N
Item_1	7.3000	2.71006	10
Item_2	8.8000	5.05085	10
Item_3	9.1000	4.74810	10
Item_4	9.4000	2.71621	10
Item_5	7.0000	3.80058	10

Inter-Item Correlation Matrix					
	Item_1	Item_2	Item_3	Item_4	Item_5
Item_1	1.000	0.679	0.351	0.827	0.022
Item_2	0.679	1.000	0.612	0.743	−0.463
Item_3	0.351	0.612	1.000	0.462	−0.129
Item_4	0.827	0.743	0.462	1.000	−0.011
Item_5	0.022	−0.463	−0.129	−0.011	1.000

We can see that SPSS reports a raw reliability coefficient of 0.633 and 0.691 based on standardized items. SPSS also reports item statistics, which include the mean and standard deviation of each item, as well as the inter-item correlation matrix, which, not surprisingly, has values of 1.0 down the main diagonal (i.e., the correlation of an item with itself is equal to 1.0).

Next, SPSS features **Item-Total Statistics**, which contains useful information for potentially dropping items and seeking to ameliorate reliability:

Item-Total Statistics				
	Scale Mean if Item Deleted	Scale Variance if Item Deleted	Corrected Item-Total Correlation	Squared Multiple Correlation	Cronbach's Alpha if Item Deleted
Item_1	34.3000	108.900	0.712	0.726	0.478
Item_2	32.8000	80.400	0.558	0.841	0.476
Item_3	32.5000	88.278	0.512	0.448	0.507
Item_4	32.2000	104.844	0.796	0.776	0.445
Item_5	34.6000	164.267	−0.228	0.541	0.824

The most relevant column of the above is the last one on the far right, "**Cronbach's Alpha if Item Deleted**." What this reports is how much alpha would change if the given item were excluded. We can see that for all items, alpha would **decrease** if the given item were excluded, but for item 5, alpha would **increase**. If we drop item 5 then, we should expect alpha to increase. We recompute alpha after removing item 5:

```
RELIABILITY
  /VARIABLES=Item_1 Item_2 Item_3 Item_4
  /SCALE('ALL VARIABLES') ALL
  /MODEL=ALPHA
  /STATISTICS=DESCRIPTIVE SCALE CORR
  /SUMMARY=TOTAL.
```

Reliability Statistics		
Cronbach's Alpha	Cronbach's Alpha Based on Standardized Items	Not Items
0.824	0.863	4

As we can see, alpha indeed did increase to 0.824 as indicated it would based on our previous output. Hence, according to coefficient alpha, dropping item 5 may be worthwhile in the hopes of improving the instrument and making its items a bit more interrelated.

Though we have provided an easy demonstration of Cronbach's alpha, it would be negligent at this point to not issue a few cautions and caveats regarding its everyday use. According to Green and Yang (2009), the regular employment of coefficient alpha for assessing reliability should be discouraged based on the fact that assumptions for the statistic are rarely ever met, and hence the statistic can exhibit a high degree of bias. What is more, according to a now classic paper by Schmitt (1996), alpha should not be used to conclude anything about **unidimensionality** of a test, and thus should not be interpreted as such. **Confirmatory factor analysis models** (Chapter 15) are typically better suited for assessing and establishing the dimensionality of a set of items. What is more, cut-offs for alpha regarding what is low versus high internal consistency can be very difficult to define, and as argued by Schmitt, low levels of alpha may still be useful. Hence, though easily computable in SPSS and other software, the reader should be cautious about the unrestricted employment of alpha in their work. For more details on how it should be used, in addition to the aforementioned sources, Cortina (1993) and Miller (1995) are very informative readings and should be read before you readily and regularly adopt alpha in your everyday statistical toolkit.

2.18 COVARIANCE AND CORRELATION MATRICES

Having reviewed the concept of covariance, we need a way to account for the covariance of many variables. For this, we write the sample covariance in matrix form:

$$
\mathbf{S} = \left(s_{jk}\right) = \begin{pmatrix} s_{11} & s_{12} & \cdots & s_{1p} \\ s_{21} & s_{22} & \cdots & s_{2p} \\ \cdot & \cdot & & \cdot \\ \cdot & \cdot & \cdots & \cdot \\ \cdot & \cdot & \cdots & \cdot \\ s_{p1} & s_{p2} & \cdots & s_{pp} \end{pmatrix}
$$

where s_{jk} are the covariances for variables j by k. The population covariance matrix Σ can be analogously defined:

$$
\sum = \sigma_{jk} = \begin{pmatrix} \sigma_{11} & \sigma_{12} & \cdots & \sigma_{1p} \\ \sigma_{21} & \sigma_{22} & \cdots & \sigma_{2p} \\ \cdot & \cdot & & \cdot \\ \cdot & \cdot & \cdots & \cdot \\ \cdot & \cdot & \cdots & \cdot \\ \sigma_{p1} & \sigma_{p2} & \cdots & \sigma_{pp} \end{pmatrix}
$$

where along the main diagonal of the covariance matrix are variances σ_{11}, σ_{22}, etc., for variables 1, 2, etc., up to σ_{pp}, the variance of the pth variable.

When we standardize the covariance matrix, dividing each of its elements by respective products of standard deviations, we obtain the **correlation matrix**:

$$\mathbf{R} = \left(r_{jk} \right) = \begin{pmatrix} 1 & r_{12} & \cdots & r_{1p} \\ r_{21} & 1 & \cdots & r_{2p} \\ \cdot & \cdot & & \cdot \\ \cdot & \cdot & & \cdot \\ \cdot & \cdot & & \cdot \\ r_{p1} & r_{p2} & \cdots & 1 \end{pmatrix}$$

where r_{12} is the correlation between variables 1 and 2, etc., and r_{1p} is the correlation between variable 1 and the pth variable.

An example of a correlation matrix (Heston, 1948) is that between different tests on the GRE (**Graduate Record Examination**):

Intercorrelations Among The G.R.E. Tests Of General Education

	Math	P.S.	B.S.	Soc.	Lit.	Arts	Exp.	Voc.
Mathematics		.55	.44	.51	.36	.35	.52	.38
Physical Science	.55		.49	.43	.20	.40	.32	.29
Biological Science	.44	.49		.57	.42	.42	.46	.50
Social Studies	.51	.43	.57		.54	.40	.61	.59
Literature	.36	.20	.42	.54		.39	.53	.54
Arts	.35	.40	.42	.40	.39		.42	.52
Effecive Expression	.52	.32	.46	.61	.53	.42		.66
Vocabulary	.38	.29	.50	.59	.54	.52	.66	

From the matrix, we can see that most correlations are low to moderate, with the correlation between Effective Expression and Vocabulary relatively large at a value of 0.66. The correlation between Physical Science and Vocabulary is relatively small, equaling 0.29.

2.19 OTHER CORRELATION COEFFICIENTS

It often happens that once we hear of Pearson's r, this becomes the **only** correlation coefficient in one's vocabulary, and too often the **concept**, rather than **calculation**, of a correlation is automatically linked to Pearson's r. Pearson r is but one of **many** correlation coefficients available at one's disposal in applied research. Recall that Pearson r captures **linear** relationships between (typically) continuous variables. If the relationship is not linear, or one or more variables are not continuous, or again if the data are in the form of ranks, then other correlation coefficients are generally more suitable. We briefly review Spearman's rho, although a host of other correlation coefficients exist that are well-suited for a variety of particular types of data.[8]

---

[8] For an overview of alternative correlation coefficients such as the biserial, point-biserial and tetrachoric coefficients, see Howell (2002) or Warner (2013).

Spearman's r_s ("rho"), named after Charles Spearman who developed the coefficient in 1904,[9] is a correlation coefficient suitable for data on two variables that are expressed in terms of **ranks** rather than actual measurements on a continuous scale. Mathematically, the Spearman correlation coefficient is equivalent to a Pearson r when the data are ranked. There are important differences between these two coefficients. Spearman's r_s can be defined as:

$$r_s = 1 - \frac{6 \sum \left(R_x - R_y\right)^2}{n(n^2 - 1)} = 1 - \frac{6 \sum d_i^2}{n(n^2 - 1)}$$

where R_x and R_y are the ranks on x_i and y_i for the ith individual in the data, d_i^2 are squared rank deviations, and n is the number of pairs of ranks (Kirk, 2008). When we compute r_s on the Galton data, we obtain:

```
> cor.test(parent, child, method = "spearman")

        Spearman's rank correlation rho

data:  parent and child
S = 76569964, p-value < 2.2e-16
alternative hypothesis: true rho is not equal to 0
sample estimates:
      rho
0.4251345
```

We see that r_s of 0.425 is slightly less than was Pearson r of 0.459.

To understand why Spearman's rank correlation and Pearson coefficient differ, consider data (Table 2.5) on the rankings of favorite movies for two individuals. In parentheses are subjective scores of "favorability" of these movies, scaled 1–10, where 1 = least favorable and 10 = most favorable.

From the table, we can see that Bill very much favors Star Wars (rating of 10) while least likes Batman (rating of 2.1). Mary's favorite movie is Scarface (rating of 9.7) while her least favorite movie is Batman (rating of 7.6). We will refer to these subjective scores in a moment. For now, we focus only on the ranks. For instance, Bill's ranking of Scarface is third, while Mary's ranking of Star Wars is third.

TABLE 2.5   Favorability of Movies for Two Individuals in Terms of Ranks

Movie	Bill	Mary
Batman	5 (2.1)	5 (7.6)
Star Wars	1 (10.0)	3 (9.0)
Scarface	3 (8.4)	1 (9.7)
Back to the Future	4 (7.6)	4 (8.5)
Halloween	2 (9.5)	2 (9.6)

Actual scores on the favorability measure are in parentheses.

[9] The coefficient appears in Spearman, C. (1904). The proof and measurement of association between two things. American Journal of Psychology, 15, 72–101.

To compute Spearman's r_s in R the "long way," we generate two vectors that contain the respective rankings:

```
> bill <- c(5, 1, 3, 4, 2)
> mary <- c(5, 3, 1, 4, 2)
```

Because the data are already in the form of ranks, both Pearson r and Spearman rho will agree:

```
> cor(bill, mary)
[1] 0.6
```

```
> cor(bill, mary, method = "spearman")
> 0.6
```

Note that by default, R returns the Pearson correlation coefficient. One has to specify `method = "spearman"` to get r_s. Consider now what happens when we correlate, instead of rankings, the actual subjective favorability scores corresponding to the respective ranks. When we plot the favorability data, we obtain:

```
> bill.sub <- c(2.1, 7.6, 8.4, 9.5, 10.0)
> mary.sub <- c(7.6, 8.5, 9.0, 9.6, 9.7)
> plot(mary.sub, bill.sub)
```

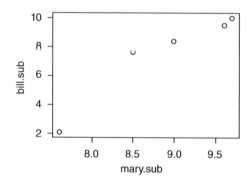

Note that though the relationship is not perfectly linear, each increase in Bill's subjective score is nonetheless associated with an increase in Mary's subjective score. When we compute Pearson's r on this data, we obtain:

```
> cor(bill.sub, mary.sub)
[1] 0.9551578
```

However, when we compute r_s, we get:

```
> cor(bill.sub, mary.sub, method = "spearman")
[1] 1
```

Spearman's r_s is equal to 1.0 because the rankings of movie preferences are perfectly **monotonically increasing** (i.e., for each increase in movie preference along the abscissa corresponds an increase in movie preference along the ordinate). In the case of Pearson's, the correlation is less than 1.0 because *r* captures the **linear** relationship among variables and not simply a monotonically increasing one. Hence, a high magnitude coefficient for Spearman's essentially tells us that two variables are "moving together," but it does not necessarily imply the relationship is a linear one. A similar test that measures rank correlation is that of Kendall's rank-order correlation. See Siegel and Castellan (1988, p. 245) for details.

2.20 STUDENT'S *t* DISTRIBUTION

The density for Student's *t* is given by (Shao, 2003):

$$f(t) = \frac{\Gamma[(v + 1)/2]}{\sqrt{v\pi}\,\Gamma(v/2)} \left(1 + \frac{t^2}{v}\right)^{-(v+1)/2}$$

where Γ is the gamma function and *v* are degrees of freedom. For small degrees of freedom *v*, the *t* distribution is quite distinct from the standard normal. However, as degrees of freedom increase, the *t* distribution converges to that of a normal density (Figure 2.11). That is, in the limit, $f(t) \rightarrow f(z)$, or a bit more formally, $\lim_{v \to \infty} f(t) = f(z)$.

The fact that *t* converges to *z* for large degrees of freedom but is quite distinct from *z* for small degrees of freedom is one reason why *t* distributions are often used for **small sample** problems. When sample size is large, and so consequently are degrees of freedom, whether one treats a random variable as *t* or *z* will make little difference in terms of computed *p*-values and decisions on respective null hypotheses. This is a direct consequence of the convergence of the two distributions for large degrees of freedom. For a historical overview of how *t*-distributions came to be, consult Zabell (2008).

2.20.1 *t*-Tests for One Sample

When we perform hypothesis testing using the *z* distribution, we assume we have knowledge of the population variance σ^2. Having direct knowledge of σ^2 is the most **ideal** and preferable of circumstances. When we know σ^2, we can compute the standard error of the mean directly as

$$\sigma_M = \frac{\sigma}{\sqrt{n}}$$

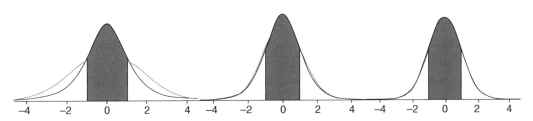

FIGURE 2.11   Student's *t* versus normal densities for 3 (left), 10 (middle), and 50 (right) degrees of freedom. As degrees of freedom increase, the limiting form of the *t* distribution is the *z* distribution.

Recall that the form of the one-sample z test for the mean is given by

$$z_M = \frac{\bar{y} - \mu_0}{\frac{\sigma}{\sqrt{n}}}$$

where the numerator $\bar{y} - \mu_0$ represents the distance between the sample mean and the population mean μ_0 under the null hypothesis, and the denominator $\frac{\sigma}{\sqrt{n}}$ is the standard error of the mean.

In most research contexts, from simple to complex, we usually do not have direct knowledge of σ^2. When we do not have knowledge of it, we use the next best thing, an estimate of it. We can obtain an unbiased estimate of σ^2 by computing s^2 on our sample. When we do so, however, and use s^2 in place of σ^2, we can no longer pretend to "know" the standard error of the mean. Rather, we must concede that all we are able to do is estimate it. Our estimate of the standard error of the mean is thus given by:

$$\hat{\sigma}_M = \frac{s}{\sqrt{n}}$$

When we use s^2 (where $\sqrt{s^2} = s$) in place of σ^2, our resulting statistic is no longer a z statistic. That is, we say the ensuing statistic is no longer **distributed** as a standard normal variable (i.e., z). If it is not distributed as z, then what is it distributed as? Thanks to William Sealy Gosset who in 1908 worked for **Guinness Breweries** under the pseudonym "Student" (Zabell, 2008), the ratio

$$t = \frac{\bar{y} - E(\bar{y})}{\hat{\sigma}_M} = \frac{\bar{y} - E(\bar{y})}{\frac{s}{\sqrt{n}}}$$

was found to be distributed as a t statistic on $n - 1$ degrees of freedom. Again, the t distribution is most useful for when sample sizes are rather small. For larger samples, as mentioned, the t distribution converges to that of the z distribution. If you are using rather large samples, say approximately 100 or more, whether you evaluate your null hypothesis using a z or t distribution will not matter much, because the critical values for z and t for such degrees of freedom (99 for the one-sample case) will be relatively alike, that practically at least, the two test statistics can be considered more or less equal. For even larger samples, the convergence is that much more fine-tuned.

The concept of convergence between z and t can be easily illustrated by inspecting the variance of the t distribution. Unlike the z distribution where the variance is set at 1.0 as a constant, the variance of the t distribution is defined as:

$$\sigma_t^2 = \frac{v}{v - 2}$$

where v are the degrees of freedom. For small degrees of freedom, such as $v = 5$, the variance of the t distribution is equal to:

$$\sigma_t^2 = \frac{5}{5 - 2} = \frac{5}{3} \approx 1.67$$

Note what happens as v increases, the ratio $\frac{v}{v-2}$ gets closer and closer to 1.0, which is the precise variance of the z distribution. For example, $v = 20$ yields:

$$\sigma_t^2 = \frac{20}{20 - 2} = \frac{20}{18} \approx 1.11$$

which is already quite close to the variance of a standardized normal variable z (i.e., 1.0). Hence, we can say more formally

$$\lim_{v \to \infty} \left(\frac{v}{v-2}\right) = 1.0$$

That is, as v increases without bound, the variance of the t distribution equals that of the z distribution, which is equal to 1.0.

We demonstrate the use of the one-sample t-test using SPSS. Consider the following small, hypothetical data on IQ scores on five individuals:

```
IQ

105
 98
110
105
 95
```

Suppose that the hypothesized mean IQ in the population is equal to 100. The question we want to ask is—**Is it reasonable to assume that our sampled data could have arisen from a population with mean IQ equal to 100?** We assume we have no knowledge of the population standard deviation, and hence must estimate it from our sample data. To perform the one-sample t-test in SPSS, we compute:

```
T-TEST
  /TESTVAL=100
  /MISSING=ANALYSIS
  /VARIABLES=IQ
  /CRITERIA=CI(.95).
```

The line /TESTVAL = 100 inputs the test value for our hypothesis test, which for our null hypothesis is equal to 100. We have also requested a 95% confidence interval for the mean difference.

One-Sample Statistics				
	N	Mean	SD	SE Mean
IQ	5	102.6000	6.02495	2.69444

We confirm from the above that the size of our sample is equal to 5, and the mean IQ for our sample is equal to 102.60 with standard deviation 6.02. The standard error of the mean reported by SPSS of 2.69 is actually not the **true** standard error of the mean. It is the **estimated** standard error of the mean, since recall that we did not have knowledge of the population variance (otherwise we would have been performing a z-test instead of a t-test).

One-Sample Test						
Test Value = 100						
					95% Confidence Interval of the Difference	
	t	Df	Sig. (2-tailed)	Mean Difference	Lower	Upper
IQ	0.965	4	0.389	2.60000	−4.8810	10.0810

We note from the above output:

- Our obtained t-statistic is equal to 0.965 and is evaluated on four degrees of freedom (i.e., $n - 1 = 5 - 1 = 4$). We lose a degree of freedom because recall that in estimating the population variance σ^2 with s^2, we had to compute a sample mean $\bar{y}$ and hence this value is regarded as "fixed" as we carry on with our t-test. Hence, we lose a single degree of freedom.
- The two-tailed p-value is equal to 0.389, which, assuming we had set our criteria for rejection at $\alpha = 0.05$, leads us to the decision to not reject the null hypothesis. The two-tailed (as opposed to **one-tailed** or **directional**) nature of the statistical test in this example means that we allow for a rejection of the null hypothesis in either direction from the value stated under the null. Since our null hypothesis is $\mu_0 = 100$, it means we were prepared to reject the null hypothesis for observed values of the sample mean that deviate "significantly" either **greater than** or **less than** 100. Since our significance level was set at 0.05, this means that we have $0.05/2 = 0.025$ area in each end of the t distribution to specify as our rejection region for the test. The question we are asking of our sample mean is—**What is the probability of observing a sample mean that falls much greater OR much less than 100?** Because the observed sample mean can only fall in one tail or the other on any single trial (i.e., we are conducting a single "trial" when we run this experiment a single time), this implies these two events are **mutually exclusive**, which by the addition rule for mutually exclusive events, we can add them. When we add their probabilities, we get $0.025 + 0.025 = 0.05$, which, of course, is our significance level for the test.
- The actual mean difference observed is equal to 2.60, which was computed by taking the mean of our sample, that of 102.6 and subtracting the mean hypothesized under the null hypothesis, that of 100 (i.e., $102.6 - 100 = 2.60$).
- The 95% confidence interval of the difference is interpreted to mean that with 95% confidence, the interval with lower bound -4.8810 and upper bound 10.0810 will capture the true parameter, which in this case is the population mean difference. We can see that 0 lies within the limits of the confidence interval, which again confirms why we were unable to reject the null hypothesis at the 0.05 level of significance. Had zero lay outside of the confidence interval limits, this would have been grounds to reject the null at a significance level of 0.05 (and consequently, we would have also obtained a p-value of less than 0.05 for our significance test). Recall that the true mean (i.e., parameter) is not the random component. Rather, the sample is the random component, on which the interval is then computed. It is important to emphasize this distinction when interpreting the confidence interval.

We can easily generate the same t-test in R. We first generate the vector of data then carry on with the one-sample t-test, which we notice mirrors the findings obtained in SPSS:

```
> iq <- c(105, 98, 110, 105, 95)
> t.test(iq, mu = 100)

        One Sample t-test

data:  iq
t = 0.965, df = 4, p-value = 0.3892
alternative hypothesis: true mean is not equal to 100
95 percent confidence interval:
  95.11904 110.08096
sample estimates:
mean of x
   102.6
```

2.20.2 *t*-Tests for Two Samples

Just as the *t*-test for one sample is a generalization of the *z*-test for one sample, for which we use s^2 in place of σ^2, the *t*-test for two independent samples is a generalization of the *z*-test for two independent samples. Recall the *z*-test for two independent samples:

$$z_M = \frac{E(\bar{y}_1) - E(\bar{y}_2)}{\sqrt{\dfrac{\sigma_1^2}{n_1} + \dfrac{\sigma_2^2}{n_2}}} = \frac{\mu_1 - \mu_2}{\sqrt{\dfrac{\sigma_1^2}{n_1} + \dfrac{\sigma_2^2}{n_2}}}$$

where $E(\bar{y}_1)$ and $E(\bar{y}_2)$ denote the expectations of the sample means $\bar{y}_1$ and $\bar{y}_2$ respectively (which are equal to μ_1 and μ_2).

When we do not know the population variances σ_1^2 and σ_2^2, we shall, as before, obtain estimates of them in the form of s_1^2 and s_2^2. When we do so, because we are using these estimates instead of the actual variances, our new ratio is no longer distributed as *z*. Just as in the one-sample case, it is now distributed as *t*:

$$t = \frac{E(\bar{y}_1) - E(\bar{y}_2)}{\sqrt{\dfrac{s_1^2}{n_1} + \dfrac{s_2^2}{n_2}}} = \frac{\mu_1 - \mu_2}{\sqrt{\dfrac{s_1^2}{n_1} + \dfrac{s_2^2}{n_2}}} \tag{2.6}$$

on degrees of freedom $v = n_1 - 1 + n_2 - 1 = n_1 + n_2 - 2$,

The formulization of *t* in (2.6) assumes that $n_1 = n_2$. If sample sizes are unequal, then **pooling** variances is recommended. To pool, we weight the sample variances by their respective sample sizes and obtain the following **estimated standard error of the difference in means**:

$$\hat{\sigma}_{diff} = \sqrt{\hat{\sigma}^2{}_{pooled}\left(\frac{1}{n_1} + \frac{1}{n_2}\right)} = \sqrt{\frac{(n_1 - 1)s_1^2 + (n_2 - 1)s_2^2}{n_1 + n_2 - 2}\left(\frac{n_1 + n_2}{n_1 n_2}\right)}$$

which can also be written as

$$\hat{\sigma}_{diff} = \sqrt{\hat{\sigma}^2{}_{pooled}\left(\frac{1}{n_1} + \frac{1}{n_2}\right)} = \sqrt{\frac{\hat{\sigma}^2{}_{pooled}}{n_1} + \frac{\hat{\sigma}^2{}_{pooled}}{n_2}}$$

Notice that the pooled estimate of the variance $\dfrac{(n_1 - 1)s_1^2 + (n_2 - 1)s_2^2}{n_1 + n_2 - 2}$ is nothing more than an averaged **weighted sum**, each variance being weighted by its respective sample size. This idea of weighting variances as to arrive at a pooled value is not unique to *t*-tests. Such a concept forms the very fabric of how MS error is computed in the analysis of variance as we shall see further in Chapter 3 when we discuss the ANOVA procedure in some depth.

2.20.3 Two-Sample *t*-Tests in R

Consider the following hypothetical data on pass-fail grades ("0" is fail, "1" is pass) for a seminar course with 10 attendees:

```
grade          studytime
0                 30
0                 25
0                 59
0                 42
0                 31
1                140
1                 90
1                 95
1                170
1                120
```

To conduct the two-sample *t*-test, we generate the relevant vectors in R then carry out the test:

```
> grade.0 <- c(30, 25, 59, 42, 31)
> grade.1 <- c(140, 90, 95, 170, 120)
> t.test(grade.0, grade.1)

        Welch Two Sample t-test

data:  grade.0 and grade.1
t = -5.3515, df = 5.309, p-value = 0.002549
alternative hypothesis: true difference in means is not equal to 0
95 percent confidence interval:
 -126.00773  -45.19227
sample estimates:
mean of x mean of y
     37.4     123.0
```

Using a **Welch adjustment** for unequal variances (Welch, 1947) automatically generated by R, we conclude a statistically significant difference between means ($p = 0.003$). With 95% confidence, we can say the true mean difference lies between the lower limit of approximately −126.0 and the upper limit of approximately −45.2. As a quick test to verify the assumption of equal variances (and to confirm in a sense whether the Welch adjustment was necessary), we can use var.test which will produce a ratio of variances and evaluate the null hypothesis that this ratio is equal to 1 (i.e., if the variances are equal, the numerator of the ratio will be the same as the denominator):

```
> var.test(grade.0, grade.1)

        F test to compare two variances

data:  grade.0 and grade.1
F = 0.1683, num df = 4, denom df = 4, p-value = 0.1126
alternative hypothesis: true ratio of variances is not equal to 1
95 percent confidence interval:
 0.01752408 1.61654325
sample estimates:
ratio of variances
        0.1683105
```

The var.test yields a *p*-value of 0.11, which under most circumstances would be considered insufficient reason to doubt the null hypothesis of equal variances. Hence, the Welch adjustment on the variances was probably not needed in this case as there was no evidence of an inequality of variances to begin with.

Carrying out the same test in SPSS is straightforward by requesting (output not shown):

```
t-test groups = grade(0 1)
    /variables = studytime.
```

A classic **nonparametric** equivalent to the independent-samples *t*-test is the **Wilcoxon rank-sum** test. It is a useful test to run when either distributional assumptions are known to be violated or when they are unknown and sample size too small for the central limit theorem to come to the "rescue." The test compares **rankings** across the two samples instead of actual scores. For a brief overview of how the test works, see Kirk (2008, Chapter 18) and Howell (2002, pp. 707–717), and for a more thorough introduction to nonparametric tests in general, see the following chapter on ANOVA in this book, or consult Denis (2020) for a succinct chapter and demonstrations using R. We can request the test quite easily in R:

```
> wilcox.test(grade.0, grade.1)

        Wilcoxon rank sum test

data:  grade.0 and grade.1
W = 0, p-value = 0.007937
alternative hypothesis: true location shift is not equal to 0
```

We see that the obtained *p*-value still suggests we reject the null hypothesis, though the *p*-value is slightly larger than for the Welch corrected parametric test.

2.21 STATISTICAL POWER

Power, first and foremost, is a **probability**. Power is the probability of rejecting a null hypothesis given that the null hypothesis is false. It is equal to $1 - \beta$ (i.e., 1 minus the type II error rate). If the null hypothesis were true, then regardless of how much power one has, one would still not be able to reject the null. We may think of it somewhat in terms of the **sensitivity** of a statistical test for detecting the falsity of the null hypothesis. If the test is not very sensitive to departures from the null (i.e., in terms of a particular alternative hypothesis), we will not detect such departures. If the test is very sensitive to such departures, then we will correctly detect these departures and be able to infer the statistical alternative hypothesis in question.

A useful analogy for understanding power is to think of a sign on a billboard that reads "H_0 is false." Are you able to detect such a sign with your current glasses or contact lenses that you are wearing? If not, **you lack sufficient power**. That is, you lack the sensitivity in your instrument (your reading glasses) to correctly detect the falsity of the null hypothesis, and in doing, be in a position to reject it. Alternatively, if you have 20/20 vision, you will be able to detect the false null with ease, and reject it with confidence. A key point to note here is that if H_0 is false, it is false **regardless of your ability to detect it**, analogous to a virus strain being present but biomedical engineers lacking a powerful enough microscope to see it. If the null is false, the only question that remains is whether or not you will have a powerful enough test to detect its falsity. If the null were not false on the other hand, then regardless of your degree of power, you will not be able to detect its falsity (because it is not false to begin with).

Power is a function of four elements, all of which will be featured in our discussion of the *p*-value toward the conclusion of this chapter:

1. The value hypothesized under the statistical alternative hypothesis, H_1. All else equal, a greater distance between H_0 and H_1 means greater power. Though "distance" in this regard is not a one-to-one concept with **effect size**, the spirit of the two concepts is the same. The greater the sci-entific effect, the more power you will have to detect that effect. This is true whether we are dealing with mean differences in ANOVA-type models or testing a null hypothesis of the sort $H_0 : R^2 = 0$ in regression. In all such cases, we are seeking to detect a deviation from the null hypothesis.

2. The **significance level**, or type I error rate (α) at which you set your test. All else equal, a more liberal setting such as 0.05 or 0.10 affords more statistical power than a more conservative setting such as 0.01 or 0.001, for instance. It is easier to detect a false null if you allow yourself more of a risk of committing a type I error. Since we usually want to minimize type I error, we typically want to regard α as fixed at a nominal level (e.g., 0.05 or 0.01) and consider it not amenable to adjustment for the purpose of increasing power. Hence, when it comes to boosting power, researchers usually do not want to "mess with" the type I error rate.

3. **Population variability**, σ^2, often unknown but estimated by s^2. All else equal, the greater the variance of objects studied in the population, the **less sensitive** the statistical test, and the less power you will have. Why is this so? As an analogy, consider a rock thrown into the water. The rock will make a definitive particular "splash" in that it will displace a certain amount of water when it hits the surface. This can be considered to be the "effect size" of the splash. If the water is noisy with wind and waves (i.e., high population variability), it will be difficult to detect the splash. If, on the other hand, the water is calm and serene (i.e., low population variability), you will more easily detect the splash. Either way, the rock made a particular splash of a given size. The **magnitude of the splash** is the same regardless of whether the waters are calm or tur-bulent. Whether we can detect the splash or not is in part a function of the variance in the population.

4. Applying this concept to research settings, if you are sampling from "noisy" populations, it is harder to see the effect of your independent variable than if you are sampling from less noisy and thus, less variable, populations. This is why research using lab rats or other equally **controllable** objects can usually detect effects with relatively few animals in a sample, whereas research studying humans on variables such as intelligence, anxiety, attitudes, etc., usually requires many more subjects in order to detect effects. A good way to boost power is to study populations that have relatively low variability before your treatment is administered. If your treatment works, you will be able to detect its efficacy with fewer subjects than if dealing with a highly variable population. Another approach is to **covary out** one or two factors that are thought to be related to the dependent variable through a technique such as the **analysis of covariance** (Keppel and Wickens, 2004), discussed and demonstrated later in the book.

5. **Sample size**, n. All else equal, the greater the sample size, the greater the statistical power. Boost-ing sample size is a common strategy for increasing power. Indeed, as will be discussed at the conclusion of this chapter, for any significance test in which there is at least **some** effect (i.e., some distance between the null and alternative), statistical significance is assured for a large-enough sample size. Obtaining large samples is a **good thing** (since after all, the most ideal goal would be to have the actual **population**), but as sample size increases, the p-value becomes an increasingly poor indicator or measure of experimental effect. **Effect sizes should always be reported alongside any significance test**.

2.21.1 Visualizing Power

Figure 2.12, adapted from Bollen (1989), depicts statistical power under competing values for detecting the population parameter θ. Note carefully in the figure that the critical value for the test remains constant as a result of our desire to keep the type I error rate constant. It is the **distance** from $\theta = 0$ to $\theta = C_1$ or $\theta = C_2$ that determines power (the shaded region in distributions (b) and (c)).

Statistical power matters so long as we have the inferential goal of rejecting null hypotheses. A study that is underpowered risks not being able to reject null hypotheses even if such null hypotheses are in reality false. A failure to reject a null hypothesis under the condition of minimal power could either mean a lack of inferential support for the obtained finding, or it could simply suggest an underpowered (and consequently poorly designed) experiment or study. Ensuring adequate statistical power **before** one engages in a research study or experiment is mandatory (Cohen, 1988).

2.22 POWER ESTIMATION USING R AND G*POWER

To demonstrate the estimation of power using software, we first use `pwr.r.test` (Champely, 2014) in R to estimate required sample size for a Pearson r correlation coefficient. As an example, we estimate required sample size for a population correlation coefficient of $\rho = 0.10$ at a significance level set to 0.05, with desired power equal to 0.90. Note that in the code that follows, we purposely leave n empty so R can estimate this figure for us:

```
> install.packages("pwr")
> library(pwr)
> pwr.r.test(n = , r = .10, sig.level = .05, power = .90)
```

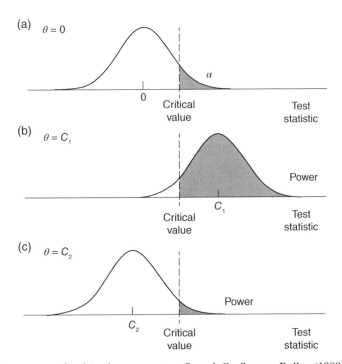

FIGURE 2.12 Power curves for detecting parameters C_1 and C_2. Source: Bollen (1989). Reproduced with permission from John Wiley & Sons, Inc.

```
approximate correlation power calculation (arctangh transformation)
            n = 1046.423
            r = 0.1
    sig.level = 0.05
        power = 0.9
  alternative = two.sided
```

We see that to detect a correlation coefficient of 0.10 at a desired level of power equal to 0.9, a sample size of 1046 is required. We could round up to 1047 for a slightly more **conservative estimate**. It is a more conservative estimate because 1047 is slightly more "generous" of a sample than R is reporting is necessary (1046). Now, in this case, the difference is extremely slight, but in general, when you provide your analysis with more subjects than what may be necessary for a given level of power, you are guarding against the possibility of obtaining smaller effects than what you believe are "out there" in your population. If in doubt, **larger samples are always preferable to smaller ones**, and thus rounding "up" on sample size requirements is usually a good idea.

Estimating in G*Power,[10] we obtain that given in Figure 2.13.

Note that our power estimate using G*Power is identical to that using R (i.e., power of 0.90 requires a sample size of 1046 for an effect size of $\rho = 0.10$). G*Power also allows us to draw the corresponding power curve. A power curve is a simple depiction of required sample size as a function of power and estimated effect size. What is nice about power curves is that they allow one to see how estimated sample size requirements and power **increase** or **decrease** as a function of effect size. For the estimation of estimated sample size for detecting $\rho = 0.10$, G*Power generates the curve in Figure 2.14 (top curve).

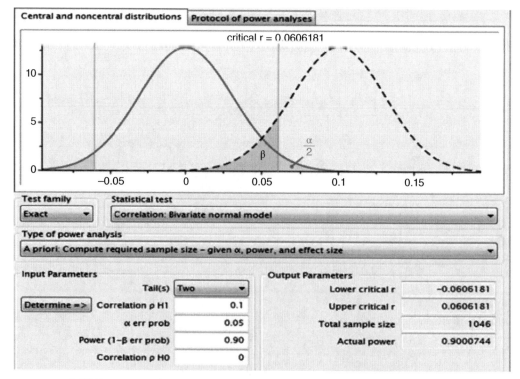

FIGURE 2.13   G*Power output for estimating required sample size for $r = 0.10$.

---

[10] G*Power is a user-friendly statistical program that can be downloaded for free at: https://www.psychologie.hhu.de/arbeitsgruppen/allgemeine-psychologie-und-arbeitspsychologie/gpower.html.

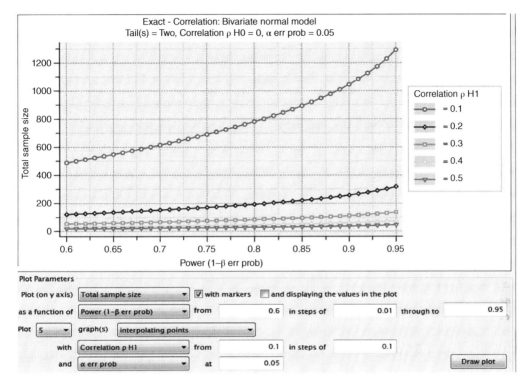

FIGURE 2.14   Power curves generated by G*Power for detecting correlation coefficients of $\rho = 0.10$ to 0.50.

Especially for small hypothesized values of ρ, the required sample size for even poor to modest levels of statistical power is quite large. For example, reading off the plot in Figure 2.14, to detect $\rho = 0.10$, at even a relatively low power level of 0.60, one requires upward of almost 500 participants. This might explain why many studies that yield relatively small effect sizes never get published. **They often have insufficient power to reject their null hypotheses**. As effect size increases, required sample size drops substantially. For example, to attain a modest level of power such as 0.68 for a correlation coefficient of 0.5, one requires only 21.5 participants, as can be more clearly observed from Table 2.6 which corresponds to the power curves in Figure 2.14 for power ranging from 0.60 to 0.69.

Hence, one general observation from this simple power analysis for detecting ρ is that **size of effect** (in this case, ρ) plays a very important role in determining estimated sample size. As a general rule, across virtually all statistical tests, **if the effect you are studying is large, a much smaller sample size is required than if the effect is weak**. Drawing on our analogy of the billboard sign that reads "H_0 is false," all else equal, if the sign is in large print (i.e., strong effect), you require less "power" in your prescription glasses to detect such a large sign. If the sign is in small print (i.e., weak effect), you require much more "power" in your lenses to detect it.

2.22.1   Estimating Sample Size and Power for Independent Samples *t*-Test

For an independent-samples *t*-test, required sample size can be estimated through R using `pwr.t.test`:

```
> pwr.t.test (n =, d =, sig.level =, power =, type = c("two.sample", "one.
sample", "paired"))
```

where, n = sample size per group, d = estimate of standardized statistical distance between means (Cohen's d), `sig.level` = desired significance level of the test, `power` = desired power level, and `type` = designation of the kind of *t*-test you are performing (for our example, we are performing a two sample test).

TABLE 2.6   Power Estimates as a Function of Sample Size and Estimated Magnitude Under Alternative Hypothesis

		Exact – Correlation: Bivariate Normal Model				
		Tail(s) = Two, Correlation ρ H0 = 0, α err prob = 0.05				
		Correlation ρ H1 = 0.1	Correlation ρ H1 = 0.2	Correlation p HI = 0.3	Correlation ρ HI = 0.4	Correlation ρ HI = 0.5
#	Power(1-β err prob)	Total Sample Size	Total Sample Size	Total Sample Size	Total Sample Size	Total Sample Size
1	0.600000	488.500	121.500	53.5000	29.5000	18.5000
2	0.610000	500.500	124.500	54.5000	30.5000	18.5000
3	0.620000	511.500	126.500	55.5000	30.5000	19.5000
4	0.630000	523.500	129.500	56.5000	31.5000	19.5000
5	0.640000	535.500	132.500	58.5000	32.5000	19.5000
6	0.650000	548.500	135.500	59.5000	32.5000	20.5000
7	0.660000	561.500	138.500	60.5000	33.5000	20.5000
8	0.670000	574.500	142.500	62.5000	34.5000	21.5000
9	0.680000	587.500	145.500	63.5000	34.5000	21.5000
10	0.690000	601.500	148500	64.5000	35.5000	22.5000

It would be helpful at this point to translate Cohen's d values into R^2 values to learn how much variance is explained by differing d values. To convert the two, we apply the following transformation:

$$\mathbf{d} = \sqrt{\frac{4r^2}{1 - r^2}}$$

Table 2.7 contains conversions for r increments of 0.10, 0.20, 0.30, etc.

To get a better feel for the relationship between Cohen's d and r^2, we obtain a plot of their values (Figure 2.15).

As can be gleamed from Figure 2.15, the relationship between the two effect size measures is not exactly linear and increases rather sharply for rather large values (the curve is somewhat exponential).

Suppose a researcher would like to estimate required sample size for a two-sample t-test, for a relatively small effect size, d = 0.41 (equal to r of 0.20), at a significance level of 0.05, with a desired power level of 0.90. We compute:

```
> pwr.t.test (n =, d =0.41, sig.level =.05, power =.90, type = c("two.sample"))

      Two-sample t test power calculation

              n = 125.9821
              d = 0.41
      sig.level = 0.05
          power = 0.9
    alternative = two.sided

NOTE: n is number in *each* group
```

Thus, the researcher would require a sample size of approximately 126. As R emphasizes, this sample size is **per group**, so the **total** sample size required is 126(2) = 252.

TABLE 2.7   Conversions for $r \rightarrow r^2 \rightarrow$ d.[11]

r	r^2	d
0.10	0.01	0.20
0.20	0.04	0.41
0.30	0.09	0.63
0.40	0.16	0.87
0.50	0.25	1.15
0.60	0.36	1.50
0.70	0.49	1.96
0.80	0.64	2.67
0.90	0.81	4.13
0.99	0.98	14.04

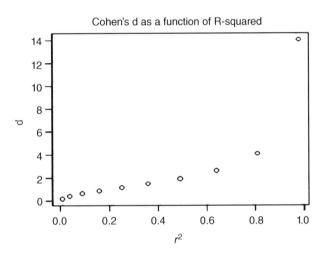

FIGURE 2.15   Relationship between Cohen's d and R-squared.

2.23   PAIRED-SAMPLES *t*-TEST: STATISTICAL TEST FOR MATCHED-PAIRS (ELEMENTARY BLOCKING) DESIGNS

Oftentimes in research, we are able to sample observations that are **matched** on one or more variables or characteristics. For instance, consider the hypothetical data in Table 2.8.

[11]
```
> r <- c(0.10, 0.20, 0.30, 0.40, 0.50, 0.60, 0.70, 0.80, 0.90, 0.99)
> r_squared <- r^2
> r_squared
 [1] 0.0100 0.0400 0.0900 0.1600 0.2500 0.3600 0.4900 0.6400 0.8100 0.9801

> d <- sqrt((4*r^2)/(1-r^2))
> d
 [1]  0.2010076  0.4082483  0.6289709  0.8728716  1.1547005  1.5000000
 [7]  1.9603921  2.6666667  4.1294832 14.0358479
```

TABLE 2.8   Matched-Pairs Design

	Treatment 1	Treatment 2
Block 1	10	8
Block 2	15	12
Block 3	20	14
Block 4	22	15
Block 5	25	24

About Table 2.8:

- In each block (1 through 5), participants **within** blocks are assumed to be more **homogeneous** on one or more variables than participants **between** blocks.
- Participants are randomly assigned to condition (i.e., treatment 1 versus treatment 2) within each block.
- Whether the blocks are naturally occurring or our sampling scheme is designed purposely to create the blocks, we can exploit the homogeneity of participants within each block by including this source in our statistical analysis as to potentially reduce the error term of our statistical test.
- The matched-pairs design is a simpler version of the full-blown **randomized block design** in which one can have more than just two levels of the independent variable (e.g., treatment 1 versus treatment 2 versus treatment 3). However, the principle behind the matched-pairs design and that of randomized block designs is the same, that of exploiting the covariance between conditions and removing it from the error term of the test statistic (t in matched-pairs, F in randomized block designs).
- In more advanced analyses such as repeated measures, longitudinal, and mixed effects modeling, we will say that subjects are **nested** within block. **A nesting structure simply implies that subjects within a block share similarity compared to subjects between blocks.** Good statistical analyses will attempt to account for this similarity, remove it from respective error terms for tests, and hence make the statistical test for effects more sensitive (i.e., more powerful).

As an example of a matched-pairs situation, suppose we are interested in evaluating the effects of melatonin[12] dose on average hours of sleep. However, we know that due to age, some people will naturally sleep longer than others irrespective of how much melatonin they receive. We do not want this natural sleep tendency due to age to confound the effect we are actually interested in studying (i.e., that of melatonin dose), and so we will match participants on their age level, or perhaps even crudely on **age group** (e.g., young, middle-aged, old), and carry out our study **within each age group**. Then, when we perform statistical analyses, we will be able to extract this variation due to age out of the error term of the analysis, and hence boost statistical power for estimating the effect we are actually interested in (melatonin dosage).

When we sample observations in pairs, as was true for the independent samples t-test, the expectation of the difference between sample means is given by:

$$E(\bar{y}_1 - \bar{y}_2) = \mu_1 - \mu_2$$

However, because observations are sampled (or "matched") in pairs, we naturally expect there to be a covariance different from zero between pairs. We can exploit this covariance and remove it from the error term of our statistical test. As given in Hays (1994, p. 339), the variance of the difference becomes

$$\sigma_{diff}^2 = \sigma_{M_1}^2 + \sigma_{M_2}^2 - 2\text{cov}(\bar{y}_1, \bar{y}_2)$$

---

[12] Melatonin is sometimes used as a non-prescription sleep aid.

with standard error equal to

$$\sigma_{diff} = \sqrt{\sigma_{diff}^2} = \sqrt{\sigma_{M_1}^2 + \sigma_{M_2}^2 - 2\text{cov}(\bar{y}_1, \bar{y}_2)}$$

Notice that we have subtracted $2\text{cov}(\bar{y}_1, \bar{y}_2)$ from the denominator of our statistic. Assuming the covariance between pairs is unequal to 0, this will serve to **lower** the standard error of our statistic, and hence, boost statistical power. In practice, this is accomplished by conducting a t-test on the **difference scores** between samples. As Hays (1994, p. 339) notes, "the matching and the consequent **dependence** within the pairs changes the standard error of the difference between the sample means."

In the classic between-subjects design where participants are not matched, the expectation is that covariance between treatments is equal to 0, and hence, we would have:

$$\begin{aligned}
\sigma_{diff}^2 &= \sigma_{M_1}^2 + \sigma_{M_2}^2 - 2\text{cov}(\bar{y}_1, \bar{y}_2) \\
&= \sigma_{M_1}^2 + \sigma_{M_2}^2 - 2(0) \\
&= \sigma_{M_1}^2 + \sigma_{M_2}^2
\end{aligned}$$

The matched-pairs design is a very important concept in statistics and design of experiments, because this simple design is the starting point to understanding more complicated designs and modeling such as mixed effects and hierarchical models.

We analyze the hypothetical data in Table 2.8 using a paired samples t-test in R by requesting `paired = TRUE`:

```
> treat <- c(10, 15, 20, 22, 25)
> control <- c(8, 12, 14, 15, 24)
> t.test(treat, control, paired = TRUE)

        Paired t-test

data:  treat and control
t = 3.2827, df = 4, p-value = 0.03042
alternative hypothesis: true difference in means is not equal to 0
95 percent confidence interval:
 0.5860324 7.0139676
sample estimates:
mean of the differences
                3.8
```

The obtained p-value of 0.03 is statistically significant at a 0.05 level of significance. We reject the null hypothesis and conclude the population means for the treatment conditions to be different.

As a **nonparametric** test, the Wilcoxon rank-sum test featured earlier can be adapted to incorporate paired observations. For our data, we have:

```
> wilcox.test(treat, control, paired = TRUE)

        Wilcoxon signed rank test

data:  treat and control
V = 15, p-value = 0.0625
alternative hypothesis: true location shift is not equal to 0
```

TABLE 2.9   Randomized Block Design

	Treatment 1	Treatment 2	Treatment 3
Block 1	10	9	8
Block 2	15	13	12
Block 3	20	18	14
Block 4	22	17	15
Block 5	25	25	24

We notice that the obtained p-value is somewhat greater for the nonparametric test than for the parametric one. In terms of significance tests, this emphasizes the fact that there is usually a cost to not being able to make parametric assumptions.

2.24   BLOCKING WITH SEVERAL CONDITIONS

We have said that in a blocking design, between treatment conditions we expect the covariance to be unequal to 0. Now, consider a design in which, once again we block, but this time on more than two treatment levels. The layout for such a design is given in Table 2.9.

Now, here is the trick to understanding **advanced modeling**, including a primary feature of mixed effects modeling. We know that we expect the covariance between treatments to be unequal to 0. This is analogous to what we expected in the simple matched-pairs design. It seems then that a reasonable assumption to make for the data in Table 2.9 is that the covariances between treatments are **equal**, or at minimum, follow some hypothesized correlational structure. In multilevel and hierarchical models, attempts are made to account for the correlation between treatment levels instead of assuming these correlations to equal 0 as is the case for classical between-subjects designs. In Chapter 6, we elaborate on these ideas when we discuss randomized block and repeated measures models.

2.25   COMPOSITE VARIABLES: LINEAR COMBINATIONS

In many statistical techniques, especially multivariate ones, statistical analyses take place not on individual variables, but rather on **linear combinations** of variables. A linear combination in linear algebra can be denoted simply as:

$$\ell_i = a_1 y_1 + a_2 y_2 + \ldots + a_p y_p$$
$$= \mathbf{a'y}$$

where $\mathbf{a}' = (a_1, a_2, \ldots, a_p)$. These values are scalars, and serve to weight the respective values of y_1 through y_p, which are the variables.

Just as we did for "ordinary" variables, we can compute a number of central tendency and dispersion statistics on linear combinations. For instance, we can compute the mean of a linear combination ℓ_i as

$$\bar{\ell} = \frac{1}{n} \sum_{i=1}^{n} \ell_i = \mathbf{a'\bar{y}}$$

We can also compute the sample variance of a linear combination:

$$s_\ell^2 = \frac{\sum_{i=1}^{n} \left(\ell_i - \bar{\ell} \right)^2}{n-1} = \mathbf{a'Sa}$$

for $\ell_i = \mathbf{a}'\mathbf{y}$, $i = 1, 2, ..., n$, and where $\mathbf{S}$ is the sample covariance matrix. Though the form $\mathbf{a}'\mathbf{S}\mathbf{a}$ for the variance may be difficult to decipher at this point, it will become clearer when we consider techniques such as **principal components** later in the book.

For two linear combinations,

$$\ell_1 = a_1 y_1 + a_2 y_2 + ... + a_p y_p = \mathbf{a}'\mathbf{y}$$

and

$$\ell_2 = b_1 y_1 + b_2 y_2 + ... + b_p y_p = \mathbf{b}'\mathbf{y}$$

we can obtain the sample covariance between such linear combinations as follows:

$$\text{cov}_{\ell_1,\ell_2} = \frac{\sum_{i=1}^{n} \left(\ell_{i1} - \bar{\ell}_1\right)\left(\ell_{i2} - \bar{\ell}_2\right)}{n-1} = \mathbf{a}'\mathbf{S}\mathbf{b}$$

The correlation of these linear combinations (Rencher and Christensen, 2012, p. 76) is simply the standardized version of $\text{cov}_{\ell_1,\ell_2}$:

$$r_{\ell_1,\ell_2} = \frac{\text{cov}_{\ell_1,\ell_2}}{\sqrt{s_{\ell_1}^2 s_{\ell_2}^2}} = \frac{\mathbf{a}'\mathbf{S}\mathbf{b}}{\sqrt{(\mathbf{a}'\mathbf{S}\mathbf{a})(\mathbf{b}'\mathbf{S}\mathbf{b})}}$$

As we will see later in the book, if r_{ℓ_1,ℓ_2} is the **maximum correlation** between linear combinations on the same variables, it is called the **canonical correlation**, discussed in Chapter 12. The correlation between linear combinations plays a central role in multivariate analysis. Substantively, and geometrically, linear combinations can be interpreted as "projections" of one or more variables onto new dimensions. For instance, in simple linear regression, the fitting of a least-squares line is such a projection. It is the projection of points such that it guarantees that the sum of squared deviations from the given projected line or "surface" (in the case of higher dimensions) is kept to a minimum.

If we can assume multivariate normality of a distribution, that is, $\mathbf{Y} \sim \mathbf{N}[\boldsymbol{\mu}, \boldsymbol{\Sigma}]$, then we know linear combinations of $\mathbf{Y}$ are also normally distributed, as well as a host of other useful statistical properties (see Timm, 2002, pp. 86–88). In multivariate methods especially, we regularly need to make assumptions about such linear combinations, and it helps to know that so long as we can assume multivariate normality, we have some idea of how such linear combinations will be distributed.

2.26 MODELS IN MATRIX FORM

Throughout the book, our general approach is to first present models in their simplest possible form using only scalars. We then gently introduce the reader to the corresponding matrix counterparts and extensions. The requirement of matrices for such models is to accommodate numerous variables and dimensions. Matrix algebra is the vehicle by which multivariate analysis is communicated, though most of the **concepts** of statistics can be communicated using simpler scalar algebra. Knowing matrix algebra for its own sake will not necessarily equate to understanding statistical concepts. Indeed, hiding behind the mathematics of statistics are the philosophically "sticky" issues that mathematics or statistics cannot, on their own at least, claim to solve. These are often the problems confronted by researchers and scientists in their empirical pursuits and attempts to draw conclusions from data. For instance, what is the nature of a "correct" model? Do latent variables exist, or are they only a consequence of generating linear combinations? The nature of a latent variable is not necessarily contingent on the linear algebra that seeks to define it. Such questions are largely philosophical, and if such interest you, you are strongly encouraged to familiarize yourself with the **philosophy of statistics and mathematics** (you may not always find answers to your questions, but you will appreciate the complexity of such questions, as they are beyond our current study here). For a gentle introduction to the philosophy of statistics, see Lindley (2001).

As an example of how matrices will be used to develop more complete and general models, consider the multivariate general linear model in matrix form:

$$\mathbf{Y} = \mathbf{XB} + \mathbf{E} \tag{2.7}$$

where $\mathbf{Y}$ is an $n \times m$ matrix of n observations on m response variables, $\mathbf{X}$ is the model or "design" matrix whose columns contain k regressors which includes the intercept term, $\mathbf{B}$ is a matrix of regression coefficients, and $\mathbf{E}$ is a matrix of errors. Many statistical models can be incorporated into the framework of (2.7). As a relatively easy application of this general model, consider the simple linear regression model (featured in Chapter 7) in matrix form:

$$
\mathbf{Y} = \begin{bmatrix} y_{i=1} \\ y_{i=2} \\ y_{i=3} \\ . \\ . \\ . \\ y_{i=n} \end{bmatrix}
\quad
\mathbf{X} = \begin{bmatrix} 1 & x_{i=1} \\ 1 & x_{i=2} \\ 1 & x_{i=3} \\ . & . \\ . & . \\ . & . \\ 1 & x_{i=n} \end{bmatrix}
\quad
\mathbf{B} = \begin{bmatrix} \alpha \\ \beta \end{bmatrix}
\quad
\boldsymbol{\varepsilon} = \begin{bmatrix} \varepsilon_1 \\ \varepsilon_2 \\ \varepsilon_3 \\ . \\ . \\ . \\ \varepsilon_n \end{bmatrix}
$$

where $y_{i=1}$ to $y_{i=n}$ are observed measurements on some dependent variable, $\mathbf{X}$ is the model matrix containing a constant of 1 in the first column to represent the common intercept term (i.e., "common" implying there is one intercept that represents all observations in our data), $x_{i=1}$ to $x_{i=n}$ are observed values on a predictor variable, α is the fixed intercept parameter, β is the slope parameter, which we also assume to be fixed, and $\boldsymbol{\varepsilon}$ is a vector of errors ε_1 to ε_n (we use $\boldsymbol{\varepsilon}$ here instead of $\mathbf{E}$).

Suppose now we want to add a second response variable. Because of the generality of (2.7), this can be easily accommodated:

$$
\mathbf{Y} = \begin{bmatrix} y_{i=1,1}, y_{i=1,2} \\ y_{i=2,1}, y_{i=2,2} \\ y_{i=3,1}, y_{i=3,2} \\ . \\ . \\ . \\ y_{i=n,1}, y_{i=n,2} \end{bmatrix}
$$

where now, a second response variable is represented in **Y** by a second column. That is, $y_{i\,=\,1,\,2}$ corresponds to individual 1 on response variable 2, $y_{i\,=\,2,\,2}$ is individual 2 on response variable 2, etc. We will at times refer to matrix representations throughout the book.

2.27 GRAPHICAL APPROACHES

Performing inferential tests to help draw conclusions about population parameters is useful, but ultimately the findings of a statistical analysis should make their way into a graph or other visualization. **Data visualization** is a field in itself, and with the advent of modern computing power, possibilities exist today that could only be dreamt of in the past. Simple visualizations such a **histograms**, **boxplots**, **scatterplots**, etc., can be useful in depicting findings but also in helping to verify assumptions that underlay the statistical model one is using. For example, since many tests of normality and equality of variances (and covariances) are relatively sensitive to the types of data to which they are applied, oftentimes researchers will generate simple plots in order to detect potential gross violations of such assumptions. We feature such techniques throughout the book.

For graphical displays meant to communicate findings (rather than test assumptions), Friendly (2000) puts the field into context:

> **Designing good graphics is surely an art, but as surely, it is one that ought to be informed by science** … **In this view, an effective graphical display, like good writing, requires an understanding of its** *purpose* **– what aspects of the data are to be communicated to the viewer. In writing, we communicate most effectively when we know our audience and tailor the message appropriately**. (p. 8)

In high-dimensional space, the challenge of graphical approaches is to summarize data into lower dimensions, while still retaining most of the information in the original data. We feature some such plots in later chapters. For a thorough account of data visualization, see datavis.ca (Friendly, 2020). For sophisticated graphics using R, consult Wickham (2009).

For now, it is useful to briefly review some basic plots for which the reader is likely already familiar.

2.27.1 Box-and-Whisker Plots

The boxplot was a contribution of John Tukey (1977) in the spirit of what is called **exploratory data analysis**, or "EDA" which encouraged scientists to spend more of their energy on descriptive techniques instead of focusing exclusively on confirmatory statistical tests. Boxplots of parent heights from Galton's data appear below:

```
> attach(Galton)
> boxplot(parent)
> library(lattice)
> bwplot(parent)
```

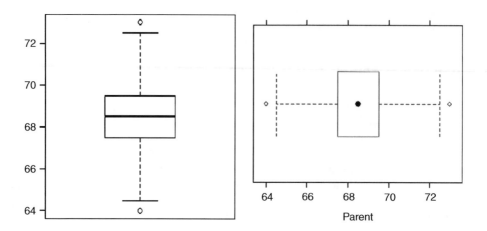

The boxplot provides what is generally known as a **five-number summary** of a distribution, of which we can obtain most of the numbers we need by the summary function in R:

```
> summary(parent)
   Min. 1st Qu.  Median    Mean 3rd Qu.    Max.
  64.00   67.50   68.50   68.31   69.50   73.00
```

Recall that the **median** is the point in the ordered data that divides the data set into two equal parts. The location of the median is computed by $(n + 1)/2$. In Galton's data, there are 928 observations, and so the location of the median is at 464.5th (i.e., (928 + 1)/2) point in the ordered data set. For parent, this value is equal to 68.50. The first and third quartiles represent the 25th and 75th percentiles and are 67.50 and 69.50 respectively. We can also compute the range as

```
> range(parent)
[1] 64 73
```

We can also generate boxplots by category. Throughout the book, we use Fisher's iris data (Fisher, 1936) in which flower characteristics such as sepal and petal length are categorized by species of flower. We plot sepal length by species:

```
> library(lattice)
> attach(iris)
> bwplot(Sepal.Length ~ Species)
```

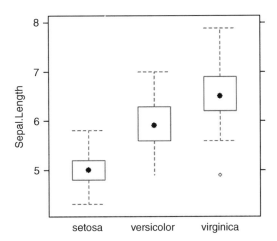

Data points falling beyond the whiskers of the plots may reveal the presence of outliers, and should be investigated (though of course, not necessarily **deleted**, see Section 7.23 for a discussion). If you are completely unfamiliar with boxplots, see Denis (2020) for an overview.

Stem-and-leaf plots are also easily produced. These visual displays are kind of "naked histograms," because they reveal the actual observations in the data while also providing information about their frequency of occurrence. In 1710, John Arbuthnot analyzed data on the ratios of males to female births in London from 1629 to 1710 and in so doing made an argument for these births being a function of a "divine being" (Arbuthnot, 1710; Shoesmith, 1987). One of his variables was the number of male christenings (i.e., baptisms) over the period 1629–1710. We generate a stem-and-leaf plot in R of these male christenings using package aplpack (Wolf and Bielefeld, 2014), for which the "leaves" are corresponding hundreds. For example, in the following plot, the first value of 2|8 would appear to represent a value of 2800 but is rounded down from the actual value in the data (which is also the minimum) of 2890. The maximum in the data is actually equal to 8426, but is represented by 8400 (i.e., 8|0012334):

```
> install.packages("aplpack")
> library(aplpack)
> library(HistData)
> attach(Arbuthnot)
> stem.leaf(Males)

1 | 2: represents 1200
 leaf unit: 100
              n: 82
    1     2. | 8
   10     3* | 011222334
   15     3. | 66777
   18     4* | 014
   25     4. | 6777899
   36     5* | 01112233444
   38     5. | 56
  (11)    6* | 00001122444
   33     6. | 5555899
   26     7* | 244
   23     7. | 5555666666778999
    7     8* | 0012334
```

2.28   WHAT MAKES A *p*-VALUE SMALL? A CRITICAL OVERVIEW AND PRACTICAL DEMONSTRATION OF NULL HYPOTHESIS SIGNIFICANCE TESTING

The workhorse for establishing statistical evidence in the social and natural sciences is the method of **null hypothesis significance testing** (or, "NHST" for short). However, since its inception with R.A. Fisher in the early 1900s, the significance test has been the topic of much debate, both statistical and philosophical. Throughout much of this book, NHST is regularly used to evaluate null hypotheses in methods such as the analysis of variance, regression, and various multivariate procedures. Indeed, the procedure is universally used in most statistical methods.

It behooves us then, before embarking on all of these methodologies, to discuss the nature of the null hypothesis significance test, and clearly demonstrate what it actually **means**, not only in a statistical context but also in how it should be interpreted in a **research** or **substantive** context.

The purpose of this final section of the present chapter is to provide a clear and concise demonstration and summary of the factors that influence the size of a computed *p*-value in virtually every statistical significance test. Understanding why statements such as "$p < 0.05$" can be reflective of even the smallest and trivial of effects is critical for the practitioner or researcher to appreciate if he or she is to assess and appraise statistical evidence in an intelligent and thoughtful manner. It is not an exaggeration to say that **if one does not understand the make-up of a *p*-value and the factors that directly influence its size, one cannot properly evaluate statistical evidence, nor should one even make the attempt to do so**. Though these arguments are not new and have been put forth by even the very best of methodologists (e.g., see Cohen, 1990; Meehl, 1978) there is evidence to suggest that many practitioners and researchers do not understand the factors that determine the size of a *p*-value (Gigerenzer, 2004). To emphasize once again—understanding the determinants of a *p*-value and what makes *p*-values distinct from effect sizes is not simply "fashionable." Rather, it is absolutely mandatory for any attempt to properly evaluate statistical evidence in a research report. Does the paper you're reading provide evidence of a successful treatment for cancer? **If you do not understand the distinctions between *p*-values and effect sizes, you will be unable to properly assess the evidence. It is that important**. As we will see, stating a result as "statistically significant" does not in itself tell you whether the treatment works or does not work, and in some cases, tells you very little at all from a scientific vantage point.

2.28.1   Null Hypothesis Significance Testing (NHST): A Legacy of Criticism

Criticisms targeted against null hypothesis significance testing have inundated the literature since at least the time Berkson in 1938 brought to light how statistical significance can be easily achieved by simple manipulations of sample size:

> **I believe that an observant statistician who has had any considerable experience with applying the chi-square test repeatedly will agree with my statement that, as a matter of observation, when the numbers in the data are quite large, the P's tend to come out small**. (p. 526)

Since Berkson, the very best and renown of methodologists have remarked that the significance test is subject to gross misunderstanding and misinterpretation (e.g., see Bakan, 1966; Carver, 1993; Cohen, 1990; Estes, 1997; Loftus, 1991; Meehl, 1978; Oakes, 1986; Shrout, 1997; Wilson, Miller, and Lower, 1967). And though it can be difficult to assess or evaluate whether the situation has improved, there is evidence to suggest that it has not. Few describe the problem better than Gigerenzer in his article **Mindless statistics** (Gigerenzer, 2004), in which he discusses both the roots and truths of hypothesis testing, as well as how its "statistical rituals" and practices have become far more of a **sociological** phenomenon rather than anything related to good science and statistics.

Other researchers have found that misinterpretations and misunderstandings about the significance test are widespread not only among students but also among their instructors (Haller and Krauss, 2002). What determines statistical significance and what is it a function of? This is an extremely important question. An unawareness of the determinants of statistical significance leaves the door open to misunderstanding and misinterpretation of the test, and the danger to potentially draw false conclusions based on its results. Too often and for too many, the finding "$p < 0.05$" simply denotes a "good thing" of sorts, without ever being able to pinpoint what is so "good" about it.

Recall the familiar one-sample *z*-test for a mean discussed earlier:

$$z_M = \frac{\bar{y} - \mu_0}{\frac{\sigma}{\sqrt{n}}}$$

where the purpose of the test was to compare an obtained sample mean $\bar{y}$ to a population mean μ_0 under the null hypothesis that $\mu = \mu_0$. Sigma, σ, recall is the standard deviation of the population from which the sample was presumably drawn. Recall that in practice, this value is rarely if ever known for certain, which is why in most cases an estimate of it is obtained in the form of a sample standard deviation *s*. What determines the size of z_M, and therefore, the smallness of *p*? There are three inputs that determine the size of *p*, which we have already featured in our earlier discussion of statistical power. These three factors are $\bar{y} - \mu_0$, σ and *n*. We consider each of these once more, then provide simple arithmetic demonstrations to emphasize how changing any one of these necessarily results in an arithmetical change in z_M, and consequently, a change in the observed *p*-value.

As a first case, consider the distance $\bar{y} - \mu_0$. **Given constant values of σ and *n*, the greater the distance between $\bar{y}$ and μ_0, the larger z_M will be.** That is, as the numerator $\bar{y} - \mu_0$ grows larger, the resulting z_M also gets larger in size, which as a consequence, decreases *p* in size. As a simple example, assume for a given research problem that σ is equal to 20 and *n* is equal to 100. This means that the standard error is equal to $20/\sqrt{100}$, which is equal to 20/10 = 2. Suppose the obtained sample mean $\bar{y}$ were equal to 20, and the mean under the null hypothesis, μ_0, were equal to 18. The numerator of z_M would thus be 20 − 18 = 2. When 2 is divided by the standard error of 2, we obtain a value for z_M of 1.0, which is not statistically significant at $p < 0.05$.

Now, consider the scenario where the standard error of the mean remains the same at 2, but that instead of the sample mean $\bar{y}$ being equal to 20, it is equal to 30. The difference between the sample mean and the population mean is thus 30 − 18 = 12. This difference represents a greater distance between means, and presumably, would be indicative of a more "successful" experiment or study. Dividing 12 by the standard error of 2 yields a z_M value of 6.0, which is highly statistically significant at $p < 0.05$ (whether for a one- or two-tailed test).

Having the value of z_M increase as a result of the distance between $\bar{y}$ and μ_0 increasing is of course what we would expect from a test statistic if that test statistic is to be used in any sense to evaluate the strength of the **scientific** evidence against the null. That is, if our obtained sample mean $\bar{y}$ turns out to be very different than the population mean under the null hypothesis, μ_0, we would hope that our test statistic would measure this effect, and allow us to reject the null hypothesis at some preset significance level (in our example, 0.05). If interpreting test statistics were always as easy as this, there would be no misunderstandings about the meaning of statistical significance and the misguided decisions to automatically attribute "worth" to the statement "$p < 0.05$." However, as we discuss in the following cases, there are other ways to make z_M big or small that do not depend so intimately on the distance between $\bar{y}$ and μ_0, and this is where interpretations of the significance test usually run awry.

Consider the case now for which the distance between means, $\bar{y} - \mu_0$ is, as before, equal to 2.0 (i.e., 20 − 18 = 2.0). As noted, with a standard error also equal to 2.0, our computed value of z_M came out to be 1.0, which was not statistically significant. However, is it possible to increase the size of z_M without changing the observed distance between means? Absolutely. Consider what happens to the size of z_M as we change the magnitude of either σ or n, or both. First, we consider how z_M is defined in part as a function of σ. For convenience, we assume a sample size still of $n = 100$. Consider now three hypothetical values for σ: 2, 10, and 20. Performing the relevant computations, observe what happens to the size of z_M in the case where $\sigma = 2$:

$$z_M = \frac{\bar{y} - \mu_0}{\frac{\sigma}{\sqrt{n}}} = \frac{20 - 18}{\frac{2}{\sqrt{100}}} = \frac{2}{0.2} = 10$$

The resulting value for z_M is quite large at 10. Consider now what happens if we increase σ from 2 to 10:

$$z_M = \frac{\bar{y} - \mu_0}{\frac{\sigma}{\sqrt{n}}} = \frac{20 - 18}{\frac{10}{\sqrt{100}}} = \frac{2}{1} = 2$$

Notice that the value of z_M has decreased from 10 to 2. Consider now what happens if we increase σ even more to a value of 20 as we had originally:

$$z_M = \frac{\bar{y} - \mu_0}{\frac{\sigma}{\sqrt{n}}} = \frac{20 - 18}{\frac{20}{\sqrt{100}}} = \frac{2}{2} = 1$$

When $\sigma = 20$, the value of z_M is now equal to 1, which is no longer statistically significant at $p < 0.05$. Be sure to note that the distance between means $\bar{y} - \mu_0$ has remained constant. In other words, and this is important, z_M **did not decrease in magnitude by altering the actual distance between the sample mean and the population mean, but rather decreased in magnitude only by a change in σ.**

What this means is that given a constant distance between means $\bar{y} - \mu_0$, whether or not z_M will or will not be statistically significant can be manipulated by changing the value of σ. Of course, a researcher would never arbitrarily manipulate σ directly. The way to decrease σ would be to sample from a population with less variability. The point is that decisions regarding whether a "positive" result occurred in an experiment or study should not be solely a function of whether one is sampling from a population with small or large variance!

Suppose now we again assume the distance between means $\bar{y} - \mu_0$ to be equal to 2. We again set the value of σ at 2. With these values set and assumed constant, consider what happens to z_M as we increase the sample size n from 16 to 49 to 100. We first compute z_M assuming a sample size of 16:

$$z_M = \frac{\bar{y} - \mu_0}{\frac{\sigma}{\sqrt{n}}} = \frac{20 - 18}{\frac{2}{\sqrt{16}}} = \frac{2}{0.5} = 4$$

With a sample size of 16, the computed value for z_M is equal to 4. When we increase the sample size to 49, again, **keeping the distance between means constant, as well as the population standard deviation constant**, we obtain:

$$z_M = \frac{\bar{y} - \mu_0}{\frac{\sigma}{\sqrt{n}}} = \frac{20 - 18}{\frac{2}{\sqrt{49}}} = \frac{2}{0.29} = 6.9$$

We see that the value of z_M has increased from 4 to 6.9 as a result of the larger sample size. If we increase the sample size further, to 100, we get

$$z_M = \frac{\bar{y} - \mu_0}{\frac{\sigma}{\sqrt{n}}} = \frac{20 - 18}{\frac{2}{\sqrt{100}}} = \frac{2}{0.2} = 10$$

and see that as a result of the even larger sample size, the value of z_M has increased once again, this time to 10. Again, we need to emphasize that the observed increase in z_M is occurring not as a result of changing values for $\bar{y} - \mu_0$ or σ, as these values remained constant in our above computations. Rather, **the magnitude of z_M increased as a direct result of an increase in sample size, n, alone**. In many research studies, **the achievement of a statistically significant result may simply be indicative that the researcher gathered a minimally sufficient sample size that resulted in z_M falling in the tail of the z distribution**. In other cases, the failure to reject the null may in reality simply indicate that the investigator had insufficient sample size. The point is that unless one knows how n can directly increase or decrease the size of a p-value, one cannot be in a position to understand, in a scientific sense, what the p-value actually means, or intelligently evaluate the statistical evidence before them.

2.28.2 The Make-Up of a *p*-Value: A Brief Recap and Summary

The simplicity of these demonstrations is surpassed only by their profoundness. In our simple example of the one-sample z-test for a mean, we have demonstrated that the size of z_M is a direct function of three elements: (1) distance $\bar{y} - \mu_0$, (2) population standard deviation σ, and (3) sample size n. A change in **any** of these while holding the others constant will necessarily, through nothing more than the consequences of how the significance test is constructed and functionally defined, result in a change in the size of z_M. The implication of this is that one can make z_M as small or as large as one would like by choosing to do a study or experiment such that the combination of $\bar{y} - \mu_0$, σ, and n results in a z_M value that meets or exceeds a pre-selected criteria of statistical significance.

The important point here is that a large value of z_M does not necessarily mean something of any **practical** or **scientific** significance occurred in the given study or experiment. This fact has been reiterated countless times by the best of methodologists, yet too often researchers fail to emphasize this extremely important truth when discussing findings:

A p-value, no matter how small or large, does not necessarily equate to the success or failure of a given experiment or study.

Too often a statement of "$p < 0.05$" is recited to an audience with the implication that somehow this necessarily constitutes a "scientific finding" of sorts. This is entirely misleading, and the practice needs to be avoided. The solution, as we will soon discuss, is to pair the p-value with a report of the effect size.

2.28.3 The Issue of Standardized Testing: Are Students in Your School Achieving More Than the National Average?

To demonstrate how adjusting the inputs to z_M can have a direct impact on the obtained p-value, consider the situation in which a school psychologist practitioner hypothesizes that as a result of an intensified program implementation in her school, she believes that her school's students, on average, will have a higher achievement mean compared to the national average of students in the same grade. Suppose that the national average on a given standardized performance test is equal to 100. If the school psychologist is correct that her students are, on average, more advanced performance-wise than the national average, then her students should, on average, score higher than the national mark of 100. She decides to sample 100 students from her school and obtains a sample achievement mean of $\bar{y} = 101$. Thus, the distance between means is equal to $101 - 100 = 1$. She computes the estimated population standard deviation s equal to 10. Because she is estimating σ^2 with s^2, she computes a one-sample t-test rather than a z-test. Her computation of the ensuing t is:

$$t = \frac{\bar{y} - \mu_0}{\frac{s}{\sqrt{n}}} = \frac{101 - 100}{\frac{10}{\sqrt{100}}} = \frac{1}{1} = 1$$

On degrees of freedom equal to $n - 1 = 100 - 1 = 99$, for a two-tailed test, we require a t statistic of $\pm$ 1.984 for the result to be statistically significant at a level of significance of 0.05. Hence, the obtained value of $t = 1$ is not statistically significant. That the result is not statistically significant is hardly surprising, since the sample mean of the psychologist's school is only 101, a single mean point higher than the national average of 100. It would seem then that the computation of t is telling us a story that is consistent with our intuition, that there is no reason to believe that the school's performance is higher than that of the national average in the population from which these sample data were drawn.

Now, consider what would have happened had the psychologist collected a larger sample, suppose $n = 500$. Using our new sample size, and still assuming an estimated population standard deviation s equal to 10 and a distance between means equal to 1, we repeat the computation for t:

$$t = \frac{\bar{y} - \mu_0}{\frac{s}{\sqrt{n}}} = \frac{101 - 100}{\frac{10}{\sqrt{500}}} = \frac{1}{0.45} = 2.22$$

What happened? The obtained value of t increased from 1 to 2.22 **simply as a result of collecting a larger sample**, nothing more. The actual distance between means remained the same ($101 - 100 = 1$). The degrees of freedom for the test have changed and are now equal to 499 (i.e., $n - 1 = 500 - 1 = 499$). Since our obtained t of 2.22 exceeds critical t, our statistic is deemed statistically significant at $p < 0.05$. What is important to realize is that we did not change the difference between the sample mean $\bar{y}$ and the population mean μ_0, it remained extremely small at only a single mean achievement point (i.e., $101 - 100 = 1$). Even with the same distance between means, the obtained t of 2.22 and it being statistically significant at $p < 0.05$ now means we will reject the null hypothesis, and infer the alternative hypothesis that $\mu \neq \mu_0$. And because scientists have historically considered the infamous statement "$p < 0.05$" to be automatically and necessarily equivalent to something meaningful or important, the obvious danger is that the rejection of the null hypothesis at $p < 0.05$ is considered by some (or even **most**) a "positive" result. When in reality, the difference, in this case, is nothing short of **trivial**.

The problem is not that the significance test is not useful and therefore should be banned. The problem is that too few are aware that the statement "$p < 0.05$," in itself, **scientifically** (as opposed to **statistically**) may have little meaning in a given research context, and at worst, may be entirely misleading if automatically assigned any degree of scientific importance by the interpreter.

2.28.4 Other Test Statistics

The factors that influence the size of a p-value are, of course, not only relevant to z- and t-tests, but are at work in essentially every test of statistical significance we might conduct. For instance, as we will see in the following chapter, the size of the F-ratio in traditional one-way ANOVA is subject to the same influences. Taken as the ratio of MS between to MS error, the three determining influences for the size of p are (1) size of MS between, which is a reflection of the extent to which means are different from group to group, (2) size of MS error, which is in part a reflection of the within-group variability, and (3) sample size (when computing MS error, we divide the sum of squares for error by degrees of freedom, in which the degrees of freedom are determined in large part by sample size). Hence, a large F-stat does not necessarily imply that MS between is absolutely large, no more than a large t necessarily implies the size of $\bar{y} - \mu_0$. **A small p-value associated with a computed F could be a result of small within-group variation and/or a large sample size. It does not necessarily mean that group-to-group mean differences are substantial, which was presumably the goal of the study or experiment**

by the investigator. That is, the goal was not to simply obtain small within-group variation. The goal was to demonstrate mean differences between groups.

These ideas for significance tests apply in even the most advanced of modeling techniques, such as structural equation modeling (see Chapter 15). The typical measure of model fit is the chi-square statistic, χ^2, which as reported by many (e.g., see Bollen, 1989; Hoelter, 1983) suffers the same interpretational problems as t and F regarding how its magnitude can be largely a function of sample size. That is, one can achieve a small or large χ^2 simply because one has used a small or large sample. If a researcher is not aware of this fact, he or she may decide that a model is well-fitting or poor-fitting based on a small or large chi-square value, without awareness of its connection with n. This is in part why other measures, as we will see, have been proposed for interpreting the fit of SEM models (e.g., see Browne and Cudeck, 1993).

2.28.5 The Solution

The solution to episodes of misunderstanding the significance test is not to drop or ban it, contrary to what some have recommended (e.g., Hunter, 1997). Rather, the solution is to supplement it with a measure that accounts for the actual distance between means and serves to convey the magnitude of the actual **scientific** finding, as opposed to **statistical** finding, should there be one. Measures of **effect size**, interpreted in conjunction with significance tests, help to communicate whether something has "happened" or "not happened" in the given study or experiment. The reader interested in effect sizes can turn to a multitude of sources (Cortina and Nouri, 1999; Rosenthal, Rosnow, and Rubin, 2000). For our purposes, it suffices to review the **principle** of an effect size measure rather than catalog the wealth of possibilities for effect sizes available. Perhaps the easiest and most straightforward way of conceptualizing an effect size is to consider a measure of **standardized statistical distance**, or Cohen's d, already featured in our computations of power.

2.28.6 Statistical Distance: Cohen's d

For a one-sample z-test, Cohen's d (Cohen, 1988) is defined as the absolute distance between the observed sample mean and the population mean under the null hypothesis, divided by the population standard deviation:

$$d = \left| \frac{\bar{y} - \mu_0}{\sigma} \right|$$

In the above, since $\bar{y}$ is serving as the estimate of μ, the numerator can also be given as $\mu - \mu_0$. However, using $\bar{y}$ instead of μ above is a reminder of where this mean is coming from. It is coming from our sample data, and we wish to compare that sample mean to the population mean μ_0 under the null hypothesis.

As an example, where $\bar{y} = 20$, $\mu_0 = 18$, and $\sigma = 2$ Cohen's d is computed as:

$$d = \left| \frac{20 - 18}{2} \right|$$
$$= 1.0$$

Cohen offered the guidelines of 0.20, 0.50, and 0.80 as representing small, medium, and large effects respectively (Cohen, 1988). However, relying on effect size guidelines to indicate the absolute size of an experimental or nonexperimental effect should only be done in the complete and absolute absence of all other information for the research area. In the end, **it is the researcher, armed with knowledge of the history of the phenomenon under study, who must evaluate whether an effect is small or large**. For instance, referring to the achievement example discussed earlier, Cohen's d would be equal to:

$$d = \left| \frac{101 - 100}{10} \right|$$
$$= 0.1$$

The effect size of 0.1 is small according to Cohen's guidelines, but more importantly, also small **substantively**, since a difference in means of 1 point is, by all accounts, likely trivial. In this case, both Cohen's guidelines and the actual substantive evaluation of the size of effect coincide. However, this is not always the case. In physical or biological experiments, for instance, one can easily imagine examples for which an effect size of even 0.8 might be considered "small" relative to the research area under investigation, since the degree of control the investigator can impose over his or her subjects is much greater. In such cases, it may very well be that Cohen's d values in the neighborhood of two or three would be required for an effect to be considered "large." The point is that only in the complete absence of information regarding an area of investigation is it appropriate to use "rules of thumb" to evaluate the size of effect. Cohen's d, or effect size measures in general, should always be used in conjunction with statements of statistical significance, since they tell the researcher what she is actually wanting to know, that of the estimated separation between sample data (often in the form of a sample mean) and the null hypothesis under investigation. Oftentimes **meta-analysis**, which is a study of the overall measure of effect for a given phenomenon, can be helpful in comparing new research findings to the "status quo" in a given field. For a thorough user-friendly overview of the methodology, consult Shelby and Vaske (2008).

2.28.7   What Does Cohen's d Actually Tell Us?

Writing out a formula and plugging in numbers, unfortunately, does not necessarily give us a feeling for what the formula actually means. This is especially true with regard to Cohen's d. We now discuss the statistic in a bit more detail, pointing out why it is usually interpreted as the **standardized difference between means**.

Imagine you have two independent samples of laboratory rats. To one sample, you provide normal feeding and observe their weight over the next 30 days. To the other sample, you also feed normally, but also give them regular doses of a weight-loss drug. You are interested in learning whether your weight-loss drug works or not. Suppose that after 30 days, on average, a mean difference of 0.2 pounds is observed between groups. How big is a difference of 0.2 pounds for these groups? If the average difference in weight among rats in the population were very large, say, 0.8 pounds, then a mean difference of 0.2 pounds is not that impressive. After all, if rats weigh very differently from one rat to the next, then really, finding a mean difference of 0.2 between groups cannot be that exciting. However, if the average weight difference between rats were equal to 0.1 pounds, then all of a sudden, a mean difference of 0.2 pounds seems more impressive, because that size of difference is **atypical** relative to the population. What is "typical?" This is exactly what the **standard deviation** reveals. Hence, when we are computing Cohen's d, we are in actuality producing a **ratio of one deviation relative to another**, similar to how when we compute a z-score, we are comparing the deviation of $y - \mu$ with the **standard** deviation σ. The extent to which observed differences are large relative to "average" differences will be the extent to which d will be large in magnitude.

2.28.8 Why and Where the Significance Test Still Makes Sense

At this point, the conscientious reader may very well be asking the following question: **If the significance test is so misleading and subject to misunderstanding and misinterpretation, how does it even make sense as a test of anything? It would appear to be a nonsensical test and should forever be forgotten**. The fact is that the significance test **does** make sense, only that the sense that it makes is not necessarily always **scientific**. Rather, it is **statistical**. To a pure theoretical statistician or mathematician, a decreasing p-value as a function of an increasing sample size makes perfect sense—as we snoop a larger part of the population, the random error we expect typically decreases, because with each increase in sample size we are obtaining a better estimate of the true population parameter. Hence, that we achieve statistical significance with a sample size of 500 and not 100, for instance, is well within that of statistical "good sense." That is, **the p-value is functioning as it should**, and likewise yielding the correct **statistical** information.

However, **statistical truth does not equate to scientific truth** (Bolles, 1962). Statistical conclusions should never be automatically equated with scientific ones. They are different and distinct **things**. When we arrive at a statistical conclusion (e.g., when deciding to reject the null hypothesis), one can never assume that this represents anything that is necessarily or absolutely scientifically meaningful. Rather, the statistical conclusion should be used as a potential **indicator** that something scientifically interesting **may** have occurred, the evidence for which must be determined by other means, which includes effect sizes, researcher judgment, and putting the obtained result into its proper interpretive context.

2.29 CHAPTER SUMMARY AND HIGHLIGHTS

- To understand advanced statistical procedures, it is necessary to have a firm grasp on the **foundations of introductory statistics**. Advanced procedures are typically extensions of **first principles**.
- **Densities** are theoretical probability distributions. The **normal univariate density** is an example.
- The **standard normal distribution** has a mean μ of 0 and a variance σ^2 of 1.
- **z-scores** are useful for comparing raw scores emanating from different distributions. **Standardization** transforms raw scores to a common scale, allowing for comparison between scores.
- **Binomial distributions** are useful in modeling experiments in which the outcome can be conceptualized as a "success" or "failure." The outcome of the experiment must be **binary** in nature for the binomial distribution to apply.
- The **normal distribution** can be used to **approximate** the binomial distribution. In this regard, we say that the **limiting form** of the binomial distribution is the normal distribution.
- The **bivariate normal density** expresses the probability of the joint occurrence of **two variables**.
- The **multivariate normal density** expresses the probability of the joint occurrence of **three or more variables**.
- The **mean, variance, skewness**, and **kurtosis** are all **moments** of a distribution.
- The **mean (arithmetic)**, the first moment of a distribution, either of a mathematical variable or a random variable, can be regarded as the **center of gravity** of the distribution such that the sum of deviations from the mean for any distribution is equal to zero.
- The **variance**, the second moment of a distribution, can be computed for either a mathematical variable or a random variable. It expresses the degree to which scores, on average, deviate from the mean in **squared units**.

- The **sample variance** with n in the denominator is **biased**. To correct for the bias, a single degree of freedom is subtracted so that the new denominator is $n - 1$.
- The **expectation** of the uncorrected version of the sample variance is not equal to σ^2. That is, $E(S^2) \neq \sigma^2$. However, the corrected version of the sample variance (with $n - 1$ in the denominator) is equal to σ^2. That is, $E(s^2) = \sigma^2$.
- **Skewness**, the third moment of a distribution, reflects the extent to which a distribution **lacks symmetry**.
- **Kurtosis**, the fourth moment of a distribution, reflects the extent to which a distribution is peaked or flat and also having much to do with a distribution's tails.
- **Covariance** and **correlation** are defined for both empirical variables and random variables. Both measure the extent to which two variables are **linearly related**. Pearson r is the standardized version of the covariance, and is dimensionless, meaning that its value is not dependent on the variance in each variable. Pearson r ranges from -1 to $+1$ in value.
- One popular use of correlation is in establishing **reliability** and **validity** of psychometric measures.
- In **multivariable** contexts, covariance and correlation **matrices** are used in place of single coefficients.
- There are numerous other correlation coefficients available other than Pearson r. One such coefficient is **Spearman's** r_s, which captures **monotonically increasing** (or decreasing) relationships. Monotonic relationships do not necessarily have to be linear.
- The issue of measurement should be carefully considered before data is collected. S.S. Stevens proposed four scales of measurement, **nominal**, **ordinal**, **interval**, and **ratio**. The most sophisticated level of measurement is that of the ratio scale where a value of zero on the scale truly means an absence of the attribute under study.
- A **random variable** is a mathematical variable that is associated with a probability distribution. More formally, it is a function from a sample space into the real numbers.
- An **estimator** is a function of a sample used to estimate a parameter in the population.
- An **interval estimator** provides a range of values within which the true parameter is hypothesized to exist.
- An **unbiased estimator** is one in which its expectation is equal to the corresponding population parameter. That is, $E(T) = \theta$.
- An estimator is **consistent** if as sample size increases without bound, the variance of the estimator approaches zero.
- An estimator is **efficient** if it has a relatively low mean squared error.
- An estimator is **sufficient** for a given parameter if the statistic tells us everything we need to know about the parameter and our knowledge of it could not be improved if we considered additional information (e.g., such as a secondary statistic).
- The concept of a **sampling distribution** is at the heart of statistical inference. A sampling distribution of a statistic is a theoretical probability distribution of that statistic. It is idealized, and hence not ordinarily empirically derived.
- The **sampling distribution of the mean** is of great importance because so many of our inferences feature means.
- As a result of $E(\bar{y}) = \mu$, we can say that $\mu_{\bar{y}} = \mu$, that is, the **mean of all possible sample means** we could draw from some specified population is equal to the mean of that population.

- The **variance of the sampling distribution of the mean** is equal to $\frac{1}{n}$ of the original population variance. That is, it is equal to $\frac{\sigma^2}{n}$.
- The square root of the sampling variance for the mean is equal to the standard error, $\sqrt{\frac{\sigma^2}{n}} = \frac{\sigma}{\sqrt{n}}$.
- The **central limit theorem** is perhaps the most important theorem in all of statistics. Though there are different forms of the theorem, in general, it states that the sum of random variables approximates a normal distribution as the size upon which each sample is based increases without bound.
- **Confidence intervals** provide a range of values for which we can be relatively certain to lay the true parameter we are seeking to estimate. Key to understanding confidence intervals is to recognize that it is the sample upon which the interval is computed that is the random component, and not the parameter we are seeking to estimate. The parameter is typically assumed to be fixed.
- **Student's t distribution**, derived by William Gosset (or "Student") in 1908, is useful when σ^2 is unknown and must be estimated from the sample. Because in the limit $f(t) = f(z)$ (i.e., $\lim_{v \to \infty} f(t) = f(z)$), for large samples, whether one uses z or t will make little difference in terms of whether or not the null hypothesis is rejected.
- The **t-test** for one sample compares an obtained sample mean to a population mean and evaluates the null hypothesis that the sample mean could have reasonably been drawn from the given population.
- As degrees of freedom increase, the **variance of the t-distribution approaches 1**, which is the same as that for a standardized normal variable. That is, $\lim_{v \to \infty} \left(\frac{v}{v-2} \right) = 1.0$.
- The **t-test for two samples** tests the null hypothesis that both samples were selected from the same population. A rejection of the null hypothesis suggests the samples arose from populations with different means.
- **Power** is the probability of rejecting a null hypothesis given that it is false. It is equal to $1 - \beta$ (i.e., 1 – type II error rate). Power is a function of four elements: (1) hypothesized value under H_1, (2) significance level, or type I error rate, α, (3) variance, σ^2, in the population, and (4) sample size.
- Experiments or studies suffering from **insufficient power** make it difficult to ascertain why the null hypothesis failed to be rejected.
- The **paired-samples t-test** is useful for **matched-pairs** (elementary blocking) designs.
- The paired-samples t-test usually results in an **increase in statistical power** because the covariance between measurements is subtracted from the error term. In general, anything that makes the error term smaller helps to boost statistical power.
- The paired-samples t-test and the matched design which it serves provides a good entry point into the discussion of the **randomized block design**, the topic of Chapter 6.
- In **multivariable** contexts, **linear combinations** of variables are generated of the form $\ell_i = a_1 y_1 + a_2 y_2 + \ldots + a_p y_p$. Means and variances of linear combinations can be obtained, as well as the covariance and correlation between linear combinations.
- Representing statistical models in matrix form is required in statistical analyses of higher dimensions than 1 (e.g., multiple regression, multivariate analysis of variance, principal components analysis, etc.). The fundamental general linear model can be given by $\mathbf{Y} = \mathbf{XB} + \mathbf{E}$.
- Understanding what makes a p-value small or large is essential if a researcher is to intelligently interpret statistical evidence is his or her field. The history of null hypothesis significance testing (NHST) is plagued with controversy, and a solid understanding of the difference between **statistical significance** and **effect size** (e.g., Cohen's d) is necessary before one attempts to interpret any research findings.

REVIEW EXERCISES

2.1. Distinguish between a **density** and an **empirical distribution**. How are they different? How are they similar?

2.2. Consider the **univariate normal density**:

$$f(x_i, \mu, \sigma^2) = \frac{1}{\sqrt{2\pi\sigma^2}} e^{-(x_i-\mu)^2/2\sigma^2}$$

Show that for a standard normal distribution, the above becomes $f(x_i, \mu, \sigma^2) = \dfrac{e^{-\frac{1}{2}x_i^2}}{\sqrt{2\pi}}$.

2.3. Explain the nature of a z-score, $z = \dfrac{x_i - \mu}{\sigma}$. Why is it also called a **standardized** score?

2.4. Using R, compute the probability of observing a standardized score of **1.0 or greater**. What is then the probability of observing a score less than 1.0 from such a distribution?

2.5. Think up a research example in which the **binomial distribution** would be useful in evaluating a null hypothesis.

2.6. Rafael Nadal, a professional tennis player, as of 2020 had won the French Open tennis championship a total of 13 times in the past 16 tournaments. If we set the probability of him winning each time at 0.5, determine the probability of winning 13 times out of 16. **Make a statistical argument that Nadal is an exceptional tennis player at the French Open**. What if we set the probability of a win at 0.1? Does this make Nadal's achievements less or more impressive? Why? Explain.

2.7. Give an example using the **binomial distribution** in which the null hypothesis would not be rejected even if observing 2 out of 10 heads on flips of a coin.

2.8. On a fair coin, what is the **probability of observing 0 heads or 5 heads**? How did you arrive at this probability, and which rules of probability did you use in your computation?

2.9. Discuss what a **limiting form** of a distribution means, and how the limiting form of the binomial distribution is that of the normal distribution.

2.10. Consider the **multivariate density**:

$$g(x_i) = \frac{1}{\left(\sqrt{2\pi}\right)^P |\Sigma|^{1/2}} e^{-(x-\mu)'\Sigma^{-1}(x-\mu)/2}$$

All else constant, what effect does an increasing value of the **determinant** ($|\Sigma|$) have on the density, and how does this translate when using real variables?

2.11. What is meant by the **expectation** of a random variable?

2.12. Compare these two products, and explain how and why they are different from one another when taking **expectations**: $y_i p(y_i)$ versus $y_i p(y_i) dy$

2.13. Why is it reasonable that the arithmetic mean is the **center of gravity** of a distribution?

2.14. What is an **unbiased estimator** of a population mean vector?

2.15. Discuss what it means to say that $E(S^2) \neq \sigma^2$, and the implications of this. What is $E(S^2)$ equal to?

2.16. Even though $E(S^2) \neq \sigma^2$, how can it be true nonetheless that $\lim_{n \to \infty} E(S^2) = \sigma^2$? Explain.

2.17. Explain why the following form of the sample variance is considered to be an **unbiased estimator** of the population variance:

$$s^2 = \frac{\sum_{i=1}^{n}(y_i - \bar{y})^2}{n-1}$$

2.18. Draw a distribution that is **positively skewed**. Now draw one that is **negatively skewed**.

2.19. Compare and contrast the **covariance** of a random variable: $\text{cov}(x_i, y_i) = \sigma_{xy} = E[(x_i - \mu_x)(y_i - \mu_y)]$ with that of the **sample covariance**: $\text{cov} = \dfrac{\sum_{i=1}^{n}(x_i - \bar{x})(y_i - \bar{y})}{n-1}$. How are they similar? How are they different? What in their definitions makes them different from one another?

2.20. What effect (if any) does **increasing sample size** n have on the magnitude of the covariance? If it does not have any effect, explain why it does not.

2.21. Explain or show how the **variance** of a variable can be conceptualized as the **covariance of a variable with itself**.

2.22. Cite three reasons why the covariance is not a pure or **dimensionless** measure of **relationship** between two variables.

2.23. Why is **Pearson r** not suitable for measuring relationships that are **nonlinear**? What is an alternative coefficient (one of many) that may be computed that is more appropriate for relationships that are nonlinear?

2.24. What does it mean to say the relationship between two variables is **monotonically increasing**?

2.25. What does a **correlation matrix** have along its main diagonal that a **covariance matrix** does not? What is along the main diagonal of a covariance matrix?

2.26. Define, in general, what it means to **measure** something.

2.27. Explain why it is that something measurable at the **ratio level of measurement** is also measurable at the **interval**, **ordinal**, and **nominal** levels as well.

2.28. Is something such as **intelligence** measurable on a **ratio scale**? Why or why not?

2.29. Distinguish between a **mathematical variable** and a **random variable**.

2.30. Distinguish between an **estimator** and an **estimate**.

2.31. Define what is meant by an **interval estimator**.

2.32. Define what is meant by the **consistency of an estimator** and what

$$\lim_{n \to \infty} \sigma_T^2 = 0$$

means in this context.

2.33. Compare the concepts of **efficiency** versus **sufficiency** with regard to estimators. How are they different?

2.34. The **sampling distribution of the mean** is an idealized distribution. However, discuss how one would generate the sampling distribution of the mean empirically.

2.35. Discuss why for a higher level of confidence, all else equal, a **confidence interval** widens rather than narrows.

2.36. Define what is meant by a **maximum-likelihood estimator**.

2.37. Discuss the behavior of the *t* **distribution** for increasing degrees of freedom. What is the limiting form of the *t* distribution?

2.38. In a research setting, under what condition(s) is a *t*-**test** usually preferred over a *z*-**test**?

2.39. Verbally interpret the nature of **pooling** in the independent-samples *t*-test. Under what condition(s) do we pool variances? Under what condition(s) should we not pool?

2.40. Discuss why an estimate of **effect size** is required for estimating **power**.

2.41. Using R, estimate required **sample size** for detecting a population correlation coefficient of 0.30 at a significance level of 0.01, with power equal to 0.80.

2.42. Repeat exercise 2.41, this time using **G*Power**.

2.43. Using R, estimate power for an **independent samples *t*-test** for a sample size of 100 per group and Cohen's d equal to 0.20.

2.44. For a value of $r^2 = 0.70$, compute the corresponding value for **d**.

2.45. Discuss how the **paired-samples *t*-test** can be considered a special case of the wider and more general **blocking design**.

2.46. Define what is meant by a **linear combination**.

2.47. Define and describe each term in the **multivariate general linear model Y = XB + E**.

2.48. Discuss the key determinants of the *p* value in a significance test

2.49. A researcher collects a sample of $n = 10,000$ observations and tells you that with such a large sample size, he is guaranteed to reject the null hypothesis. Explain why the researcher's claim is false.

2.50. A researcher collects a sample size of $n = 5$, computes z_M and rejects the null hypothesis. Argue on the one hand for why this might be impressive scientifically, then argue why it may not be.

2.51. Consider once more Galton's data on heights (only the first 10 observations are shown):

```
> library(HistData)
> attach(Galton)
> Galton
      parent   child
   1    70.5    61.7
   2    68.5    61.7
   3    65.5    61.7
   4    64.5    61.7
   5    64.0    61.7
   6    67.5    62.2
   7    67.5    62.2
   8    67.5    62.2
   9    66.5    62.2
  10    66.5    62.2
```

(a) Compute a **histogram** of parent height, as well as an index of **skewness** and **kurtosis**. What do your measures of skewness and kurtosis suggest about the distribution?

(b) Transform the distribution of child heights to *z*-**scores**. What effect did such a transformation have on the mean and variance of the original distribution? Second, did it change its shape at all? Why or why not?

(c) Compute the **covariance** between **parent height** and **child height**. Does the sign of the covariance suggest a positive or negative relationship?

(d) Standardize the covariance by computing **Pearson** *r*. Interpret the obtained correlation coefficient, and test it for statistical significance using either SPSS or R.

2.52. Consider the following data on whether a student passed or failed a mathematics course (grade = 0 is "failed" and grade = 1 is "passed"), along with that student's study time for the course, in average minutes per day for the duration of the course:

grade	studytime
0	30
0	25
0	59
0	42
0	31
1	140
1	90
1	95
1	170
1	120

Conduct an **independent-samples** *t*-**test** on this data using SPSS and R. Verify that the assumption of **homogeneity of variances** is met in SPSS.

2.53. A researcher is interested in conducting a **two-sample** *t*-**test** between a treatment group and a control group. The researcher anticipates an effect size of approximately d = 1.5 and wishes to test the null hypothesis $\mu_1 = \mu_2$ at a significance level of 0.05. Estimate required **sample size** assuming the researcher wishes to attain **power** of at least 0.90 for her test of the null hypothesis.

Further Discussion and Activities

2.54. As discussed in this chapter, null hypothesis significance testing (NHST) has been critically evaluated and dissected as a means for drawing scientific inferences in the social and natural sciences. Rozeboom (1960) quite nicely summarized the main criticisms in **The Fallacy of the Null-Hypothesis Significance Test**. Read the article and discuss Rozeboom's distinction between **decisions** versus **degrees of belief**. Why is such a distinction important for a scientist to understand the difference between **statistical** versus **scientific** inference? Rozeboom's article can be downloaded from Christopher D. Green's **Classics in the History of Psychology** website: http://psychclassics.yorku.ca/Rozeboom/

2.55.   R.A. Fisher, the modern "father of statistics" wrote in 1956:

> **"… no scientific worker has a fixed level of significance at which from year to year, and in all circumstances, he rejects hypotheses; he rather gives his mind to each particular case in the light of his evidence and his ideas."**

Many writers and researchers, however, have found that since the inception of the significance test in the early 1900s, scientists, both social and otherwise, routinely employ the 0.05 level of significance in rejecting null hypotheses. Read **Mindless Statistics** by Gigerenzer (2004), and discuss the dangers and risks, both practical and theoretical, of allowing the "null ritual" to dominate in science.

3

ANALYSIS OF VARIANCE: FIXED EFFECTS MODELS

The prime objective of this book is to put into the hands of researchworkers, [sic] and especially of biologists, the means of applying statistical tests accurately to numerical data accumulated in their own laboratories or available in the literature. Such tests are the result of solutions of problems of distribution, most of which are but recent additions to our knowledge and have so far only appeared in specialised [sic] mathematical papers.

(Fisher, 1925, p. 4, Statistical Methods for Research Workers)

Suppose a researcher is interested in knowing whether melatonin, a popular sleep aid, is effective at helping individuals fall asleep at night. The researcher samples 75 individuals at random, and assigns 25 to a control group receiving no melatonin, 25 to a treatment group receiving 1 mg of melatonin, and 25 to another treatment group receiving 3 mg of melatonin nightly. These **specific** doses of 1 and 3 mg are of interest to the researcher, since it is hypothesized that the greater dose of 3 mg will be more effective at promoting sleep compared to the lesser dose of 1 mg, which will, in turn, be more effective than receiving no melatonin at all.

Monitoring EEG levels of all participants in a sleep lab, the researcher measures the time it takes from ingestion of the melatonin to the time the participant reaches NREM ("non-rapid eye movement") sleep. The "time until NREM," measured in minutes, is generally known as **sleep onset latency**. To assess whether the melatonin has an effect on sleep onset, the researcher wishes to compare mean sleep latencies across groups to discern any treatment effect that may be present (see Figure 3.1). Such a research design calls for a **one-way fixed effects analysis of variance**.

Applied Univariate, Bivariate, and Multivariate Statistics: Understanding Statistics for Social and Natural Scientists, With Applications in SPSS and R, Second Edition. Daniel J. Denis.
© 2021 John Wiley & Sons, Inc. Published 2021 by John Wiley & Sons, Inc.
Companion Website: www.wiley.com/go/denis/appliedstatistics2e

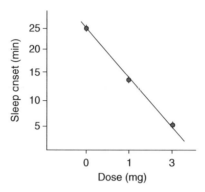

FIGURE 3.1   Sleep onset as a function of melatonin dose (hypothetical). Circles represent means for each dose. Dotted line suggests a negative relationship between sleep onset and dose.

3.1   WHAT IS ANALYSIS OF VARIANCE? FIXED VERSUS RANDOM EFFECTS

The analysis of variance (ANOVA) is the workhorse of experimental research across the social and natural sciences. The methodology is generally attributed to R.A. Fisher who wrote in 1925 **Statistical Methods for Research Workers**, which provided scientists with a novel quantitative method for partitioning sources of variance in a set of data and making inferences about effects in the population from which sample data were drawn. Estimation of parameters in the analysis of variance usually boils down to obtaining least-squares solutions analogous to what is done in regression (see Chapter 7), but as remarked by Eisenhart (1947), Fisher's primary contribution was in how he **packaged** the analysis of variance procedure:

> With respect to the problems of estimation belonging to this class [class of estimating fixed effects], analysis of variance is simply a form of the method of least squares: the analysis-of-variance solutions are the least-squares solutions. The cardinal contribution of analysis of variance to the actual procedure is the **analysis-of-variance table** devised by R.A. Fisher, which serves to simplify the arithmetical steps and to bring out more clearly the significance of the results obtained. (p. 3)

Fisher also published in 1935 **Design of Experiments** in which he elucidated principles of research methodology that continue to this day to be the bedrock of modern experimental design. The type of analysis of variance model one specifies depends in large part on the assumptions that go into the model and what conclusions one wishes to make regarding observed effects. In a **fixed effects** analysis of variance, the investigator is interested in testing null hypotheses of the sort:

$$H_0 : \mu_1 = \mu_2 = \mu_3 = \mu_J$$

where the particular, **exact** levels of the independent variable chosen by the experimenter are of specific interest. That is, the investigator would like to draw conclusions about those particular levels chosen for the study and is not interested in generalizing conclusions to a wider population of levels. The levels of the independent variable are **fixed** in advance of performing the analysis, and conclusions drawn are about those levels and those levels **alone** in the fixed effects model.

In a **random effects** model, the investigator is interested in generalizing his findings not only to the levels chosen for the experiment but also to the **population** of levels from which the experimental levels were drawn. In this model, the researcher is not specifically interested in the **particular** levels of the independent level chosen for the given study. He is most interested in what these randomly chosen levels might suggest about the population of levels from which he randomly sampled the ones appearing in the given experiment. In the random effects model, the investigator is interested not in **mean population differences**, but rather in the extent to which **variance** in the dependent variable can be explained or accounted for by changing levels of the independent variable. When both fixed and random effects are present in the same model, we have the **mixed model analysis of variance**. We discuss random effects and mixed models in Chapter 5.

Note carefully that in all of these models under discussion we seek to infer conclusions drawn from samples to respective population parameters. This is not what distinguishes one model from the other. What does distinguish models is the extent to which conclusions about **sampled factor levels** are generalizable to the **population of factor levels**. Many times, students, attempting to distinguish fixed versus random effects, mistakenly conclude that fixed effects are somehow not as "inferential" as random effects, in the sense that if we are dealing with a fixed effect that somehow we are no longer interested in inferential statistics. But this is entirely incorrect. In fixed effects models, we **do** make inferences, only the inferences of treatment effects are specific to the levels actually chosen by the researcher, and not to the population of levels of which the chosen ones are but a random sample, as one would have in a random effects models.

Hays (1994), Kirk (1995), and Winer, Brown, and Michels (1991) are all classic resources on ANOVA. Maxwell and Delaney (2004) also provide a very readable overview of ANOVA models. A more technical and advanced treatment that assumes a grounding in matrix algebra is Scheffé (1999). Federer (1955), Snedecor and Cochran (1967), and Edwards (1985) are also excellent sources.

3.1.1   Small Sample Example: Achievement as a Function of Teacher

A motivating example will help set the stage for discussing the one-way fixed effects ANOVA and will be extended to two-way models in the following chapter. We also refer to this example when we discuss random effects and mixed models in Chapter 5.

Consider Table 3.1, featuring hypothetical data for students' standardized mathematics achievement scores as a function of teacher. In this design, students were assigned, at random, one of four mathematics teachers for the course of a full school year. At the end of the year, students were evaluated on their mathematics achievement (scores range from 0 to 100) through standardized testing. Students were screened beforehand to ensure they possessed an approximately equivalent degree of mathematical skill before being randomly assigned.

Some features of the data include:

- There are a total of six observations per group for a total of 24 data points. It is a **balanced** design, meaning that in each group there are an equal number of data points (when we study factorial ANOVA in the following chapter, a design will be balanced if there are an equal number of data points in each **cell**).

TABLE 3.1   Achievement as a Function of Teacher

		Teacher	
1	2	3	4
70	69	85	95
67	68	86	94
65	70	85	89
75	76	76	94
76	77	75	93
73	75	73	91
$M = 71.00$	$M = 72.5$	$M = 80.0$	$M = 92.67$

- The dependent or response variable is student achievement score on a standardized test (range from 0 to 100).
- The last row of the table contains the means for each group (71.00, 72.50, 80.00, 92.67).

3.1.2   Is Achievement a Function of Teacher?

We would like to know whether a student's mathematics achievement score is dependent on what teacher they were randomly assigned to for the school year. Recall what we mean in general by a function statement. When we ask the question "**Is achievement a function of teacher?**" what we are essentially asking is if I tell you one's teacher, are you able to predict, with some degree of certainty, their achievement score? Even if the assignment of teacher is related to achievement, we should not expect it to be a function of the kind $f(x) = x$. Such would imply that given one's teacher, we could predict their mathematics achievement **perfectly**. That is, it would imply we are working with a **deterministic** rather than a **probabilistic** or **stochastic** model. Most models in the social, medical, and other sciences are not deterministic. Rarely can we ever expect a perfect functional relation between two or more variables. This very idea, historically, of prediction but with a measure of uncertainty, is what set into motion the evolution (and revolution) of probability and of statistical modeling in the sciences.

The following is a subset of our data:

```
> achiev <- read.table("achievement.txt", header = T)
> library(car)
> some(achiev)

   ac teach
1  70      1
2  67      1
7  69      2
```

We visualize the data to get a better sense of whether mean differences may exist:

```
> achiev$teach
 [1] 1 1 1 1 1 1 2 2 2 2 2 2 3 3 3 3 3 3 4 4 4 4 4 4
> with (achiev, boxplot(ac ~ teach))
```

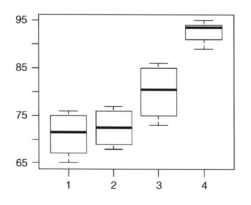

Though boxplots feature the **median** as a measure of central tendency, they still help us get a first-glance visualization both regarding potential mean differences as well as whether the assumption of **homogeneity of variance** (to be discussed) is satisfied. As we can see from the boxplots, it would appear that achievement, in general, increases across teachers. We will address the assumption of homogeneity of variance later, but for now, it would appear that the dispersion of scores within each "teacher group" is approximately similar, though there appears to be noticeably less variability of scores for the 4th teacher compared to the others. Observing boxplots such as these is useful as an exploratory aid before conducting more formal inferential statistics. As a general rule, **always explore your data before conducting formal analyses. Get to "know" your data very well**.

If achievement is (at least imperfectly) a function of teacher, then we would expect achievement means to differ by teacher. Our sample averages definitely do differ. There is no doubt that sample means 71.00 versus 72.50 versus 80.0 versus 92.67 **are** numerically different from one another. However, these are only **sample** means. They are not **population** means. Why are they not population means? Because presumably, we are not specifically interested in only these sample data points when discerning whether teacher is related to student achievement. More likely we are interested in using this sample data to draw **inferences** to the population from which these data were drawn. If we were only interested in drawing **descriptive** conclusions about the data in Table 3.1 and making no inferences to a wider population, then these data could indeed constitute population data. Remember, one researcher's sample is another researcher's population. We must always ask ourselves whether the data in front of us are considered the complete set of observations or whether they are considered a **subset** of a larger set.

Hence, we wish to evaluate a null hypothesis that the population means are equal across teacher. We can state this null hypothesis as:

$$H_0 : \mu_1 = \mu_2 = \mu_3 = \mu_4$$

Our statistical alternative hypothesis is that somewhere in the set of population means there is at least one difference in means. Of course, even if we end up rejecting the null hypothesis, we do not immediately know **where** the difference(s) lie. A couple of possibilities for statistical alternatives include:

$$H_1 : \mu_1 \neq \mu_2 = \mu_3 = \mu_4$$
$$H_1 : \mu_1 = \mu_2 \neq \mu_3 = \mu_4$$

Note that for convenience and ease of visualization of mean differences, we are notating possible alternatives as $H_1 : \mu_1 \neq \mu_2 = \mu_3 = \mu_4$ and $H_1 : \mu_1 = \mu_2 \neq \mu_3 = \mu_4$ to indicate equality or differences between group means, where in the case of $H_1 : \mu_1 \neq \mu_2 = \mu_3 = \mu_4$ is meant to read "mean 1 is different from mean 2, but mean 2 is equal to means 3 and 4." Though this makes for a convenient visual for looking at the alternative hypothesis, to be more formal about it, it would be more correct to write $H_1 : \mu_1 \neq \mu_2, \mu_2 = \mu_3 = \mu_4$. No harm is done in writing it as $H_1 : \mu_1 \neq \mu_2 = \mu_3 = \mu_4$ so long as one understands what is being communicated.

We will use procedures such as **contrasts** and **post-hoc tests** to help in discerning where mean differences may lie given a rejected null hypothesis. We discuss contrasts and post-hocs later in the chapter.

3.2   HOW ANALYSIS OF VARIANCE WORKS: A BIG PICTURE OVERVIEW

How do we go about testing a null hypothesis of the kind in our example, that of $H_0 : \mu_1 = \mu_2 = \mu_3 = \mu_4$? We could compare the sample means directly, those of 71.00 versus 72.50 versus 80.00 versus 92.67, and since they are not identical, conclude that mean differences in the population exist. But as mentioned, this would be a grossly **incorrect** way of proceeding. Since these are only sample means, any relatively small differences between means can most likely be explained by **sampling error** or **chance**. That is, we must ask the question:

> **Do sample mean differences of the kind 71.00 versus 72.50 versus 80.00 versus 92.67 actually reflect a mean difference in the population? Or, are these sample differences small enough to be simply attributable to differences generated by the simple process of sampling (i.e., "sampling error" or "chance")?**

To begin to address the above question, we must ask ourselves a related question:

> **If we sampled repeatedly an infinite number of times from this population, what is the probability of observing differences of the kind 71.00 versus 72.50 versus 80.00 versus 92.67 if the null hypothesis H_0 were actually true? That is, if $H_0 : \mu_1 = \mu_2 = \mu_3 = \mu_4$ really does represent reality, what is the probability of obtaining mean differences of the magnitude that we have in our current sample?**

If mean differences of the magnitude that we are observing happen frequently in repeated sampling under the null model of equal population means (i.e., $H_0 : \mu_1 = \mu_2 = \mu_3 = \mu_4$), then it is probably safe to at least tentatively conclude that the observed mean difference is most easily explained by sampling error or chance. In other words, since these kinds of differences happen so often in repeated sampling even when the null is true, we would have no reason to start believing the null is false and to start inferring the alternative hypothesis.

However, if mean differences of the kind we are observing in our sample turn out to be **unlikely** under the null hypothesis, then we might start thinking the null hypothesis does **not** represent reality after all. At that point, if the probability of the observed data under the null hypothesis is low enough, then we have reason to reject the null hypothesis and make an inference toward the statistical alternative hypothesis, that at least somewhere among means, there are mean differences in the population. This is the essence of how hypothesis-testing works in ANOVA.

3.2.1 Is the Observed Difference Likely? ANOVA as a Comparison (Ratio) of Variances

The next question becomes one of asking how to determine whether the observed mean difference is likely or unlikely under the null hypothesis. To help better appreciate this question, imagine if the data turned out to be as in Table 3.2 instead of how they actually are in Table 3.1.

Notice that under this idealized (and quite unrealistic) situation, every observation within its respective group is equal to the sample mean for that group. Notice that **between groups**, we still have the same mean differences. However, **within groups**, there is no variation. We ask the question we posed earlier—**What is the probability of obtaining mean differences of the kind 71.00** versus **72.50** versus **80.00** versus **92.67 if the null hypothesis were true?**

It seems intuitive that the probability of obtaining the mean differences we observed is much **lower** for the data in Table 3.2 (condition of no within-group variability) than it is for the data in Table 3.1 (condition of within-group variability) **if the null hypothesis were actually true**. In other words, in Table 3.2, all of the variation occurring is attributable to between-group differences. In Table 3.1, all of the variation occurring is attributable to not only between-group differences but also within-group differences. What we need now is a way to compare these sources of variation in some systematic and statistically correct fashion.

Suppose we could obtain a measure of just how much variance in a data set is attributable to **between-group** differences and how much is attributable to **within-group** differences. If most of the variance were attributable to within-group differences, then it would suggest that any between-group differences we are observing could probably be best explained by **random variation in the**

TABLE 3.2 Hypothetical Achievement Data

	Teacher		
1	2	3	4
71.00	72.50	80.00	92.67
71.00	72.50	80.00	92.67
71.00	72.50	80.00	92.67
71.00	72.50	80.00	92.67
71.00	72.50	80.00	92.67
$M = 71.00$	$M = 72.50$	$M = 80.00$	$M = 92.67$

Between-group variation but no within-group variation.

data, that is, of the same kind that is happening within groups. However, if most of the variance is attributable to between-group differences, then it would suggest that any between-group differences we are observing is not easily explained by within-group variation. That is, the between-group differences we are observing might actually represent a **real difference** of population means in the population that generated the sample data, and not simply sampling error or chance, the very type of variation we are witnessing within groups.

Our goal then is to make the following comparison in the form of a ratio:

$$\frac{\sigma^2_{Between}}{\sigma^2_{Within}}$$

where $\sigma^2_{Between}$ represents "variance between groups" and σ^2_{Within} represents "variance within groups." If $\sigma^2_{Between} > \sigma^2_{Within}$ to such an extent that we can **exclude** sampling error as being "responsible" for this inequality, then we will reject the null hypothesis $H_0 : \mu_1 = \mu_2 = \mu_3 = \mu_4$ in favor of the statistical alternative hypothesis, H_1. If, on the other hand, $\sigma^2_{Between}$ is more or less equal to σ^2_{Within}, then it would suggest that any observed mean differences in our sample are most easily explained by chance or sampling error. In other words, we have no evidence to conclude or argue that H_0 is actually false. This ratio that compares $\sigma^2_{Between}$ to σ^2_{Within} is called the **F-ratio** (or **F statistic**), named in honor of R.A. Fisher, and constitutes the overall **omnibus test of significance** in the analysis of variance model.

3.3 LOGIC AND THEORY OF ANOVA: A DEEPER LOOK

Having presented a brief overview of how ANOVA works, we now develop the theory at a slightly deeper level, essentially "unpacking" and elaborating on the brief discussion aforementioned. We begin first by drawing on previous exposure to the independent-samples t-test. In this respect, we present ANOVA as an **extension** of the independent-samples t-test where we are interested now in testing null hypotheses on more than two independent samples (i.e., $H_0 : \mu_1 = \mu_2$). We wish to generalize the null hypothesis to reflect a test of J population means, $H_0 : \mu_1 = \mu_2 = \mu_3 = \mu_J$. After presenting the parallels between independent-samples t-tests and ANOVA, we go into a more thorough discussion of how the ANOVA model itself is built and conceptualized, starting with the idea of modeling a randomly chosen observation in observed data, right up to the derivation of the sums of squares. We then take expectations of mean squares (i.e., expected values of sums of squares divided by their respective degrees of freedom) which lead us to generating the F ratio.

3.3.1 Independent-Samples *t*-Tests Versus Analysis of Variance

Most statistical methods are usually based on the same fundamental principles learned in a first course on statistics. The advanced methods simply constitute different and many times more complex ways of arranging these fundamental tools. If you truly understand the logic of how an independent-samples *t*-test works (see Chapter 2), then understanding the nuts and bolts of ANOVA will not be that difficult.

Recall that we have said we would like to test a null hypothesis of the kind:

$$H_0 : \mu_1 = \mu_2 = \mu_3 = \mu_J$$

To help us appreciate just how we will go about testing the null hypothesis for ANOVA, let us briefly review the form of the independent samples *t*-test, searching for some insight or ideas on how we might tackle our ANOVA problem. Recall the independent samples *t* of Chapter 2 (2.7):

$$t = \frac{E(\bar{y}_1) - E(\bar{y}_2)}{\sqrt{\dfrac{s_1^2}{n_1} + \dfrac{s_2^2}{n_2}}} = \frac{\mu_1 - \mu_2}{\sqrt{\dfrac{s_1^2}{n_1} + \dfrac{s_2^2}{n_2}}}$$

where $E(\bar{y}_1)$ and $E(\bar{y}_2)$, the expectations of $\bar{y}_1$ and $\bar{y}_2$, are equal to μ_1 and μ_2, s_1^2 and s_2^2 are unbiased estimators of their population counterparts σ_1^2 and σ_2^2, and n_1 and n_2 are the sample sizes in each group. Recall that (2.7) essentially has two parts to it. In the numerator is expressed a difference in means of the kind $\mu_1 - \mu_2$. In the denominator is a sum of sample variances, each weighted by the sample size n on which it was computed:

$$\sqrt{\frac{s_1^2}{n_1} + \frac{s_2^2}{n_2}} \tag{3.1}$$

Collectively, (3.1) formed an estimate of the **standard error of the difference in means**. That is, the denominator of the *t*-test gave us an idea of how much variance in sample mean differences $\bar{y}_1 - \bar{y}_2$ we could **expect** to see if we sampled infinitely from the given population under consideration. The job of (2.7) was to compare an observed mean difference to the variability we might expect to see if we were able to sample mean differences an infinite number of times from the population. When $n_1 \neq n_2$, we used a pooled estimator of the population variance,

$$s_{pooled}^2 = \frac{(n_1 - 1)s_1^2 + (n_2 - 1)s_2^2}{n_1 + n_2 - 2}$$

making the independent-samples *t*-test equal to

$$t = \frac{E(\bar{y}_1) - E(\bar{y}_2)}{\sqrt{\dfrac{(n_1 - 1)s_1^2 + (n_2 - 1)s_2^2}{n_1 + n_2 - 2} \left(\dfrac{n_1 + n_2}{n_1 n_2}\right)}}$$

In the analysis of variance, we will arrive at a ratio quite similar to that of the independent-samples *t*-test, one which effectively compares a numerator term expressing mean difference to a denominator term that effectively represents an estimate of population variability. If mean differences in the numerator are large relative to expected overall variability in the denominator, then it will suggest that the mean difference in the sample may not be due simply to chance or sampling error alone. In other words, it will be grounds for establishing statistical significance and rejecting the null hypothesis.

The question for us right now is how to conceptualize ANOVA so that we can actually run a similar test as we did for the independent-samples t-test but with more than two means. How do we conceive of our numerator and denominator for our test? For this, we need to focus our attention on how the analysis of variance model is conceptualized through a **model equation**. That is, we need to consider how the ANOVA model **arises**. We start at the beginning with fundamental ideas concerning what constitutes a statistical model. These ideas will not only be useful for understanding the current model we are dealing with but will also be helpful in understanding other statistical models presented in this book and elsewhere. Taking the time to learn and understand the material to follow will pay dividends when extending your knowledge base to more complex models such as **repeated-measures**, **random effects**, and **multilevel** or **hierarchical modeling**. It is well worth your time to understand the ANOVA model very well.

3.3.2 The ANOVA Model: Explaining Variation

Recall the concept of a **model** introduced in Chapter 1. The idea of a model is to think up an equation that best accounts for how observed data were generated.[1] For our achievement example, we ask questions of the kind—**Why was the score for a randomly drawn observation in our data equal to y_i?** Why was another observation equal to y_{i+1} (i.e., a different y_i in our data?). To "explain" these observations, we need to come up with a **theory** as to why they are what they are. This is the essence of most traditional statistical model-building, **to come up with a mathematical equation that best accounts for observed data, and to use that equation for making inferences toward the population**. The search is for an equation to account for observed data. In psychology, theoreticians seek out narratives to explain human behavior. In medicine, researchers theorize causes to cancer. In finance, investors theorize predictors of stock growth. In statistics, these narratives go by the name of **statistical models**.

For instance, Sigmund Freud (1856–1939) used a model for the **id**, **ego**, and **superego** to help explain human behavior, to explain the data he observed. His predictions based on his theory were not always correct. Even with what Freudians would argue was a solid theory to draw upon, he could still not explain **all** behavior (some would even argue, very little of it), and had to admit that sometimes, his theory failed, it was in **error**. As another example, recall that B.F. Skinner was able to predict behavior in the pigeon, he had a theory as to why the pigeon responded as it did. It predicted a lot of observations successfully, but others it did not predict so well. Sometimes, his theory failed too in its predictions of behavior.

In ANOVA, just as in virtually all statistical modeling, we will put forth a theory that attempts to explain observed data, and likewise, sometimes the theory will predict accurately, but other times, it will not. The times it does not predict accurately we will denote as **errors of prediction**. How the technique of analysis of variance partitions variability into **predictable** versus **unpredictable** components is the topic of this chapter. Indeed, most statistical models do something strikingly similar, that of attempting to separate the "signal" from the "noise."

Referring again to our data in Table 3.1, we notice that there exists variability in the sample achievement data, that is, $s^2 > 0$. We will define the **grand mean** of all the data as equal to

$$\bar{y} = \sum_{i=1}^{n} \frac{y_i}{N}$$

---

[1] Kirk (1995) calls the model equation an **experimental design model equation** (p. 32). Though the current text is not about experimental design per se, the fields of applied statistics and experimental design are necessarily intimately (and historically) linked. One cannot make intelligent selections of statistical analyses without a keen awareness of experimental design issues.

or, since these data are **balanced** (i.e., equal numbers per group), we can calculate the grand mean as the "mean of means," where J designates the number of groups:

$$\bar{y}.. = \sum_{j=1}^{J} \frac{\bar{y}_j}{J}$$

The grand mean for this data is equal to

$$\sum_{j=1}^{J} \frac{\bar{y}_j}{J} = \frac{71.00 + 72.50 + 80.00 + 92.67}{4} = 79.04$$

Given that we have at least some variability in the data, we can express each observation y_{ij} as being somewhat "off" from the grand mean, and calculate a deviation score for each observation. If we let any given observation i in a given group j be represented by y_{ij}, and the mean of all observations to equal $\bar{y}..$, then we can express the deviation for any given score as $y_{ij} - \bar{y}..$. For instance, we observe the following deviations for the first few data points in each teacher group (70, 69, 85, 95):

$$y_{ij} - \bar{y}.. = y_{11} - \bar{y}.. = 70.00 - 79.04 = -9.04$$
$$y_{ij} - \bar{y}.. = y_{12} - \bar{y}.. = 69.00 - 79.04 = -10.04$$
$$y_{ij} - \bar{y}.. = y_{13} - \bar{y}.. = 85.00 - 79.04 = 5.96$$
$$y_{ij} - \bar{y}.. = y_{14} - \bar{y}.. = 95.00 - 79.04 = 15.96$$

To give more examples of what we are doing, consider the deviations for the last observations in each age group (73, 75, 73, 91):

$$y_{ij} - \bar{y}.. = y_{61} - \bar{y}.. = 73.00 - 79.04 = -6.04$$
$$y_{ij} - \bar{y}.. = y_{62} - \bar{y}.. = 75.00 - 79.04 = -4.04$$
$$y_{ij} - \bar{y}.. = y_{63} - \bar{y}.. = 73.00 - 79.04 = -6.04$$
$$y_{ij} - \bar{y}.. = y_{64} - \bar{y}.. = 91.00 - 79.04 = 11.96$$

We could continue to do this for the entire data set. The important point to note (so far) is that each observed score in the data set can be expressed as a **deviation from the grand mean**.

3.3.3 Breaking Down a Deviation

Now that we have reasoned that we can represent any single score as a deviation from the mean (even if the score is equal to the mean, the deviation is then equal to 0), our next point of interest is to break down the deviation further. That is, we are interested in the following important question:

Why does any given score in our data deviate from the overall mean?

We need to think about the possible **reasons** why a given deviation, of the kind $y_{ij} - \bar{y}..$, might exist in a set of data. This is an equivalent question to asking why a given score in our data **is what it actually is**, only now, we are asking this question in terms of the given score's deviation from the overall mean.

Consider again the first observation, that of $y_{ij} = y_{11} = 70.00$. What "explanations" or "reasons" can you come up with for why this observation is not equal to the overall mean of 79.04? One reason you might come up with is that quite simply, even if the grand mean is equal to 79.04, it does not imply that all scores are going to be equal to the mean, **and for no particular good reason**. That is, you might theorize that 70.00 is different from the overall mean of 79.04 as a simple artifact of the data, out of pure and simple variability, nothing more. However, is this explanation enough to account for the given observation? Maybe not. For one, we have a grouping factor in our data, which is the teacher assigned to that particular individual. We must reason that it is possible that a given data point differs from the overall mean not only because of an artifact of the data (or chance, random variability), but because it is **in a particular group (i.e., teacher) and not another group**. When we hypothesize that scores are the way they are because they are in one group and not another, we reflect this by the deviation:

$$\bar{y}_j - \bar{y}.$$

which represents "between group" variability. That is, if the data point 70.00 differs from the overall mean because it received the 1st teacher rather than the other three teachers, then it would seem of interest to calculate the sample mean for this group and subtract the overall mean to reflect this deviation. If there is an "effect" of being assigned the 1st, 2nd, 3rd, or 4th teacher, then this should be reflected in the deviation $\bar{y}_j - \bar{y}.$. We can express the deviation $y_{ij} - \bar{y}.$ as being "made up" or "composed" of two parts. In fact, we can say further that the deviation $y_{ij} - \bar{y}.$ is equal to the **sum** of two parts, a part representing variability **within** a given group, $(y_{ij} - \bar{y}_j)$, and a part representing variability **between** groups, $(\bar{y}_j - \bar{y}.)$. The entire sum is thus:

$$\left(y_{ij} - \bar{y}.\right) = \left(y_{ij} - \bar{y}_j\right) + \left(\bar{y}_j - \bar{y}.\right) \tag{3.2}$$

Equation (3.2) is a fundamental identity in the analysis of variance. It expresses the make-up or composition of any randomly chosen observation in a one-way layout as a sum of two parts. The overriding goal of the analysis of variance is to learn whether the deviations in a set of data are better **explained** or **accounted for by within-group** deviations of the kind $(y_{ij} - \bar{y}_j)$, or **between-group** deviations of the kind $(\bar{y}_j - \bar{y}.)$. You may be able to foresee where this discussion is headed. If it turns out that deviations of the kind $(y_{ij} - \bar{y}.)$ are better explained by between-group deviations than they are by within-group deviations, it would suggest that our samples may have been drawn from **distinct** and **unique** populations. Experimentally, this would make good sense, since this is presumably why we did the study in the first place, to seek out mean differences between such treatment groups. In other words, if we found, overall, that deviations of the kind $\bar{y}_j - \bar{y}.$ were large relative to deviations of the kind $y_{ij} - \bar{y}_j$, then it might suggest our treatment was effective. We will return to this point later after we have more fully developed the logic behind ANOVA. Before we do anything more, we need to give names to each of the deviations in our fundamental identity of (3.2).

3.3.4   Naming the Deviations

Let us consider the first deviation of (3.2), that of $(y_{ij} - \bar{y}_j)$ to the immediate right of the equal sign. Why might such a deviation arise? That is, why would your data exhibit a deviation of the form, $(y_{ij} - \bar{y}_j)$? They are all in the same group, are they not? Therefore, it cannot be due to a "grouping" effect of any kind. After all, they were all collected and treated the same way. The best we can do to explain this deviation is to call it "error," or to say that the score y_{ij} deviates from $\bar{y}_j$ due simply to "chance," or to "random factors" that we cannot immediately account for or explain. In brief, **we**

do not know why one score deviates from another within a given group. Hence, we will call the term $(y_{ij} - \bar{y}_j)$ by the name of error and denote it as e_{ij}, which represents the error for any given individual i in a given group j. Substituting e_{ij} for $(y_{ij} - \bar{y}_j)$ we rewrite the identity in (3.2) as:

$$(y_{ij} - \bar{y}.) = e_{ij} + (\bar{y}_j - \bar{y}.)$$

Now, we ask the following question—**Why would the second deviation, that of the form, $(\bar{y}_j - \bar{y}.)$, exist?** That is, why would one group's mean differ from the overall mean of all the observations? A sensible explanation is that there is some kind of "effect" of being in that particular group versus being in another group, and that is why a given group mean is different from the overall mean. We will name this deviation by "a_j" and let it represent the **sample effect** or **treatment effect** of being in a particular group. Hence, when we further substitute a_j for $(\bar{y}_j - \bar{y}.)$, we can write our equation for the deviation $(y_{ij} - \bar{y}.)$ in (3.2) as,

$$(y_{ij} - \bar{y}.) = e_{ij} + a_j$$

or, more commonly, we will reorder a_j before the error effect:

$$(y_{ij} - \bar{y}.) = a_j + e_{ij} \qquad (3.3)$$

In summary then, what we have done thus far is to reason that a given deviation of the kind $(y_{ij} - \bar{y}.)$ can be composed of two "things." Either it is due to an effect of being in one sample versus another, which we call by the name of **sample effect**, a_j, or, it is due simply to "error" which we designate as e_{ij}.

The analysis of variance partitions variability in this way such that we can eventually test (through an F ratio) the assumption that deviations are due to "error" alone, or equivalently, that the sample effects we have observed in our sample are not large enough to begin to doubt that the population effects (which we will denote by α_j, the population counterparts to a_j) are actually equal to zero.

3.3.5   The Sums of Squares of ANOVA

We have concluded that any deviation from the grand mean can be said to be represented by, or "composed of" $(y_{ij} - \bar{y}.) = a_j + e_{ij}$. However, when summing any deviations about a mean, we know that the sum of the deviations will equal zero. That is, if we did take the sum of deviations $(y_{ij} - \bar{y}.)$, we know that $\sum(y_{ij} - \bar{y}.) = 0$ would be true for any data set we deal with, real or hypothetical. Though calculating the sum of absolute deviations of the form $\sum |y_{ij} - \bar{y}.|$ is a possibility for avoiding the sum of zero, the solution historically adopted for this problem has been to **square the deviations**, then sum them up. We will apply the same principle of squaring deviations to our model equation $(y_{ij} - \bar{y}.) = a_j + e_{ij}$.

We omit here the actual derivation of the sums of squares. For details, see Hays (1994) or Kirk (1995). After deriving the sums of squares, we arrive at the following identity:

$$\sum_{j=1}^{J} \sum_{i}^{n} (y_{ij} - \bar{y}.)^2 = \sum_{j}^{J} n_j (\bar{y}_j - \bar{y}.)^2 + \sum_{j=1}^{J} \sum_{i}^{n} (y_{ij} - \bar{y}_j)^2 \qquad (3.4)$$

Equation (3.4) is referred to as the **partition of the sums of squares** for a one-way fixed effects between-subjects analysis of variance. Notice that it is made up of three parts, which we detail now:

SS total: $\sum\limits_{j=1}^{J} \sum\limits_{i}^{n} \left(y_{ij} - \bar{y}.\right)^2$—This is the **total sum of squares** for the entire data set. It is the sum of squared deviations of every individual value in a sample of data from the grand mean of all observations. This term is also equal to calculating NS^2 (see Hays, 1994, for details).

SS between: $\sum\limits_{j}^{J} n_j \left(\bar{y}_j - \bar{y}.\right)^2$—This is the sum of squares representing variation due to a potential treatment effect. Notice that in our derivation, we picked up the "multiplier" n_j. This is simply equal to the number of observations per group, which for balanced designs, we assume to be equal.

SS within: $\sum\limits_{j=1}^{J} \sum\limits_{i}^{n} \left(y_{ij} - \bar{y}_j\right)^2$—This is the sum of squares representing **error** or **within-group** variation. It is the sum of squared deviations for all observations in each group from its respective group mean. It is a measure of error because if all observations are in the same group (i.e., they were all treated the same way), we would expect scores to be more or less the same, and would attribute any differences to chance or unexplainable variability (which we call **error**).

3.4 FROM SUMS OF SQUARES TO UNBIASED VARIANCE ESTIMATORS: DIVIDING BY DEGREES OF FREEDOM

Recall how we calculated a variance—we produced a sum of squares, and then divided this sum of squares by an appropriate denominator:

$$S^2 = \frac{\sum\limits_{i}^{n} (y_i - \bar{y})^2}{N} \quad \text{or} \quad s^2 = \frac{\sum\limits_{i}^{n} (y_i - \bar{y})^2}{N - 1} \tag{3.5}$$

The denominator for the uncorrected variance was simply N, and for the corrected variance, $N - 1$. The reason for dividing by $N - 1$ was to obtain an **unbiased** estimator of the population variance σ^2. When we divide by either denominator, we are in essence producing a "mean" of the squares, only that in one case, we are basing the mean on N pieces of information, and in the other case, basing it on $N - 1$ pieces of information. However, the concept of generating an "average" is the same in both contexts.

The important element then is the selection of appropriate denominators for our various sums of squares. We do the exact same thing in ANOVA, only that now, instead of having only one sum of squares to be concerned about, we have three, **SS total**, **SS between,** and **SS within**. The question boils down to deciding what degrees of freedom are appropriate for each sum of squares in generating suitable variances.

We saw that SS between is calculated as $\sum\limits_{j}^{J} n_j \left(\bar{y}_j - \bar{y}.\right)^2$. Notice that we are subtracting the grand mean from group means. The degrees of freedom for SS between are equal to one less than the number of groups we have. This is because in our calculation of the sums of squares, $\bar{y}.$ is implicitly serving as an estimate of μ. What this means is that this value can be considered fixed, and implies that one of the group means is not free to vary (recall the "Beautiful Triangle" of Chapter 2). Hence, we lose one degree of freedom. For instance, for three treatment groups, the degrees of freedom are equal to $J - 1$, which, in this case, is $3 - 1 = 2$.

Recall that SS within is calculated as $\sum\limits_{j=1}^{J} \sum\limits_{i}^{n} \left(y_{ij} - \bar{y}_j\right)^2$. Notice that we are subtracting the group mean from individual scores within the given group. In this case, we are fixing the given group mean, since the group mean $\bar{y}_j$ is implicitly being used as an estimate of μ_j, so we lose one degree of freedom **per group**. For example, for three treatment groups of $n = 10$ observations per group, the degrees of freedom for SS within are $(10 - 1) + (10 - 1) + (10 - 1) = 9 + 9 + 9 = 27$. Alternatively, we could have also computed these degrees of freedom as $N - J$, that is, the total number of observations minus the number of groups. Losing one degree of freedom per group or computing $N - J$ on the entire sample amounts to the same thing for a balanced design.

Finally, although we will not be deriving any mean square estimates using SS total, it is nonetheless useful to know that the degrees of freedom for SS total are equal to one less than the total number of observations in the entire data. For instance, if there are $N = 30$ observations, then since each deviation for SS total consists of subtracting the grand mean (and hence, we are constrained by it), we will lose 1 degree of freedom, giving us $30 - 1$, or more generally, $N - 1$ degrees of freedom for **SS total**. Notice that this is simply the "ordinary" corrected variance we started out with.

In dividing by appropriate degrees of freedom, we transform our sums of squares into mean squares, one for between-group variance,

$$\text{MS Between} = \frac{\sum_{j}^{J} n_j \left(\bar{y}_j - \bar{y}_. \right)^2}{J - 1}$$

and one for within-group variance,

$$\text{MS Within} = \frac{\sum_{j=1}^{J} \sum_{i}^{n} \left(y_{ij} - \bar{y}_j \right)^2}{N - J}$$

Note that while the sums of squares are **additive** in that **SS total = SS between + SS within**, the mean squares are generally not. Because we are dividing by degrees of freedom, mean squares vary depending on the given experiment and on such things as the operationalization of levels of the independent variable. The "breakdown" is simply not the same as for the sums of squares. This is why you will not typically find MS total figures in an ANOVA table.

Having developed the necessary mean squares for computing variances, our next task is to learn what these mean squares actually estimate in the population. That is, we need to take **expectations** of these mean squares.

3.5 EXPECTED MEAN SQUARES FOR ONE-WAY FIXED EFFECTS MODEL: DERIVING THE *F*-RATIO

As a recap, we have seen how a given deviation from the grand mean of the form $y_{ij} - \bar{y}_.$ can be said to be made up of two parts. The first part reflects deviations between sample means and the grand mean, $\bar{y}_j - \bar{y}_.$. The second part reflects deviations between single observations in each group from their respective group means, $y_{ij} - \bar{y}_j$. We also saw how to produce sums of squares to account for the various sources of variation, and how to divide by appropriate degrees of freedom to obtain unbiased estimators of variance, the so-called "mean squares." We obtained a mean squares between (MS between) and a mean squares within (MS within).

What are the expectations of these mean squares? Recall that when we derived the sample variance, we were interested in its expectation. We found that the expectation of the corrected version of the sample variance was equal to the population variance σ^2. That is, $E(s^2) = \sigma^2$. We were interested in the expected value because we wanted to know that over an infinite number of potential samples, and by the algebra and rules of expectations, the value of the sample variance would equal that of the population variance. That is, we wanted some comfort and assurance in knowing that s^2 was actually estimating the correct quantity, that of σ^2. If a statistic we have computed is not estimating the population parameter we are actually interested in knowing about, it usually does us little good (though biased estimators at times are useful, such as in ridge regression). We are just as curious about the values of MS between and MS within. We would like to know their **expectations**. The reason why we need to know what quantities they are estimating is so we know how to generate appropriate and relevant F-ratios.[2]

Expectations for MS between and MS within can be found in Hays (1994) or Kirk (1995). For our purposes, we cut to the chase without derivation. Based on the exercise of taking expected mean squares, it can be shown that for between and within, respectively, they are equal to:

$$E(\text{MS Between}) = \frac{\sum_j \left(n_j \alpha_j^2 \right)}{J-1} + \sigma_e^2 \text{ and } E(\text{MS Within}) = \sigma_e^2$$

When we take a ratio of MS between to MS within, we find that under the circumstance where there is a **complete absence of sample effects**, the ratio should equal approximately

$$\frac{\sigma_e^2}{\sigma_e^2} \approx 1.0$$

When sample effects are present, we expect the ratio to be greater than 1.0.

We summarize this partition of the sums of squares in what is known as the **Analysis of Variance Summary Table**, given in Table 3.3.

TABLE 3.3   Summary Table for One-Way Fixed Effects Analysis of Variance

Source	Sum of Squares	df	Mean Squares	F-Ratio	p-Value
Between	$\sum_j^J n_j \left(\bar{y}_j - \bar{y}. \right)^2$	$J-1$	$\dfrac{\sum_j^J n_j \left(\bar{y}_j - \bar{y}. \right)^2}{J-1}$	$\dfrac{\sum_{j=1}^J n_j \left(\bar{y}_j - \bar{y}. \right)^2 / (J-1)}{\sum_{j=1}^J \sum_i^n \left(y_{ij} - \bar{y}_j \right)^2 / (N-J)}$	Evaluate obtained ratio as F-stat on $J-1$ and $N-J$ degrees of freedom
Within	$\sum_{j=1}^J \sum_i^n \left(y_{ij} - \bar{y}_j \right)^2$	$N-J$	$\dfrac{\sum_{j=1}^J \sum_i^n \left(y_{ij} - \bar{y}_j \right)^2}{N-J}$	—	—
Total	$\sum_{j=1}^J \sum_{i=1}^n \left(y_{ij} - \bar{y}. \right)^2$	$N-1$	$s^2 = \dfrac{\sum_{j=1}^J \sum_{i=1}^n \left(y_{ij} - \bar{y}. \right)^2}{N-1}$ $(N-1)(s^2) = \sum_{j=1}^J \sum_{i=1}^n \left(y_{ij} - \bar{y}. \right)^2$	—	—

[2] Though we are computing the F statistic as a ratio of $\sigma^2_{Between}/\sigma^2_{Within}$, the actual F density distribution (i.e., the actual F statistic that we compare the F-ratio to) is given by

$$F = \frac{\chi_1^2 / v_1}{\chi_2^2 / v_2}$$

where χ_1^2 and χ_2^2 are independently distributed chi-square variables on v_1 and v_2 degrees of freedom. For details, see Hays (1994) or Kirk (1995).

3.6   THE NULL HYPOTHESIS IN ANOVA

Having conceptualized the ANOVA model and computed expected mean squares, the next task is to get on with testing null hypotheses. There are two common ways we can state the null hypothesis in the one-way fixed effects ANOVA. The test of both hypotheses will suggest the same decision on H_0. If MS between is equal to MS within, then this suggests they are each estimating the same variance. That is, each term is measuring error variance, σ_e^2. Recall that the expected mean square for MS between is equal to

$$E(\text{MS between}) = \sigma_e^2 + \frac{\sum_j n_j \alpha_j^2}{J-1}$$

If there are no sample effects present in a given analysis, then it suggests that all corresponding population effects α_j of the form $\alpha_j = \mu_j - \mu.$ are equal. If they are all equal, then the sum of $n_j \alpha_j^2$ must be 0, giving us the following for the expected mean squares for between:

$$E(\text{MS between}) = \sigma_e^2 + \frac{\sum_j n_j (0)_j^2}{J-1}$$
$$= \sigma_e^2 + 0$$
$$= \sigma_e^2$$

Notice that under the condition that all effects α_j are equal to 0, the mean squares between is estimating the same as the mean squares within, that of simply unexplainable or unaccounted for deviation of scores within their respective groups, that is, σ_e^2. When both MS between and MS within are estimating the same quantity, the expectation for F is approximately 1.0.[3] Recall that while it is true that the expectation for F is equal to approximately 1.0, we will rarely if ever obtain this in practice in our sample even if the null hypothesis were true. Sampling error always makes its way into things, so we will usually deviate slightly from expectation even under a true null hypothesis. **The question is always whether our deviation from expectation is enough to cause us to reject the null hypothesis of equal population means**.

Hence, one way of positing the null hypothesis for the one-way ANOVA is that all population effects are equal to 0. More formally, we could state the null as

$$H_0 : \alpha_j = 0, \text{ for all populations } j.$$

If at least one of the group means does differ from the grand mean, then we have a **sample effect** for the given group. We are interested in knowing whether the sample effect is large enough to suggest an effect unequal to zero in the population. The alternative hypothesis, H_1, can be stated as

$$H_1 : \alpha_j \neq 0, \text{ for at least some populations } j.$$

If there are no population effects in a one-way ANOVA, then this implies that all population means are equal. Because of this, we can also state the null hypothesis as:

$$H_0 : \mu_j = \mu. \text{ for all populations } j. \tag{3.6}$$

The null hypothesis in (3.6) reads that all population means are equal to the grand mean of all the populations. If this is true, then it implies that there cannot be any differences in means between

---

[3] The expectation of F turns out to not equal 1.0 exactly. Under the null hypothesis of equal population means, as noted in Howell (2002, p. 331), $E(F) = \dfrac{df \text{ error}}{df \text{ error} - 2}$.

populations. The alternative hypothesis would be that for at least one population, its mean does not equal the grand mean of all the populations. That is,

$$H_1 : \mu_j \neq \mu. \text{ for at least some } j \text{ populations.}$$

We see then that whether we state the null in terms of population effects or population means, it amounts to the same null hypothesis under test. Usually, however, the null hypothesis is expressed as simply the equality of population means. The hypothesis $H_0 : \mu_j = \mu.$ for all populations j, however, gives us an actual idea of how this former hypothesis is being evaluated.

Remember always that the null and alternative hypotheses are about **parameters**, and not **sample** statistics, which is why we are using the notation α_j and $\mu.$ or μ_j to represent respective population effects and population means. Recall also that in our sample, we fully expect inequality to some degree, for instance, $\bar{y}_1 \neq \bar{y}_2 \neq \bar{y}_3$. That our sample means are not exactly equal to one another is hardly a shocking result or momentous finding. What we are really interested in, is in knowing whether such deviations are large enough relative to what we would expect simply to due sampling error. Generally, in research, we are usually not all that interested in sample statistics. We are most interested in **parameters**. Statistics usually simply serve as a means of estimating these parameters.

3.7 FIXED EFFECTS ANOVA: MODEL ASSUMPTIONS

Any mathematical model, whether statistical or otherwise, comes with it a set of assumptions on which the model is based. If these assumptions are not satisfied, especially to a substantial degree, it could cast into doubt the very correctness and utility of the model you are fitting to your data. It should be noted as well at the outset that when one does not engage in the process of estimating parameters, that is, of **statistical inference**, the analysis of variance itself is, as Fisher put it, simply a way of "**arranging the arithmetic**." One does not require assumptions for arranging this arithmetic. Eisenhart described this very idea quite eloquently as well:

> … when the formulas and procedures of analysis of variance are used merely to summarize properties of the data in hand, no assumptions are needed to validate them. On the other hand, when analysis of variance is used as a method of statistical inference, for inferring properties of the "population" from which the data in hand were drawn, then certain assumptions, about the "population" and the sampling procedure by means of which the data were obtained, must be fulfilled if the inferences are to be valid.
>
> (Eisenhart, 1947, p. 8)

Hence, when we use the arithmetic of ANOVA to make **inferences**, we require assumptions. The assumptions for the one-way fixed effects ANOVA can be summarized into the following:

- $E(\varepsilon_{ij}) = 0$, the expectation of the error term is equal to 0. We use ε_{ij} here in place of e_{ij} to denote the population parameter.
- ε_{ij} are $NI(0, \sigma_e^2)$, the errors are normally distributed (N) and independent (I) of one another having a mean equal to 0 and variance equal to σ_e^2. In general, if one can assume that the errors within each population are normally distributed, then this implies that the observations on the dependent variable in each of the populations are also normally distributed (Kirk, 1995). Normality can be tested using graphical methods such as histograms, residual plots, and Q–Q plots, whereas **independence of error**, a much more difficult assumption to verify, can be investigated at least somewhat via residual plots, but is usually ensured by the method of data collection and random assignment for the given experiment.

- $\sigma_{e_{ij}}^2 < \infty$, the variance of the errors is some finite number (which simply implies that it is less than infinity).

- $\text{Cov}(\varepsilon_{ij}, \varepsilon_{i'j}) = 0$, the covariance between errors is equal to 0. If errors are indeed independent, then this assumption is already implied. Independence and an absence of covariance are not equivalent properties, however. As we will see when we study block designs, these assumptions will typically be violated.

- $\sigma_{j=1}^2 = \sigma_{j=2}^2 = \sigma_{j=J}^2$, the variances across populations as operationalized by the independent variable are equal (often called the **homogeneity** assumption). This can be tested using a **variance ratio test** in R where the largest variance is compared to the smallest. Levene's test or Bartlett's test is also useful in verifying this assumption.

- Measurements on the dependent variable are observed values of a random variable that are distributed about true mean values that are fixed constants. This assumption (adapted from Eisenhart, 1947, p. 9) is equivalent to the assumption that the levels of the independent variable used for the given experiment constitute the **only** levels the researcher wishes to generalize to in the population. This is precisely what defines the fixed effects model as **fixed**. Recall that if this is not the case, and the experimenter wishes to generalize these levels to a **population** of levels of which the levels appearing in the experiment are but a random sample, then the correct model is not that of a fixed effects model, but rather that of a **random** effects model.

When we perform an analysis of variance in R and SPSS toward the end of this chapter, we will briefly demonstrate how one can go about verifying some of these assumptions using inferential tests and graphical displays. Light to moderate departures from these assumptions is usually not a major concern (other than that for independence of errors, which is a serious concern if violated) since ANOVA is quite robust against violations (generally implying the type I error rate and power will remain relatively stable even in the face of violations). However, there are remedies for violations if they get to be severe. For instance, one can perform power transformations to help establish a sense of normality in the dependent variable. These often take the form of square root or logarithmic transformations or others. We do not cover transformations in any detail here, because usually, they are not required except for rather extreme violations. In the event that you do wish to transform to **near normality**, you are encouraged to consult any of the excellent resources on this topic. Fox (1997) is especially good. The function boxcox (named for Box-Cox transformations) in the MASS package (Venables and Ripley, 2002) in R offers some options in helping one decide on the most optimal transformation for a set of data.

If you suspect a violation of the assumption of equal variances, so-called **heterogeneity of variance**, options such as the **Brown–Forsythe test** (see Kirk, 1995) or the **Welch procedure** (1951) can be used for adjusting the obtained F from ANOVA so it better incorporates a possible violation. These tests will typically be more **conservative** than the omnibus ANOVA test. Since ANOVA is quite robust to violations of this assumption, we do not cover these tests to any extent in this book (though we do demonstrate the Welch test in SPSS in our ANOVA example). Howell (2002) does an excellent job at summarizing their contributions and is highly recommended. In cases where either sample sizes are very small or even the prospect of satisfying assumptions in ANOVA seems impossible, a **nonparametric** test may be a better choice. Nonparametric tests make fewer assumptions about the population from which the sample data were drawn. The counterpart to the one-way fixed effects analysis of variance is the **Kruskal–Wallis one-way analysis of variance test** and is available in most statistical software packages. For details, see Rice (1995, p. 453).

Another assumption that is more or less implicit in the ANOVA model is that the model equation

$$y_{ij} = \mu + \alpha_j + \varepsilon_{ij}$$

of which the sample equivalent is given by

$$y_{ij} = \bar{y}_. + a_j + e_{ij}$$

contains all the relevant sources of variation for the given experiment. That is, we assume the model is correctly **specified**. Of course, no model is ever completely perfectly specified, but the point of this assumption is to say that there are no **obvious** sources of variation that were omitted from the model. For example, since it is generally known that there are gender differences in depression rates (e.g., see Salk, Hyde, and Abramson, 2017), if one did not include gender in a model of predicting mean differences on depression, one could easily argue that the model is misspecified. Detecting specification errors is sometimes a skill more honed by experience in a given research area than anything else, in that experienced researchers are often well-familiar with the "big player" predictors in their respective fields. If those predictors are for some reason left out of an analysis, such folks will surely be the first to call you out on a specification error. We revisit this topic when we discuss the regression model in Chapter 7.

3.8   A WORD ON EXPERIMENTAL DESIGN AND RANDOMIZATION

Up to now, we have not commented much on the actual process of experimentation or the randomization of subjects to treatment groups. The process of randomization is that of administering subjects randomly to levels of the treatment factor, with the goal of eliminating as much as possible any source of **bias** that could potentially confound findings. For instance, in our melatonin example discussed at the start of the chapter, randomly assigning subjects to dosage levels (control, 1 mg, 3 mg) is our best assurance (though by no means **guarantee**) that the infinite number of "nuisance factors" are evenly dispersed among our treatment groups.

   Nuisance factors are all those things that could theoretically be acting on the dependent variable but that we have not accounted for or measured in our experiment or study. For example, again referring to the melatonin study, surely some people are more predisposed to falling asleep with ease compared to other people. If we do not account for this in our design (e.g., through **blocking** or **analysis of covariance**, for example), then we are relegated to hoping that randomization "balances things out" and that there will not be any systematic bias built up in any group on the said nuisance factor.

   Randomization is the ideal "gold standard" for experimental design. In many studies, however, it is either impossible or unethical to randomly assign participants to treatment conditions. For instance, if we wanted to learn whether mammography screening reduces the risk of death from breast cancer, it would be somewhat unethical to randomly assign some participants to mammography screening while others to a control group, especially if we have prior knowledge that screening is effective. Likewise, it would be unethical to randomly assign some participants to a "smoking group" and others to a "non-smoking" group to observe the effects of cigarette consumption over time. The experimenter simply cannot have such a level of **control** over his or her subjects. In many cases, we have to take subjects **as they come**. The best we can do often is record whether a subject has or has not received mammography screening and associate that with their later cancer risk or survival. Likewise, we often have to take smokers as they come, and compare them to nonsmokers. But this means we are no longer randomly assigning participants to treatments. Designs such as this where we are unable, for whatever reason, ethical or otherwise, to randomly assign participants to treatment conditions are generally known as **quasi-experimental** designs (e.g., see Eliopoulos et al., 2005). If there is absolutely no control imposed at any level, these designs can sometimes be considered similar to **correlational designs** (Campbell and Stanley, 1963).

As a guideline, whenever you do an experiment or study, it is usually best to try for randomization at least at some level. If such is not attainable, then resorting to a quasi-experimental design might be considered as a second option. A randomized design should usually be your **first** attempt or choice, however.

3.9   A PREVIEW OF THE CONCEPT OF NESTING

The curious reader may have noticed an important element in our discussion of mammography screening (and that of smoking behavior) just mentioned. Women who receive mammographies may be more likely to be **alike** than women who do not receive them. That is, they may share characteristics (other than mammography screening) that women who do not receive mammographies do not share. Perhaps those who receive mammographies are more concerned with their health than those who do not. Perhaps they are more educated, have better health insurance, or share numerous other similarities. This idea, which we briefly introduced in the previous chapter in the context of the matched-samples design, generally goes by the name of **nesting**.

In many designs, observations are naturally nested within a given group. A classic example is that of school children nested within classrooms, and classrooms nested within schools. That is, children sharing the same classroom (and thus, the same teacher) may be similar in ways compared to children in another classroom (and thus, with another teacher). Likewise, classrooms in the same school may be more similar than classrooms in different schools. We only briefly mention the topic here as a preview to our further discussion of it when we consider random effects and randomized block designs in chapters to follow. The concept of nesting in this manner forms the basis for such modeling as **hierarchical modeling and multilevel modeling**, topics that are well beyond the scope of the current text, but have gained popularity in the social and behavioral sciences in the last 30 years or so. We will recap and extend this discussion of nesting when we consider randomized block and repeated-measures models in Chapter 5.

3.10   BALANCED VERSUS UNBALANCED DATA IN ANOVA MODELS

In all of our discussion of ANOVA thus far, we have assumed that group sizes have an equal number of subjects. These data layouts are referred to as **balanced**. Layouts in which groups do not have the same number of measured objects per group (or cell, in the case of factorial ANOVA, the topic of the following chapter) are referred to as **unbalanced**. Balanced data are generally preferred to unbalanced data for the reason that effects in a balanced design are **orthogonal**, which typically implies that associated tests are independent of one another. That is, when data are unbalanced, the possibility arises that main effects and interactions will no longer be independent of one another, which also translates to the fact that sums of squares may not be **additive**. As Tabachnick and Fidell (2007) note, the problem of unequal sample sizes is more relevant if the groups with **small sample size also exhibit relatively high variance**. This could potentially lead to an inflated type I error rate. Orthogonality of factors also helps to ensure that comparisons of one factor at levels of the other factor will not be unduly influenced by groups on one factor having more "information" (i.e., in terms of objects studied) than another. For a brief discussion of unbalanced designs, see Steinhorst (1982). As mentioned, in most of our examples of ANOVA and beyond, for convenience, we generally assume balanced designs.

3.11   MEASURES OF ASSOCIATION AND EFFECT SIZE IN ANOVA: MEASURES OF VARIANCE EXPLAINED

Obtaining a statistically significant F statistic literally means that the statistic we have obtained is relatively **rare** assuming that it arose from an F sampling distribution specified by $J - 1$ and $N - J$ degrees of freedom. However, as we have already discussed in relation to z and t-tests, statistically significant statistics do not necessarily suggest a **large** difference between means. That is, **statistical significance does not necessarily imply a large or meaningful effect size**. This is because, as summarized in Chapter 2, there are other things in a test statistic's "DNA" that influence its magnitude, such as sample size, variance, and, considered jointly, its standard error. Note that when we produce the F statistic by dividing MS between by MS within, we risk getting a large F simply as a result of MS within being small. The smaller MS within is, the larger will be our resulting F statistic, all else equal. Similarly, one can see that by simply increasing sample size, $N - J$, the degrees of freedom for SS within will get larger and larger. As we increase the degrees of freedom, MS within necessarily gets smaller, since we are dividing SS within by a larger and larger number.

Do not misunderstand. Having a small MS within is always a good thing **statistically**. **The inferential statistic is doing its job**. A small error term suggests we have a good degree of precision in our estimation. But scientifically, the small error term does not in itself guarantee that anything **important** or **practical** has happened in the experiment or study. **The distinction between statistical significance and effect size is one you must understand in order to evaluate scientific evidence in an intelligent manner. A misunderstanding of this distinction can lead to serious misunderstandings in the global interpretation of evidence**.

The F-test then, or any inferential test statistic for that matter, will not reflect a pure measure of the obtained sample effects. How do we solve this problem? One option is to take a ratio of SS between to SS total before these sums get converted into mean squares. That way, we are not having our F statistic unduly influenced by sample size since we are not yet dividing by $N - J$ in the denominator. This ratio of SS between to SS total goes by the name of **Eta-Squared**, symbolized as η^2. We now discuss this important statistic.

3.11.1 η^2 Eta-Squared

As discussed, it seems intuitive that if we wanted a more pure measure of the difference in means, we should consider further the magnitude of the **sample effects**, without necessarily requiring an inferential statement about them. Recall that any potential sample effects are included in SS between:

$$\sum_{j}^{J} n_j \left(\bar{y}_j - \bar{y}. \right)^2$$

whereas deviations of the kind $\bar{y}_j - \bar{y}.$ grow larger and larger, this is indicative of an increasingly larger difference between sample means. And since SS total is a measure of **total variation** in a set of data, it seems sensible to take the ratio of SS between to SS total as our measure of effect size:

$$\eta^2 = \frac{\sum_{j}^{J} n_j \left(\bar{y}_j - \bar{y}. \right)^2}{\sum_{j=1}^{J} \sum_{i=1}^{n} \left(y_{ij} - \bar{y}. \right)^2}$$

where η^2 can range from 0 to 1. A value of 0 suggests that sample effects are accounting for no variance in the dependent variable and the only source of variation that is "at work" in our data is random error. A proportion of 1 suggests that the total variation in our data is accounted for entirely by our obtained sample effects. Otherwise said, a measure of 1 suggests that all variation in our data is attributable to between-group differences (i.e., which are in effect, the obtained sample effects).

To reiterate then, η^2 represents the **proportion of variance in the dependent variable that is accounted for by the independent variable**. In the case of a one-way fixed effects ANOVA, that independent variable is the grouping factor.

3.11.2   Omega-Squared

It is well known that values of η^2 tend to report an overly "optimistic" picture of the magnitude of effect. This is in part because η^2 is a **descriptive measure** of effect size in the particular sample on which it is computed and assumes that the population regression line (if we were to know it) passes through the group means on the independent variable (Howell, 2002, p. 353). It does not accurately estimate what the actual **true** effect might be in the population from which data were drawn.

Omega-squared, ω^2, is a less-biased estimate of effect size and serves as an estimate of the actual population effect size. An estimate of ω^2 for the one-way fixed effects analysis of variance can be obtained by:

$$\hat{\omega}^2 = \frac{\text{SS between} - (J-1)\,\text{MS within}}{\text{SS total} + \text{MS within}}$$

where the values of SS between and MS within are obtained from the analysis of variance table. As noted by Kirk (1995, p. 178), $\hat{\omega}^2$ can also be computed as

$$\hat{\omega}^2 = \frac{(J-1)(F-1)}{(J-1)(F-1) + nJ}$$

where F is that obtained from the overall ANOVA, n is the sample size per group (we assume equal n per group), and J is the number of levels on the independent variable. This formulation is especially useful for situations in which you wish to compute omega but do not have access to a researcher's ANOVA summary table (and are only provided with F).

η^2 and ω^2 are by far the most popular effect size measures used to contextualize findings in the analysis of variance. However, relatively recently, attention has been drawn to the fact that these measures do not incorporate the potential influence that design features might have on the effect size estimate, especially for the factorial designs of the following chapter. One recommendation given to overcome these deficiencies is to compute **generalized eta squared** and **generalized omega squared** statistics. These statistics, in part, incorporate the influence of design features into their estimates. Though we do not discuss these effect size measures here, the interested reader is encouraged to consult Olejnik and Algina (2003) for a discussion of such measures.

3.12   THE F-TEST AND THE INDEPENDENT SAMPLES t-TEST

Recall that ANOVA can be conceptualized as an extension of the independent samples t-test. Given this, it stands that we should be able to conduct an ANOVA on a two-sample problem and translate obtained F into a t statistic.

TABLE 3.4 Hypothetical Data on Two Independent Samples

Sample 1	Sample 2
2	7
1	6
3	8
2	9
Mean = 2.0	Mean = 7.5

Consider the small hypothetical data set in Table 3.4. Suppose we wished to test the null hypothesis that $\mu_1 = \mu_2$. Both ANOVA and the independent samples t-test can be used to evaluate the tenability of this null.

An independent samples t-test on these data yields a t-statistic of $|7.20|$. Evaluated on $(n_1 - 1) + (n_2 - 1)$ degrees of freedom, we find t to be statistically significant at the 0.05 level of significance.

Suppose now that instead of the t-test, we wish to perform a one-way fixed effects ANOVA on these same data. If we square our obtained t statistic, it will equal the obtained F that we get in the ANOVA. That is, $(7.20)^2 = 51.84 = F$. Likewise, we can go the other way. The square root of F will equal t, that is, $\sqrt{F} = \sqrt{51.84} = 7.2 = t$.

Hence, if an F statistic is statistically significant at a given significance level α on $J - 1$ and $N - J$ degrees of freedom, then the corresponding value of $t = \sqrt{F}$ will be statistically significant at the same α level on degrees of freedom $(n_1 - 1) + (n_2 - 1)$ in a two-tailed test. If the statistical alternative to the null hypothesis is one-sided (also known as "directional"), then the sign of the t statistic must be taken into consideration.

3.13 CONTRASTS AND POST-HOCS

The overall F statistic computed in the analysis of variance tests the general null hypothesis of equality among population means. It is the so-called **omnibus** test of equality among population means. Often-times in research, however, we have **planned** hypotheses that we would like to test that reduce the omnibus null hypothesis to a series of two-group comparisons. Each comparison uses up a single degree of freedom, and so they are sometimes called **single-degree-of-freedom** contrasts.

For example, referring once again to the achievement data of Table 3.1, suppose the researcher was interested in specifically comparing achievement means on teachers 1 and 2 **taken together** with the achievement means on teachers 3 and 4, also considered simultaneously. Notice that, in this case, we are not so much interested in a general mean difference as much as we are interested in a **specific** mean difference between the first and second and the third and fourth teachers.

Such a hypothesis calls for a **population comparison** among means. We can define a population comparison as the following linear combination:

$$C_i = c_1\mu_1 + c_2\mu_2 + \cdots + c_J\mu_J = \sum_{j=1}^{J} c_j\mu_j \tag{3.7}$$

where c_j is a set of real numbers, not all zero, and μ_j is the relevant population means. For an example in which we have three means, the population comparison would be defined as:

$$C_i = c_1\mu_1 + c_2\mu_2 + c_3\mu_3$$

For our example, since we are interested in comparing μ_1 and μ_2 taken as a **set** with μ_3 and μ_4, we will weight the first two population means with an identical weight compared to the last two population means. The following assignment of weights would work:

$$C_i = c_1\mu_1 + c_2\mu_2 + c_3\mu_3 + c_4\mu_4$$
$$= (1)\mu_1 + (1)\mu_2 + (-1)\mu_3 + (-1)\mu_4$$

Note carefully how we assigned the weights. The first two means received weights of "1" while the last two means received weights of "−1." Weighting the means this way has the effect of comparing $\mu_1 + \mu_2$ to $\mu_3 + \mu_4$. Because our linear combination has weights that sum to zero, that is,

$$\sum_{j=1}^{J} c_j = 0$$

the linear combination C_i is given a special name. It is called a **contrast**. A contrast is simply a linear combination of the form (3.7) for which $\sum_{j=1}^{J} c_j = 0$.

Of course, as usual, we rarely if ever have population means at our disposal. When we reject the null hypothesis in the ANOVA F-test, we are implying that there is at least one statistically significant comparison of the type

$$\hat{C}_i = c_1\bar{y}_1 + c_2\bar{y}_2 + \cdots + c_J\bar{y}_J$$
$$= \sum_{l=1}^{J} c_j\bar{y}_j$$

where $\hat{C}_i$ is the estimate for the population comparison C_i. When we take the expectation of $\hat{C}_i$, we find that

$$E(\hat{C}_i) = E\left(\sum_j c_j\bar{y}_j\right)$$
$$= \sum_j c_j E(\bar{y}_j)$$
$$= C_i$$

That is, $\hat{C}_i$ is an unbiased estimator of C_i.

Recall that we do not immediately know the nature of the comparison when we reject an omnibus null hypothesis in ANOVA. For instance, the population comparison could be μ_1 versus μ_2, μ_3 or it could be μ_1, μ_2 versus μ_3, etc. There are a variety of possible comparisons one could make. As noted by Hsu (1996), "to consider multiple comparisons as to be performed only if the F-test for homogeneity [i.e., equality of population means] rejects is a mistake" (Hsu, 1996, p. 178). Hence, it behooves us to consider contrasts quite carefully, since we may wish to make them even without a rejection of omnibus F.

It is very important to also note that whether C_i or $\hat{C}_i$, when computing a comparison, we are computing a **weighted sum of means**. That is, when we speak of a **value** for C_i or $\hat{C}_i$, **we are speaking of one and only one value which is equal to the weighted sum of means we are computing**. Oftentimes comparisons can seem confusing until it is realized that they are, in the end, reduced to a single number, C_i. They are linear combinations (see Chapter 2), and even for the most complex of linear combinations, in the end, they still boil down to a **single number**. Concepts of linear combinations preview the study of multivariate methods in later chapters in this book.

An appropriate null hypothesis for a population comparison is the following:

$$H_0 : C_i = 0$$

since as mentioned, $E(\hat{C}_i) = 0$. A two-sided alternative would be: $H_1 : C_i \neq 0$.

There are many types of "canned" contrasts available in software. These include **simple contrasts** and **Helmert contrasts**, among others. Each contrast-type differs in the comparisons of means it tests. For example, Helmert contrasts feature the comparison of each level of a factor against the average of subsequent levels. That is, in a three-population ANOVA, Helmert contrasts would compare the first mean with a combination of the second and third means, then the second mean with the combination of the third and successive means (of which, in this case, there are none). One should be aware that depending on how software defines the Helmert contrast, the output may differ somewhat from how we have defined them. For instance, the function `contr.helmert()` in R contrasts the second level with the first, the third with the average of the first two, and so forth.

We demonstrate a very simple comparison using the achievement data of Table 3.1. Again, suppose we wished to contrast teachers 1 and 2 with 3 and 4. That is, we wish to estimate values for the following population contrast:

$$C_i = c_1\mu_1 + c_2\mu_2 + c_3\mu_3 + c_4\mu_4$$

Recall that to make it a legitimate contrast, we must select c_1, c_2, c_3 and c_4 such that their sum is equal to 0. To set up the contrast, we can use weights 1, 1 and −1, −1, giving us the estimated contrast value of −29.17:

$$\begin{aligned}
\hat{C}_i &= c_1\bar{y}_1 + c_2\bar{y}_2 + c_3\bar{y}_3 + c_4\bar{y}_4 \\
&= 1(71.00) + 1(72.50) + (-1)(80.0) + (-1)(92.67) \\
&= 71.00 + 72.50 - 80.00 - 92.67 \\
&= -29.17
\end{aligned}$$

We notice immediately that the value of our estimated comparison $\hat{C}_i$ is not equal to 0, which 0 is what we would have expected under the null hypothesis $H_0 : C_i = 0$. How might we interpret this contrast, even before testing it for statistical significance? Since we chose to compare teachers 1 and 2 with 3 and 4 and obtained a **negative** value for our estimated contrast, we can say, without even making an inferential statement yet, that the sum of average achievement for the first two teachers in the sample is 29.17 units **less** than the achievement of students assigned to teachers 3 and 4. We can conclude this by how we assigned the weights (i.e., the two "positive" means came before the negative ones). Note, however, that we have concluded nothing yet about population parameters. We are merely observing our descriptive linear combination. The task of gambling whether these sample results suggest a rejection of the null that $H_0 : C_i = 0$ is the task of statistical inference. But to test such a null, we will need, as is true for any inferential test statistic, an estimated standard error.

We can write the estimated variance for a sample comparison $\hat{C}_i$ by

$$\sigma_{\hat{C}_i}^2 = \sigma_e^2 \sum_j \frac{c_j^2}{n_j}$$

Of course, we do not know the population variance σ_e^2, but we can obtain an unbiased estimate of it, $\hat{\sigma}_e^2$, in the form of MS error. That is, upon making the relevant substitution, we get

$$\hat{\sigma}_{\hat{C}_i}^2 = \text{MS error} \sum_j \frac{c_j^2}{n_j}$$

We know that to get a standard deviation from a variance, it is a simple matter to take the square root of the variance. Likewise, to get the standard error from $\hat{\sigma}_{\hat{C}_i}^2$, we take the square root of $\hat{\sigma}_{\hat{C}_i}^2$:

$$\hat{\sigma}_{\hat{C}_i} = \sqrt{\hat{\sigma}_{\hat{C}_i}^2}$$

Now that we have obtained a standard error for our statistic, we are now in a position to test $\hat{C}_i$ for statistical significance. Recall that we are testing $H_0 : C_i = 0$ against the alternative hypothesis that $H_1 : C_i \neq 0$. For this, we can use a t-test,

$$t = \frac{\hat{C}_i - C_i}{\hat{\sigma}_{\hat{C}_i}}$$

evaluated on $N - J$ degrees of freedom.

In SPSS, we compute the contrast for the achievement data, comparing teachers 1 and 2 to 3 and 4:

```
ONEWAY ac BY teach
/CONTRASTS=1 1-1-1
```

Contrast Coefficients				
Contrast		Teach		
1	1.00	2.00	3.00	4.00
	1	1	−1	−1

Contrast Tests	Contrast	Value of Contrast	Std. Error	t	df	Sig. (2-tailed)
ac Assume equal variances	1	−29.1667	3.54417	−8.229	20	0.000
Does not assume equal variances	1	−29.1667	3.54417	−8.229	15.034	0.000

The contrast, both for equal variances and unequal variances assumed, suggests we reject the null hypothesis.

3.13.1 Independence of Contrasts

When we speak of the pairwise independence of contrasts, we are speaking, substantively, of whether each comparison provides independent and unique **information**. To determine the independence of two contrasts, we need simply to verify the product of respective weights for the two contrasts, that is, $\sum_j c_{1j} c_{2j} = 0$. This requirement holds if sample sizes are equal (for the case of unequal sample sizes, see Hays, 1994, p. 435), and also generally assumes samples come from normal populations with typically equal variances (Hays, 1994, p. 434). C_1 and C_2 are considered to be **orthogonal contrasts**. The number of orthogonal contrasts for a set of groups is always equal to $J - 1$.

Note carefully that when considering comparisons, "orthogonal" and "independent" mean the same thing only when populations are normal with preferably homogeneous variances (Hays, 1994, p. 434). Be sure to note as well that when speaking of independence or orthogonality of comparisons, we are usually referring to a very **specific set of comparisons**, and to ensure orthogonality among the set, we require that all pairwise comparisons be orthogonal to one another. It needs to be noted as well that simply because two or more comparisons are independent does not necessarily guarantee that the t- or F-tests on these comparisons are likewise independent. It simply means that each contrast is providing us with unique information with regards to tested hypotheses. For a discussion of this issue, see Hays (1994).

3.13.2 Independent Samples t-Test as a Linear Contrast

The observant reader may have noticed at this point that an independent-samples t-test is actually a special case of a more general linear contrast. This intuition is correct. To demonstrate such, we perform a t-test **and** a linear contrast on the hypothetical data featured in Table 3.5.

The contrast of interest to us in an independent-samples t-test is to compare group 1 (coded as 0) on X to group 2 (coded as 1). Our contrast is thus of the kind, $\hat{C}_i = (\bar{y}_1 - \bar{y}_2)$, where the corresponding weights we will assign to these means are 1 and -1 respectively. Any other positive and negative balance of weight coefficients would have worked as well such as 2, -2, 3, -3, etc. We compute the mean for group 1 to be 3.2 and the mean for group 2 to be 12.4. Weighting the two means, we obtain:

$$\hat{C}_i = c_1 \bar{y}_1 + \ldots + c_J \bar{y}_J = \sum_{j=1}^{J} c_j \bar{y}_j$$
$$= (1)(3.2) + (-1)(12.4)$$
$$= -9.2.$$

Our obtained value of the contrast is -9.2. We next evaluate as t:

$$t = \frac{\hat{C}_i - C_i}{\hat{\sigma}_{\hat{C}_i}}$$

TABLE 3.5 Hypothetical Data on Dependent Variable Y and Independent Variable X

Y	X
1	0
6	0
4	0
2	0
3	0
8	1
9	1
10	1
15	1
20	1

for which $\hat{\sigma}^2_{\hat{C}_i}$ and $\hat{\sigma}_{\hat{C}_i}$ are equal to:

$$\hat{\sigma}^2_{\hat{C}_i} = \sigma^2_e \sum_j \frac{c^2_j}{n_j}$$

$$= \frac{14.5\left[(1)^2 + (-1)^2\right]}{5}$$

$$= 29/5 = 5.8$$

$$\hat{\sigma}_{\hat{C}_i} = \sqrt{5.8}$$

$$= 2.408.$$

and where

$$\text{SS error} = \sum_j \sum_i \left(y_{ij} - \bar{y}.\right)^2$$

$$= 116.0$$

and so

$$\text{MS error} = \frac{\text{SS error}}{N - J} = \sigma^2_e = \frac{116}{8} = 14.5$$

To compare the above contrast to t, we now compute a t statistic, for which our estimated standard error of the difference, $\hat{\sigma}_{diff}$, once computed, is equal to 2.408. Our t statistic is therefore equal to

$$t = \frac{(\bar{y}_1 - \bar{y}_2)}{\hat{\sigma}_{diff}}$$

$$t = \frac{(3.2 - 12.4)}{2.408}$$

$$t = -3.82.$$

We evaluate t on $N - J$ degrees of freedom, which for this problem are equal to $10 - 2 = 8$. The critical value for t at a significance level of 0.05 is 2.306 (two-tailed test). Since we do not care about the **sign** of the mean difference for the purpose of the contrast (our ordering of coefficients was arbitrary, we could have just as easily reordered our coefficients as -1 and 1), we consider the absolute value of our obtained t, which is equal to 3.82. Since obtained t exceeds the critical value, we reject the null hypothesis and conclude a statistically significant difference between the sample means. In other words, we have evidence to suggest that in the population from which these means were drawn, we indeed have a mean difference.

3.14   POST-HOC TESTS

The contrasts that we have briefly studied are typically useful in situations in which you have strong **a priori** suspicion of where mean differences may lay in your data or a theory guiding you on which contrasts to perform. However, oftentimes, we do not have theory guiding us regarding which contrasts to perform and would like to run as many as we can in order to "snoop" the data to see where pairwise differences may lie.

Recall, however, that with each comparison or contrast we undertake, there is a risk of committing a type I error. This error rate is known generally as the **per comparison type I error rate**, generally denoted as α_{PC}. Obviously, when we perform many comparisons on the same data, the per comparison error rate will add up. The total error rate for a family of comparisons then is known as the **family-wise error rate**, generally denoted as α_{FW}. As given in Howell (2002, p. 371), assuming comparisons are independent, the relationship between α_{PC} and α_{FW} is the following:

$$\alpha_{FW} = 1 - (1 - \alpha_{PC})^k$$

where k is the number of comparisons we are carrying out. The relationship is not quite equal to a direct sum but is still relatively high. For instance, for the situation in which we are carrying out $k = 10$ comparisons, each at $\alpha_{PC} = 0.05$, α_{FW} is estimated to be

$$\begin{aligned} \alpha_{FW} &= 1 - (1 - \alpha_{PC})^k \\ &= 1 - (1 - 0.05)^{10} \\ &= 0.40 \end{aligned}$$

What the number of 0.40 is telling us is that across 10 comparisons, each performed at $\alpha_{PC} = 0.05$, the probability of committing **at least one type I error** in this family is equal to 0.40. Clearly, this error rate is unacceptably high.

What we would like to be able to do is run our 10 comparisons, but keep α_{FW} at a **nominal level** such as 0.05. How can this be done? One easy way to ensure this is to simply perform each pairwise test at a lower level of significance by simply dividing α_{FW} by the number of comparisons c we wish to perform. That is:

$$\frac{\alpha_{FW}}{c}$$

Notice that what we have done is slice up α_{FW} into c component parts. For instance, for the case in which we are performing 10 comparisons, our computation would be

$$\frac{\alpha_{FW}}{c} = \frac{0.05}{10} = 0.005$$

This would mean that we would be testing each comparison at 0.005. This adjustment to α_{PC} in which we divide a nominal α_{FW} by the number of comparisons is known as the **Bonferroni correction**.

Note that for our example at least, our corrected α_{PC} yields an extremely small α level, and hence for each comparison, we have very little power to reject a null hypothesis under this modification. If the number of comparisons were much smaller, say 3, then applying a Bonferroni correction would still keep α_{FW} at a nominal level yet not at the expense of that significant of a decrease in power to reject null hypotheses for each comparison, since,

$$\frac{\alpha_{FW}}{c} = \frac{0.05}{3} = 0.0167$$

Hence, it is clear that this intuitive way of keeping α_{FW} at a nominal level by dividing by c is probably best applied in situations where the number of comparisons is relatively small or one desires setting α_{PC} at a very low and conservative level in the case of many comparisons. For this reason, the Bonferroni correction should be used judiciously and with some judgment.

3.14.1 Newman–Keuls and Tukey HSD

There exist a significant number of post-hoc tests one may use as data snooping procedures following the analysis of variance. We certainly cannot discuss all of them, nor is doing so a productive use of our time unless we aspire ourselves to be post-hoc experts. What is useful, however, is to survey a few post-hocs for the purpose of learning how these procedures generally work.

The next post-hoc test we survey is known as the **Newman–Keuls method**. The Newman–Keuls generally does not keep the family-wise error rate at a nominal level, and though the test is recommended by very few and there are generally better post-hocs available, we survey it anyway because in it is the general logic of how many post-hoc tests function. Also, as we will see, a test that is recommended by many is the Tukey HSD (**Honestly Significant Difference**), which is related to the Newman–Keuls procedure.

To illustrate the Newman–Keuls, consider the analysis of variance summary table for the achievement data, obtained by `fit <- aov(ac ~ f.teach, data = achiev)`, where we generate a factor for teach by `f.teach <- factor(teach)`, and `summary(fit)` gives us:

```
            Analysis of Variance Table

            Response: ac
                      Df  Sum Sq Mean Sq F value    Pr(>F)
            f.teach    3 1764.13  588.04   31.21 9.677e-08 ***
            Residuals 20  376.83   18.84
```

Suppose now we produced a table of ordered pairwise differences between means on the teacher factor. These are listed in Table 3.6.

Table 3.6 is read as follows:

- Contained in each cell are the pairwise mean differences between groups. For instance, in the cell representing the joint occurrence of teacher 1 and teacher 2, the number 1.5 is the mean absolute difference between teacher 1 and teacher 2 (i.e., 72.50–71.00).
- We note that the largest pairwise difference occurs between teacher 1 and teacher 4 (i.e., a mean difference of 21.67)

TABLE 3.6 Pairwise Differences Between Achievement Means for Respective Teacher Assignments

	Teacher 1	Teacher 2	Teacher 3	Teacher 4	
	71.00	72.50	80.00	92.67	Layer
Teacher 1 (71.00)	0	1.5	9.0	21.67	Layer 3; $k = 4$
Teacher 2 (72.50)		0	7.5	20.17	Layer 2; $k = 3$
Teacher 3 (80.00)			0	12.67	Layer 1; $k = 2$
Teacher 4 (92.67)				0	

- "Layer 1" represents a mean difference of two steps (i.e., $k = 2$), from teacher 3 to teacher 4 (each level is counted as a step).
- "Layer 2" represents a mean difference of three steps (i.e., $k = 3$) from teacher 2 to teacher 4 (again, each level is counted as a step, which is why there are 3 steps here, teacher 2 (step 1) to teacher 3 (step 2) to teacher 4 (step 3).

The logic of the Newman–Keuls test is that **pairwise mean differences that are greater steps apart should be tested against a more stringent significance level than mean differences that are lesser steps apart**. In the opinion of the Newman–Keuls test, if means are more steps apart than less, their comparison needs to "pay the price" in terms of being harder to claim as statistically significant. That is, they need to be tested against a more stringent significance level than means that are less distant. For each layer of the test, a different critical value is computed. These critical values are computed from the **studentized range distribution** (a statistic called "q") and also incorporates α and degrees of freedom. The critical value for a given layer of the test is computed:

$$\text{layer } k - 1 = q_{(\alpha, k, v)} \sqrt{\frac{\text{MS error}}{n}}$$

where $q_{(\alpha, k, v)}$ is the critical value for q at significance level α, number of steps k, and degrees of freedom for the ANOVA MS error v. All means that are a given number of steps apart are tested at the critical value for the given layer. One moves diagonally across the table to locate these pairwise differences that are k steps apart. For example, pairwise differences for teacher 3 versus 4, 2 versus 3, and 1 versus 2 would all be tested against the same critical value at layer 1. Likewise, pairwise differences for teacher 2 versus 4 and 1 versus 3 would also be tested against the same critical value, this time at layer 2.

To demonstrate the computation of the critical value for layer 1, we have:

$$\text{layer } 2 - 1 = q_{(\alpha, k, v)} \sqrt{\frac{\text{MS error}}{n}} = q_{(0.05, 2, 20)} \sqrt{\frac{18.84}{6}} = 2.950(1.772) = 5.23$$

The critical value for q for the above is 2.950 (which was found by seeking it out in a sampling distribution of q, available in many introductory statistics texts). The value for the layer is equal to 5.23. If our obtained pairwise difference meets or exceeds a value of 5.23, we may deem it statistically significant at the 0.05 level. From our table, we see that 12.67 does exceed 5.23, and hence, the pairwise difference between teacher 3 and teacher 4 is considered statistically significant. Moving up the table diagonally, we note as well that the pairwise difference between teacher 2 and teacher 3 (7.5) is also statistically significant, but that the pairwise difference between teacher 1 and teacher 2 is not (1.5 does not exceed 5.23).

3.14.2 Tukey HSD

We have seen that the Newman–Keuls test specifies a different critical value dependent on the number of steps means are apart. Tukey HSD tests each mean comparison as though they were the **maximum** steps apart. This produces a much more conservative test than the Newman–Keuls, but as many would argue, is a **better** test. For the data in Table 3.6, the Tukey HSD tests each difference at layer $= 3$ ($k = 4$).

We demonstrate the Tukey test in R for the achievement data:

```
> TukeyHSD(fit)
  Tukey multiple comparisons of means
    95% family-wise confidence level

Fit: aov(formula = ac ~ f.teach)

$f.teach
        diff        lwr       upr     p adj
2-1  1.50000 -5.5144241  8.514424 0.9313130
3-1  9.00000  1.9855759 16.014424 0.0090868
4-1 21.66667 14.6522425 28.681091 0.0000002
3-2  7.50000  0.4855759 14.514424 0.0334428
4-2 20.16667 13.1522425 27.181091 0.0000006
4-3 12.66667  5.6522425 19.681091 0.0003278
```

We can see that the mean difference between teacher 2 and teacher 1 is 1.5, which is not statistically significant ($p = 0.93$) at a conventional level. All other mean differences between teachers are statistically significant at $p < 0.05$. For a detailed discussion of the Tukey test, see Montgomery (2005).

3.14.3   Scheffé Test

The **Scheffé** test is one of the more stringent, conservative tests of the post-hoc family. Recall what it means for a test to be conservative. Pragmatically, it means that if you are able to find statistical significance using the Scheffé, there is a good bet a difference in means **truly** exists in the population. It is definitely the test for hardliners. However, along with its stringent quality comes lower statistical power. As noted by Kirk (1995), Scheffé controls the type I error rate at or less than α_{FW} across all number of contrasts, not only pairwise.

We demonstrate the Scheffé test in SPSS on the teacher factor:

```
ONEWAY ac BY teach
  /MISSING ANALYSIS
  /POSTHOC=SCHEFFE ALPHA(0.05).
```

Multiple Comparisons

Dependent Variable: ac Scheffé

(I) f.teach	(J) f. Teach	Mean Difference (I–J)	Std. Error	Sig.	95% Confidence Lower Bound	Interval Upper Bound
1.00	2.00	−1.50000	2.50610	0.948	−9.1406	6.1406
	3.00	−9.00000[a]	2.50610	0.017	−16.6406	−1.3594
	4.00	−21.66667[a]	2.50610	0.000	−29.3073	−14.0261
2.00	1.00	1.50000	2.50610	0.948	−6.1406	9.1406
	3.00	−7.50000	2.50610	0.056	−15.1406	0.1406
	4.00	−20.16667[a]	2.50610	0.000	−27.8073	−12.5261
3.00	1.00	9.00000[a]	2.50610	0.017	1.3594	16.6406
	2.00	7.50000	2.50610	0.056	−0.1406	15.1406
	4.00	−12.66667[a]	2.50610	0.001	−20.3073	−5.0261
4.00	1.00	21.66667[a]	2.50610	0.000	14.0261	29.3073
	2.00	20.16667[a]	2.50610	0.000	12.5261	27.8073
	3.00	12.66667[a]	2.50610	0.001	5.0261	20.3073

[a]The mean difference is significant at the 0.05 level.

We see that by the Scheffé, the mean difference between teacher 2 and 3, which was found to be statistically significant using the Tukey, is no longer statistically significant at 0.05 ($p = 0.056$). This is a consequence of Scheffé being a more conservative test.

3.14.4 Other Post-Hoc Tests

In addition to the Newman–Keuls, Tukey, and Scheffé tests discussed, there are a host of other post-hocs available to researchers. A distinguishing feature of these tests is their power for pairwise and linear contrasts and their ability to minimize type I error rates across numerous comparisons. Other relatively popular tests include the **Holm test**, the **Ryan test** (REGWQ), and **Dunnett's test**. We do not review these tests here, though the interested reader is encouraged to consult Howell (2002) for a succinct discussion of the benefits and drawbacks to using these tests. Montgomery (2005) also provides further discussion. If you understand the **logic** of post-hoc procedures such as the Newman–Keuls and Tukey HSD, what is meant by a more **conservative** versus more **liberal** test, and what it means to protect an error rate against a multitude of comparison possibilities, then you are in a good position to confront any post-hoc test you may come across and know which types of questions to ask of it. An **understanding** of post-hoc tests is more important than memorizing a catalog of them. For the reader interested in a much more thorough and deeper discussion of post-hocs, consult Hsu (1996) or Miller (1981).

3.14.5 Contrast versus Post-Hoc? Which Should I Be Doing?

After learning about contrasts and post-hocs, students are often still unsure why contrasts are generally recommended if one has a theoretical **planned** prediction about a mean difference, but that post-hocs should be resorted to if one does not have such strong predictions. To help clarify, consider the following three hypothetical sample means:

$$\bar{y}_1 = 10 \quad \bar{y}_2 = 40 \quad \bar{y}_3 = 41$$

Now, after **seeing** these sample means, if I gave you the opportunity to test **one** pairwise mean difference such that you wished to **maximize** your chance of finding statistical significance, which would it be? You would probably choose the comparison $\bar{y}_1 = 10$ versus $\bar{y}_3 = 41$ since these are the most distant means. However, did you really make just **this** comparison? No. Cognitively, when viewing the means, you made a lot more than just one comparison. Implicitly, you compared $\bar{y}_1$ to $\bar{y}_2$, $\bar{y}_1$ to $\bar{y}_3$, etc. So when you decided to test $\bar{y}_1 = 10$ versus $\bar{y}_3 = 41$, it would be incorrect to assume this is the **only** comparison you would be making. You undoubtedly mentally made a lot more comparisons. The job of post-hoc tests is to help guard against these many comparisons you have made, even "below the radar" that could unduly increase the family-wise type I error rate.

Now, pretend for a moment that you had not yet collected the above data, yet based on your **theoretical prediction** and experience in the research area, decided that once the data became available, you would like to compare $\bar{y}_1$ to $\bar{y}_2$. Note that this is an **informed** comparison, it is based on your expertise in predicting which means will be different. In such a comparison, the type I error rate is equal to whatever significance level you set for the comparison. This is because you are making the prediction without first looking at the means and so you are not "punished" for snooping the data and potentially inflating the type I error rate.

To summarize, the critical distinction between **a priori** contrasts and post-hocs is that if you are able to make strong theoretical predictions **before looking at the data**, then contrasts are a suitable option, so long as you are not exhausting the number of contrasts you do on the same data (otherwise, you are more of a "snooper" than a prediction-focused scientist). If you look at the data first, then it must be assumed that you are making a whole lot more comparisons than any specific comparison that you **do** choose to make. In such a case, post-hoc tests are needed to hold you "accountable" for making so many comparisons and help regulate, or keep, the type I error rate under control. Note that the key distinction is not **when the data are collected**. The key feature is whether the data were **observed** before comparisons were made or predicted. If one collected data ten years ago but did not look at the data, then performing a contrast on such data is still acceptable if it is governed by theoretical prediction. As soon as one looks at the means, however, then one has to assume that virtually all (or at least, many) comparisons have been made (mentally), and thus enter post-hoc tests to help control the type I error rate.

The issue of a priori versus post-hoc comparisons is a general theme of scientific credibility. If one is able to predict an outcome before seeing that outcome, the finding is quite impressive. This is the idea of **a priori** comparisons and contrasts. On the other hand, if one observes an outcome and then simply remarks that it occurred, the skill of the scientist is not as apparent. If I hold an apple in my hand and have a theory that if I let it go mid-air it will drop to the floor, that is one thing. If I let go of the apple in mid-air and simply record that it fell, without having any **a priori** prediction that it **would** fall, that is quite another.

3.15   SAMPLE SIZE AND POWER FOR ANOVA: ESTIMATION WITH R AND G*POWER

The concept of power was briefly introduced in Chapter 2. Recall that power is the probability of rejecting a null hypothesis given that it is false. In general, increasing sample size per cell (i.e., replicating units) serves to increase statistical power since it increases degrees of freedom for error (pause for a moment to consider why this statement is true). We now consider how to estimate statistical power using software for the one-way fixed effects analysis of variance model. We demonstrate using R and G*Power.

3.15.1   Power for ANOVA in R and G*Power

Suppose a researcher is interested in testing a balanced one-way fixed effects analysis of variance. The dependent variable is a continuous variable. The independent variable has five levels. The researcher sets the type I error rate at 0.05 and desires a minimal level of power equal to 0.90. In R, we can use `pwr.anova.test` to estimate power for this situation:

```
> library(pwr)
> pwr.anova.test(k= , n= , f= , sig.level= , power= )
```

where, `k` = number of levels on the independent variable (in the current case, equal to 5), `n` = sample size per group; recall for a **balanced** design, there are an equal number of observations per cell, `f` = expected or minimally-desired effect size, `sig.level` = significance level for the omnibus F-test, and `power` = desired or computed power level for the test.

3.15.2   Computing f

Interpreting f values is awkward and unintuitive. Much more intuitive is to convert these to R^2 values. The following is the conversion we need (see Table 3.7):

$$f^2 = \frac{R^2}{1 - R^2}$$

TABLE 3.7 $R^2 \to f^2 \to f$ **Conversions**[4]

R^2	f^2	f
0.10	0.11	0.33
0.20	0.25	0.50
0.30	0.43	0.65
0.40	0.67	0.82
0.50	1.00	1.00
0.60	1.50	1.22
0.70	2.33	1.53
0.80	4.00	2.00
0.90	9.00	3.00
0.99	99.00	9.95

[4]These conversion values were computed in R by the author as follows:

```
> r_squared <- c(.10, .20, .30, .40, .50, .60, .70, .80, .90, .99)
> f_squared <- (r_squared)/(1 - r_squared)
> f_squared
 [1]  0.1111111 0.2500000 0.4285714 0.6666667 1.0000000 1.5000000
 [7]  2.3333333 4.0000000 9.0000000 99.0000000

> f <- sqrt(f_squared)
> f
 [1]  0.3333333 0.5000000 0.6546537 0.8164966 1.0000000 1.2247449 1.5275252
 [8]  2.0000000 3.0000000 9.9498744
```

Suppose for our example that the researcher specifies a minimal effect of interest of $R^2 = 0.10$, which enters `pwr.anova.test` as an f of 0.33. The computation for sample size is the following:

```
> pwr.anova.test(k = 5, n = , f = .33, sig.level = .05, power = .90)

     Balanced one-way analysis of variance power calculation

              k = 5
              n = 29.25818
              f = 0.33
      sig.level = 0.05
          power = 0.9
NOTE: n is number in each group
```

The required sample size **per group** is equal to 29.26. Of course, obtaining "fractions" of participants or subjects can be somewhat difficult, so we will round **up** (not down) for an estimated sample size of 30 participants per group. Even though customarily the rounding of 0.25818 would suggest we settle on 29 participants per group, recall it is good practice to always **round up when estimating sample size**. It is the more **conservative estimate** for the desired level of power (i.e., it will give you slightly **more** power than you have requested if the effect size does turn out as you have estimated).

We perform the same computations in G*Power (Figure 3.2). Notice the identical entries of f, α, desired power and number of groups on the left-hand side. On the right-hand side, G*Power computes the representative noncentrality parameter, along with the critical F statistic required for rejection of the null hypothesis. A total sample size of 150 is the output, which is, within rounding error, equal to our computation using R for 30 participants per group ($150 = 5(30)$). G*Power also computes for us the representative degrees of freedom, 4 for numerator ($J - 1 = 5 - 1 = 4$) and $N - J = 150 - 5 = 145$.

For demonstration, we generate power curves for effect size, f, values of 0.33, 0.63, 0.93, 1.23, and 1.53 (see Figure 3.3).

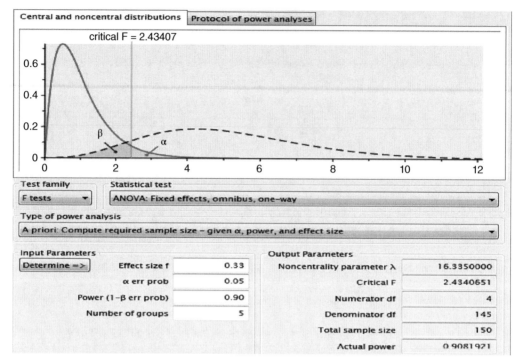

FIGURE 3.2   Power analysis for fixed effects analysis of variance.

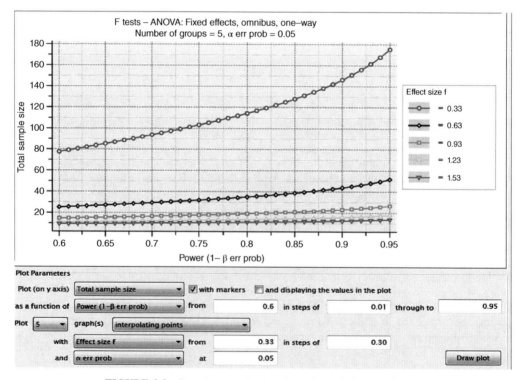

FIGURE 3.3   Power curves for fixed effects analysis of variance.

Notice also that we requested power estimates from 0.6 to 0.95 in increments of 0.01, and effect size f values beginning at 0.33 in increments of 0.30. For relatively large effect sizes of $f = 1.23$ and 1.53 and higher, sample size requirements for a given level of power are relatively constant, though still increasing. For a more moderate effect size of $f = 0.63$, sample size requirements increase slightly more steeply for higher levels of desired power than for lower levels. For a relatively small effect size of $f = 0.33$, a much larger sample size is required for even low levels of power (e.g., 0.6), and the curve increases rather dramatically as increased power is desired. Linearity of the curve is sometimes implicitly assumed when researchers are considering increasing sample size from say, 30 participants to 60. It is often mistakenly believed that such a doubling of sample size equates to a doubling of power for a given effect size. But as power curves demonstrate, this is generally not the case.

Recall that anything that significantly lowers MS error usually results in an increase in statistical power. Hence, in addition to increasing sample size or decreasing the variance of the population under study, one might also adopt the strategy of including additional factors in the design in an effort to reduce MS error. Variables that are added into the model for the sole purpose of boosting power and reducing MS error are generally known as **covariates**. The **analysis of covariance** (ANCOVA) is an extension of the ANOVA model in which covariates are included in an effort to boost statistical power rather than in specifically studying their effect on the dependent variable. We delay our discussion of ANCOVA until Chapter 9, where it will be seen that ANOVA and ANCOVA are both best conceptualized as special cases of the more general linear regression model.

3.16 FIXED EFFECTS ONE-WAY ANALYSIS OF VARIANCE IN R: MATHEMATICS ACHIEVEMENT AS A FUNCTION OF TEACHER

We now conduct a full fixed effects ANOVA on the achievement data of Table 3.1. We designate teacher as a factor having levels 1 through 4:

```
> achiev <- read.table("achievement.txt", header = T)
> f.teach <- factor(teach)
> f.teach
 [1] 1 1 1 1 1 1 2 2 2 2 2 2 3 3 3 3 3 3 4 4 4 4 4 4
Levels: 1 2 3 4
```

We obtain the mean achievement scores for each of the four teachers:

```
> tapply(ac, f.teach, mean)
        1        2        3        4
71.00000 72.50000 80.00000 92.66667
```

The grand mean of the data, or, equivalently, because this is a balanced design, the mean of all means, is computed as:

```
> mean(ac)
[1] 79.04167
```

We next obtain the summary table for our ANOVA using aov:

```
> fit <- aov(ac ~ f.teach, data = achiev)
> summary(fit)
            Df Sum Sq Mean Sq F value   Pr(>F)
f.teach      3 1764.1   588.0   31.21 9.68e-08 ***
Residuals   20  376.8    18.8
---
Signif. codes:  0 '***' 0.001 '**' 0.01 '*' 0.05 '.' 0.1 ' ' 1
```

The above summary table was built under the assumption that population variances are equal. With a reported F stat of 31.21 evaluated on 3 and 20 degrees of freedom yielding a p-value of 9.68e−08, we reject the null hypothesis of equal population means.

We can obtain sample (or treatment) effects from model.tables in R:

```
> model.tables(fit)

Tables of effects

 f.teach
f.teach
      1       2       3       4
 -8.042  -6.542   0.958  13.625
```

We can also use the plot.design function (see Crawley, 2013, p. 238) to visualize the means relative to the overall grand mean:

```
> plot.design(ac ~ f.teach)
```

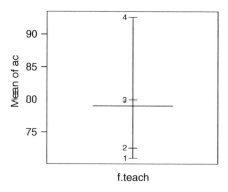

We can see from the plot that the means for teachers 1 and 2 are relatively close, whereas the means for teachers 3 and 4 are quite distant. The horizontal bar just below the 3rd mean is the grand mean of all observations, which recall is equal to 79.04.

3.16.1 Evaluating Assumptions

Since we have very small numbers per group, it would be very difficult to even attempt to test the assumption of normality within each level of teacher in any formal way. We will instead generate a Q–Q plot (Teetor, 2011) for the entire sample to get a rough idea as to whether achievement appears to be at least approximately normally distributed:

```
> qqnorm (ac)
> qqline(ac)
```

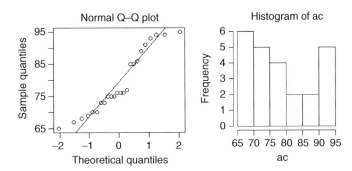

Next to the Q–Q plot is also a histogram (right), obtained using `hist(ac)`. As we can see from both plots, a perfect normal distribution is not evidenced by any means, though the deviation from normality, in this case, is likely nothing to be too concerned about as it relates to pushing forward with the ANOVA (recall again that in a true evaluation of normality, we would have to evaluate each group, or the residuals of the model; our evaluation here on the total sample is simply for demonstration). Though graphical displays are often sufficient for detecting gross violations, we could have also performed the **Shapiro–Wilk** test to evaluate the tenability of the null hypothesis that these sample data arose from a normal population:

```
> shapiro.test(ac)

        Shapiro-Wilk normality test

data:  ac
W = 0.9057, p-value = 0.02842
```

The test yields a *p*-value of 0.028, which is statistically significant tested at 0.05, but not statistically significant when tested at a more stringent significance level, such as 0.01. Since ANOVA is relatively robust to violations of normality, we carry on with the analysis.

R provides several options for verifying the homogeneity of variances assumption. One can use the **Fligner–Killeen test**, **Bartlett's test,** or **Levene's test**. The Fligner–Killeen, which is a nonparametric test, is recommended by some (e.g., Crawley, 2013) over Bartlett's and Levene's because it is quite robust against departures of normality:

```
> fligner.test(ac ~ f.teach, data = achiev)

        Fligner-Killeen test of homogeneity of variances

data:  ac by f.teach
Fligner-Killeen:med chi-squared = 10.8128, df = 3, p-value = 0.01278
```

The test rejects the null hypothesis that population variances are equal, which would suggest an **inequality** of population variances.

In comparison, we now proceed with Bartlett's test:

```
> bartlett.test(ac ~ f.teach, data = achiev)

        Bartlett test of homogeneity of variances

data:  ac by f.teach
Bartlett's K-squared = 3.8962, df = 3, p-value = 0.2729
```

A *p*-value of 0.2729 indicates insufficient evidence against the null hypothesis of equal population variances. Consequently, we would not reject the null and could tentatively assume equality of variance, or at minimum, proceed with our ANOVA.

Next is Levene's test where we specify `center = mean` to denote the fact that we want the test based on means rather than another measure of central tendency (such as medians):

```
> leveneTest(ac, f.teach, center = mean)

Levene's Test for Homogeneity of Variance (center = mean)
      Df F value   Pr(>F)
group  3   7.671 0.001327 **
       20
---
Signif. codes:  0 '***' 0.001 '**' 0.01 '*' 0.05 '.' 0.1 ' ' 1
```

A *p*-value of 0.0013 leads us to reject the null hypothesis of equal variances. Hence, two of the three tests performed suggest we may have a problem with variances. As mentioned, since ANOVA is quite robust to such a violation, we will proceed with performing the ANOVA with the assumption that the condition of variances is satisfied, and then will compare results to a model in which we assume variances are unequal. Even under cases where homogeneity of variance is questionable, R gives us the option of carrying on with the ANOVA by requesting `var.equal = FALSE` in the `oneway.test` function. To demonstrate, we first run the test under the assumption that variances are equal by specifying `var.equal = TRUE`:

```
> oneway.test(ac ~ f.teach, var.equal = TRUE)

        One-way analysis of means

data:  ac and f.teach
F = 31.2096, num df = 3, denom df = 20, p-value = 9.677e-08
```

The observed *p*-value for the analysis is extremely small (i.e., 9.677e−08). We now run the same ANOVA, but this time, under the assumption that the equality of variance assumption is not satisfied:

```
> oneway.test(ac ~ f.teach, var.equal = FALSE)

        One-way analysis of means (not assuming equal variances)

data:  ac and f.teach
F = 57.3175, num df = 3.000, denom df = 10.419, p-value = 8.982e-07
```

Notice that for these data, the *p*-value increased slightly as a result of the assumption not being recognized (i.e., it rose from 9.677e−08 to 8.982e−07). Hence, even when incorporating a violation in variances, because we have such a large effect, our ANOVA is still reporting an extremely small *p*-value.

3.16.2 Post-Hoc Tests on Teacher

We already performed the Tukey earlier on these data in our discussion of post-hoc tests (see Section 3.14.2). For convenience, we reproduce the results of the Tukey HSD test on the teacher factor:

```
> fit <- aov(ac ~ f.teach)
> TukeyHSD(fit)


  Tukey multiple comparisons of means
    95% family-wise confidence level


Fit: aov(formula = ac ~ f.teach)


$f.teach
        diff        lwr        upr      p adj
2-1  1.50000 -5.5144241  8.514424 0.9313130
3-1  9.00000  1.9855759 16.014424 0.0090868
4-1 21.66667 14.6522425 28.681091 0.0000002
3-2  7.50000  0.4855759 14.514424 0.0334428
4-2 20.16667 13.1522425 27.181091 0.0000006
4-3 12.66667  5.6522425 19.681091 0.0003278
```

The above results reveal that virtually all pairwise differences may be of interest (i.e., they are associated with relatively low p-values) except for teacher 2 versus teacher 1 which yields a value of $p = 0.93$.

We can easily observe mean differences by plotting our TukeyHSD post-hoc findings through 95% confidence intervals, where we see the confidence interval for teacher 2 versus 1 (i.e., the first interval below) includes 0, indicating the null hypothesis should not be rejected, which is consistent with the p-value we obtained above.

```
> plot(TukeyHSD(fit))
```

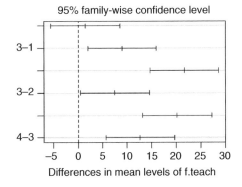

3.17   ANALYSIS OF VARIANCE VIA R's lm

We could have also analyzed the achievement data using R's lm ("linear model") function:

```
> fit.lm <- lm(ac ~ f.teach)
> summary(fit.lm)

Coefficients:
            Estimate Std. Error t value Pr(>|t|)
(Intercept)   71.000     1.772   40.066  < 2e-16 ***
f.teach2       1.500     2.506    0.599  0.55620
f.teach3       9.000     2.506    3.591  0.00183 **
f.teach4      21.667     2.506    8.646 3.44e-08 ***
---
Signif. codes:  0 '***' 0.001 '**' 0.01 '*' 0.05 '.' 0.1 ' ' 1

Residual standard error: 4.341 on 20 degrees of freedom
Multiple R-squared:  0.824,     Adjusted R-squared:  0.7976
F-statistic: 31.21 on 3 and 20 DF,  p-value: 9.677e-08
```

The output supplies us with all of the information provided by aov but also much more, including mean contrasts of interest:

- The intercept value of 71.00 is the mean achievement for students assigned to the first teacher. R takes this to be the "baseline" group since it is the first category of teacher.
- f.teach2 represents a mean difference between the first teacher and the second teacher, of 72.5 − 71.00 = 1.50.
- f.teach.3 represents a mean difference between the first teacher and the third teacher, of 80.00 − 71.00 = 9.00.
- f.teach.4 represents a mean difference between the first teacher and the fourth teacher, of 92.67 − 71.00 = 21.67.
- The obtained p-value for the model is identical to that obtained using aov (both are equal to 9.677e − 08).
- R-squared for the model, which for ANOVA, in this case, amounts to η^2, is equal to 0.824, indicating that a whopping 82.4% of the variance in achievement can be explained by mean differences between these **particular teachers** (recall it's a **fixed effects** model) chosen for the experiment. We postpone a discussion of **Adjusted R-squared** until Chapter 7.

3.18   KRUSKAL–WALLIS TEST IN R AND THE MOTIVATION BEHIND NONPARAMETRIC TESTS

Most of the statistical models surveyed in this book, in one way or another, make distributional assumptions. For example, in the prior chapter featuring t-tests, as well as this chapter discussing ANOVA, it has been generally assumed that population distributions and sampling distributions have been at least approximately normally distributed with no serious violations. As we have seen, the central limit theorem aids in assuring in most cases the normality of sampling distributions, and often the assumption of normality of populations is at least feasible for t-tests and ANOVA. We have also had to assume such things as equality of variances, and later in the book, we will learn that regression models come with them their own set of parametric assumptions that need to verified and satisfied for inferences on parameters to be justified.

There are situations, however, where we can be fairly confident parametric assumptions are not satisfied, which could, in turn, cause problems for drawing inferences from samples to populations. For example, sometimes distributions of sample data may be severely **abnormal**, or have **heavy tails** that stretch out to extremes, generating highly **skewed** data. In such cases, even data transformations may be hopeless in helping ensure assumptions are met. In other cases, sample size may be exceedingly small, such as having 5–10 subjects in a t-test, in which case it becomes virtually impossible to assure ourselves with any kind of certainty that the assumptions underlying the parametric test could ever be justified, unless of course, we have a priori intimate knowledge of the underlying population distributions. In most cases, especially with **small samples** in particular, we do not have this knowledge, and hence we are left conducting inferences on the parametric model with a high degree of skepticism that all is in place for inferences to be valid.

There is, however, a potential solution out of this dilemma, and that is to employ in place of a parametric test, a **nonparametric** method. Nonparametric tests carry with them the advantage of making virtually no assumptions about the underlying population, and thus are especially useful for sample data that is severely misbehaving, or data that are very small in number such that we could never know anyway whether parametric assumptions are satisfied. And though nonparametric tests are themselves unique in their own way, many parametric tests have nonparametric "equivalents" for which researchers can employ either alongside or in replacement of the parametric test. Nonparametric tests can be especially useful in situations where data (especially for small samples) are in the from of **ranks**, and in many cases, nonparametric tests actually work by first translating measurement data into simpler ranks as part of its procedure.

Though nonparametric tests are useful and we feature demonstrations of a few of them in this book, it should be recognized that if assumptions of the test are satisfied, then parametric tests will typically have more **power** over nonparametric ones, and are usually, therefore, preferred over nonparametric methods in most, though certainly not all cases (Howell, 2002). Nonparametric tests are also a bit more "crude" in that, as mentioned, they usually convert continuous data into such things as ranks, which necessarily causes a **loss of information** in the data. As an example, if we consider two temperature ratings of 30 degrees and 20, that distance of 10 degrees between values may be extremely important for data analysis and be of primary interest to the investigator. However, many nonparametric tests will simply treat the value of 30 as "first" and 20 as "second" and assign an **ordinal ranking** to the values instead of appreciating and recognizing the **magnitude** of their difference. This is one potential drawback to using nonparametric tests, and so before using one, a researcher should seriously contemplate whether the (pseudo) continuity in one's data is especially valuable, or whether a more crude ranking of values is sufficient (or even preferable).

As a quick example of how a nonparametric alternative test can be applied to the achievement data, we test whether ac is a function of f.teach using the **Kruskal–Wallis test**, which can be considered the nonparametric equivalent to the parametric ANOVA featured in this chapter. For details of the test, see Howell (2002). We conduct the test using `kruskal.test`:

```
> kruskal.test(ac ~ f.teach)

        Kruskal-Wallis rank sum test

data:  ac by f.teach
Kruskal-Wallis chi-squared = 16.2665, df = 3, p-value = 0.0009999
```

Though reporting a larger p-value than the parametric ANOVA run earlier, clearly, we still have evidence to reject the null hypothesis that the samples arose from the same population ($p = 0.00099$).

We can perform a nonparametric post-hoc using the **Tukey and Kramer (Nemenyi)** test to follow up on the Kruskal–Wallis. We use the PMCMR package (Pohlert, 2014) in R to conduct the test:

```
> library(PMCMR)
> posthoc.kruskal.nemenyi.test(ac, f.teach, method = "Tukey")

    Pairwise comparisons using Tukey and Kramer (Nemenyi) test
            with Tukey-Dist approximation for independent samples

data:  ac and f.teach

   1       2       3
2 0.9658  -       -
3 0.3054  0.5849  -
4 0.0014  0.0074  0.2117

P value adjustment method: none
Warning message:
In posthoc.kruskal.nemenyi.test(ac, f.teach, method = "Tukey") :
  Ties are present, p-values are not corrected.
```

What appears in the above table are p-values, not mean differences. Comparisons between teachers 1 versus 4 and 2 versus 4 yield small p-values (0.0014 and 0.0074, respectively). These represent more conservative findings when compared to the parametric counterpart post-hoc (Tukey) performed earlier.

3.19   ANOVA IN SPSS: ACHIEVEMENT AS A FUNCTION OF TEACHER

We now present select output for the analysis performed in SPSS. We only briefly discuss the results, as they for the most part parallel those generated by R. Entered into SPSS, our data file appears as:

	ac	teach
1	70.00	1.00
2	67.00	1.00
3	65.00	1.00
4	75.00	1.00
5	76.00	1.00
6	73.00	1.00
7	69.00	2.00
8	68.00	2.00
9	70.00	2.00
10	76.00	2.00
11	77.00	2.00
12	75.00	2.00
13	85.00	3.00
14	86.00	3.00
15	85.00	3.00
16	76.00	3.00
17	75.00	3.00
18	73.00	3.00
19	95.00	4.00
20	94.00	4.00
21	89.00	4.00
22	94.00	4.00
23	93.00	4.00
24	91.00	4.00

We perform the analysis using the following syntax:

```
UNIANOVA ac BY teach
  /METHOD=SSTYPE(2) * requests Type II sums of squares (Type III, the
default in SPSS, would have produced the same output for this analysis)
  /POSTHOC=teach(TUKEY SCHEFFE)
  /EMMEANS=TABLES(teach) * requests estimated marginal means (i.e.,
the means of each group, in this case)
  /PRINT=ETASQ HOMOGENEITY * requests Eta-squared and a test of
homogeneity of variance (Levene's test)
  /CRITERIA=ALPHA(.05) * sets the significance level for the F-test at 0.05
```

Levene's test suggests the same finding as that found in R, that there is a difference in variances in the population:

Levene's Test of Equality of Error Variances[a]

Dependent Variable: ac

F	df1	df2	Sig.
7.671	3	20	0.001

Tests the null hypothesis that the error variance of the dependent variable is equal across groups.

[a]Design: Intercept + teach

Tests of Between-Subjects Effects

Dependent Variable: ac

Source	Type II Sum of Squares	Df	Mean Square	F	Sig.	Partial Eta-Squared
Corrected model	1764.125[a]	3	588.042	31.210	0.000	0.824
Intercept	149 942.042	1	149 942.042	7958.003	0.000	0.997
teach	1764.125	3	588.042	31.210	0.000	0.824
Error	376.833	20	18.842			
Total	152 083.000	24				
Corrected total	2140.958	23				

[a]R-Squared = 0.824 (Adjusted R-Squared = 0.798)

The resulting ANOVA table parallels that generated by R (we do not reproduce the Tukey and Scheffé tests here).

To run a more robust test of means, one less sensitive to model assumptions, we could have run the Welch test (1951):

```
ONEWAY ac BY teach
  /STATISTICS WELCH
  /MISSING ANALYSIS.
```

Robust Tests of Equality of Means ac

	Statistic[a]	df1	df2	Sig.
Welch	57.318	3	10.419	0.000

[a]Asymptotically F distributed.

Just as we found in R, the null hypothesis is rejected even under the more robust test.

3.20 CHAPTER SUMMARY AND HIGHLIGHTS

- The **analysis of variance**, or ANOVA for short, is a statistical method useful for partitioning variability in a sample for the purpose of testing null hypotheses about the equality of population means (fixed effects) or null hypotheses about the extent to which one or more factors account for variance in a dependent variable (random effects).

- The **one-way analysis of variance** is defined to have a single categorical independent variable and a single continuous dependent (or "response") variable.

- ANOVA models are usually distinguished between **fixed effects**, **random effects**, and **mixed models**.

- In **fixed effects models**, the researcher is specifically interested in the levels of the independent variable chosen for the particular experiment. The specific levels were **deliberately** chosen.

- In **random effects** models, the researcher is not specifically interested in the levels of the independent variable chosen for the particular experiment but is rather most interested in **generalizing these levels** to the population of levels from which the sample levels were drawn.

- **Mixed models** contain a blend of both fixed and random effects.

- The fact that **sample means** may differ in a data set is not itself evidence against the null hypothesis. What we ask of the data is the likelihood of such differences in the sample under the null hypothesis. If such differences are unlikely under the null, then we have reason to reject the null hypothesis and conclude there to be population mean differences.

- The inferential test for ANOVA essentially boils down to a comparison of variances in terms of a ratio. If **between-group variance** is large relative to **within-group variance**, then this may be taken as evidence against the null hypothesis. The expectation for F under the null hypothesis is approximately equal to 1.0.

- **Fixed effects analysis of variance** can be understood as an extension of the independent-samples t-test, or, the independent-samples t-test can be understood as a special case of the wider ANOVA model.

- When we break down a deviation into its constituent parts, the essential goal of ANOVA is obtaining an answer to the question—**Why does any given score in our data deviate from the overall mean?** The extent to which these deviations are due to **between-group effects** rather than **within-group variability** is the extent to which we gather evidence against the null hypothesis.

- When we square respective deviations, we find that **SS total** can be partitioned into **SS between** + **SS within**.

- The **expected mean squares** for both between and within suggest that when squared population effects equal 0 (i.e., $\alpha_j^2 = 0$), the appropriate denominator for the F-test is that of **MS within**.

- The **ANOVA summary table** is a convenient way of representing the results of the analysis of variance.

- The **assumptions of fixed effects ANOVA**, in addition to the fixed nature of the levels chosen for the experiment, include $E(\varepsilon_{ij}) = 0$, ε_{ij} are $NI(0, \sigma_e^2)$, $\sigma_{e_{ij}}^2 < \infty$, $Cov(\varepsilon_{ij}, \varepsilon_{i'j'}) = 0$, and $\sigma_{j=1}^2 = \sigma_{j=2}^2 = \sigma_{j=J}^2$. An additional assumption is that the model is correctly specified, which means that the model at least reasonably accounts for the major sources of variation in the response variable.

- In an **experimental design** featuring random assignment of subjects to groups, individuals within each group are not expected to be similar a priori the randomization. However, in nonexperimental studies, individuals in existent groups usually share characteristics that are similar. That is, individuals within groups are usually more alike compared to individuals across groups. This concept generally goes by the name of **nesting** and is the motivation behind such relatively advanced techniques as **hierarchical** and **multilevel** modeling.

- Obtaining a **statistically significant** F in ANOVA in no way guarantees a meaningful scientific finding. **Effect size** measures are required to assess the degree to which the independent variable explains variance in the response variable.

- **Eta-squared** is a traditional effect size computed by taking the ratio of **SS between** to **SS total**. A value of 0 indicates zero variance explained. A value of 1.0 indicates 100% of variance explained.

- **Omega-squared** is an effect size measure used to help correct the overly optimistic estimates typically provided by Eta-squared. Omega-squared attempts to better estimate the corresponding effect size in the population, and thus is typically less than Eta-squared.

- Computing a **t-test via ANOVA** is a useful exercise to appreciate the similarities between the two procedures by noting the relation $t = \sqrt{F}$.

- **Contrasts** are useful in providing custom hypothesis tests between pairs of population means.

- The independent-samples t-test can be interpreted as an example of a **linear contrast**.

- **Post-hoc tests** are used to snoop the data following a statistically significant F in ANOVA. The objective of a post-hoc test is to help control the **family-wise error rate**, that is, the error rate generated by successive tests across the "family" of comparisons. Good post-hoc tests are generally those that keep the error rate at a nominal level but not at the expense of a significant loss of **power**.

- The **Bonferroni correction** divides the family-wise error rate across the number of pairwise comparisons one wishes to make. The test quickly loses power as the number of means (and thus comparisons) increases.

- The **Newman–Keuls method**, though somewhat unpopular because of its failure to protect family-wise error, is nonetheless useful for describing the general logic of a layered test. The Tukey HSD test is a more common test than the Newman–Keuls and is also more conservative. It is highly recommended for most cases.

- The **Scheffé test** is a very conservative post-hoc test that protects not only against pairwise comparisons but also against all linear contrasts. If one finds a sample difference with the Scheffé, one can be relatively confident that the difference exists in the population.

- **Sample size** and **power** can be estimated with relative ease using R or G*Power.

REVIEW EXERCISES

3.1. Give a definition for the **fixed effects** analysis of variance.

3.2. Compare and contrast a **fixed effect** versus a **random effect**.

3.3. Explain how models in virtually all sciences are not **deterministic** but rather **probabilistic**. In the achievement example discussed in the chapter, what would it mean to say that mathematics achievement is a true **function** of teacher? Why is such an ideal likely virtually impossible in practice?

3.4. Explain why observing **differences in sample means** does not alone constitute evidence against a **null hypothesis** tested in ANOVA. What more information do we require?

3.5. Compare the equation for an **independent-samples t-test** to that of an **F-test in ANOVA**, and comment on their similarities and differences. How do they both essentially answer a similar question?

3.6. Discuss the importance and significance of the **identity**

$$\left(y_{ij} - \bar{y}.\right) = \left(y_{ij} - \bar{y}_j\right) + \left(\bar{y}_j - \bar{y}.\right)$$

as it pertains to the logic of ANOVA. Identify each component.

3.7. Discuss the effect of **squaring deviations** in the identity $\left(y_{ij} - \bar{y}.\right) = \left(y_{ij} - \bar{y}_j\right) + \left(\bar{y}_j - \bar{y}.\right)$.

3.8. Verbally interpret and discuss the following **identity**:

$$\sum_{j=1}^{J} \sum_{i}^{n} \left(y_{ij} - \bar{y}.\right)^2 = \sum_{j=1}^{J} \sum_{i}^{n} \left(y_{ij} - \bar{y}_j\right)^2 + \sum_{j}^{J} n_j \left(\bar{y}_j - \bar{y}.\right)^2$$

3.9. Briefly discuss what role dividing by **degrees of freedom** has on the sums of squares of ANOVA. What is the purpose of dividing by degrees of freedom?

3.10. Explain why **sums of squares** are generally **additive** for balanced designs, but **means squares** are not.

3.11. What is the approximate **expectation of F** under a true null hypothesis? Why is this so?

3.12. State two ways in which the **null hypothesis** for ANOVA can be operationalized.

3.13. List the **assumptions** of the ANOVA model.

3.14. Define what is meant by a **nuisance factor** and comment on why **randomization** does not guarantee that nuisance factors will be evenly dispersed among treatment groups.

3.15. Distinguish between an **experimental design** versus a **quasi-experimental design**.

3.16. For an experiment in which virtually all **variance** is accounted for by the **treatment effect**, what value of η^2 would you expect to obtain? Why?

3.17. For an experiment in which virtually none of the **variance** is accounted for by the **treatment effect**, what value of η^2 would you expect to obtain? Why?

3.18. Discuss the **difference** between η^2 and ω^2.

3.19. Discuss the purpose of **contrasts** in ANOVA.

3.20. Distinguish between a **linear combination** and a **contrast**.

3.21. Derive a data set for which the dependent variable is continuous and the independent variable consists of a three-level grouping variable. Generate the data for which there is **much within-group variability**, but very little **between-group variability**. In such a case, what decision on the null hypothesis $H_0: \mu_1 = \mu_2 = \mu_3$ would likely result? Why? Explain.

3.22. The analysis of variance was developed primarily to address problems in agriculture, genetics, and biology. Consider data from R.A. Fisher's **Statistical Methods for Research Workers** published in 1925, the book credited with the first comprehensive introduction to the analysis of variance. In Table 41, p. 217 (1934 edition), Fisher presents data on **soil bacteria** in which soil data was separated into four samples. On each sample, seven plates were inoculated, and the number of colonies recorded on each plate. The data are reproduced in Table 3.8.

TABLE 3.8   Number of Bacteria Colonies by Plate and Sample (Fisher, 1925/1934)

Plate	Sample	Number of Colonies
1	1	72
1	2	74
1	3	78
1	4	69
2	1	69
2	2	72
2	3	74
2	4	67
3	1	63
3	2	70
3	3	70
3	4	66
4	1	59
4	2	69
4	3	58
4	4	64
5	1	59
5	2	66
5	3	58
5	4	62
6	1	53
6	2	58
6	3	56
6	4	58
7	1	51
7	2	52
7	3	56
7	4	54

Source: Fisher (1925, 1934).

Answer the following questions with regards to Fisher's data:

(a) Is there evidence to suggest that the **mean number of colonies differs by plate**? Conduct a one-way fixed effects analysis of variance.

(b) Is there evidence to suggest that the **mean number of colonies differs by sample**? Conduct a one-way fixed effects analysis of variance.

Further Discussion and Activities

3.23. The majority of statistical procedures can be represented through concepts of **covariance** and **correlation**. The analysis of variance, though focusing on mean differences, can nonetheless be expressed through simple correlational analysis. An excellent and relatively easy read that describes these ideas is given in Levin et al. (1989). Read the paper and summarize the essential ideas of how **ANOVA** can be conceptualized in terms of **correlational theory**.

4

FACTORIAL ANALYSIS OF VARIANCE: MODELING INTERACTIONS

The assignable sources of variation in a manufacturing process may be divided into two categories. First, there are those factors which introduce variation in a random way. Lack of control at some stage of production very often acts in this manner, and the material itself usually exhibits an inherent random variability. The other type of factor gives rise to systematic variation.

(Daniels, 1939, p. 187, *The Estimation of Components of Variance*)

The researcher of Chapter 3 who studied the effect of melatonin dosage on sleep onset is interested now in learning whether these effects are consistent across ambient noise levels present during sleep. For this experiment, the researcher again randomly assigns 25 individuals to a control group, 25 more to a group receiving 1 mg of melatonin, and 25 more to a group receiving 3 mg of melatonin. In addition, within each of these conditions, half of the participants receive either no ambient noise or a low amount of ambient noise at the moment of melatonin ingestion and lasting throughout the night (for instance, a slight buzzing sound). The researcher would like to test whether sleep onset is a function of dosage, ambient noise, and a potential **combination** of the two factors. That is, the researcher is interested in detecting a potential **interaction** between dose and noise level. He is only interested in generalizing his findings to these particular doses of melatonin and to these particular noise levels. Such a research design calls for a **two-way fixed effects factorial analysis of variance**.

4.1 WHAT IS FACTORIAL ANALYSIS OF VARIANCE?

In the one-way ANOVA of the previous chapter, we tested null hypotheses about equality of population means of the kind:

$$H_0 : \mu_1 = \mu_2 = \mu_3 = \mu_J$$

Applied Univariate, Bivariate, and Multivariate Statistics: Understanding Statistics for Social and Natural Scientists, With Applications in SPSS and R, Second Edition. Daniel J. Denis.
© 2021 John Wiley & Sons, Inc. Published 2021 by John Wiley & Sons, Inc.
Companion Website: www.wiley.com/go/denis/appliedstatistics2e

In the two-way and higher-order analysis of variance, we have more than a single factor in our design. As we did for the one-way analysis, we will test similar **main effect** hypotheses for each individual factor, but we will also test a new null hypothesis, one that is due to an **interaction** between factors.

In the two-factor design on melatonin and ambient noise level, we are interested in the following effects:

- **Main effect** due to drug dose in the form of mean sleep differences across dosage levels.
- **Main effect** due to ambient noise level in the form of mean sleep differences across noise levels.
- **Interaction** between drug dose and noise level in the form of mean sleep differences on drug not being consistent across noise levels (or vice versa).

It does not take long to realize that science is about the discovery not of main effects, but of interactions. Yes, we are interested in whether melatonin has an effect, but we are even more interested in whether melatonin has an effect **differentially** across noise levels. And beyond this, we may be interested in even higher-order effects, such as **three-way interactions**. Perhaps melatonin has an effect, but mostly at lower noise levels, and mostly for those persons aged 40 and older. This motivates the idea of a three-way interaction, drug dose by noise level by age. One will undoubtedly remark the tone of **conditional probability** themes in the concept of an interaction.

As another example of an interaction, consider Table 4.1 and corresponding Figure 4.1. The plot features the achievement data of the previous chapter, only that now, in addition to students being randomly assigned to one of four teachers (f.teach), they were also randomly assigned to the study of one of two mathematics textbooks (f.text).

What we wish to know from Figure 4.1 is whether textbook differences (1 versus 2) are **consistent** across levels of teacher. For instance, at teacher = 1, we ask whether the same textbook "story" is being told as at teachers 2, 3, and 4. What this "story" is, are the distances between cell means, as emphasized in part (b) of the plot. Is this distance from textbook 1 to textbook 2 consistent across teachers, or do such differences depend in part on which teacher one has? These are the types of questions we need to ask in order to ascertain the presence or absence of an interaction effect. And though it would appear that mean differences are not equal across teacher, the question we really need to ask is whether these sample differences across teacher are large enough to infer population mean differences. These questions will be addressed by the test for an **interaction effect** in the two-way fixed effects analysis of variance model.

TABLE 4.1 Achievement as a Function of Teacher and Textbook

Textbook	Teacher			
	1	2	3	4
1	70	69	85	95
1	67	68	86	94
1	65	70	85	89
2	75	76	76	94
2	76	77	75	93
2	73	75	73	91

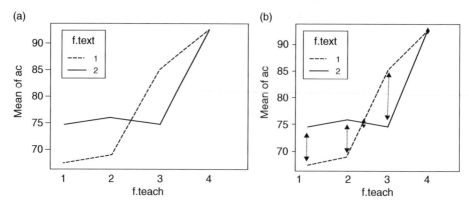

FIGURE 4.1   (a) Cell means for teacher∗textbook on achievement. (b) Distances between cell means as depicted by two-headed arrows. (where `f.text` is the factor name for textbook and `f.teach` is the factor name for teacher).

4.2   THEORY OF FACTORIAL ANOVA: A DEEPER LOOK

As we did for the one-way analysis of variance, we develop the theory of factorial ANOVA from fundamental principles which will then lead us to the derivation of the sums of squares. The main difference between the simple one-way model and the two-way model is the consideration of **cell effects** as opposed to simply **sample effects**. Consider, in Table 4.2, what the two-way layout might look like for our melatonin example in the factorial design.

We are interested in both **row** mean differences, summing across melatonin dose, as well as **column** mean differences, summing across noise level. We ask ourselves the same question we asked in the previous chapter for the one-way model:

Why does any given score in our data deviate from the mean of all the data?

Our answer must now include four possibilities:

- An effect of being in one melatonin-dose group versus others.
- An effect of being in one noise level versus others.
- An effect due to the combination (**interaction**) of dose and noise.

TABLE 4.2   Cell Means of Sleep Onset as a Function of Melatonin Dose and Noise Level (Hypothetical Data)

Noise Level	Melatonin Dose			
	0 mg	1 mg	3 mg	Row Means
High	15	11	8	11.3
Low	12	10	4	8.7
Column means	13.5	10.5	6.0	10.0

Column

Row		1	2	3	•	•	•	•	•	*j*	•	•	*c*	*Row means*
	1	X_{11}	X_{12}	X_{13}	•	•	•	•	•	X_{ij}	•	•	X_{1c}	$X_{1.}$
	2	X_{21}	X_{22}	X_{23}	•	•	•	•	•	X_{2j}	•	•	X_{2c}	$X_{2.}$
	•	•	•	•	•	•	•	•	•	•	•	•	•	•
	•	•	•	•	•	•	•	•	•	•	•	•	•	•
	•	•	•	•	•	•	•	•	•	•	•	•	•	•
	i	X_{i1}	X_{i2}	X_{i3}	•	•	•	•	•	X_{ij}	•	•	X_{ic}	$X_{i.}$
	•	•	•	•	•	•	•	•	•	•	•	•	•	•
	•	•	•	•	•	•	•	•	•	•	•	•	•	•
	r	X_{i1}	X_{i2}	X_{i3}	•	•	•	•	•	X_{rj}	•	•	X_{rc}	$X_{r.}$
	Col means	$X_{.1}$	$X_{.2}$	$X_{.3}$	•	•	•	•	•	$X_{.j}$	•	•	$X_{.c}$	$X_{..}$

FIGURE 4.2 Generic two-way analysis of variance layout. The two-way factorial analysis of variance has row effects, column effects, and interaction effects. Each value within each cell represents a data point. Row and column means are represented by summing across values of the other factor. Source: Eisenhart (1947). Reproduced with permission from John Wiley & Sons.

- Chance variation that occurs within each **cell** of the design. Notice that this 4th possibility is now the **within-group** variation of the previous one-way model of Chapter 3, only that now, the "within group" is, in actuality, **within cell**. The error variation occurs within the cells of a factorial design.

In the spirit of history, we show an earlier and more generic layout of the two-way model diagramed by Eisenhart (1947) and reproduced in Figure 4.2, where entries in the cells depict data points for each row and column combination. Note the representation of row means and column means. These will aid in the computation of main effects for each factor.

4.2.1 Deriving the Model for Two-Way Factorial ANOVA

We now develop some of the theory behind the two-way factorial model. As always, it is first helpful to recall the essentials of the one-way model, then extend these principles to the higher-order model. Recall the one-way fixed effects model of the previous chapter:

$$y_{ij} = \bar{y}_{.} + a_j + e_{ij}$$

where the sample effect a_j was defined as $a_j = (\bar{y}_j - \bar{y}.)$. The sample effect a_j denoted the effect of being in one particular sample in the layout. Recall that in the one-way layout, $\sum_j n_j a_j = 0$, which in words meant that the sum of weighted sample effects, where n_j was the sample size per group, summed to zero. For this reason, we squared these treatment effects, which provided us with a measure of the sums of squares between groups:

$$\text{SS between} = \sum_j n_j a_j^2$$

TABLE 4.3   Cell Means Layout for 2 × 3 Factorial Analysis of Variance

	Factor 2		
Factor 1	Level 1	Level 2	Level 3
Level 1	$\bar{y}_{jk}$	$\bar{y}_{jk}$	$\bar{y}_{jk}$
Level 2	$\bar{y}_{jk}$	$\bar{y}_{jk}$	y_{jk}

It turned out as well that the sample effect a_j was an **unbiased estimator** of the corresponding population effect, α_j. That is, the expectation of a_j is equal to α_j, or, more concisely, $E(a_j) = \alpha_j$. Recall that the sample effect represents the effect or influence of group membership in our design. For instance, for an independent variable having three levels, we had three groups ($J = 3$) on which to calculate respective sample effects. In the factorial two-way analysis of variance, we will have more than J groups because we are now **crossing** two variables with one another. For example, the layout for a 2×3 (i.e., 2 rows by 3 columns) design is given in Table 4.3.

Notice that now, we essentially have six "groups" in the 2×3 factorial model, where each combination of factor levels generates a mean $\bar{y}_{jk}$, where j designates the row and k designates the column. The "groups" that represent this combination of factor 1 and factor 2 we will refer to as **cells**. This is why we have been putting "groups" in quotation marks, because what these things really are in the factorial design are cells. **The heart of partitioning variability in a factorial design happens between cells.** In addition to defining the sample effects associated with each factor (i.e., a_j and b_k), we will now also need to define what is known as a **cell effect**.

4.2.2   Cell Effects

A **sample cell effect** (Hays, 1994, p. 477) is defined as:

$$[ab]_{jk} = \left(\bar{y}_{jk} - \bar{y}..\right)$$

and represents a measure of variation for being in one cell and not others. Notice that to compute the cell effect, we are taking each cell mean $\bar{y}_{jk}$ and subtracting the grand mean, $\bar{y}..$ (we carry **two periods** as subscripts for the grand mean now to denote the summing across j rows and k columns). But why do this? We are doing this similar to why we took the group mean and subtracted the grand mean in a simple one-way analysis of variance. In that case, in which we computed $a_j = \left(\bar{y}_j - \bar{y}.\right)$, we were interested in the "effect" of being in one group versus other groups (which was represented by subtracting the overall mean).

Likewise, in computing cell effects, we are interested in the effect of being in one cell versus other cells, because now, in the two-way factorial model, in addition to both main effects for row and column, it is the cell effect that will represent our interests in there possibly existing an interaction between the two factors. We will need to compute an interaction effect to do this, but getting the cell effect is the first step toward doing so.

As it was true that the sum of sample effects in the one-way model was equal to 0, $\sum_j n_j a_j = 0$, it will also be true that the sum of cell effects is equal to 0 for any given sample. That is,

$$\sum_j \sum_k [ab]_{jk} = \sum_j \sum_k \left(\bar{y}_{jk} - \bar{y}..\right) = 0$$

TABLE 4.4 Deviations Featured in One-way and Two-way Analysis of Variance

Deviation	In Words	Solution is Squaring Deviations
$\sum_{i=1}^{n}(y_i - \bar{y}.) = 0$	The sum of score deviations around a mean equals 0	$\sum_{i=1}^{n}(y_i - \bar{y}.)^2 > 0$
$\sum_{i=1}^{n}(\bar{y}_j - \bar{y}..) = 0$	The sum of row sample mean deviations around a grand mean equals 0	$\sum_{i=1}^{n}(\bar{y}_j - \bar{y}..)^2 > 0$
$\sum_{i=1}^{n}(\bar{y}_k - \bar{y}..) = 0$	The sum of column sample mean deviations around a grand mean equals 0	$\sum_{i=1}^{n}(\bar{y}_k - \bar{y}..)^2 > 0$
$\sum_{i=1}^{n}(\bar{y}_{jk} - \bar{y}..) = 0$	The sum of cell mean deviations around a grand mean equals 0	$\sum_{i=1}^{n}(\bar{y}_{jk} - \bar{y}..)^2 > 0$

In each case, the sum of deviations equals to 0.

where the double summation represents the summing across k columns first, then across j rows. We can easily demonstrate this by computing the cell effects for Table 4.2 across each row of noise level. For the first cell mean of 15 in row 1, column 1, the cell effect is computed as $15 - 10 = 5$. For row 1, column 2, the cell effect is $11 - 10 = 1$. The remaining cell effects are computed analogously (-2, 2, 0, -6). The sum of these cell effects is easily demonstrated to be zero ($(5 + 1 + (-2) + 2 + 0 + (-6) = 0$). But why would this be true? It is true for the same reason why summing sample effects equals 0. We are taking deviations from the grand mean, and by definition, the grand mean is the "center of gravity" of all means (in a balanced design). So, it is reasonable then that the sum of deviations around that value should be equal to 0. To avoid this, just as we did for the ordinary variance and for the variances derived in the one-way analysis of variance, we square deviations.

To better conceptualize deviations from means across the one-way and two-way factorial designs, it is helpful to compare and contrast the four scenarios featured in Table 4.4.

We can see from Table 4.4 that the solution in each case is to **square** respective deviations. This is precisely why in the case of cell effects, as we did for single deviations and mean deviations, we will likewise square them. We will call this sum of squared cell effects by the name of **SS AB cells**:

$$\text{SS AB cells} = \sum_{j}\sum_{k} n\left([ab]_{jk}\right)^2$$

where n is the number of observations per cell, which we assume to be equal for our purposes.

4.2.3 Interaction Effects

Having defined the meaning of a cell effect, we are now ready to define what is meant by an **interaction effect**. These interaction effects are the reason why we computed the cell effects in the first place. The **sample interaction effect** for each cell jk is given by

$$
\begin{aligned}
(ab)_{jk} &= \text{interaction effect of cell } jk \\
&= \text{cell effect for cell } jk - \text{effect for row } j - \text{effect for column } k \\
&= [ab]_{jk} - a_j - b_k \\
&= \bar{y}_{jk} - \bar{y}.. - (\bar{y}_j - \bar{y}..) - (\bar{y}_{.k} - \bar{y}..) \\
&= \bar{y}_{jk} - \bar{y}_j. - \bar{y}_{.k} + \bar{y}..
\end{aligned}
$$

A few things to remark about sample interaction effects:

- A sample interaction effect $(ab)_{jk}$ exists for each cell in the design.
- The sample interaction effect is defined by the cell effect minus the row and column effects (i.e., $[ab]_{jk} - a_j - b_k$); this makes sense, since it is reasonable that we are interested in the effect of being in a particular cell over and above the corresponding row and column effects.
- The sample interaction effect can also be defined as taking the mean of each cell, $\bar{y}_{jk}$, and subtracting out row means and column means (i.e., $\bar{y}_{.j.}$ and $\bar{y}_{.k}$, respectively), then adding on the grand mean, $\bar{y}_{..}$.

As we did for sample effects, we will square the interaction effects so that they do not always sum to zero:

$$\text{SS A} \times \text{B interaction} = \sum_j \sum_k n(ab)_{jk}^2$$

4.2.4 Cell Effects Versus Interaction Effects

It is useful at this point to emphasize an important distinction and to clarify something that may at first be somewhat confusing. We have introduced the ideas of cell effects and interaction effects. It is important to recognize that these are not the same things, as evidenced by their different computations. To help clarify, let's compare the two concepts:

$$\text{Cell Effect } [ab]_{jk} = \left(\bar{y}_{jk} - \bar{y}..\right) \text{ versus Interaction Effect } (ab)_{jk} : [ab]_{jk} - a_j - b_k$$

Notice that the interaction effect $(ab)_{jk}$ uses the cell effect in its computation. In our operationalization of the two-way ANOVA, the cell effect is just the starting point to computing the interaction effect. The cell effect simply measures the deviation of a cell mean from the grand mean. It is the **interaction effect** that takes this deviation value and then subtracts further the row and column effects. **Be sure not to confuse cell effects and interaction effects as they are not one and the same**.

4.2.5 A Model for the Two-Way Fixed Effects ANOVA

Having now defined the sample interaction effect, which again, is the distinguishing feature between a one-way fixed effects model and a two-way fixed effects model, we can now state a general linear model for the two-way, one that includes an interaction term:

$$y_{ijk} = \bar{y}.. + a_j + b_k + (ab)_{jk} + e_{ijk}$$

where a_j is the sample effect of membership in row j, b_k is the sample effect of membership in column k, $(ab)_{jk}$ is the interaction effect associated with the cell jk, and e_{ijk} is the error associated with observation i in cell jk. In words, what the model says is that any given randomly selected observation from the two-way layout, represented by y_{ijk}, individual i in cell jk, can be theorized to be a function of the grand mean of all observations, $\bar{y}..$, an effect of being in a particular row j, a_j, an effect of being in a particular column k, b_k, the effect of being in a particular cell combination, jk, which is expressed via the interaction effect $(ab)_{jk}$, and an effect unique to individuals within each cell jk, e_{ijk}, for which we either did not account for in our design, or, we concede is due to random variation which we will call by the name of "chance." Either way, e_{ijk} represents our inability to model y_{ijk} perfectly in a truly functional manner. Just as was true for the one-way model, e_{ijk} is the effect that makes our model truly probabilistic.

4.3 COMPARING ONE-WAY ANOVA TO TWO-WAY ANOVA: CELL EFFECTS IN FACTORIAL ANOVA VERSUS SAMPLE EFFECTS IN ONE-WAY ANOVA

It is pedagogical at this point to compare, side by side, the one-way model of the previous chapter to the two-way model of the current chapter. Recall the overall purpose of writing out a model equation. It is an attempt to "explain," in as functional a way as possible, the makeup of a given observation. In the one-way model, we attempted to explain observations by theorizing a single grouping factor along with within-group variability. Our sample model was

$$y_{ij} = \bar{y}_. + a_j + e_{ij}$$

Notice that for such a model, it was not appropriate to append the additional subscript k to y_{ij} such as in y_{ijk}, because we did not have "cells" in the one-way ANOVA. Defining the idea of a "cell" did not make a whole lot of sense, since we were simply dealing with a single grouping variable. Subscripting y_{ij} to represent individual i in group j was enough. Indeed, if we were to "pretend" for a moment that we were dealing with cells, we could write the one-way model as

$$y_{ij} = \bar{y}. + \left(\bar{y}_j - \bar{y}.\right) + e_{ij}$$
$$y_{ij} = \bar{y}. + [ab]_j + e_{ij} \tag{4.1}$$

Nothing has changed in (4.1) except for equating "groups" with "cells." Why do this? Simply to note how the factorial model compares with that of the one-way model. Notice that the difference between the one-way model and the two-way model in terms of cell effects is that instead of hypothesizing y_{ijk} to be a function of $a_j = \bar{y}_j - \bar{y}.$, we are now hypothesizing y_{ijk} to be a function of $\bar{y}_{jk} - \bar{y}...$ In both cases, whether $a_j = \bar{y}_j - \bar{y}.$ for the one-way model or $[ab]_{jk} = \bar{y}_{jk} - \bar{y}..$ for the two-way model, **the total systematic variation in the data is represented by either of these, depending on whether there is one factor or two. Sample effects represent the systematic variation in a one-way model, and cell effects represent the systematic variation in a two-way model.** If you understand this concept, then generalizing these ANOVA models to higher-order models (e.g., three-way, four-way, and potentially higher) will not be intimidating, because you will realize at the outset that the systematic variation in the entire model is "housed" in the cell effects, regardless of the complexity of the model. To reiterate, we can say as a general principle of fixed effects analysis of variance models that

> In the fixed effects analysis of variance model, the systematic variation is housed in the cell effects. In the special case where we have only a single independent variable, the cell effects are equivalent to the sample (group) effects.

4.4 PARTITIONING THE SUMS OF SQUARES FOR FACTORIAL ANOVA: THE CASE OF TWO FACTORS

Just as we did for the one-way model, we will now work out the partition of the sums of squares for the two-way factorial model. Remember, the reason why we are partitioning "sums of squares" and not simply unsquared effects, is because if we attempted to partition unsquared effects (e.g.,

$a_j = \bar{y}_j - \bar{y}.$ or $[ab]_{jk} = \bar{y}_{jk} - \bar{y}..$), these effects would always sum to 0 (unless of course there is no variation in the data, then whether squared or not, they **will** sum to 0 regardless).

When we partitioned the sums of squares for the one-way model, we started out by hypothesizing what any single observation in our data, y_i, could be a function of. After a process of deliberate reasoning, we concluded that y_i was a function of **between** variation and **within** variation. Upon squaring deviations, we arrived at the identity:

$$\sum_{j=1}^{J} \sum_{i}^{n} (y_{ij} - \bar{y}.)^2 = \sum_{j}^{J} n_j (\bar{y}_j - \bar{y}.)^2 + \sum_{j=1}^{J} \sum_{i}^{n} (y_{ij} - \bar{y}_j)^2$$

which we called the partition of sums of squares for the one-way fixed effects analysis of variance model. We called it an "**identity**" simply because it holds true for virtually any given data set having a continuously measured dependent variable and a categorically-defined independent variable.

Likewise, in the two-way factorial model, we again want to consider how the partition of the sums of squares works out and can be derived. As we did for the one-way model, we follow a very logical process in determining this partition.

4.4.1   SS Total: A Measure of Total Variation

Just as we did in deriving the total sums of squares for the one-way model, instead of simply considering the makeup of y_{ijk}, we will consider the makeup of deviations of the form $y_{ijk} - \bar{y}..$, which when we incorporate into the model, we obtain, quite simply:

$$y_{ijk} = \bar{y}.. + [ab]_{jk} + e_{ijk}$$
$$y_{ijk} - \bar{y}.. = [ab]_{jk} + e_{ijk}$$

Notice that similar to how we did for the one-way model, in which $(y_{ij} - \bar{y}.) = a_j + e_{ij}$ was true, for the two-way model, we likewise claim that the makeup of any given observation is of two "things," systematic variation as represented by $[ab]_{jk}$ (in the one-way model the systematic variation was represented by a_j), and random variation as represented by e_{ijk} (in the one-way model the random variation was represented by e_{ij} — note the subscripts, we did not have **cells** in the one-way, so we did not need to append the subscript k). Instead of squaring $a_j + e_{ij}$ as is done in the one-way model, we will square $[ab]_{jk} + e_{ijk}$. When we take these squares and sum them, as given in Hays (1994, p. 481), we get:

$$\text{SS total} = \sum_i \left([ab]_{jk} + e_{ijk} \right)^2$$

$$= \sum_j \sum_k \sum_i \left([ab]_{jk}^2 + 2[ab]_{jk} e_{ijk} + e_{ijk}^2 \right)$$

$$= \sum_j \sum_k \sum_i \left([ab]_{jk}^2 + 2 \sum_j \sum_k [ab]_{jk} \sum_i e_{ijk} + \sum_j \sum_k \sum_i e_{ijk}^2 \right)$$

$$= \sum_j \sum_k n[ab]_{jk}^2 + \sum_j \sum_k \sum_i e_{ijk}^2$$

Notice that the term $2\sum_{j}\sum_{k}[ab]_{jk}\sum_{i}e_{ijk}$ dropped out of the above summation (3rd line of the equation). What happened to this term? Since the cell effects $[ab]_{jk}$ sum to zero and the errors within any given cell $\sum_{i}e_{ijk}$ sum to 0, the term $2\sum_{j}\sum_{k}[ab]_{jk}\sum_{i}e_{ijk}$ drops out of the derivation, since $2\sum_{j}\sum_{k}[ab]_{jk}\sum_{i}e_{ijk} = 0$. Hence, we are left simply with:

$$SS\ total = \sum_{j}\sum_{k}n[ab]_{jk}^{2} + \sum_{j}\sum_{k}\sum_{i}e_{ijk}^{2}$$

What we have just found is that the total variation in the two-factorial model is a function of the sum of **squared cell effects** and **random variation**. Once we have accounted for the systematic variation in $[ab]_{jk}$, then whatever is leftover must be random error, or otherwise denoted, the variation **within** the cells. Also, because the cell effects, $[ab]_{jk}$, contain all **systematic** variation, it makes sense that within these cell effects will be "hidden" a main effect for A, main effect for B, and interaction effect, A × B. That is, if you take the sums of squares for a cell effect which by itself contains all the systematic variation, it seems reasonable that we could break this down further into the **SS for factor A**, **SS for factor B**, and the **SS for the A x B** interaction, such that:

SS AB cells = SS factor A + SS factor B + SS A × B interaction

If we put these two partitions together, we end up with the following identities:

SS total = SS AB cells + SS within cells

SS total = SS factor A + SS factor B + SS A × B interaction + SS within

In considering now the main effects for the two-way factorial model, as in the one-way ANOVA, the sample main effect of any level j of the row factor A is given by $a_{j} = \bar{y}_{.j.} - \bar{y}_{...}$, where a_{j} as before represents the effect for a particular row, and $\bar{y}_{.j.} - \bar{y}_{...}$ represents the given row mean minus the grand mean of all observations. As in the one-way, the sum of the fixed sample main effects for factor A will be 0, $\sum_{j}a_{j} = 0$. Notice again here we are specifying the word "fixed." This is because for a fixed effects model, the sum of effects for a main effect sum to 0. However, in the following chapter, when we consider **random** and **mixed models**, we will see that this is not necessarily the case for certain factors. This will have important implications in how we construct F-ratios.

The sums of squares for factor A is thus $\sum_{j}Kn(a_{j})^{2}$, where K is the number of columns, and n is the number of observations per cell. For the column main effect (i.e., factor B), the sample main effect is $b_{k} = \bar{y}_{..k} - \bar{y}_{...}$, where $\bar{y}_{..k}$ is the sample mean corresponding to a particular column k. As with the sample effects for a_{j}, the sum of the column sample effects, b_{k}, will also be 0, $\sum_{k}b_{k} = 0$. The sums of squares for factor B is thus $\sum_{k}Jn(b_{k})^{2}$, where J is the number of rows.

4.4.2 Model Assumptions: Two-Way Factorial Model

The assumptions for a two-way fixed effects analysis of variance are similar to those of the one-way analysis of variance model, only now, because we have cells in our design, these are the "groups" about which we have to make assumptions when involving the interaction term:

- $E(\varepsilon_{ijk}) = 0$, that is, the expectation of the error term is equal to 0. Note the extra subscript on e_{ijk} to reflect not only the jth population but also the jk^{th} cell.
- ε_{ijk} are $NI(0, \sigma_e^2)$, that is, the errors are normally distributed and independent of one another. Just as we did for the one-way, we are using ε_{ijk} to denote the corresponding population parameter of the sample quantity e_{ijk}.
- $\sigma_{e_{ijk}}^2 < \infty$, that is, the variance of the errors is some finite number (which, as was true in the one-way model, implies that it is less than infinity).
- $\sigma_{jk=1}^2 = \sigma_{jk=2}^2 = \sigma_{jk=JK}^2$, that is, the variances across cell populations are equal (recall this is called the **homoscedasticity** assumption).
- Measurements on the dependent variable are observed values of a random variable that are distributed about true mean values that are **fixed** constants. This is the same assumption made for the one-way model in which we were interested in the fixed effects. This assumption will be relaxed when we contemplate **random effects** models in chapters to come.

We could also add the assumption, as we did for the one-way model, that the model is correctly **specified**, in that there are reasonably no other sources acting on the dependent variable to an appreciable extent. If there were, and we did not include them in our model, we would be guilty of a **specification error** or of more generally misspecifying our model.

4.4.3 Expected Mean Squares for Factorial Design

In deriving F-ratio tests for the various effects in the two-way ANOVA, just as we did for the one-way ANOVA, we need to derive the expectations for the various sums of squares, and then divide these by the appropriate degrees of freedom to produce a mean square for the given factor or interaction. Hence the phrase, "expected mean squares." We adapt the following derivations from Hays (1994), Kempthorne (1975), and Searle, Casella, and McCulloch (1992). We begin with the expected mean squares for within cells (Hays, 1994, p. 485):

$$
\begin{aligned}
E(\text{SS within cells}) &= E\left[\sum_k \sum_j \sum_i \left(y_{ijk} - \bar{y}_{jk}\right)^2\right] \\
&= \sum_k \sum_j E\left[\sum_i \left(y_{ijk} - \bar{y}_{jk}\right)^2\right] \\
&= \sum_k \sum_j (n-1)\sigma_e^2 \\
&= JK(n-1)\sigma_e^2
\end{aligned}
\tag{4.2}
$$

Why does $\sum_k \sum_j E\left[\sum_i \left(y_{ijk} - \bar{y}_{jk}\right)^2\right]$ equal $\sum_k \sum_j (n-1)\sigma_e^2$? To understand this, recall in the one-way layout:

$$
E(\text{SS within}) = E\left[\sum_j \sum_i \left(y_{ij} - \bar{y}_j\right)^2\right]
$$

However, for any given sample group j, we know that we have to divide SS by $n - 1$ in order to get an **unbiased** estimate of the error variance. That is, we know that $E\left[\sum_j \sum_i (y_{ij} - \bar{y}_j)^2\right]$ does not "converge" to σ_e^2, but that $E\left[\sum_i \frac{(y_{ij} - \bar{y}_j)^2}{n_j - 1}\right]$ does. So, we can rearrange this slightly to get

$$E\sum_i (y_{ij} - \bar{y}_j)^2 = (n_j - 1)\sigma_e^2$$

Finally, how did we go from $\sum_k \sum_j (n-1)\sigma_e^2 = JK(n-1)\sigma_e^2$ in the final term of (4.2)? By the rules of summation, $\sum_j^J y = Jy$, and so $\sum_k^K \sum_j^J y = JKy$, in which in our case $(n-1)\sigma_e^2$ acts as "y."

Now that we have the expectation for SS error, that of $JK(n-1)\sigma_e^2$ of (4.2), let us consider what we have to divide this sum of squares by to get MS error. That is, we need to determine the **degrees of freedom for error**. Since there are $J \times K$ cells, we will lose 1 degree of freedom **per cell**, which gives us degrees of freedom $= JK(n-1)$. So, MS error is equal to:

$$\begin{aligned} \text{MS error} &= \frac{\text{SS error}}{JK(n-1)} \\ &= \frac{JK(n-1)\sigma_e^2}{JK(n-1)} \\ &= \sigma_e^2 \end{aligned}$$

That is, as was the case in the one-way ANOVA, MS error is simply equal to the error variance alone in a two-way fixed effects ANOVA.

What about the mean square for factor A? When determining an appropriate mean square for any term, recall that it is essential to consider what goes into the numerator. For the error term, as we just saw, all that goes into the calculation of error is simply σ_e^2. When considering the effect for factor A, we need to recall that in any given row J, both the column effects b_k and the interaction effects sum to 0. That is, $\sum_k b_k = 0$ and $\sum_k (ab)_{jk} = 0$. Notice that we are summing over k columns to get the row effect. Why is this important? It is important because it tells us what we can leave out of the mean square for factor A. Because we know $\sum_k b_k = 0$ and $\sum_k (ab)_{jk} = 0$, we become aware that these terms will not be part of the mean square for factor A. If you prefer, we might say they **will** still be part of the term, but since they sum to 0, why include them in the mean square for factor A at all? Both ways of thinking about it gets us to the same place in that we do not have to incorporate them when computing our mean squares.

Recall that the sums of squares for factor A are given by

$$Kn\sum_j a_j^2 = Kn\sum_j (\bar{y}_{j.} - \bar{y}_{..})^2$$

Given this, and the fact that $\sum_k b_k = 0$ and $\sum_k (ab)_{jk} = 0$, the expectation for MS factor A in which factor A is fixed, is:

$$E(\text{MS A}) = \sigma_e^2 + \frac{Kn\sum_j \alpha_j^2}{J-1}$$

In words, the expectation is equal to error variance, σ_e^2, plus a term containing variability due to factor A, $\frac{Kn\sum_j \alpha_j^2}{J-1}$. Given the expected mean square, we would like to produce an F-ratio to test the main effect for factor A of $H_0 : \alpha_j = 0$ versus $H_1 : \alpha_j \neq 0$ for at least some population as specified by the levels of factor A. If there is absolutely no effect, we will have:

$$E(\text{MS A}) = \sigma_e^2 + \frac{Kn\sum_j 0}{J-1}$$

and hence

$$E(\text{MS A}) = \sigma_e^2$$

And so it is easy to see that the following F-ratio will be a suitable one for testing the effect due to factor A:

$$F = \frac{\text{MS A}}{\text{MS error}}$$

on $J-1$ and $JK(n-1) = N - JK$ degrees of freedom. That is, in the two-way fixed effects analysis of variance, MS error is the correct error term for testing the effect of factor A.

A similar argument applies to the factor B mean square. Since $\sum_j a_j = 0$ and $\sum_j (ab)_{jk} = 0$, we will only expect variability due to that in columns when considering factor B, since the effects for A and interaction effects will both sum to 0 in the fixed effects model we are currently considering (they will not necessarily in random and mixed models of the following chapter). Therefore, the relevant expectation is:

$$E(\text{MS B}) = \sigma_e^2 + \frac{Jn\sum_k \beta_k^2}{K-1}$$

where similar to the case for factor A, the term $Jn\sum_k \beta_k^2$ simply comes from the derivation of the sums of squares for factor B, that of:

$$\text{SS B} = Jn\sum_k b_k^2 = Jn\sum_j (\bar{y}_{.k} - \bar{y}_{..})^2$$

Under the null hypothesis, it will be the case that $\beta_k^2 = 0$, and so we are left with σ_e^2. Hence, the appropriate F ratio is:

$$F = \frac{\text{MS B}}{\text{MS error}}$$

TABLE 4.5   ANOVA Summary Table for Two-Way Factorial Design

Source	Sums of Squares	df	Mean Squares	F
A (rows)	SS A	$J-1$	SS A/$J-1$	MS A/MS error
B (columns)	SS B	$K-1$	SS B/$K-1$	MS B/MS error
A × B	SS AB cells−SSA−SSB	$(J-1)(K-1)$	SS A × B/$(J-1)(K-1)$	MS A × B/MS error
Error	SS total−(SSA + SSB + SS A × B)	$N-JK$	SS error/$(N-JK)$	
Total	SS total	$N-1$		

on $K-1$ and $JK(n-1) = N - JK$ degrees of freedom. That is, in the two-way fixed effects analysis of variance, MS error is the correct error term for testing the effect of factor B.

Finally, what of the expected mean squares for interaction? In generating the mean square, we follow a similar argument as when producing the terms for factor A and factor B. That is, we ask ourselves, **what went into the interaction term?** Well, we know that for the sample cell effect, $[ab]_{jk}$, we saw that it was composed of variability due to factor A, factor B, and the A × B interaction. What goes into the interaction term $(ab)_{jk}$ is simply variability due to an interaction between factor A and factor B. Thus, for the interaction, we have:

$$E(\text{MS interaction}) = \sigma_e^2 + \frac{n\sum_j\sum_k (\alpha\beta)_{jk}^2}{(J-1)(K-1)}$$

If the interaction effects end up being 0, that is, if $n\sum_j\sum_k (\alpha\beta)_{jk}^2 = 0$, then we will wind up with simply σ_e^2.

Hence, the appropriate F-ratio is MS interaction/MS error on $(J-1)(K-1)$ and $JK(n-1) = N - JK$ degrees of freedom. The summary table for the two-way factorial design is given in Table 4.5.

4.4.4   Recap of Expected Mean Squares

Recall that the practical purpose behind deriving expected mean squares, whether in the one-way or higher-order ANOVA models, is to be able to generate meaningful F-ratios and test null hypotheses of interest to us. In our discussion of mean squares, we have justified the use of F-ratios for testing the main effect of A, main effect of B, and the interaction of A × B. Notice that in each case, MS error is the appropriate denominator in the **fixed effects model of analysis of variance**. When we consider **random** and **mixed effects models** in chapters to follow, we will see that, and more importantly understand why, MS error is not always the appropriate denominator for testing effects.

4.5   INTERPRETING MAIN EFFECTS IN THE PRESENCE OF INTERACTIONS

Typically, if one has found evidence for an interaction in an ANOVA, one can still interpret main effects, so long as one realizes that the main effects no longer "tell the whole story." As noted by Kempthorne (1975, p. 483), however, "the testing of main effects in the presence of interaction, without additional input, is an exercise in fatuity."

As an illustration, suppose the researcher investigating the effect of melatonin did find an effect, but that the drug was only truly effective in conditions of very low noise. If ambient noise is elevated, melatonin no longer reduces sleep onset time. In other words, an interaction is present. In light of this interaction, if we interpreted **by itself the effect of dosage** without also including noise level in our "story," then we would be potentially misleading the reader who may mistakenly conclude taking melatonin could help him get to sleep faster even if in a college dormitory (which is relatively noisy, even at night). **The take-home message is clear—if you have evidence for an interaction in your data, it is the interaction that should be interpreted first**. Interpreting main effects second is fine, so long as you caution your reader that they do not tell the **whole story**. The more complete story is housed in the interaction term.

4.6   EFFECT SIZE MEASURES

Recall that for the one-way fixed effects analysis of variance model, we computed

$$\eta^2 = \frac{\sum_{j}^{J} n_j \left(\bar{y}_j - \bar{y}.\right)^2}{\sum_{j=1}^{J} \sum_{i=1}^{n} \left(y_{ij} - \bar{y}.\right)^2}$$

as a measure of effect size in the sample. It revealed the proportion of variance in the dependent variable that was accounted for by knowledge of the independent variable.

In the factorial design, we can likewise compute η^2, but this time for each factor and interaction. That is, we will have, for respective main effects and interaction,

$$\eta_A^2 = \frac{SS\ A}{SS\ total} \quad \eta_B^2 = \frac{SS\ B}{SS\ total} \quad \eta_{A \cdot B}^2 = \frac{SS\ A \times B}{SS\ total}$$

Each of these, as was true for the one-way model, will give us an estimate of the variance explained in the dependent variable given the particular source of variation. As was true for the fixed effects model, these measures of η^2 are all **descriptive** measures of what is going on in the particular sample. Measures of η^2 are **biased upward**, and hence the true strength of association in the corresponding population parameters is usually less than what values of η^2 suggest.

In factorial designs, since we are modeling more than a single effect, one can also compute $\eta_{Partial}^2$, defined as:

$$\eta_{Partial}^2 = \frac{SS\ effect}{SS\ effect + SS\ error}$$

A look at $\eta_{Partial}^2$ reveals that the denominator contains not the **total variation** as in η^2, but rather SS for the effect we are considering in addition to what is "left over" from the ANOVA in terms of error. For the one-way ANOVA, $\eta^2 = \eta_{Partial}^2$. Some authors (e.g., see Tabachnick and Fidell, 2007) recommend the reporting of $\eta_{Partial}^2$ for the reason that the size of η^2 will depend on the **complexity** of the model. That is, for a given effect, η^2 will typically be smaller in a model containing many effects than in a simpler model as a result of the total variation being larger in the former case. In the case of $\eta_{Partial}^2$, we are not allowing all of these effects to be a part of our denominator, and so $\eta_{Partial}^2$, all else equal, will be greater than η^2.

Analogous to the one-way model, ω^2 can also be computed in factorial models such that it provides a better approximation of the strength of association in the population and yields a more accurate estimate compared to η^2. Estimates of ω^2 can be obtained for both main effects and interactions, though ω^2 is less common in most software than is η^2 and $\eta^2_{Partial}$. For derivation and computation details, see Vaughan and Corballis (1969).

4.7   THREE-WAY, FOUR-WAY, AND HIGHER MODELS

The cases of three or more independent variables are a natural extension of the case for two. The only difference in terms of the partition is that in higher-order models, in addition to subtracting out SS A and SS B, etc., (depending on how many factors we have) from the cells term, we also need to subtract out all **two-way** interaction terms as well, since they are also naturally "part" of the cells term. Hence, for a three-way model, we would have:

$$\textbf{SS A} \times \textbf{B} \times \textbf{C} = (\textbf{SS ABC cells}) - (\textbf{SS A}) - (\textbf{SS B}) - (\textbf{SS C}) - (\textbf{SS A} \times \textbf{B}) - (\textbf{SS A} \times \textbf{C}) - (\textbf{SS B} \times \textbf{C})$$

This is nothing new. The principle is the same as for the two-way. Because cell terms contain all systematic effects in an experiment, we need to subtract all effects that may have "gone into" this term. This includes main effects and two-way interactions, which is why we include them in the subtraction.

4.8   SIMPLE MAIN EFFECTS

Given the presence of an interaction, the examination of **simple main effects** allows us to study the effect associated with some level of a given factor when the level of another factor is prespecified. We will usually want to perform simple effects analysis for any statistically significant interaction, and the precise number of simple effects we perform should align at least somewhat with our theoretical predictions as to not unduly inflate type I error rates (or at minimum, we could use a Bonferroni-type correction on α_{FW} to attempt to keep the family-wise error rate at a nominal level).

To understand simple main effects, we begin first by reconsidering factor A with J levels. Recall that the main effect associated with this factor in a two-way factorial model is $a_j = \bar{y}_{j.} - \bar{y}...$ That is, the effect a_j is defined as the difference between the mean for that particular row, $\bar{y}_{j.}$ and the grand mean of $\bar{y}..$ (Recall that the periods following the letters are simply used as "placeholders" for columns k when considering $\bar{y}_{j.}$ and for rows j and columns k when considering the grand mean, $\bar{y}..$). In the presence of a two-way interaction, if we chose only one level k of factor B, and examined only the effects of factor A **within a given level of factor B**, each of these effects would be called **simple main effects**. They are analogously derived for column effects. They are effects (usually main effects, but as we will see, they can also be **interaction effects** in the case of a three-way or higher ANOVA) of a factor at one level of another factor. They allow us to "tease apart" an interaction to learn more about what generated the interaction in the first place.

As a visualization to better understand the concept of a simple main effect, consider once more Figure 4.1 given at the outset of this chapter, only now, with a simple main effect indicated at the level of the first teacher (Figure 4.3). It is the simple main effect of mean achievement differences on textbook at the first teacher.

We can define the simple main effect in Figure 4.3 as:

$$\bar{y}_{jk} - \bar{y}_{.k}$$

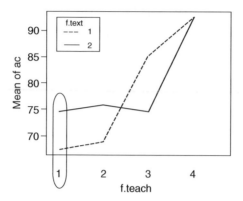

FIGURE 4.3   A simple main effect: Mean difference of textbook at level 1 of teacher.

where $\bar{y}_{jk}$ is the mean for a given textbook cell and $\bar{y}_{.k}$ is the mean for teacher 1, collapsing across textbooks. We can define a number of other simple main effects:

- textbook 1 versus textbook 2 @ teacher 2
- textbook 1 versus textbook 2 @ teacher 3
- textbook 1 versus textbook 2 @ teacher 4

We could also define simple main effects the other way (though not as easily visualized in Figure 4.3):

- teacher 1 versus teacher 2 versus teacher 3 versus teacher 4 @ textbook 1
- teacher 1 versus teacher 2 versus teacher 3 versus teacher 4 @ textbook 2

We carry out analyses of simple main effects in software toward the conclusion of the chapter, where much of this will likely make more sense in the context of a full analysis.

4.9   NESTED DESIGNS

Up to this point in the chapter, our idea of an interaction for the achievement data has implied that all teachers were **crossed** with all textbooks. The layout of 2×4 (i.e., 2 textbooks by 4 teachers) of both Table 4.1 and Figure 4.1 denotes the fact that all combinations of textbook and teacher are represented and analyzed in the ANOVA.

Nesting in experimental design occurs when **particular levels of one or more factors appear only at particular levels of a second factor**. For example, using the example of teachers and textbooks, if only teachers 1 and 2 used the first textbook but teachers 3 and 4 used the second textbook, then we would say the **factor teacher is nested within the factor textbook** (Table 4.6). These types of designs are sometimes referred to as **hierarchical designs** (e.g., see Kirk, 1995, p. 476). Though we do not consider nested designs in any detail in this book, it is important to understand how such designs (should you be confronted with one) differ from the classical factorial design in which all levels are crossed. For further details on nested designs, see Casella (2008), Kirk (1995), Mead (1988), and Montgomery (2005).

TABLE 4.6   Nested Design: Teacher is Nested Within Textbook

Textbook 1		Textbook 2	
Teacher 1	Teacher 2	Teacher 3	Teacher 4
70	69	85	95
67	68	86	94
65	70	85	89
75	76	76	94
76	77	75	93
73	75	73	91
Mean = 71.0	Mean = 72.5	Mean = 80.0	Mean = 92.7

4.9.1   Varieties of Nesting: Nesting of Levels Versus Subjects

It is well worth making another point about nesting. Recall that in our brief discussion of Chapter 3, nesting was defined as a **similarity** of objects or individuals within a given group, whether it be those women receiving mammographies or those exhibiting smoking behavior, or those children within the same classroom, classrooms within the same school, etc. It should be noted at this point that the nesting featured in Table 4.6 in relation to factor levels, other than for a trivial similarity, is not of the same kind of nesting as that of subjects within groups. The word "nesting" is used interchangeably in both circumstances, and much confusion can result from equating both designs.

To illustrate the important distinction, let us conceptualize a design in which the same subject is measured successively over time. These are so-called **repeated-measures designs**, to be discussed at some length in Chapter 6. Consider the data in Table 4.7 in which rats 1 through 6 were each measured a total of three times, once for each trial of a learning task. For this hypothetical data, rats were tested to measure the elapsed time it took to press a lever in an operant conditioning chamber. The response variable is the time (measured in minutes) it took for them to learn the lever-press response. We would expect that if learning is taking place, the time it takes to press the lever should **decrease** across trials.

In such a layout, it is often said that "trials are nested within subject" (in this case, the rats). That is, **measurements from trial 1 to 3 are more likely to be similar within a given rat than between rats**. If a rat performs poorly at trial 1, even if it improves by trials 2 and 3, we could probably still expect a relatively lowered performance overall. On the other hand, if a rat performs very well at trial 1, this information probably will tell us something about its performance at trials 2 and 3. That is, because observations occur **within rat**, we expect trials to be **correlated**.

TABLE 4.7   Learning as a Function of Trial (Hypothetical Data)

Rat	Trial			Rat Means
	1	2	3	
1	10.0	8.2	5.3	7.83
2	12.1	11.2	9.1	10.80
3	9.2	8.1	4.6	7.30
4	11.6	10.5	8.1	10.07
5	8.3	7.6	5.5	7.13
6	10.5	9.5	8.1	9.37
Trial means	$M = 10.28$	$M = 9.18$	$M = 6.78$	

This is one crucial difference when we speak of nesting. On the one hand, we have nested designs in which factor levels of one factor are nested within factor levels of a second factor. This is the nesting featured in Table 4.6. On the other hand, we have **nested measurements**, in which factor levels usually remain the same from subject to subject (or "block to block" as we will see in Chapter 6), but that several measurements are made on each subject. These two types of nesting are not quite the same. The only way the two types of nesting do converge is if we consider **subject to be simply another factor**. In hierarchical and multilevel models, for instance, we say that students are nested within class-room. But what are students? In the sense of nesting, students are but another factor of which we sample many different **levels** (i.e., many different subjects). Likewise, different classrooms have different students, and if there is more similarity among students within the same classroom than between, then we would like this similarity to be taken into account in the statistical analysis. Nesting of this sort is a characteristic of **randomized block designs** and **multilevel sampling**. We discuss this topic further when we survey random effects and mixed models in the next two chapters. For now, it is enough to understand that when the word "nesting" is used, it is important to garner more details about the design to learn exactly how it applies. Half of the battle in understanding statistical concepts is often in appreciating just how the word is being used in the given context.

4.10   ACHIEVEMENT AS A FUNCTION OF TEACHER AND TEXTBOOK: EXAMPLE OF FACTORIAL ANOVA IN R

Having surveyed the landscape of factorial analysis of variance, we now provide an example to help motivate the principles aforementioned. We once more use the hypothetical achievement data for our illustration. As discussed, instead of only randomly assigning students to one of four teachers, we also randomly assign students to one of two textbooks. We are only interested in generalizing our findings to these four teachers and these two textbooks, making the fixed effects model appropriate.

Our data of Table 4.1 appears below in R:

```
> achiev.2 <- read.table("achievement2.txt", header = T)
> achiev.2
> some(achiev.2)

   ac teach text
1  70     1    1
2  67     1    1
3  65     1    1
```

First, as usual, we identify teacher and text as factors:

```
> attach(achiev.2)
> f.teach <- factor(teach)
> f.text <- factor(text)
```

We proceed with the 2×2 factorial ANOVA:

```
> fit.factorial <- aov(ac ~ f.teach + f.text + f.teach:f.text,
data = achiev.2)
```

```
> summary(fit.factorial)

              Df Sum Sq Mean Sq F value     Pr(>F)
f.teach        3 1764.1   588.0 180.936 1.49e-12 ***
f.text         1    5.0     5.0   1.551    0.231
f.teach:f.text 3  319.8   106.6  32.799 4.57e-07 ***
Residuals     16   52.0     3.3
—
Signif. codes:  0 '***' 0.001 '**' 0.01 '*' 0.05 '.' 0.1 ' ' 1
```

We note that the main effect for teacher is statistically significant, while the main effect for text is not. The interaction between teacher and text is statistically significant ($p = 4.57\text{e-}07$). The identical model can be tested in SPSS (output not shown) using:

```
UNIANOVA ac BY teach text
  /METHOD=SSTYPE(3)
  /INTERCEPT=INCLUDE
  /CRITERIA=ALPHA(0.05)
  /DESIGN= teach text teach*text.
```

To look at means more closely, we may use the package phia (Rosario-Martinez, 2013), and request cell means for the model:

```
> library(phia)
> (fit.means <- interactionMeans(fit.factorial))

  f.teach f.text adjusted mean
1       1      1      67.33333
2       2      1      69.00000
3       3      1      85.33333
4       4      1      92.66667
5       1      2      74.66667
6       2      2      76.00000
7       3      2      74.66667
8       4      2      92.66667
```

We reproduce the cell means in Table 4.8.

Remember, when trying to discern whether an interaction exists, we ask ourselves the following question—**At each level of one independent variable, is the same "story" being told at each level**

TABLE 4.8   Achievement Cell Means Teacher∗Textbook

| | Teacher | | | | |
Textbook	1	2	3	4	Row Means
1	$\bar{y}_{jk} = \bar{y}_{11} = 67.33$	$\bar{y}_{jk} = \bar{y}_{12} = 69.00$	$\bar{y}_{jk} = \bar{y}_{13} = 85.33$	$\bar{y}_{jk} = \bar{y}_{14} = 92.67$	$\bar{y}_{j.} = \bar{y}_{1.} = 78.58$
2	$\bar{y}_{jk} = \bar{y}_{21} = 74.67$	$\bar{y}_{jk} = \bar{y}_{22} = 76.00$	$\bar{y}_{jk} = \bar{y}_{23} = 74.67$	$\bar{y}_{jk} = \bar{y}_{24} = 92.67$	$\bar{y}_{j.} = \bar{y}_{2.} = 79.50$
Column Means	$\bar{y}_{.k} = \bar{y}_{.1} = 71.00$	$\bar{y}_{.k} = \bar{y}_{.2} = 72.5$	$\bar{y}_{.k} = \bar{y}_{.3} = 80.0$	$\bar{y}_{.k} = \bar{y}_{.4} = 92.67$	$\bar{y}_{..} = 79.04$

of the other independent variable? What such a question begs us to do is look at means at the level of one factor **conditioned** on levels of the other factor.

For example, examine the mean teacher differences at textbook 1 in Table 4.8. We note the means to be 67.33, 69.00, 85.33, and 92.67 for the first, second, third, and fourth teachers, respectively. Notice how these means represent a continuous increase from teachers one through four. This is what we mean by the "story" being told at the level of textbook = 1. The actual "story" is not the actual **values** of the means, but rather the **differences** between means. That is, the story is the **magnitude and direction on which these cell means differ**. We can see the story for textbook = 2 is similar, yet not the same as for textbook = 1 (for example, from teacher 2 to 3 denotes a mean **decrease**, not an **increase**).

Trying to discern all this in a table of cell means is quite difficult, and we are better off graphing these cell means, which we can do via an **interaction plot** in R as we did in Figure 4.1 to open this chapter. We reproduce the plot here:

```
> interaction.plot(f.teach, f.text, ac)
```

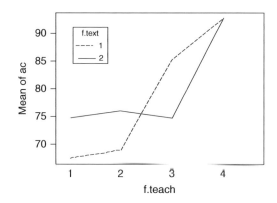

Be sure you are able to match up the interaction plot with the cell means in Table 4.8. The plot provides a much better picture of what is really going on in the achievement data than a table of numbers only could ever reveal. Is the same mean difference story of textbook differences on achievement being told at each level of teacher? The plot helps to answer such questions. It would appear from the plot that for the first and second teachers, textbook 2 is more effective than textbook 1. But for teacher 3, textbook 1 is more effective than textbook 2. That is, there is a **reversal** of means from teacher 2 to teacher 3. For teacher 4, it appears that achievement is equal regardless of which textbook is used.

Of course, visualizing mean differences in a plot is one thing and provides strong evidence for an interaction in the **sample data**. However, simply because we are seeing that mean differences of teacher across textbook are not equal is not reason in itself to reject the null hypothesis of no interaction and infer the alternative hypothesis that there is one in the **population** from which these data were drawn. We need to conduct the formal test of significance to know if rejecting the null of no interaction is warranted.

Always remember that differences and effects in sample data may not generalize to actual differences and effects in the populations from which the sample data were drawn. This is the precise point of the inferential significance test and associated *p*-value, to make a decision as to whether observed differences or effects potentially seen in the sample can be inferred to the population.

Recall also that **as sample size $n \to \infty$, that is, as it grows without bound, even for miniscule sample effects or sample interaction effects, statistical significance is assured**. This may make it sound like it is sample size that is dictating whether we "find something or not." And this is precisely true if we are foolish enough to consider the p-value as the "be all and end all" of things. As we pointed out in Chapter 2, when interpreting statistical and scientific evidence, the p-value should be used as only **one** indicator of the potential presence of a scientific finding. The other indicator is **effect size**.

To reiterate and emphasize, **distinguishing between statistical significance and effect size is not only a good idea, it is essential if you are to evaluate scientific evidence in an intelligent manner**. If you are of the mind that p-values, and p-values alone, should be used in the evaluation of scientific evidence, then you should not be interpreting scientific evidence in the first place. Being able to distinguish between what a p-value tells you and what an effect size tells you is **that** mandatory. It is not merely a preferred or "fashionable" custom, it is absolutely necessary for quality interpretation of scientific findings.

Another way to visualize the interaction is through R's `plot.design`, where we notice that means across teacher are quite disperse and means across textbook are quite close to one another:

```
> plot.design(ac ~ f.teach + f.text + f.teach:f.text, data = achiev.2)
```

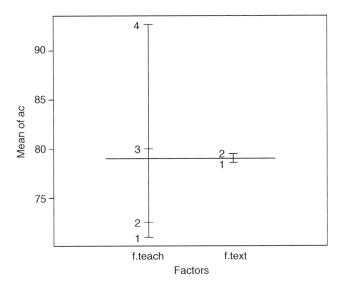

The plot allows us to see the main effects for teacher and textbook. Recall, however, that in the presence of an interaction effect, it is the interaction effect that should be emphasized in interpretation, not the main effects, as these latter effects do not tell us the "whole story."

4.10.1 Comparing Models Through AIC

A model is considered nested within another model if it estimates a subset of the parameters estimated in the larger model. **Akaike's information criteria**, introduced in Chapter 2, is a useful measure when comparing the fit of nested models. It can also be used for comparing the fit of non-nested models as well, however, it is commonly used for comparing nested models. Because the main-effects-only model can be considered a model nested within the higher-order interaction model, computing AIC for each model can also give us a measure of improvement in terms of how much "better" the inter-action model is relative to the main-effects-only model. We first compute AIC for the main-effects model:

```
> fit.main <- aov(ac ~ f.teach + f.text, data = achiev.2)
> AIC (fit.main)
[1] 145.8758
```

We next compute AIC for the model containing the interaction term:

```
> fit.int <- aov(ac ~ f.teach + f.text + f.teach:f.text, data = achiev.2)
> AIC (fit.int)
[1] 104.6656
```

Recall that a **decrease** in AIC values denotes an improvement in model fit. The AIC value for the main-effects-only model is 145.88, while AIC for the model containing the interaction term is 104.67, which helps statistically substantiate our obtained evidence for an interaction effect.

Collapsing across cells, the sample means for teacher are computed:

```
> library(phia)
> interactionMeans(fit.factorial, factors = "f.teach")
  f.teach adjusted mean std. error
1       1      71.00000  0.7359801
2       2      72.50000  0.7359801
3       3      80.00000  0.7359801
4       4      92.66667  0.7359801
```

As before, these means for teacher are found by summing across the means for textbook. Are there mean differences for teacher? Our **sample** definitely shows differences, and based on our obtained p-value for teacher, we also have statistical evidence to infer this conclusion to the population from which these data were drawn. Suppose we decided to **not** control for **per comparison error rate** and decided to simply run independent samples t-tests. In R, we can use the `pairwise.t.test` function and for `p.adj`, specify "none" to indicate that we are not interested in adjusting our per comparison error rate:

```
> pairwise.t.test(ac, f.teach, p.adj = "none")

        Pairwise comparisons using t tests with pooled SD

data:  ac and f.teach

    1       2       3
2 0.5562  -       -
3 0.0018  0.0072  -
4 3.4e-08 1.1e-07 6.1e-05

P value adjustment method: none
```

What is reported in the table are the *p*-values associated with the pairwise differences. We note the *p*-value for comparing teacher 1 to teacher 2 is equal to 0.5562, which is not statistically significant at the 0.05 level. The *p*-value for comparing teacher 1 to teacher 4 is equal to 3.4e-08, and hence, is statistically significant. The *p*-value for comparing teacher 2 to teacher 3 is equal to 0.0072 and is also statistically significant. The remaining *p*-values for comparing teacher 2 to 4 and 3 to 4 are likewise very small and hence the differences are statistically significant.

We now perform the same comparisons, but this time using a **Bonferroni correction** to adjust the per comparison error rate. We do this by requesting p.adj = "bonf":

```
> pairwise.t.test(ac, f.teach, p.adj = "bonf")

        Pairwise comparisons using t tests with pooled SD

data:   ac and f.teach

  1       2        3
2 1.00000 -        -
3 0.01095 0.04316  -
4 2.1e-07 6.4e-07  0.00036

P value adjustment method: bonferroni
```

Though we notice all pairwise differences that were statistically significant (at 0.05) without using a correction are still significant after using a Bonferroni correction, we note the increase in *p*-values for each comparison. Comparison 2 versus 3 now yields a *p*-value of 0.04316, which for instance, would no longer be statistically significant if evaluated at the 0.01 level of significance. This is because the Bonferroni, through its adjustment of the significance level for each comparison, is making it a bit "harder" to reject null hypotheses in an effort to keep the overall type I error rate across comparisons at a nominal level.

We can also obtain means for the textbook factor:

```
> library(phia)
> interactionMeans(fit.factorial, factors = "f.text")
  f.text adjusted mean std. error
1      1      78.58333  0.5204165
2      2      79.50000  0.5204165
```

Since there are only two levels to the textbook factor, conducting a post-hoc test on it makes no sense. There is no type I error to adjust since there is only a single comparison. The problem of "multiple comparisons" does not exist.

4.10.2 Visualizing Main Effects and Interaction Effects Simultaneously

A very nice utility in the phia package is its ability to generate a graph for which one can visualize both main effects and potential interaction effects simultaneously. We obtain this with plot (fit.means):

```
> library(phia)
> plot(fit.means)
```

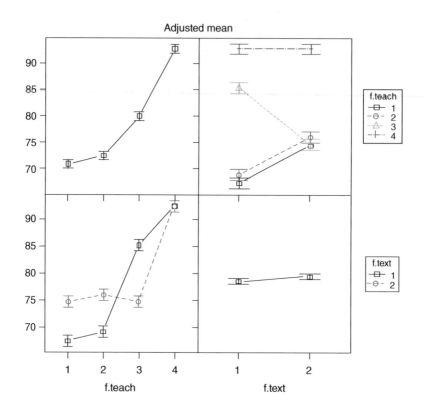

In the quadrants running from top left to lower right are shown the main effects for teacher and textbook, respectively. In the quadrants running from top right to lower left are shown the sample interaction effects for teacher*textbook. Both of the interaction graphs are yielding the same essential information but in the one case (lower left), teacher is plotted on the *x*-axis while in the other (upper right), textbook is plotted on the *x*-axis. In both graphs, an interaction effect is evident.

4.10.3 Simple Main Effects for Achievement Data: Breaking Down Interaction Effects

Recall that the purpose of conducting simple main effects is to break an interaction effect down into components to better understand it, to learn what is promoting there to be an interaction in the first place. They are essentially reductions of the sample space in order to zero in on analyses that tease apart the interaction effect.

Ideally, a researcher should usually only test the simple main effects of **theoretical** or **substantive** interest. Otherwise, the exercise becomes not one of scientific hypothesis-testing but rather one of data-mining and exploration (and potentially, "fishing"). Data mining and exploration are not "bad" things by any means, only be aware that if you do "exploit" your data, you increase the risk of committing inferences that may turn out to be wrong if replication (or cross-validation) is not performed. If you do decide to test numerous simple main effects, then using a correction on the type I error rate (e.g., Bonferroni) is advised. At minimum, you owe it to your audience to tell them which findings resulted from your predictions, and which were stumbled upon in exploratory searches. From a scientific perspective, especially when working with messy high-variability data, the two are not one and the same.

We evaluate mean differences of textbook across teacher:

```
> library(phia)
> testInteractions(fit.factorial, fixed = "f.teach", across = "f.text")

F Test:
P-value adjustment method: holm
            Value Df Sum of Sq      F     Pr(>F)
1         -7.3333  1    80.667 24.820 0.0004071 ***
2         -7.0000  1    73.500 22.615 0.0004299 ***
3         10.6667  1   170.667 52.513 7.809e-06 ***
4          0.0000  1     0.000  0.000 1.0000000
Residuals          16    52.000
—
Signif. codes:  0 '***' 0.001 '**' 0.01 '*' 0.05 '.' 0.1 ' ' 1
```

R generates the **Holm test**, which is a multistage test, similar in spirit to the Bonferroni, but in splitting up α per comparisons c, adjusts c depending on the number of null hypotheses remaining to be tested (see Howell, 2002, pp. 386–387 for details). The Holm test is thus generally more powerful than the Bonferroni. The value of the first contrast is the mean difference between textbook 1 versus textbook 2 at teacher 1 (i.e., $67.33 - 74.67 = -7.33$), and is statistically significant. The value of the second contrast is the mean difference between textbook 1 versus textbook 2 at teacher 2 (i.e., $69.00 - 76.00 = -7.00$), also statistically significant. The third contrast is the mean difference between textbook 1 versus textbook 2 at teacher 3 (i.e., $85.33 - 74.67 = 10.67$), and the fourth contrast is the mean difference between textbook 1 versus textbook 2 at teacher 4 (i.e., $92.67 - 92.67 = 0.00$). The last of these, of course, is not statistically significant.

Simple main effects of text differences within each teacher can also be tested in SPSS using:

```
UNIANOVA
ac BY teach text
/METHOD = SSTYPE(3)
/INTERCEPT = INCLUDE
/EMMEANS = TABLES(teach*text) COMPARE (text) ADJ(BONFERRONI)
/CRITERIA = ALPHA(.05)
/DESIGN = teach text teach*text.
```

One could also test for corresponding teacher differences within each textbook by adjusting the above code appropriately (i.e., COMPARE (teach)).

4.11   INTERACTION CONTRASTS

Whereas simple main effects analyze mean differences on one factor at a single level of another factor, **interaction contrasts** constitute a comparison, not of means, but of **mean differences** (i.e., a contrast of contrasts). That is, they compare a **mean difference on one factor to a mean difference on a second factor**. We can obtain values for all interaction contrasts in one large set:

```
> testInteractions(fit.factorial)
```

```
F Test:
P-value adjustment method: holm
                 Value Df Sum of Sq        F    Pr(>F)
1-2 : 1-2   -0.3333    1      0.083   0.0256  0.8747843
1-3 : 1-2  -18.0000    1    243.000  74.7692  1.196e-06 ***
1-4 : 1-2   -7.3333    1     40.333  12.4103  0.0084723 **
2-3 : 1-2  -17.6667    1    234.083  72.0256  1.278e-06 ***
2-4 : 1-2   -7.0000    1     36.750  11.3077  0.0084723 **
3-4 : 1-2   10.6667    1     85.333  26.2564  0.0004079 ***
Residuals             16     52.000
---
Signif. codes:  0 '***' 0.001 '**' 0.01 '*' 0.05 '.' 0.1 ' ' 1
```

The value of the first contrast is the difference between mean differences teacher 1 and teacher 2 for textbook 1 ($67.33 - 69.00 = -1.67$) and teacher 1 and teacher 2 for textbook 2 ($74.67 - 76.00 = -1.33$). That is, it is the difference $-1.67 - (-1.33) = -0.33$. This comparison is not statistically significant ($p = 0.87$). The value of the second contrast is the difference between mean differences teacher 1 versus teacher 3 for textbook 1 ($67.33 - 85.33 = -18.00$) and teacher 1 versus teacher 3 for textbook 2 ($74.67 - 74.67 = 0$). That is, it is the difference $-18.00 - 0 = -18.00$. This comparison is statistically significant ($p = 1.196e-06$). Remaining contrasts are interpreted in an analogous fashion.

4.12   CHAPTER SUMMARY AND HIGHLIGHTS

- **Factorial analysis of variance** is a suitable statistical method to test both **main effects** and **interactions** in a model where the dependent variable is continuous and the independent variables are categorical.
- The benefit of using factorial ANOVA over separate one-way ANOVAs is the ability to test for **interactions** between factors.
- Whereas **sample effects** constituted the basis of the one-way ANOVA model, **sample cell effects** constitute the systematic variation in the factorial ANOVA model.
- **Interaction effects** are computed by subtracting row and column effects from the cell effect.
- It is important to understand that **cell effects** are not equal to **interaction effects**. Rather, cell effects are used in the computation of interaction effects.
- Just as was true in the one-way model, the **error term** ε_{ijk} accounts for variability not explained by effects in the model. In the case of a two-way factorial, the error term corresponds to **within-cell** unexplained variation.
- A comparison of the **one-way model** to the **two-way model** is useful so that one can appreciate the conceptual similarities between **sample effects** and **cell effects**.
- In a **two-way model**, the sums of squares for cells partition into row, column, and interaction effects.
- The **assumptions** of the two-way factorial model parallel those of the one-way model, except that now, errors ε_{ijk} are distributed within cells, hence the requirement of the additional subscript k.
- **Expected mean squares** for factors A, B, and A × B reveal that MS error is a suitable denominator for all F-ratios.
- Interpreting **main effects** in the presence of **interaction effects** is permissible so long as one is clear to the fact that an interaction was also detected. Ideally, interaction terms should be interpreted before any main effect findings are discussed.

- A suitable **effect size** measure for terms in a factorial model is η^2, though it suffers from similar problems in the factorial model as it does in the one-way model. For a less biased estimate, ω^2 is usually recommended.
- A **simple main effect** is the effect of one factor at a particular level of another factor. Simple main effects are useful in following up a statistically significant interaction effect.
- **Interaction contrasts** can also be tested in factorial designs. These are comparisons of mean differences on one factor to mean differences on a second factor. They are "contrasts of contrasts."
- Factorial analysis of variance can be very easily performed using R or SPSS. Using the `phia` package in R, one can generate useful interaction graphs to aid in the interpretation of findings.

REVIEW EXERCISES

4.1. Define what is meant by a **factorial analysis of variance**, and discuss the purpose(s) of conducting a factorial ANOVA.

4.2. Explain, in general, what are meant by **main effects** and **interaction effects** in factorial ANOVA.

4.3. Invent a research scenario where a **two-way factorial ANOVA** would be a useful and appropriate model.

4.4. In a **two-way factorial ANOVA**, explain the four reasons why a given randomly sampled data point might differ from the **grand mean** of all the data.

4.5. Define what is meant by a **cell effect**, and why summing cell effects will always result in a **sum of zero**. What do we do to cell effects so that they do not sum to zero for every data set?

4.6. Define an **interaction effect**.

4.7. What is the difference between a **cell effect** and an **interaction effect**?

4.8. To help make the conceptual link between the **one-way model** and the **two-way**, why is it permissible (and perhaps helpful) to think of a_j as cell effects in $y_{ij} = \bar{y}. + a_j + e_{ij}$? Explain.

4.9. What is the **expected mean squares** for **MS within** in the two-factor model? Does this expectation differ from the one-way model? Why or why not?

4.10. What are the **expected mean squares** for **factor A** and **factor B** in the two-way factorial model? How do these compare to the expectations for the one-way model?

4.11. What is the **expected mean squares** for the **interaction term** in the two-way model? Under the null hypothesis of no interaction effect, what do you expect MS interaction to be?

4.12. In constructing F**-ratios**, what are the correct error terms for factor A, B, and A × B in the two-way model? What argument says that this is correct?

4.13. Given the presence of an interaction effect in a two-way model, **argue for** and **against** the interpretation of main effects.

4.14. Define what is meant by a **simple main effect**.

4.15. Discuss how an **interaction graph** can display a **sample interaction**, but that evidence might not exist to infer a **population interaction effect**.

4.16. Suppose a researcher wants to test all **simple main effects** in his or her data. Discuss potential problems with such an approach, and how that researcher might go about protecting against such difficulties.

4.17. In our computation of **interaction contrasts**, we interpreted two of them. Interpret the remaining interaction contrasts for the achievement analysis:

```
1-4 : 1-2  -7.3333  1    40.333 12.4103 0.0084723 **
2-3 : 1-2 -17.6667  1   234.083 72.0256 1.278e-06 ***
2-4 : 1-2  -7.0000  1    36.750 11.3077 0.0084723 **
3-4 : 1-2  10.6667  1    85.333 26.2564 0.0004079 ***
```

4.18. In our analysis of the achiev.2 data, we computed the **simple main effects** of textbook across teacher. Compute and interpret the simple main effects of teacher across textbook.

4.19. One way to conceptualize the testing of an interaction effect in ANOVA is to compare **nested models**. Recall a model is considered nested within another if it estimates a subset of parameters of the first model. For the achiev.2 data, though the significance test for interaction indicated the presence of an interaction, compare the **main-effects-only models** to that of the model containing an **interaction term** through the following:

(a) Test the main-effects-only model for teacher. Name the object `main.effects.teacher` in R.

(b) Test the main-effects-only model for teacher and textbook. Name the object `main.effects.textbook` in R.

(c) Test the interaction model. Name the object `interaction.effect` in R.

(d) Compare the models in R using: `anova(main.effects.teacher, main.effects.textbook, interaction.effect)`. Was adding the textbook and interaction effect worth it to the model?

5

INTRODUCTION TO RANDOM EFFECTS AND MIXED MODELS

This class includes all problems of estimating, and testing to determine whether to infer the existence of, components of variance ascribable to random deviation of the characteristics of individuals of a particular generic type from the mean values of these characteristics in the "population" of all individuals of that generic type, etc. In a sense, this is the true analysis of variance, and the estimation of the respective components of the over-all [sic] variance of a single observation requires further steps beyond the evaluations of the entries of the analysis-of-variance table itself.

(Eisenhart, 1947, p. 4)

The researcher of the previous two chapters, having discovered an effect of melatonin dosage on sleep onset, now ponders the following question:

Is sleep onset a function not only specific doses, but of melatonin dosage in general? That is, if we randomly sampled 3 dosages from a population of potential doses, would these differing doses account for variation in sleep onset?

In this situation, the researcher is not interested specifically in any particular set of doses. Rather, the researcher would like to draw the conclusion that **differing dose level is associated with differing sleep onset**. The effect for dose in this case would be considered a **random effect**, since levels of dose are randomly drawn from a wider population of possible doses. The subset of dosages randomly sampled for the given experiment is used to make a generalization to the population of dosage levels. This type of design calls for the **random effects analysis of variance model**.

Upon further thought, not only is the researcher interested in randomly sampling three dosage levels for use in his experiment, but just as he did for the two-way model of the previous chapter, he also wants to include ambient noise as a factor in his design. For this factor, he is only interested in

Applied Univariate, Bivariate, and Multivariate Statistics: Understanding Statistics for Social and Natural Scientists, With Applications in SPSS and R, Second Edition. Daniel J. Denis.
© 2021 John Wiley & Sons, Inc. Published 2021 by John Wiley & Sons, Inc.
Companion Website: www.wiley.com/go/denis/appliedstatistics2e

comparing levels **no noise** to **some noise** and hence keeps the factor **fixed**. He is not interested in generalizing to the population of noise levels. Hence, the researcher will now have one random factor (dose) and one fixed factor (noise) in his experiment. This type of design calls for a **mixed effects factorial analysis of variance model**.

In this chapter, we survey the random effects and mixed effects analysis of variance models. As we did for one-way and factorial fixed effects, we develop the conceptual basis and then move on to a consideration and development of suitable F-ratios to test effects. As we saw in previous chapters, in a fixed effects model, whether one-way, two-way, or higher-order, expected mean squares revealed that MS error was the correct error term for testing main effects and interactions. As we will see in the random effects and mixed models, **MS error is not always the most suitable error term for testing effects**. We will survey some of the theories as to why other error terms are more suitable in these situations. We also provide software examples of random effects and mixed effects models in R. For fitting mixed models in R, readers should consult Gelman and Hill (2007). Pinheiro and Bates (2000) provide an excellent treatment of the wider mixed effects model in S-Plus. Demidenko (2004) provides a very technical treatment along with some applications.

5.1   WHAT IS RANDOM EFFECTS ANALYSIS OF VARIANCE?

Recall that in the fixed effects models studied in previous chapters, what made the effects in these models "fixed" was the fact that over theoretical repetitions of the experiment, levels of the independent variable were to remain **constant**. For example, in the melatonin experiment, the fixed factor of dosage was so named because the researcher had a specific interest in the dosages tested. The idea of a random effects model is that over theoretical repetitions of the experiment, treatment effects are no longer assumed to remain fixed. Rather, treatment effects are considered to be random, and hence over numerous theoretical replications of the experiment (i.e., if we were to perform them), it is reasonable to assume that we will obtain different treatment levels when sampling each time. In a random effects model then, **the levels of a random factor are randomly sampled from a population of possible levels that could have been included in the given experiment**. When a factor is a random factor, it implies that there is a probability distribution of levels associated with that factor, and what you are using in your experiment is but a **sample** of levels from a wider set of potential levels that could have been used. In the language of sets, the levels randomly sampled are but a **proper subset** of the wider set of population levels. As Casella (2008) noted:

> … by the very nature of a random factor, we are not really interested in estimating the levels of the factor that are in the experiment. Why? Because if the factor is truly random, the levels in the experiment are nuisance parameters, and only the variance of the factor is meaningful for inference. (p. 101)

Historically, nobody better described the concept of a random effects model than Eisenhart (1947):

> … when an experimenter selects two or more treatments, or two or more varieties, for testing, he rarely, if ever, draws them at random from a population of possible treatments or varieties; he selects those that he believes are most promising. Accordingly Model I [fixed effects] is generally appropriate where treatment, or variety comparisons are involved. On the other hand, when an experimenter selects a sample of animals from a herd or a species, for a study of the effects of various treatments, he can insure that they are a random sample from the herd, by introducing randomization into the sampling procedure, for example, by using a table of random numbers. But he may consider such a sample to be a random sample from the species, only by making the assumption that the herd itself is a random sample from the species. In such a case, if several herds (from the same species) are involved, Model II [random effects] would clearly be appropriate with respect to the variation among the animals from each of the respective herds, and might be appropriate with respect to the variation of the herds from one another. (p. 19)

The random effects model has sometimes historically been called a **components of variance** model (Searle, Casella, and McCulloch, 1992) because unlike the fixed effects model in which the primary interest is in testing null hypotheses about specific differences between population means, the primary interest in a random effects model is in estimating **variance** in the dependent variable that can be attributed to main effects or interactions. This estimate of variance accounted for will apply not only to the levels actually sampled but to the larger set of possible levels (i.e., population) from which our sample was drawn. Hence, in random effects models, our primary goal is to estimate **components of variance** rather than test null hypotheses about equalities among population means as was the case in the fixed effects model.

5.2 THEORY OF RANDOM EFFECTS MODELS

Insight into the random effects model can be gleamed from a brief discussion of its assumptions, and then by comparing these assumptions to those made in the previously studied fixed effects model. Recall the one-way analysis of variance model of Chapter 3:

$$y_{ij} = \mu + \alpha_j + \varepsilon_{ij}$$

where μ is the grand mean, α_j is a population effect estimated by the sample effect $\bar{y}_j - \bar{y}_.$, and ε_{ij} is the error associated with observation i in group j. We first list the assumptions for the one-way random effects model that parallel those of the fixed effects model:

- For any treatment j, the errors ε_{ij} are normally distributed, with a mean of 0 (i.e., $E(e_{ijk}) = 0$) and variance σ_e^2, which is identical for each possible treatment j. That is, $N(0, \sigma_e^2)$. Notice that this assumption parallels the assumption of normality in the fixed effects model.
- The values of the random variable e_{ijk} are all independent (as was also assumed in the fixed effects model). In cases of naturally-occurring, or imposed hierarchical nesting structures, errors within groups may be related (see Section 5.17 and Chapter 6 for details on blocking and nesting).
- $\sigma_{e_{ijk}}^2 < \infty$, that is, the variance of the errors is some finite number (which, as was true in the one-way and two-way models, implies that it is less than infinity).
- $\sigma_{jk=1}^2 = \sigma_{jk=2}^2 = \sigma_{jk=JK}^2$, that is, the variances across cell populations are equal (recall this is called the **homoscedasticity** assumption and is essentially the same as in the fixed effects models studied previously).

Where the random effects model differs from the fixed effects model is in the following assumptions:

- a_j is a **random variable** having a distribution with mean 0 and variance σ_A^2. That is, unlike the fixed effects model, the sample treatment effects a_j are no longer considered to be **constant** across replications. Analogous to how we can reach into a bag and take a sample of 10 objects and calculate a sample mean on them, the sample mean can be considered to be a random variable that can vary from experiment to experiment. We now need to treat a_j as possibly fluctuating from sample to sample or from experiment to experiment. They are no longer **fixed** as they were in the fixed effects model.
- The values of the random variable a_j occurring in the experiment are all independent of each other (Hays, 1994).

- Each pair of random variables a_j and e_{ij} are independent. That is, the sample effects are independent of error (or if you wish, the error **effects**).

Note that the assumptions for a random effects model are for two different distributions, one for the distribution of the random variable a_j, and the other for e_{ij}. In the fixed effects model, we only made a probability assumption about e_{ij}, since we assumed a_j to be fixed across theoretical replications. Since the sample effects were assumed to be fixed, it made no sense to associate them with a probability distribution.

5.3   ESTIMATION IN RANDOM EFFECTS MODELS

There have been, historically, several different methods of estimating parameters in random effects and mixed models. The classic method in which one computes **expected values of mean squares** is historically known as **ANOVA estimation** (Wu, Yu, and Liu, 2009). This methodology has some flaws and drawbacks, and in part because of the advances in computing power, other methods of estimation have come into vogue, which include **maximum-likelihood (ML)**, **restricted maximum-likelihood (REML)**, and **minimum norm quadratic unbiased estimation**. Of these, ML and REML are dominant today in the estimation of variance components in both random effects and mixed models. These methods of estimation, however, are quite complex and require **iteration** for their solution.

As we did in prior chapters, we focus on the method of taking expectations (ANOVA estimation), largely because under certain conditions, results of ANOVA estimation match those of the iterative methods. Also, a brief study of expectations in ANOVA models, I believe, goes a long way to demystifying the theory behind estimation in general, and opens the door for the reader to understand more complex methods for estimating parameters.

In what follows then, we begin with the principles developed in previously studied fixed effects models and derive expected mean squares for random effects models. Our discussion and derivation is based largely on the work of Hays (1994), Kempthorne (1975), Searle, Casella, and McCulloch (1992), and Scheffé (1999), who all present thorough accounts of random effects ANOVA.

5.3.1   Transitioning from Fixed Effects to Random Effects

Recall the quantities of **MS Between** and **MS Within** as first derived in the fixed effects model of Chapter 3:

$$\text{MS between} = \frac{\text{SS between}}{J-1} = \frac{\sum_j n_j \left(\bar{y}_j - \bar{y}.\right)^2}{J-1}$$

$$\text{MS within} = \frac{\text{SS within}}{N-J} = \frac{\sum_j \sum_i \left(y_{ij} - \bar{y}_j\right)^2}{N-J}$$

Should we expect derived EMS on these values to be the same in a random effects model? Not necessarily. The reason is that now we are randomly selecting the J different factor levels. They are no longer fixed. Because of this, as we will see, our expected mean squares will change. They will change because we are no longer interested in population mean differences. We are interested, rather, in estimating **variances**.

Because we are randomly sampling the levels of our factor in a random effects model, we can write the mean of the sample random effects as

$$\bar{a} = \frac{\sum\limits_{i} a_j}{J}$$

where a_j is, as before, the sample effect $(\bar{y}_j - \bar{y}.)$ for a given group $J = j$. This is the mean of the sample effects for the given experiment we are conducting. Theoretically, if we were to conduct the experiment again, and obtain new levels, we would obtain another $\bar{a}$ for that particular experiment, and so on for additional repetitions of the experiment. What is key to understand here is that **this mean will surely vary from sample to sample due to sampling error** (i.e., the error generated simply by the process of sampling) inherent in the random effect. This is why a_j has now become a random variable. We can have some certainty, however, that the mean effect over **all infinite samples that could be drawn from the population** will equal to zero. More formally, we say that the expected value of $\bar{a}$ will be 0, $E(\bar{a}) = 0$. However, the value of $\bar{a}$ in any given sample need not be equal to the long-run expectation. That is, as noted by Hays (1994, p. 530), "... although the mean of the effects **over all the possible treatments** [emphasis added] must be 0, the mean $\bar{a}$ of the sample effects present in a given set of data need not be 0."

Theoretically then, in any particular experiment, the value of $\bar{a}$ is not constrained to equal 0 as it was in the fixed effects model. The major point is that in any given model with a random effects term (other than the obvious e_{ij} effect, which is indeed also a random effect), we must somehow deal with the fact that these **treatment effects a_j are now random. Being random, their values will undoubtedly change from experiment to experiment**. This change in assumption figures prominently in the derivation of the expected mean squares. We will see that because of this random quality of the sample effects, the expected mean squares in the random effects model are quite different than in the fixed. Likewise, **null hypotheses** will be defined differently as well.

5.3.2 Expected Mean Squares for MS Between and MS Within

Recall once more the reason for taking expectations of mean squares. It is to learn what parameter our given mean squares is estimating. By calculating EMS, we can then use these to generate suitable F-ratios to test various effects of interest, whether they be main effects or interactions.

As Hays (1994) does, we begin our derivation by conceptualizing the mean of the errors for any group j in a one-way random effects ANOVA as

$$\bar{e}_j = \frac{\sum\limits_{i} e_{ij}}{n}$$

where $\bar{e}_j$ is the mean error for a given group, $\sum\limits_{i} e_{ij}$ is the sum of errors across all groups j, and n is the sample size per group (as before, we are assuming a balanced design). If we take this for the entire sample across J groups, we will have

$$\bar{e} = \frac{\sum\limits_{j}\sum\limits_{i} e_{ij}}{N} = \frac{\sum \bar{e}_j}{J}$$

which means that the average overall error is equal to the mean error, $\bar{e}_j$, per group. Given this, and just as we did in previous chapters where we wrote out model equations, we can write the deviation of any group mean $\bar{y}_j$ from the grand sample mean $\bar{y}.$ as

$$(\bar{y}_j - \bar{y}.) = (a_j - \bar{a}) + (\bar{e}_j - \bar{e}) \qquad (5.1)$$

Why does it make sense to write the deviation of a group mean from the grand mean as in (5.1)? This makes sense, because we just mentioned that we can calculate a "mean of errors" term over all groups. If this is the case, then it stands to reason that for a given group j, the mean error for that particular group minus the overall mean error for the entire data will give us the "effect" of error for that particular group, just as $(\bar{y}_j - \bar{y}.)$ gives us the sample "effect" of being in a particular group j. Notice that the sum of effects for $(a_j - \bar{a})$ will sum to 0, and the sum of effects for $(\bar{e}_j - e)$ will also sum to 0; so, as usual, we take the **squared deviations**, otherwise the entire right-hand side of (5.1) will always sum to zero (this idea of the sum of unsquared effects always equaling zero should be becoming familiar territory by now). Squaring (5.1), summing, and taking expectations, we get (Hays, 1994, p. 531):

$$E\left[\sum_j (\bar{y}_j - \bar{y}.)^2\right] = E\left[\sum_j (a_j - \bar{a})^2\right] + E\left[\sum_j (\bar{e}_j - \bar{e})^2\right] \tag{5.2}$$

From (5.2), we have the expected mean squares for SS between:

$$E\left[\sum_j (\bar{y}_j - \bar{y}.)^2\right] = E\left[\sum_j (a_j - \bar{a})^2\right] + E\left[\sum_j (\bar{e}_j - \bar{e})^2\right]$$
$$E(\text{MS between}) = n\sigma_A^2 + \sigma_e^2 \tag{5.3}$$

where n is the number of subjects (or objects) per group, σ_A^2 is the variance attributable to varying levels of factor A, and σ_e^2 is the variance of the error. That is, the sum of squares for between is equal to a source of variability for factor A, $n\sigma_A^2$, and a source of variability represented by the error term, σ_e^2.

The **expectation for error**, as was true for the fixed effects model, is the average error per group:

$$E(\text{MS within}) = \sum \frac{\sigma_e^2}{J} = \sigma_e^2$$

That is, **MS within**, just as was the case for the fixed effects ANOVA, is an unbiased estimate of error variance, and **only** error variance.

5.4   DEFINING NULL HYPOTHESES IN RANDOM EFFECTS MODELS

In the random effects model, null hypotheses are stated differently than in a fixed effects model. A null hypothesis in a random effects model is not really about means. It is more about **variances**. Or to be even more precise, **variance components**. The null hypothesis for the one-way random effects model is given by

$$H_0 : \sigma_A^2 = 0$$

where σ_A^2 is the variance attributable to differing levels of factor A. If changing levels of the factor is not associated with any change in the dependent variable in our sample, then it stands that the variance explained, sampling error aside, should equal to 0. And since the purpose of conducting the investigation is usually to show that varying levels of the factor is associated with variance explained in the dependent variable, our alternative hypothesis is given by:

$$H_1 : \sigma_A^2 > 0$$

Notice that the alternative hypothesis is specified in terms of a **positive** value. The **greater than** sign denotes that σ_A^2 cannot be zero or negative given a rejection of the null hypothesis. This is reasonable, since we know variance, by definition, is a positive quantity. If there are treatment effects, either for those treatments sampled or across all treatment levels in the population, we would expect the variance attributable to our factor to be greater than 0. For the one-way random effects model then, there are two "**components of variance**" that need to be obtained. One is σ_A^2, the other is σ_e^2. Both of these components add up to the total variance σ_y^2 in the dependent variable. That is, $\sigma_y^2 = \sigma_A^2 + \sigma_e^2$. We will discuss shortly why this is the case.

5.4.1 F-Ratio for Testing H_0

How do we come up with a suitable ratio for testing $H_0 : \sigma_A^2 = 0$? We do so by considering the derived expected mean squares. As was the case for the fixed effects model, we want to isolate that part of the expected mean squares that represents the "effect" we are interested in. In $n\sigma_A^2 + \sigma_e^2$, that part is $n\sigma_A^2$. That is, if our experimental treatment "worked," (in some sense) we would expect $n\sigma_A^2$ to be large relative to σ_e^2. Notice that once we have isolated the part we are interested in, as was true for the fixed effects models of the previous chapters, the correct error term quite naturally reveals itself. Since we do not want our effects to be "polluted" by σ_e^2, we will divide $n\sigma_A^2 + \sigma_e^2$ by σ_e^2. But what is σ_e^2? This is the expectation of MS within. Hence, the F-ratio we want to produce is one which takes $n\sigma_A^2 + \sigma_e^2$ in the numerator and divides it by σ_e^2. That is, our F-ratio for the one-way random effects model is:

$$F = \frac{n\sigma_A^2 + \sigma_e^2}{\sigma_e^2}$$

At first glance, it may appear that we can simply cross out σ_e^2 in the numerator and σ_e^2 in the denominator. However, recall from the rules of algebra that we cannot do this since the numerator is a **sum** and not a **product**. Had the numerator been $\left(n\sigma_A^2\right)\left(\sigma_e^2\right)$, where the parentheses denote multiplication, then crossing out σ_e^2 would have worked. But since we are dealing with addition, we cannot eliminate σ_e^2 in this way.

Returning to our F-ratio, we can appreciate why it makes good sense to construct it as we did. If there are no treatment effects for our factor, then $n\sigma_A^2$ will be 0, since σ_A^2 would equal 0, and any n (i.e., sample size per group in a balanced design) multiplied by 0 will equal 0. Under this condition, we are simply left with σ_e^2 in the numerator, and our F-ratio would be equal to approximately

$$F = \frac{n\sigma_A^2 + \sigma_e^2}{\sigma_e^2} = \frac{n(0) + \sigma_e^2}{\sigma_e^2} = \frac{\sigma_e^2}{\sigma_e^2} = 1$$

That is, we can state more formally that under H_0,

$$E\left(\frac{n\sigma_A^2 + \sigma_e^2}{\sigma_e^2}\right) \approx 1$$

If, on the other hand, H_0 is false, then this implies the alternative hypothesis, $\sigma_A^2 > 0$, and so $n\sigma_A^2$ will be some quantity **larger** than 0. Our expectation then for our ensuing F-ratio would be

$$E\left(\frac{n\sigma_A^2 + \sigma_e^2}{\sigma_e^2}\right) > 1$$

under a false null hypothesis. As was the case for the fixed effects model, we evaluate F on $J - 1$ and $N - J$ degrees of freedom. A statistically significant F-statistic suggests that the variance attributable (or "accounted for") by our factor (i.e., either the levels represented in the sample or by the population of levels) is not equal to 0 in the population from which these data were drawn. That is, a rejection of the null hypothesis implies that the variance in our dependent variable that is accounted for by our factor is greater than 0.

5.5 COMPARING NULL HYPOTHESES IN FIXED VERSUS RANDOM EFFECTS MODELS: THE IMPORTANCE OF ASSUMPTIONS

It would do well at this point to emphasize and reiterate the fact that a rejection of the null hypothesis in a random effects analysis of variance tells us something different than a rejection of a null hypothesis in the fixed effects models of the previous chapters. In the fixed effects model, we tested hypotheses about **means**. In the random effects model, we are testing hypotheses about **variances**. A rejection of the null hypothesis in a fixed effects model hints to us that somewhere among the population means, it looks like there is a mean difference. A rejection of the null hypothesis in the random effects model tells us that changing levels of the independent variable has the effect of explaining or accounting for variance in the dependent variable. These two null hypotheses are **not the same**.

What we have noticed, however, is that the error terms used for testing both hypotheses in the one-way fixed effects and one-way random effects model are the same. In both cases, **MS error is the correct error term**. Why are they the same? They are the same (so far) because in both cases, the one-way fixed and one-way random effects, MS error "gets the job done" in terms of isolating the term in the numerator that we are interested in. Recall that in the one-way fixed effects model, the **expectation for MS between** was equal to

$$E(\text{MS between}) = \sigma_e^2 + \frac{\sum_j n_j \alpha_j^0}{J - 1}$$

The **expectation for MS within** was equal to σ_e^2, and so because we were interested in isolating

$$\frac{\sum_j n_j \alpha_j^2}{J - 1}$$

since it contained any treatment effects present, it made sense to use σ_e^2 as the error term. I want to emphasize that this is why we used MS within as the error term, because it made sense to do so in terms of what we wished to isolate in the numerator. This is the general logic of choosing error terms in ANOVA, whether in simple designs or more complex. Deciding on a correct error term is not a "mysterious" process once you have the expected mean squares at your disposal (on the other hand, **deriving** EMS can be somewhat difficult).

The expectation for MS within is again equal to σ_e^2 in our current random effects model, and so because we are interested in isolating $n\sigma_A^2$, it again makes sense to use MS error as the error term. Also, be sure to note that the phrase **error term** and **MS error** are not synonymous with one another. Under our current discussion, MS error is the appropriate error term. As we will see for the two-factor random effects model, the correct error term will be other than MS error. **It is extremely important to not get into the habit of automatically associating "error term" with "MS error." MS error is, under many circumstances and models, the appropriate error term, but under other models, it no longer is. In those cases, we will seek an error term other than MS error.**

5.6 ESTIMATING VARIANCE COMPONENTS IN RANDOM EFFECTS MODELS: ANOVA, ML, REML ESTIMATORS

Once we have computed the analysis of variance, whether in the one-way or two-way (to be discussed shortly) or higher-order analyses, our next job is to estimate **variance components** for such models. Note that in our computations of analysis of variance so far, we have not yet addressed just how quantities such as σ_A^2 and σ_e^2 are estimated. All we have considered thus far is how to use these quantities to help us derive suitable F-ratios. We first consider **ANOVA estimators**, and then move on to a brief consideration of **maximum likelihood** and **restricted maximum likelihood**.

5.6.1 ANOVA Estimators of Variance Components

ANOVA estimators are easily computed, and in some cases can be used as starting values to other forms of estimation. They are also the most historically relevant in the evolution of variance component estimation. Recall once more the expectation for MS between found in (5.3). We can solve for σ_A^2 and get an unbiased estimate of σ_A^2:

$$E(\text{MS between}) = n\sigma_A^2 + \sigma_e^2$$
$$n\sigma_A^2 + \sigma_e^2 = E(\text{MS between})$$
$$n\sigma_A^2 = E(\text{MS between}) - \sigma_e^2$$

We can then obtain our estimate of the variance component σ_A^2 quite simply:

$$\hat{\sigma}_A^2 = \frac{\text{MS between-MS within}}{n}$$

where MS between and MS within are obtained from the ANOVA, and n is the sample size per group in a balanced design. The next question is how to use this component. By itself, it simply represents a quantity of variance. What we would like to obtain is a **proportion of variance** attributable to our factor relative to the total variance in our dependent variable. To obtain this estimate, we need to know that the variance of our dependent variable y can be written as a function of two components in the one-way random effects model. The first component is σ_A^2, while the second component is σ_e^2. That is,

$$\sigma_y^2 = \sigma_A^2 + \sigma_e^2$$

This tells us that the total variance in a population for a one-factor experiment is composed of variability due to our factor, σ_A^2, and variability not due to our factor, which is relegated to the error component, σ_e^2.

The question now becomes how to estimate the total variance σ_y^2 in the random effects model. We do so by (Hays, (1994, p. 534)):

$$\hat{\sigma}_y^2 = \hat{\sigma}_A^2 + \hat{\sigma}_e^2$$
$$= \frac{\text{MS between} + (n-1)\text{MS within}}{n}$$

where $\hat{\sigma}_y^2$, $\hat{\sigma}_A^2$, and $\hat{\sigma}_e^2$ are respective estimates of variances σ_y^2, σ_A^2, and σ_e^2. Having estimated the respective components of variance, we can now assess the proportion of variance due to, or accounted for, by our factor. We take the following ratio, called the **intraclass correlation coefficient**:

$$\hat{\rho} = \frac{\hat{\sigma}_A^2}{\hat{\sigma}_A^2 + \hat{\sigma}_e^2} = \frac{\hat{\sigma}_A^2}{\hat{\sigma}_y^2} \tag{5.4}$$

The intraclass correlation coefficient measures the proportion of variance due to the grouping factor, and like all proportions, ranges from 0 to 1. As noted by Kirk (1995), it is generally considered to be the most popular measure of **effect size** for random effects.

A second, related interpretation of the intraclass correlation, is that it is the **bivariate correlation coefficient between any two randomly selected observations within a given level of the independent variable** (Fox, 2016). That is, we can define ρ as

$$\rho = \mathrm{cor}\left(y_{ij}, y_{ij'}\right)$$

where y_{ij} and $y_{ij'}$ are two distinct observations in a given group j. Intraclass correlations are useful in measuring proportions of variance explained in applications of random effects and mixed models of the current chapter as well as blocking and repeated measures models of the following chapter.

5.6.2   Maximum Likelihood and Restricted Maximum Likelihood

As discussed by Searle, Casella, and McCulloch (1992), ANOVA estimation has some weaknesses, including the fact that **negative variance estimates** are possible. According to Casella (2008, p. 143), negative variance estimates are often the fault of the estimation procedure rather than the model. Further, Casella notes that a negative variance component should not in itself imply a conclusion that $\sigma_A^2 = 0$, and that when negative estimates occur, one should try a better estimation procedure, such as **restricted maximum likelihood** (REML), which is a variation of **maximum likelihood** (ML).

Maximum likelihood estimation has its recent history beginning with a paper by Hartley and Rao (1967) in which ML equations were derived, but required iterative calculations to estimate variance components. At first, these computations were quite laborious, but with the advent of high-speed computing, iterations are now performed with relative ease and speed. Closed-form solutions for ML estimation are usually heavily dependent on normality assumptions.

Restricted maximum likelihood estimation focuses on maximizing the likelihood which is invariant (i.e., does not change) to the fixed effects of the model (called the **location parameters** of the model). REML estimates variance components as a function of **residuals** that are left over after estimating the fixed effects by least-squares (Searle, Casella, and McCulloch, 1992). For balanced data, REML solutions are identical to ANOVA estimators. For unbalanced data, ML and REML are generally preferable over ANOVA estimators (Searle, Casella, and McCulloch, 1992). Choosing between ML and REML is not straightforward, and our best advice is to follow the recommendation of Searle, Casella, and McCulloch (1992):

> As to the question "ML or REML?" there is probably no hard and fast answer. Both have the same merits of being based on the maximum likelihood principle – and they have the same demerit of computability requirements. ML provides estimators of fixed effects, whereas REML, of itself, does not. But with balanced data REML solutions are identical to ANOVA estimators which have optimal minimum variance properties – and to many users this is a sufficiently comforting feature of REML that they prefer it over ML. (p. 255)

5.7   IS ACHIEVEMENT A FUNCTION OF TEACHER? ONE-WAY RANDOM EFFECTS MODEL IN R

Recall once more the `achiev` data of Chapters 3 and 4 (reproduced in Table 5.1). In those chapters, we designated teacher as a **fixed effect**. In the current analysis, we will consider it to be a **random effect**.

Imagine the following scenario—You are the parent of Taylor, an 11-year old child in sixth grade elementary education. Taylor is not performing as well as you would like in school, and based on a few verbal reports from your daughter and parents of other children, you suspect it may have something to do with Taylor's teacher. The principal of the school, however, comes to the teacher's defense and makes the following claim to you: "Student achievement is not associated with teacher. Whether a student has one teacher or another makes no difference in how the child performs."

In advancing your argument, you would like to accumulate some evidence to help substantiate that teacher does play a "role" in academic achievement. You **randomly sample** four teachers from your city and obtain mathematics achievement scores from the children in those classes, scored from 0 to 100, where higher scores are indicative of greater achievement. Ideally, children would also be randomly assigned to teacher, but for now, our focus is simply on understanding how teacher can be considered a random effect. Even if not by experimental design, it is most likely that children were randomly assigned to teacher from the outset (i.e., unless of course a school designates particular students for particular teachers, in which case, random assignment is not taking place). For our purposes here, again, we focus simply on teachers being randomly selected from a wider set of teachers.

Notice that your hypothesis calls for a **one-way random effects model**, since levels of teacher were randomly sampled. Surely, you are not interested in showing differences (i.e., mean differences) between **these particular teachers you have sampled**. Rather, you would like to draw the conclusion that variance in achievement is a function of different teachers, of which these four in your design constitute a random sample of teachers for the given study. We thus have the perfect setup for a one-way random effects model. Should your study be "successful" in that you obtain evidence that variance in achievement accounted for by teacher is greater than 0, you would be in a position to respond to the principal of the school arguing that **varying teachers is associated with variance explained in achievement**, which would stand contrary to the principal's initial claim that regardless of teacher, students achieve to the same degree.

We run the model using the function `lmer` (linear mixed effects models) in the package `lme4` (Bates et al., 2014) specifying **teacher** as a **random effect**. To request maximum likelihood estimation, we include the statement `REML = FALSE` (i.e., by default, `lmer` will run REML):

TABLE 5.1   Achievement as a Function of Teacher

	Teacher		
1	2	3	4
70	69	85	95
67	68	86	94
65	70	85	89
75	76	76	94
76	77	75	93
73	75	73	91
$M = 71.00$	$M = 72.5$	$M = 80.0$	$M = 92.67$

```
> library(lme4)
> fit.random <- lmer(ac ~ 1 + (1|f.teach), achiev, REML = FALSE)
```

About the above model specification:

- ~ 1 fits an intercept to the model.
- (1|f.teach) specifies f.teach as a random factor.
- achiev is the name of the dataframe in which the data are contained (i.e., the .txt file we loaded into R).
- REML = FALSE tells R to bypass the default estimation method (REML) and to fit the model by maximum likelihood.

```
> fit.random
Linear mixed model fit by maximum likelihood  ['lmerMod']
Formula: ac ~ 1 + (1 | f.teach)
   Data: achiev
     AIC       BIC    logLik deviance df.resid
157.1869  160.7211  -75.5935 151.1869       21
Random effects:
 Groups    Name        Std.Dev.
 f.teach   (Intercept) 8.388
 Residual              4.341
Number of obs: 24, groups:  f.teach, 4
Fixed Effects:
(Intercept)
      79.04
```

Features of the output include the following:

- **AIC** is equal to 157.19, and recall is useful for comparing models. Lower values of AIC indicate a better-fitting model than do larger values. Recall that AIC jointly considers both the goodness-of-fit as well as the number of parameters required to obtain the given fit, essentially "penalizing" for increasing the number of parameters unless they contribute to model fit. If we were to build on the current model by potentially adding terms, then we could observe the extent to which AIC changes and use this in our global assessment of model fit.
- **BIC** yields a value of 160.72, which is also useful for comparing models. Lower values of BIC are also generally indicative of a better-fitting model than are larger values. As was true for AIC, if we were to fit additional parameters to the model, we would want to see a drop in BIC values to justify, on a statistical basis, the addition of the new parameters.
- **Deviance** of 151.19, defined as $-2[\log_e L_{Model} - \log_e L_{Saturated}]$, where L_{Model} is the likelihood of the current model and $L_{Saturated}$ is the likelihood of the saturated model. Here we assume $\log_e L_{Saturated}$ is equal to 0, hence we can also write the deviance as $-2[\log_e L_{Model}]$. Smaller values than not are indicative of better fit.
- The variance component for f.teach is equal to the square of the standard deviation. That is, $(8.388)^2 = 70.36$.

- The variance component for residual is equal to the square of the standard deviation. That is, $(4.341)^2 = 18.84$.
- The only fixed effect for this model is the intercept term, and is equal to 79.04. This is the grand mean of achievement for all observations and is not of immediate interest to us.

We could also request a summary of the fitted model (`summary(fit.random)`), which will provide us with similar output as above, with the exception that variance components are included (so we do not have to square the standard deviations ourselves).

5.7.1   Proportion of Variance Accounted for by Teacher

Having fit the model, we can now compute the proportion of variance accounted for by `f.teach`. Recall that the variance component for `f.teach` was equal to 70.36, while the variance component for residual was equal to 18.84. It is important to emphasize that these are **variance components**, they are not **proportions of variance** (that they are not proportions should be evident in itself since proportions range from 0 to 1).

Since $\sigma_y^2 = \sigma_A^2 + \sigma_e^2$, we can compute the estimated proportion of variance accounted for by our independent variable, the **intraclass correlation**, as:

$$\frac{\hat{\sigma}_A^2}{\hat{\sigma}_y^2} = \frac{\hat{\sigma}_A^2}{\hat{\sigma}_A^2 + \hat{\sigma}_e^2} = \frac{70.36}{70.36 + 18.84} = \frac{70.36}{89.20} = 0.79$$

That is, approximately 79% (we rounded up) of the variance in achievement is accounted for by teacher.

Of course, this is an extremely large measure of association for data of this kind. If you actually did find such an effect for teacher, what would it suggest? Consistent with our interpretation of the random effects model, it would imply that 79% of students' achievement variance in school is associated with varying teachers, either those teachers selected in the sample or those in the population from which the sampled levels were drawn. Does this mean that one's teacher is somehow **responsible** for one's achievement? Surely not, at least not so based on our statistical analysis.

Still, the finding of 79%, if it were actually true, could serve as a strong counter-argument against that of the principal's who claimed that teacher had no "impact" on students' achievement. Again, we must be cautious with our interpretation, because we certainly have no evidence for anything remotely close to **causal**. The word "impact" is used purposely in quotes here. Concluding that teachers "impact" student achievement implies a directional causal-like claim, and hence must be used with great care, if used at all.

However, such data are still rather strong evidence that changing teachers might be a good idea for Taylor given that she is struggling in school. And the benefit of conducting a random effects model instead of a fixed effects one is that our inferences are not restricted to generalizing to only the levels sampled for the given analysis. We can generalize to the **population** of levels of which the ones featured in the given analysis were merely a random sample. Because you conducted a random effects model, the principal cannot rebuke your evidence by accusing you of "handpicking" certain teachers over others. Your finding of 79% is generalizable to the population of teachers of which the ones you tested were but a random sample. This is what gives random effects their power to draw rather far-reaching generalizations, not unlike when we randomly sample subjects, of which the particular subjects you obtained in your experiment are but a sample of a larger population. Because of the way subject "levels" were sampled, we feel more confident about generalizing to a wider population of subjects.

5.8   R ANALYSIS USING REML

We now fit the one-way random effects model using REML estimation, and briefly compare the output to the previous analysis using ML. To fit by REML, simply exclude the statement REML = FALSE from our previous model statement (fit.random <- lmer(ac ~ 1 + (1|f.teach), achiev, REML = FALSE)):

```
> fit.random.reml <- lmer(ac ~ 1 + (1|f.teach), achiev)
> summary(fit.random.reml)

REML criterion at convergence: 146.3

Scaled residuals:
    Min       1Q   Median       3Q      Max
-1.6056  -0.8696   0.2894   0.7841   1.3893

Random effects:
 Groups    Name        Variance  Std.Dev.
 f.teach   (Intercept)  94.87     9.740
 Residual               18.84     4.341
Number of obs: 24, groups:  f.teach, 4

Fixed effects:
            Estimate Std. Error t value
(Intercept)    79.04       4.95   15.97
```

We see that the output using REML is very similar to that using ML. The variance components for teacher and residual are 94.87 and 18.84 respectively, for a proportion of variance due to teacher equal to 0.83 (i.e., $94.87/(94.87 + 18.84) = 94.87/113.71 = 0.83$), a figure slightly higher than that using maximum likelihood. We could have also obtained the standard deviations by VarCorr (fit.random.reml).

5.9   ANALYSIS IN SPSS: OBTAINING VARIANCE COMPONENTS

We now conduct the identical analysis using SPSS's VARCOMP function. We will demonstrate using both maximum likelihood (ML) and restricted maximum likelihood (REML), and briefly compare our results to those obtained using R.

To run the one-way random effects model using ML, we request in SPSS:

```
VARCOMP ac BY teach
 /RANDOM=teach
 /METHOD=ML
```

The remainder of the syntax should include a limit on the number of times you wish the algorithm to iterate (for our example, we have chosen 50), the criteria for convergence (choosing a relatively small number is recommended, or just use the default in SPSS as we have done), and the history of the iteration:

```
/CRITERIA = ITERATE(50)
/CRITERIA = CONVERGE(1.0E-8)
/PRINT = HISTORY(1)
```

Select output from the `VARCOMP` procedure follows. As we can see, much of it is essentially analogous to that obtained using R (ML).

Iteration History			
Iteration	Log-Likelihood	Var(teach)	Var(Error)
0	−83.415	98.007	89.207
1	−76.353	31.190	18.842
2	−75.593	70.365	18.842
3	−75.593[a]	70.365	18.842

Dependent variable: ac.

Method: maximum-likelihood estimation.

[a]Convergence achieved.

First, we see the **iteration history**, showing the number of times the algorithm took to converge on a log-likelihood statistic having requested convergence criteria (recall our criteria was 1.0E-8). Though the numbers are rounded, we can see that from iteration 2 to iteration 3 the difference between log-likelihood statistics is extremely small (too small to be noticeable in SPSS's report due to rounding, both values are equal to −75.593 in the output). We can also see that SPSS settled on variance components of 70.365 for teach and 18.842 for error. These are the same as those estimated in R.

Next, SPSS reports the variance component estimates that appeared at the last stage of the iteration (i.e., under iteration 3 above):

Variance Estimates	
Component	Estimate
Var(teach)	70.365
Var(Error)	18.842

Dependent variable: ac.

Method: maximum-likelihood estimation.

As we did in the analysis via R, we can compute the **proportion of variance** explained by teacher by $70.365/(70.365 + 18.842) = 70.365/89.207 = 0.79$, which is the same figure we obtained in our analysis using R.

We next briefly demonstrate the syntax and output for the same model fit in SPSS, this time fit by REML. To conserve space, only the final variance component estimates are given:

```
VARCOMP ac BY teach
 /RANDOM=teach
 /METHOD=REML [note the change from ML to REML]
/CRITERIA = ITERATE(50)
/CRITERIA = CONVERGE(1.0E-8)
/PRINT = HISTORY(1)
```

| Variance Estimates |
Component	Estimate
Var(teach)	94.867
Var(Error)	18.842

Dependent variable: ac.

Method: restricted maximum-likelihood estimation.

Using REML as our method of estimation, we see that teacher accounts for approximately 83% of the variance in achievement (i.e., $94.867/(94.867 + 18.842) = 94.867/113.709$). These results parallel those found in R using REML.

5.10   FACTORIAL RANDOM EFFECTS: A TWO-WAY MODEL

Having discussed the one-way random effects model and having come to the conclusion through expected mean squares that the correct error term was indeed MS error, we now turn to consideration of the **two-way random effects model**. In this case, both factors are random, which again implies that the levels for a given experiment are sampled levels from a wider population of levels. As was true for the one-way model, we are not interested specifically in mean differences. Rather, we are interested in variance in the dependent variable attributable to each factor, and potentially also to their interaction.

For example, suppose that instead of merely hypothesizing an association between teacher and achievement, we hypothesize that hours of homework is also related to achievement. However, as was the case for teacher, we are not interested in only **particular** hours of homework (levels), but rather would like to randomly sample a few hours (levels) in an effort to generalize our findings to a **population** of homework hours. Such would designate hours of homework to be a random effect. In this model then, both teacher and homework would be random effects, giving us the two-way random effects model:

$$y_{ijk} = \mu + \alpha_j + \beta_k + (\alpha\beta)_{jk} + \varepsilon_{ijk}$$

where μ is the population grand mean, α_j (i.e., a_j, its estimate) is the random variable for row sample effects, β_k (i.e., b_k) is the random variable for column sample effects, $(\alpha\beta)_{jk}$ (i.e., $(ab)_{jk}$) is the random interaction effect for a given cell jk, and ε_{ijk} (i.e., e_{ijk}), as before, is the error component, this time for a given individual i in a given cell jk. Notice that the only part of the model that is not random in the two-way random effects model is the grand mean (Hays, 1994). The rest of the model consists of random variables, including the error component ε_{ijk}.

The assumptions for the two-way random effects model parallel those of the one-way random model, though we now have to generally assume interaction effects, $(\alpha\beta)_{jk}$, to be normally distributed with mean 0 and variance σ_{AB}^2, as well as assuming α_j, β_k, $(\alpha\beta)_{jk}$, and ε_{ijk} are all pairwise independent (Hays, 1994).

In terms of partitioning variability, the arithmetical computations for the two-way analysis of variance under the random effects model are exactly the same as those for the two-way analysis of variance under the fixed effects model. However, as was true for the one-way model, the mean squares will be different. Consequently, this will imply that we construct our F-ratios differently than in the fixed effects model. As we will see, and for very good theoretical reasons, the **error term for each factor in the two-way random effects model will be MS interaction**, and no longer MS error. This may seem counterintuitive at first consideration, but our derivation of the EMS will prove our intuition wrong.

We begin by considering the expected mean squares. As was true for the one-way random effects model mean squares, our starting point for considering these for the two-way model begins with recalling features of the fixed effects model. Recall that in the two-way fixed effects model, the row and column effects each summed to 0, that is, $\sum_j a_j = 0$ and $\sum_k b_k = 0$. The interaction effects, $(ab)_{jk}$, also summed to zero across rows, columns, and cells. What this meant is that in the fixed effects models, when considering relevant row and column effects, we did not need to concern ourselves with interaction effects being "picked up" along the way in our computation of row or column effects, since they summed to 0 in each case. The only thing that was being accumulated in our summation was the usual error term, $\bar{e}_j$. For instance, a given row effect a_j could be written as follows:

$$\left(\bar{y}_j - \bar{y}.\right)^2 = \left(a_j + \bar{e}_j - \bar{e}\right)^2 \tag{5.5}$$

The major point of (5.5) is to emphasize that when taking squared deviations from the grand mean in the fixed effects model, the deviation reflects only the fixed effect a_j and mean error (i.e., $\bar{e}_j - \bar{e}$). **Notice that the interaction effect does not contribute to the sums of squares for rows, because the sum of the interaction effects equals 0 in the fixed effects model as we sum across columns**. Or, again, if you prefer, one could say that the interaction effect **is** included in the sum of squares for the fixed effect a_j, but that since it equals 0, it drops out of the fixed effect term. A similar situation applies to columns. There is simply no interaction effect (i.e., the interaction effect will equal to 0) included in the column effect. This is an extremely important point to grasp in order to understand the random effects model under discussion, and the mixed model to be surveyed later. When generating F-ratios for fixed effects, we were not "picking up" interaction variance, and hence had no need to consider interaction in generating suitable F-ratios to test main effects. That is, they did not figure in the expected mean squares.

5.11   FIXED EFFECTS VERSUS RANDOM EFFECTS: A WAY OF CONCEPTUALIZING THEIR DIFFERENCES

As an aside and prior to our development of the two-way model, there is a way to understand the difference between a fixed effect and a random one, and that is in drawing on our knowledge of an "effect" we are already very much familiar, that of e_{ijk}.

Recall that the effects a_j in any given sample will not necessarily equal their long-run expectation in a random effects model. Yes, while it is true that $E(\bar{a}_j) = 0$, when we simply take a random sample from the set of all possible levels, there is no guarantee, theoretically, that a given sample will match that long-run expectation. A similar situation applies for the b_k column effects. Likewise, the sample values for interaction effects $(ab)_{jk}$, because they are now too **random**, do not have to match their expected values in the sample of levels selected for the given experiment.

If you compare this with the behavior of the error term, e_{ijk}, you will notice that the error term behaves in a similar fashion. Yes, the long-run expectation of the error is equal to 0, that is, the mean of the error over an infinite number of repeated samples is expected to be 0. However, in any given experiment, in any given **sampling** of e_{ijk}, there is no reason to suspect that e_{ijk} will equal that long-run expectation. This is why e_{ijk} is quite naturally regarded as a **random effect** (even before we knew what random effects were!). Its "levels" (i.e., the values of e_{ijk} occurring in a given experiment) are randomly sampled from a larger population of potential "levels" (i.e., from a larger population of potential errors).

As we will see, it is this element of randomness of both a_j and b_k that will have influential consequences on ensuing expected mean squares and generation of suitable F-ratios to test effects of interest.

5.12 CONCEPTUALIZING THE TWO-WAY RANDOM EFFECTS MODEL: THE MAKE-UP OF A RANDOMLY CHOSEN OBSERVATION

To explain how things work in a two-way random effects model, we begin with the idea that we have been tracing since our first look at the one-way fixed effects ANOVA in Chapter 3, that of the "make-up" of a given observation for the model under consideration. We again borrow quite heavily from the work of Hays (1994), Kempthorne (1975), Kirk (1995), Searle, Casella, and McCulloch (1992), and Scheffé (1999) in what follows.

For the two-way model, we begin by conceiving that the grand sample mean $\bar{y}_{..}$ will consist of average row effects, column effects, interaction effects, and mean error:

$$\bar{y}_{..} = \bar{a}_{..} + \bar{b}_{..} + \overline{(ab)}_{..} + \bar{e}_{..}$$

The mean $\bar{y}_{j.}$ of any row will consist of the effect of that row, the mean of the column effects (because we are summing across columns), the mean of the interaction effects within that row, and the mean error in that row:

$$\bar{y}_{j.} = a_j + \bar{b}_{..} + \overline{(ab)}_{j.} + \bar{e}_{j.}$$

That is, notice that to calculate the mean of any row, $\bar{y}_{j.}$, aside from a row effect, a_j (which is what we actually **want** to obtain), we are also "picking up" mean column effects, mean interaction effects, and mean error. As Hays (1994, p. 542) notes, the difference between the row mean and the grand mean (which we want to calculate as usual to get a row effect, $\bar{y}_{j.} - \bar{y}_{..}$), will not include any column effects (we will see that it drops out of the equation), but it does include average interaction effects as well as row effects and error:

$$\begin{aligned}
\bar{y}_{j.} - \bar{y}_{..} &= a_j + \bar{b}_{..} + \overline{(ab)}_{j.} + \bar{e}_{j.} - \bar{a}_{..} - \bar{b}_{..} - \overline{(ab)}_{..} - \bar{e}_{..} \\
&= (a_j - \bar{a}_{..}) + \left[\overline{(ab)}_{j.} - \overline{(ab)}_{..}\right] + (\bar{e}_{j.} - \bar{e}_{..}).
\end{aligned} \tag{5.6}$$

Notice that when we take deviations from the grand mean, of the form $\bar{y}_{j.} - \bar{y}_{..}$, which by the above is the quantity "$a_j + \bar{b}_{..} + \overline{(ab)}_{j.} + \bar{e}_{j.}$" minus "$\bar{a}_{..} + \bar{b}_{..} + \overline{(ab)}_{..} + \bar{e}_{..}$," this difference does not include any column effects, because in (5.6), $\bar{b}_{..}$ dropped out of the final solution. It canceled out, since $\bar{b}_{..} - \bar{b}_{..} = 0$. The final solution does, however, contain row effects and interaction effects. That is, to get a row effect $\bar{y}_{j.} - \bar{y}_{..}$, we also get the "unwanted" interaction effects. We will need a way of dealing with these unwanted effects when we build our F-ratio. In the fixed effects models, we did not have to worry about picking up "nuisance effects" (other than error) when computing row or column effects. Why not? Because these nuisance factors did not exist in fixed effects models (or equivalently, they did exist, but were equal to 0).

Similarly, for the deviation of any column mean from the grand mean, we can define a column effect as containing an effect for that particular column, $b_{.k}$, the mean of the row effects, $\bar{a}_{..}$ (because we are summing this time across rows), a mean interaction effect, $\overline{(ab)}_{.k}$, and the mean error in that column, $\bar{e}_{.k}$:

$$\bar{y}_{.k} = b_{.k} + \bar{a}_{..} + \overline{(ab)}_{.k} + \bar{e}_{.k}$$

Therefore, when we take $\bar{y}_{.k}$ deviations about the grand mean, $\bar{y}_{.k} - \bar{y}..$, we end up with

$$\bar{y}_{.k} - \bar{y}.. = b_{.k} + \bar{a}.. + \overline{(ab)}_{.k} + \bar{e}_{.k} - \bar{a}.. - \bar{b}.. - \overline{(ab)}.. - \bar{e}..$$

$$= (b_{.k} - \bar{b}..) + \left[\overline{(ab)}_{.k} - \overline{(ab)}.. \right] + (\bar{e}_{.k} - \bar{e}..) \tag{5.7}$$

That is, a column deviation from the grand mean contains a column effect, average interaction effects, and average error, but no row effect, because similar to (5.6) when the column effect dropped out of the equation for the row effect, here, the row effect $\bar{a}..$ drops out of the equation. Notice that $\bar{a}.. - \bar{a}.. = 0$ in (5.7).

In summary then, we need to find a way to produce our *F*-ratios such that the interaction in the row and column effects is accounted for. As we will see, **for the two-way random effects model, this will call for a test of main effects MS against the interaction term instead of the MS error term as in the fixed effects ANOVA**. To understand why this is so, however, we need to once more consider the expected mean squares.

5.13 SUMS OF SQUARES AND EXPECTED MEAN SQUARES FOR RANDOM EFFECTS: THE CONTAMINATING INFLUENCE OF INTERACTION EFFECTS

Let us see how the interaction involvement of (5.6) and (5.7) will influence the sums of squares for rows in the two-way random effects factorial model. Recall we derived, for the two-way fixed effects model, the effect for row to be

$$SS\,A = SS \text{ between rows} = \sum_j Kn \left(\bar{y}_{j.} - \bar{y}.. \right)^2$$

Now, when we substitute $\left(\bar{y}_{j.} - \bar{y}.. \right)$ with

$$(a_j - \bar{a}..) + \left[\overline{(ab)}_{j.} - \overline{(ab)}.. \right] + (\bar{e}_{j.} - \bar{e}..)$$

of (5.6), we obtain

$$SS\,A = SS \text{ between rows} = \sum_j Kn \left\{ (a_j - \bar{a}..) + \left[\overline{(ab)}_{j.} - \overline{(ab)}.. \right] + (\bar{e}_{j.} - \bar{e}..) \right\}^2 \tag{5.8}$$

which we can now reduce to, in terms of expected mean squares:

$$E(MS\,A) = E(MS \text{ between rows}) = Kn\sigma_A^2 + n\sigma_{AB}^2 + \sigma_e^2 \tag{5.9}$$

We notice (5.9) contains the interaction term, $n\sigma_{AB}^2$. What this means is that when we consider the construction of a suitable *F*-ratio to isolate σ_A^2, we are going to need a denominator that includes $n\sigma_{AB}^2$ so that we can account for it being a part of the numerator of our *F*-test. Likewise, for factor B (columns), in terms of EMS, we have:

$$E(MS \text{ between columns}) = Jn\sigma_B^2 + n\sigma_{AB}^2 + \sigma_e^2 \tag{5.10}$$

Again, the term $n\sigma_{AB}^2$ appears in (5.10), whereas in the fixed effects model, this term did not appear (or, again, if you like, it did appear, but was equal to 0). Analogous to our test of the row effect, this will call for a different F-ratio for testing the column effect than what we had in the fixed effects model. In the fixed effects model of the previous chapter, we simply did not have to deal with the "contamination" of $n\sigma_{AB}^2$.

Finally, the expectation mean squares for the interaction term ends up being $n\sigma_{AB}^2 + \sigma_e^2$, and as usual, the expectation for MS error is σ_e^2. See Searle, Casella, and McCulloch (1992) for how this expectation is obtained.

5.13.1 Testing Null Hypotheses

As was true for the one-way random effects model, the null for factor A is given by $H_0 : \sigma_A^2 = 0$. This null hypothesis, if "true," would imply that $Kn\sigma_A^2 = 0$, and so all that is left from the expected mean squares is

$$Kn\sigma_A^2 + n\sigma_{AB}^2 + \sigma_e^2$$
$$0 + n\sigma_{AB}^2 + \sigma_e^2$$

What if we naively decided to use good 'ol MS error as our error term for testing this effect? Under the null hypothesis that $\sigma_A^2 = 0$, we would have:

$$\frac{n\sigma_{AB}^2 + \sigma_e^2}{\sigma_e^2}$$

Notice that had we used MS error, we would still have an interaction term unaccounted for in the numerator, which would mean that even if there are no effects for factor A, we might still obtain an F appreciably greater than 1. This would be because interaction variance, $n\sigma_{AB}^2$ is making its way into the numerator and we are not effectively isolating $Kn\sigma_A^2$. Therefore, this calls for us to use a new error term to test the main effect for such a random effect. Which error term shall we choose to "get rid of" $n\sigma_{AB}^2 + \sigma_e^2$? We notice that this term is actually the **mean square for interaction**, since recall that this is what we found the expectation for interaction to be.

Now, everything should be beginning to fall into place. The test for factor A must be against MS interaction as it allows us to isolate the effect of interest in the numerator:

$$F = \frac{\text{MS A}}{\text{MS A} \times \text{B interaction}} = \frac{Kn\sigma_A^2 + n\sigma_{AB}^2 + \sigma_e^2}{n\sigma_{AB}^2 + \sigma_e^2}$$

We lose a degree of freedom for row and one for column, so the degrees of freedom on which the above F will be tested are equal to $(J-1)$ and $(J-1)(K-1)$.

Likewise, for factor B, to evaluate the null hypothesis $H_0 : \sigma_B^2 = 0$, since there is interaction variance again "contaminating" the effect, $E(\text{MS between columns}) = Jn\sigma_B^2 + n\sigma_{AB}^2 + \sigma_e^2$, the appropriate denominator for testing this effect (on $(K-1)$ and $(J-1)(K-1)$ degrees of freedom) is once more $n\sigma_{AB}^2 + \sigma_e^2$:

$$F = \frac{\text{MS B}}{\text{MS A} \times \text{B interaction}} = \frac{Jn\sigma_B^2 + n\sigma_{AB}^2 + \sigma_e^2}{n\sigma_{AB}^2 + \sigma_e^2}$$

If $H_1 : \sigma_B^2 > 0$, then the term $Jn\sigma_B^2$ will reflect this effect, and the F-statistic will be appreciably greater than 1.0. Otherwise, we will be left with simply

$$F = \frac{Jn\sigma_B^2 + n\sigma_{AB}^2 + \sigma_e^2}{n\sigma_{AB}^2 + \sigma_e^2}$$

$$= \frac{0 + n\sigma_{AB}^2 + \sigma_e^2}{n\sigma_{AB}^2 + \sigma_e^2}$$

$$= \frac{n\sigma_{AB}^2 + \sigma_e^2}{n\sigma_{AB}^2 + \sigma_e^2}$$

and our expectation for F would be approximately 1.0 under the null hypothesis $\sigma_B^2 = 0$.

What is the appropriate denominator for testing $H_0 : \sigma_{AB}^2 = 0$? This one is easy. Since we found the expected mean squares to be $n\sigma_{AB}^2 + \sigma_e^2$, it is quite evident that the correct denominator in this case actually is **MS error**, evaluated on $(J-1)(K-1)$ and $JK(n-1)$ degrees of freedom. That is,

$$F = \frac{\text{MS interaction}}{\text{MS error}} = \frac{n\sigma_{AB}^2 + \sigma_e^2}{\sigma_e^2}$$

In summary then, we have found that in the two-way random effects model, both random effects are to be tested against MS interaction, while the interaction term is to be tested against MS error.

5.14 YOU GET WHAT YOU GO IN WITH: THE IMPORTANCE OF MODEL ASSUMPTIONS AND MODEL SELECTION

Even if you should never venture into models with random effects (other than, of course, the **error term** in a fixed effects model, which is virtually always present), a survey of random effects is pedagogically instructive because it serves to illustrate that the conclusions one draws from an analysis of data are very much contingent on the **assumptions** and **sampling** one enters with into the model-building process. The actual arithmetic of the ANOVA may very well be the same in many cases, but **the construction of F-ratios will differ based on the assumptions you make at the very beginning of your experiment**. We summarize this idea with the following:

> **If you use a fixed effects model, when really, you are interested in interpreting a random effects model, you will be restricted to making inferences only about the levels of the independent variable that are present in your experiment. Your substantive conclusions are intimately tied to the model you have tested.**

There are many research papers across the sciences where researchers, after conducting a fixed effects analysis of variance, regularly, and perhaps inadvertently, generalize their findings to levels of the independent variable(s) not tested in the model. As emphasized by Searle, Casella, and McCulloch (1992, p. 22), "**Users of computer packages that have F-values among their output must be totally certain that they know precisely what the hypothesis is that can be tested by each such F-value.**"

Let us shed a bit more perspective on Searle et al.'s warning. Consider the following scenario: As a researcher in sensation and perception, suppose you are interested in the variability explained in pupil size (i.e., dependent variable) when looking at various playing cards. If you select two playing cards, say a king of spades and a jack of hearts, measure pupil size, and find there is a statistically significant difference between pupil size for king of spades versus pupil size for jack of hearts, under the fixed effects ANOVA, you will only be able to conclude mean differences for **these two card-types only**, since you are assuming that in replications of the experiment, only these two cards would be used again and again. Now, had you used a random effects model, **and randomly sampled these two cards from the deck**, you could have concluded that differences in cards, either those selected randomly for the given experiment or those in the population of potential cards that could have been selected, accounts for a given amount of variance in pupil size. That is, you would be able to make a more **general** statement in the random effects model. You would be able to say something about playing cards **in general**, rather than just the two kinds you selected.

As a general guideline, when you interpret an ANOVA, always ask yourself whether the investigator is assuming a fixed or random effects model, and then critically evaluate whether the data were analyzed and interpreted in correspondence with these assumptions. Be sure to verify whether conclusions outlined in results and discussion sections agree with the model actually analyzed. If they line up, then great. If they do not, then at least you will have a sense of the limitations imposed by the analysis in relation to the potentially much more broad conclusions drawn in the discussion of the paper. **Researchers often like to overstate conclusions in discussion sections despite the fact that their statistical analyses do not support such conclusions.**

5.15 MIXED MODEL ANALYSIS OF VARIANCE: INCORPORATING FIXED AND RANDOM EFFECTS

Suppose that instead of merely wanting to demonstrate that teacher is associated with variance in achievement, you also wanted to show that the lesson plan used by the teacher is also associated with achievement. Suppose you were interested in specifically comparing five different lesson plans. Hence, teacher remains **random**, but lesson is now **fixed**. When we have a mix of fixed and random factors, we have the **mixed model analysis of variance**. Pinheiro and Bates (2000) do a nice job of summarizing the applied rationale of a mixed model:

> Mixed-effects models are primarily used to describe relationships between a response variable and some covariates in data that are grouped according to one or more classification factors. Examples of such **grouped data** include **longitudinal data**, **repeated measures data**, **multilevel data**, and **block designs**. By associating common random effects to observations sharing the same level of a classification factor, mixed-effects models flexibly represent the covariance structure induced by the grouping of the data. (p. 3)

Purely random effects models are relatively rare. Fixed effects models are much more common across the social, economic, and medical sciences. However, a study of random effects such as we have undergone is quite useful, not only because it provides an understanding of the random effects model itself, but also because it serves as a "bridge" to the mixed model, which is quite popular.

As we did for both the fixed effects and random effects models, we consider the expected mean squares for the mixed model. When we obtain effects for the fixed factor, we will need to sum across a **random factor**. Just as we summed across random factors in the two-way random effects model, we will once again conclude that this factor (i.e., the **fixed** one, not the random one) be tested against **MS interaction** and not MS error.

To help better understand the denominators we will use for testing fixed and random effects, consider the layout in Table 5.2. In this layout, the fixed factor, represented by rows, has six levels, and the random factor, represented by columns, has three levels.

TABLE 5.2 Cell Layout for 6 × 3 Mixed Model Analysis of Variance

		Random Factor (B)			
		I	II	III	Row Means
Fixed Factor (A)	I	y_{ijk}	y_{ijk}	y_{ijk}	$\bar{y}_{j.}$
	II	y_{ijk}	y_{ijk}	y_{ijk}	$\bar{y}_{j.}$
	III	y_{ijk}	y_{ijk}	y_{ijk}	$\bar{y}_{j.}$
	IV	y_{ijk}	y_{ijk}	y_{ijk}	$\bar{y}_{j.}$
	V	y_{ijk}	y_{ijk}	y_{ijk}	$\bar{y}_{j.}$
	VI	y_{ijk}	y_{ijk}	y_{ijk}	$\bar{y}_{j.}$
	Column means	$\bar{y}_{.k}$	$\bar{y}_{.k}$	$\bar{y}_{.k}$	$\bar{y}_{..}$

In the layout of Table 5.2, we will have the following effects for the fixed factor and random factor:

- **Row effects**, denoted by $\bar{y}_{j.} - \bar{y}..$, represent the effect of being in one row versus being in other rows on levels of the fixed factor.
- **Column effects**, denoted by $\bar{y}_{.k} - \bar{y}..$, represent the effect of being in one column versus being in other columns on levels of the random factor.

The questions we need to ask ourselves about Table 5.2 are the following:

- What kind of information went into producing the row effects, $\bar{y}_{j.} - \bar{y}..$? Notice that to get these row effects, we need to sum across a **random** factor. How will this summing across a random factor impact the makeup of the given row effect?
- What kind of information went into producing the column effects, $\bar{y}_{.k} - \bar{y}..$? Notice that to get these column effects, we need to sum across a **fixed** factor. How will this summing across a fixed factor impact the makeup of the given column effect?

To get a given row effect, $\bar{y}_{j.} - \bar{y}..$, because we are needing to sum across a random effect, we have every reason to believe that **the sum of interaction effects, $(ab)_{jk}$, will not equal to 0** (Hays, 1994). Hence, we will need to account for this source of variation when constructing our F-ratio. That is, within any row of the fixed effect, we can expect there to be an average interaction effect, unequal to zero (and possibly different from row to row), that we are "picking up" as we sum across the given row. These row totals then, and their corresponding effects, will not only reflect row effects, but rather will also be reflective of average **interaction effects**. To the contrary, to get a given column effect, $\bar{y}_{.k} - \bar{y}..$, because we are summing across a **fixed** effect, we have good reason to believe that the sum of interaction effects, $(ab)_{jk}$, **will** equal to 0. Hence, we do not need to account for this source of variation when constructing our F-ratio (or equivalently, we can account for it, but it will be equal to zero each time).

How are the expected mean squares impacted by all this? For the fixed effect, factor A, EMS is equal to

$$E(\text{MS A}) = \sigma_e^2 + n\sigma_{AB}^2 + \frac{Kn\left(\sum_j \alpha_{j.}^2\right)}{J-1}$$

Notice that included in this EMS is interaction variance, $n\sigma_{AB}^2$, which is **unwanted**. For the random effect, EMS is equal to

$$E(\text{MS B}) = \sigma_e^2 + Jn\sigma_B^2$$

Notice that the only **unwanted** variation in this EMS is that of σ_e^2. The EMS for the interaction term ends up being, quite simply

$$E(\text{MS AB}) = \sigma_e^2 + n\sigma_{AB}^2$$

We now have all the information necessary to build our F-ratios. For the fixed effect, under the null hypothesis of no effect, we get

$$E(\text{MS A}) = \sigma_e^2 + n\sigma_{AB}^2 + \frac{Kn\left(\sum_j \alpha_{j.}^2\right)}{J-1}$$

$$= \sigma_e^2 + n\sigma_{AB}^2 + \frac{Kn\left(\sum_j (0)_{j.}^2\right)}{J-1}$$

$$= \sigma_e^2 + n\sigma_{AB}^2$$

which suggests that the correct denominator for testing the fixed effect must be MS interaction:

$$F = \frac{\text{MS A}}{\text{MS AB}} = \frac{\sigma_e^2 + n\sigma_{AB}^2 + \frac{Kn\left(\sum_j (0)_{j.}^2\right)}{J-1}}{\sigma_e^2 + n\sigma_{AB}^2}$$

Notice that it is the fixed factor (not the random factor) that is tested against the interaction term in the mixed model.

Under the hypothesis of no column effect (random factor), $\sigma_B^2 = 0$, since $E(\text{MS B}) = Jn\sigma_B^2 + \sigma_e^2$ we end up with simply σ_e^2. Thus, the F-ratio for the random factor is given by

$$F = \frac{Jn\sigma_B^2 + \sigma_e^2}{\sigma_e^2}$$

Notice that it is the random factor (not the fixed factor) that is tested against MS error in the mixed model.

As a recap of what we have done, we have seen that in a two-way mixed model, to produce the F-test for the random effect, we divide by MS error. The reason for this is that to produce the column means, we have to sum across the fixed factor. Those respective sums are not expected to contain anything but variability due to levels of the random factor along with error.

For the fixed effect, however, what went into the sums for rows? That is, when we produce the sum (or the mean) for each row (fixed effect, in our layout), what kind of variability went into each of these row sums? There is surely (hopefully) variability due to the effect of being in that particular row and not other rows, and there is variability due to error, as usual. But, there is another source of variability, and that is interaction variance. Why? Because when we tally up the cell totals for a level of the fixed factor, we are summing across only a **sample** of possible levels of the random factor. Hence, if we were to do the experiment over, and presumably sampled different levels of the random factor, the effect we would obtain for the given level of the fixed effect might change by the very nature of summing across the random factor in question. Hence, we have "unwanted" interaction variance in the rows and have to account for this when generating the corresponding F-ratio. If we produced our F-ratio by dividing by MS error, we would still have an interaction effect left over in the numerator, and thus we would have failed to isolate the effect of interest (i.e., row effect). We would have failed to test our null hypothesis of interest.

5.15.1 Mixed Model in R

Having laid out some of the theory for mixed models, we now estimate a mixed model on the achievement data, this time specifying textbook as a fixed factor and teacher as a random effect (Table 4.1). Of course, there is much more to the fitting of a mixed model than shown here (e.g., plots, diagnostics to verify assumptions, etc.). Our purpose here is only to briefly demonstrate how such a model can be fit in R.

We use the package `nlme` (Pinheiro et al., 2014), and fit our model using REML (partial output shown below):

```
> library(nlme)
> mixed <- lme(ac ~ f.text, data = achiev, random = ~1 | f.teach)
> summary(mixed)

Random effects:
 Formula: ~1 | f.teach
         (Intercept) Residual
StdDev:    9.733736 4.423571

Fixed effects: ac ~ f.text
                Value Std.Error DF   t-value p-value
(Intercept) 78.58333  5.031607 19 15.617940  0.0000
f.text2      0.91667  1.805915 19  0.507591  0.6176
```

In the code, `random = ~1 | f.teach)` designates the random effect. The coefficient for `f.text2` is a mean contrast between the first and second textbooks (i.e., 79.50 − 78.58 = 0.92). The effect for textbook is not statistically significant ($p = 0.6176$). The variance component for `f.teach` is equal to the square of 9.73, which is 94.67. Since the square of the residual is equal to 19.57, the proportion of variance accounted for by `f.teach` is 94.67/(94.67 + 19.57) = 94.67/114.24 = 0.83 (rounded up from 0.829). Confidence intervals for effects can also be obtained via `intervals(mixed)`.

5.16 MIXED MODELS IN MATRICES

Having briefly introduced the mixed model for the simplest case, we now briefly consider the mixed model in its most general matrix form:

$$\mathbf{Y} = \mathbf{XB} + \mathbf{ZU} + \mathbf{E} \tag{5.11}$$

where, $\mathbf{Y}$ is a response matrix, $\mathbf{X}$ is a model matrix associated with the fixed effects in $\mathbf{B}$, $\mathbf{B}$ is a vector of parameters corresponding to the fixed effects, $\mathbf{Z}$ is the model matrix associated with the random effects in $\mathbf{U}$, and $\mathbf{E}$ is a vector of errors, what is left over from the model after prediction of $\mathbf{Y}$. We assume that $\mathbf{U} \sim N(\mathbf{0}, \mathbf{\Sigma}_z)$ and $\mathbf{E} \sim N(\mathbf{0}, \mathbf{\Sigma}_\varepsilon)$, where $\mathbf{\Sigma}_z$ is the covariance matrix for the random effects and $\mathbf{\Sigma}_\varepsilon$ is the covariance matrix for the errors contained in $\mathbf{E}$. This formulation of the model often goes by the name of the **Laird-Ware form**, after the seminal paper "**Random-Effects Models for Longitudinal Data**" (Laird and Ware, 1982) that provided the very general form of the mixed model. Because of $\mathbf{Y}$, the model in (5.11) can also accommodate more than a single response variable, giving us the **multivariate mixed model** (Timm, 2002), of which all other univariate mixed models can be considered special cases.

5.17 MULTILEVEL MODELING AS A SPECIAL CASE OF THE MIXED MODEL: INCORPORATING NESTING AND CLUSTERING

Our study of the mixed model lends itself well to introducing a class of modeling methodologies that is increasing in popularity in the social and natural sciences, that of **multilevel** or **hierarchical modeling**. As we discuss in the chapter to follow, mixed models are also useful for addressing problems of **repeated measurements**, which usually can also be conceptualized as having a "multilevel" or "hierarchical" structure.

The topic of multilevel modeling is beyond the scope of this book. Our goal here is to simply conclude this chapter with a **foot-in-the-door** commentary as to how these models can be conceptualized as a special case of the more general mixed model. Indeed, as Pinheiro and Bates (2000) note:

> This model with two sources of variation, b_i and ε_{ij}, is sometimes called a **hierarchical** model ... or a multilevel model. The b_i are called **random** effects because they are associated with the particular experimental units [...] that are selected at random from the population of interest. They are **effects** because they represent a deviation from an overall mean. ...Because observations made on the same [level of the independent variable] share the same random effect b_i, they are correlated. The covariance between observations on the same [level] is σ_b^2 corresponding to the correlation of $\sigma_b^2 / (\sigma_b^2 + \sigma^2)$. (p. 8)

To properly discuss the multilevel model, it helps first to recall where we have been. Recall the one-way fixed effects analysis of variance model of Chapter 3:

$$y_{ij} = \bar{y} + a_j + e_{ij}$$

In this model, we assumed the treatment effects a_j to be fixed and e_{ij} to be random and normally distributed. In specifying a_j as fixed, it implied that we were only interested in mean differences as represented by the factor levels actually included in the given experiment. If we were interested in the population of levels of which the ones showing up in our experiment constituted a random sample, then we specified a_j as random, and had the one-way random effects model, which is the same as the fixed effects model, only that now, sample effects are considered randomly sampled from a larger population.

This type of model in which we allow a_j to be random instead of fixed can, in many cases, actually be conceived as a very simple version of what is known as the **multilevel** or **hierarchical** model. What are the levels of the "hierarchy?" The observations y_{ij} constitute **level 1**, and the "grouping" random treatment effect a_j constitutes **level 2**. We say that observations y_{ij} are **nested** within level 2.

For instance, suppose that in our achievement example, instead of randomly assigning students to teacher, we simply sampled students **as they were**, and **as already associated with a given teacher**. In such a case, school children y_{ij} would be considered **nested** within teacher. If we then randomly sampled a number of teachers (say, four, as in our previous example), but wished to generalize to a wider population of teachers, then teacher becomes a random effect. But how is this also a multilevel or hierarchical model? Such models emphasize the fact that observations often occur in a **natural hierarchy** or as a result of one being imposed through a sampling plan (such as **blocking**). For our student observations, there is expected to be a **likeness** about students who share the same teacher. **Observations "within teacher" are more likely to be similar than observations between teachers, not necessarily because of any external treatment condition imposed, but simply because these students share the same teacher**. And having the same teacher means they share the same teaching style, etc., and all of the other infinite innumerable (and potentially even **immeasurable**) elements that may be related to sharing the same teacher. And though there is nothing technically inherent in the definition of "multilevel modeling" that prevents us from designating all effects as **fixed effects** (e.g., studying and generalizing to mean differences between teachers), when we speak of multilevel or hierarchical models, we are usually implicitly invoking the idea that we have one or more **random effects**. For our example, we are usually interested in generalizing to more teachers than we have sampled for our study, making it, as we have seen, a random effect.

Our point is that **multilevel structures** are often analyzed via mixed models. There is nothing inherent in such a hierarchical structure that "demands" such data be analyzed as such, but for reasons of both wanting to account for **likeness** of observations within levels of the hierarchy as well as generalizing to levels of the treatment effect, these typically necessitate the use of such models. For a classic introduction to multilevel and hierarchical data, see Raudenbush and Bryk (2002). Snijders and Bosker (1999) also provide a very readable treatment.

5.18 CHAPTER SUMMARY AND HIGHLIGHTS

- In the traditional **fixed effects model**, the specific levels of the independent variable(s) chosen by the experimenter are of interest, and population inferences are made about those, and only those, levels used in the experiment. Null hypotheses are tested of the sort $H_0 : \mu_1 = \mu_2 = \mu_3 \cdots = \mu_J$.

- In the **random effects model**, the experimenter is not interested specifically in the levels chosen for the particular experiment. Instead, the levels chosen are merely regarded as a **random sample of potential levels** that could have been chosen. The experimenter is interested in testing a null hypothesis that the variance in the dependent variable accounted for by the given factor is equal to 0, that is, $H_0 : \sigma_A^2 = 0$.

- The conceiving of **sample effects** as **random** rather than **fixed** has important implications for the construction of F-ratios.

- In the one-way random effects model, **MS error** is a suitable **error term** for constructing the F-ratio for a test of the random effect. **Variance components** may be estimated using ANOVA estimation, ML, or REML. REML is often the estimator of choice in random effects and mixed models.

- In the two-way random effects model, because each effect is computed by summing across a random effect, the expected mean squares dictate **MS interaction** to be the correct error term for each effect in the generation of F-ratios.

- When a model has a mixture of fixed and random effects (in addition, naturally, to the error term), the model is a **mixed model**. EMS for a two-way mixed model reveals that it is the **fixed effect** that is tested against **MS interaction**. The random effect is tested against MS error.

- Understanding that ε_{ijk} is always a **random effect**, whether in fixed effects, random effects, or mixed models, helps one to better appreciate the nature of random effects in general, realizing that their behavior will be governed by similar random processes as is true of ε_{ijk}.
- An understanding of basic mixed model theory coupled with the idea of **nesting structures** lends itself to conceiving the **multilevel** or **hierarchical linear model**.
- **Random effects** and **mixed models** can be fit in R using `lme4` or `nlme`. SPSS's `VARCOMP` can also be used to estimate variance components.

REVIEW EXERCISES

5.1.   Discuss why a researcher may wish to conduct a **random effects analysis of variance** instead of a **fixed effects** ANOVA.

5.2.   Elaborate on the statement "**Random effects ANOVA is not about means, it is about variances.**"

5.3.   Distinguish between a **random effects** model and a **mixed effects** model.

5.4.   Give an example of three research scenarios that would necessitate the fitting of a **random effects** model.

5.5.   Give an example of three research scenarios that would necessitate the fitting of a **mixed effects** model.

5.6.   Distinguish the **assumptions** for a one-way **fixed effects** model from those of a one-way **random effects** model. How are they similar? Different?

5.7.   How are a_j and e_{ij} similar in a **random effects** model but different in a **fixed effects** model?

5.8.   How can it be said that, technically, **virtually all ANOVA models are either random effects or mixed models**, and that purely fixed effects models rarely exist?

5.9.   What are three common ways of estimating parameters in a **random effects model**?

5.10. What is the **expected mean squares** for the **random factor** in a one-way random effects model? What implication does this EMS have on the construction of the corresponding F-ratio?

5.11. How does the **null hypothesis** for a one-way **random effects** model differ from that of a one-way fixed effects model?

5.12. Given the F-ratio for a one-way **random effects model**, what is the expectation for F under the null hypothesis, and why?

5.13. Define the **intraclass correlation** coefficient, its meaning, and its purpose.

5.14. In the chapter example of **achievement as a function of teacher**, explain how the interpretation of findings would have changed had teacher been regarded as a **fixed effect** rather than a random one. Would this have helped or hindered the cause of the parent in responding to the principal's claim? How so?

5.15. Consider the following hypothetical data in Table 5.3 on **factor A** (three levels) and **factor B** (six levels). Factor A is a **fixed factor** while factor B is a **random factor**. Within each cell is a single observation.

TABLE 5.3   Cell Layout of Data on Factors A and B

Factor B (R)	Factor A (F)			Means
	1	2	3	
1	11	27	57	31.67
2	12	29	45	28.67
3	14	31	65	36.67
4	16	26	95	45.67
5	51	36	54	47.00
6	24	35	46	35.00
Means	21.33	30.67	60.33	37.44

Estimate a two-way mixed model in R using REML. How much variance is accounted for by the random effect?

6

RANDOMIZED BLOCKS AND REPEATED MEASURES

The analysis of variance is not a mathematical theorem, but rather a convenient method of arranging the arithmetic.

(Fisher, in Wishart (1934))

In the typical between-subjects experimental design, the purpose of randomly assigning subjects to treatment conditions is so that all of the "nuisance" factors associated with subjects that we are **not** interested in studying hopefully "balance out" across treatment groups. For instance, again referring to our melatonin example, if we were to test the effectiveness of differing doses of melatonin on sleep, we might generate treatment groups of control, 1 mg and 3 mg. When randomly selecting a subject from the population then randomly assigning that subject to, say, the 1 mg group, that subject carries with him or her all characteristics **unique** to that individual that could, theoretically, be related to, or have an influence on the dependent variable we are studying (in this case, sleep onset latency). Perhaps it is true that someone with a very healthy immune system will naturally respond better to low melatonin doses than someone who is not quite as healthy. What if, just by chance, the healthy individual winds up in the 1 mg group while the weaker individual winds up in the 3 mg group? Of course, ideally, one would hope none of these nuisance effects would "pile up" in one group or the other. But if they did, it could have serious consequences on the interpretation of findings. One way to overcome this potential confound is to implement what is known as a **randomized block design**.

In this chapter, we survey the **randomized block design** and **repeated measures model**. When we generate blocks, we produce **homogeneous subsets** of subjects before administering levels of the independent variable within each block. **The goal of both the randomized block design and that of repeated measures is to account for the source of variability that is housed within blocks.** We study these two statistical techniques in the same chapter because they are very much intimately

Applied Univariate, Bivariate, and Multivariate Statistics: Understanding Statistics for Social and Natural Scientists, With Applications in SPSS and R, Second Edition. Daniel J. Denis.
© 2021 John Wiley & Sons, Inc. Published 2021 by John Wiley & Sons, Inc.
Companion Website: www.wiley.com/go/denis/appliedstatistics2e

related. As we will see, the **subject** factor in a repeated measures is, in actuality, the **blocking** factor in a randomized block design. We also briefly survey how the repeated measures model can be interpreted as the **multilevel** or **hierarchical** model discussed in the conclusion of the previous chapter. These models take into consideration the **nesting** structure implicit in the data, which, in the case of repeated measures, are the repeated measurements nested within individual.

There are a number of excellent sources on block designs, repeated measures, and longitudinal models. See Kirk (1995, Chapter 7) for a good introduction to randomized block designs. Casella (2008), Dean and Voss (1999) give good overviews of a variety of designs, which includes advanced features such as **confounding** in blocks and **fractional factorial** designs. Winer, Brown, and Michels (1991) is perhaps still, the "bible" of experimental design and should also be consulted. Singer and Willett (2003) provide extensive coverage of longitudinal models, including time-varying covariates, and applications to nonlinear and multilevel structures. Mead (1988) provides extensive coverage of blocking and repeated measures designs.

6.1   WHAT IS A RANDOMIZED BLOCK DESIGN?

The primary purpose of a randomized block design is to reduce the error term estimated in an analysis of variance through an attempt to account for one or more nuisance factors. The logic of a blocking design is produce groups of **participants who are alike** (or "**homogeneous**") on one or more nuisance variables, and then carry out the random assignment of subjects to conditions **within** each of these blocks. If blocking is successful, the resulting MS error term from the analysis of variance will typically be **smaller** than it otherwise would have been without blocking. Such is the logic of blocking, to account for additional sources of variation by including a source of variation called "blocks" into the ANOVA that would otherwise be relegated to the error term. When the error term is reduced in this fashion, we hopefully allow for a more **sensitive** (i.e., powerful) test of the null hypothesis we are interested in. Why "hopefully?" Because when we block, we typically lose degrees of freedom for generating the error term for the F-ratio (more on this shortly). Hence, it is possible that blocking simply is not "worth it" in terms of **trading off degrees of freedom for a reduction in MS error**. As summarized by Larsen and Marx (2001, p. 675), "If the block sums of squares is significantly large, as determined by the F test…the loss of degrees of freedom to blocks was a good investment." However, if SS due to blocks is not that large, then losing degrees of freedom may result in a less powerful test of the primary hypotheses in the ANOVA. The moral of the story is clear: if you are going to block, block on one or more nuisance factors that you know will "steal" a great deal from the error term. Otherwise, it may simply not be worth it (including the trouble and expense of initiating a blocking protocol in your experiment). Experienced researchers usually know what to block on, because they are intimately familiar with the literature in their area of investigation.

6.2   RANDOMIZED BLOCK DESIGNS: SUBJECTS NESTED WITHIN BLOCKS

Recall our brief discussion of the previous chapter regarding the concept of "nesting" as naturally occurring in multilevel structures and often analyzed through mixed models. In randomized block designs (and repeated measures designs, as we will see), a nesting effect is also present. However, in the typical block or repeated measures design, **we impose the nesting structure by generating blocks**. If we planned our blocks successfully, it stands that individuals **within blocks** will be more **alike** compared to individuals between blocks. It is in this sense then that **subjects are nested within blocks**. As we will see, we may treat block as either a fixed effect or a random effect. Blocks are usually decided upon and generated **beforehand** to increase the homogeneity within blocks in a deliberate fashion. For example, if we thought that IQ level might be related to our dependent variable of interest, but we had no interest in studying IQ specifically through a factorial design, we could block on **low**, **medium**, and **high** IQ. In this case, we, as investigators, are generating the blocks by pretesting individuals and classifying them into one of three blocking groups. In other instances, blocks will be more naturally occurring, such as **classrooms nested within school**. The distinction between whether we "choose" the blocking factor or whether it is inherent in the data is nicely summarized by Casella (2008):

> Thus, if we are in a situation where blocks can be chosen, it makes sense to choose them as disparate as possible. This also makes good common sense, in that we want to verify our treatment comparisons on as wide a variety of situations as possible … If the variation in blocks is not controllable … where we block on subjects, but their use is dictated by the inherent design, then we just hope that the variation removed due to blocking is a large piece (and it typically is). (p. 108)

In this chapter, we present only minimal theory regarding the randomized block design. The reason for not elaborating too much on the model is because our primary reason for discussing it at all is to use it as a precursor and introduction to the repeated measures ANOVA model, which can be conceptualized as **blocking taken to the extreme**. In a repeated measures model, we again have a blocking effect, though as mentioned, the blocks will actually be **subjects**. In such models, where subjects are tested more than a single time, we will say that **measurements are nested within subject**. Repeated measures models and longitudinal models are omnipresent in social research, so our primary focus of this chapter is to provide a reasonable, if not still quite brief, introduction to such models, and how they deal, on a statistical level, with these nested structures.

We summarize our main points so far:

- In the one-way randomized block design, subjects are nested within blocks, meaning that subjects within one level of the blocking factor will be more **alike** or **similar** than subjects between blocks. Statistically, the blocking factor could be regarded as **fixed** or **random**, for the same reasons why experimental factors were considered fixed or random in fixed, random, or mixed models of the previous chapters.
- In the **one-way repeated measures design**, measurements are nested within subject, meaning that measurements within each subject are more likely to be **alike** than measurements across subjects. Though subject can be regarded as a fixed effect, it makes much more sense to designate it as a random effect, since we are usually not specifically interested in the given subjects sampled.

The idea of blocking, though formally introduced in this chapter, is not new to us. In the **paired samples *t*-test** in which subjects served under a pretest then posttest condition, we essentially had a one-way repeated measures design where measurements were nested within subjects. **The matched-pairs design is, in fact, a basic block design**. If one truly understands the differences between independent and paired samples *t*-tests, one can quite easily grasp the idea of blocking and nesting, which opens the door to even more advanced modeling, including hierarchical and multilevel, as well as many other ANOVA and regression models. As always, most so-called "advanced" concepts are usually rooted in essential introductory concepts. That does not make the introductory concepts "easy" by any means, it simply means that "advanced statistics" usually features difficult introductory concepts that take a long time to truly master and understand. For instance, there is nothing "introductory" about understanding *z*-scores, their "whys" and their "hows," regardless of how they may first appear to an undergraduate in a first course. Statistical concepts are deceptively very slippery.

Recall Table 2.8 from Chapter 2 (reproduced here as Table 6.1). The "matching" is meant to generate subjects within each block sharing more similarity with one another than subjects between blocks.

The problem with these types of designs is that they violate the **assumption of independence** that is required for "ordinary" between-subjects analysis of variance. As a result of the nesting structure,

TABLE 6.1 Matched-Pairs Design

	Treatment 1	Treatment 2
Block 1	10	8
Block 2	15	12
Block 3	20	14
Block 4	22	15
Block 5	25	24

either naturally occurring or imposed by the design, participants within groups (i.e., blocks) usually cannot be considered independent of one another. Special considerations have to be taken into account when analyzing data that have such a nesting or **correlational** structure.

6.3 THEORY OF RANDOMIZED BLOCK DESIGNS

We now briefly discuss the theory behind randomized block designs. As we will see, having already studied the random effects and mixed models of the previous chapter, proposing a statistical model for the randomized block design is somewhat of a review of concepts already learned, since these models are ideally suited to handle the analysis of block designs.

It is first, as always, helpful to start with earlier, simpler models, then build our way up to the model under consideration. Recall once more the one-way fixed effects model:

$$y_{ij} = \mu + \alpha_j + \varepsilon_{ij}$$

where y_{ij} is the score of individual i in group j, μ is the grand mean of all observations (or of all group means, in the case of a balanced design), α_j is the population treatment effect associated with group j representing the mean difference between that particular group and the grand mean (i.e., the sample effect is $\bar{y}_j - \bar{y}.$), and ε_{ij} is the error associated with individual i in group j and represents within-group variation. Also recall that in the fixed effects model, since the investigator is only interested in making conclusions about the **specific** populations operationalized by the independent variable (rather than the population of **potential levels** that could have been included in the given experiment), we assumed that the sum of treatment effects equaled 0 (i.e., $\sum \alpha_j = 0$).

As we learned in the previous chapter, the one-way random effects model is similar to the fixed, only that in the random model, α_j (a_j being its sample estimate) is regarded as a random effect, and thus a rejection of the null hypothesis now implied that varying levels of our independent variable is associated with variance in the dependent variable. This idea of explaining variance was reflected in our null and alternative hypotheses for the random effects model, which recall were $H_0 : \sigma_A^2 = 0$ and $H_1 : \sigma_A^2 > 0$, respectively. Analogous null hypotheses were proposed for the two-way random effects model as well.

We then studied the **mixed model**, in which for the two-way case, we designated one factor as **fixed** and one factor as **random**. That model was given by:

$$y_{ijk} = \mu + \alpha_j + \beta_k + (\alpha\beta)_{jk} + \varepsilon_{ijk} \tag{6.1}$$

where, as for the two-way random effects model, we had to expand our subscript on y_{ijk} to reflect observation i in cell jk. As before in the completely fixed effects model, μ was still the grand mean of all observations and α_j was the treatment effect associated with the fixed factor. Our additional term, β_k, was a random effect, which also implied that the interaction $(\alpha\beta)_{jk}$ was also a random effect. As usual, the model also contained an error component ε_{ijk}, representing within-cell variation that is unexplained by the systematic portion of the model.

What, of the above models, is suitable for a blocking design? The blocking factor will usually (but not always) be regarded as a random effect, since when we block on a nuisance variable, we are usually not interested only in the **particular** blocks we have included in the experiment. We are usually interested in generalizing to the **population** of blocks of which our chosen blocks are but a mere sample. Hence, **assuming our treatment factor is fixed, this implies that our randomized block model will be a mixed model**. Thus, in general, we can say:

> In models for randomized block designs, the blocking factor is usually regarded as random. Assuming the other factor is fixed, this combination of a random factor and a fixed factor gives rise to the mixed model analysis of variance as a suitable model for analyzing randomized block designs.

In what follows, we consider two specific models. The first is the **nonadditive** randomized block design, which is another way of saying that the model contains the factor x block interaction term. The second is the **additive model**, which means that the model does not contain the factor x block interaction.

6.3.1   Nonadditive Randomized Block Design

The nonadditive randomized block design is usually best suited for the two-way mixed model in which there is a main effect for both factor A and factor B, as well as an interaction. The model for the nonadditive randomized block design can be given by (6.1):

$$y_{ijk} = \mu + \alpha_j + \beta_k + (\alpha\beta)_{jk} + \varepsilon_{ijk}$$

where now β_k is the effect associated with the blocking factor, usually considered to be random, $(\alpha\beta)_{jk}$ is the interaction effect for treatment by block, and ε_{ijk} is, as before, the error i associated with cell jk. Again, the inclusion of the interaction term is what makes the model nonadditive. One point worth noting is that through our use of subscripts, we are implying that within each treatment-block combination, we are able to derive an MS error term, which implies that we have a design of $n > 1$ per cell, the so-called **replicated** design. In situations in which we have only a single observation per treatment-block combination, our model can be written as:

$$y_{jk} = \mu + \alpha_j + \beta_k + (\alpha\beta)_{jk} + \varepsilon_{jk} \tag{6.2}$$

Notice that in (6.2), we have dropped the subscript i in y_{jk}, $(ab)_{jk}$ and e_{jk} to indicate that within each cell, there is only a **single** observation (i.e., $n = 1$ per cell). That is, with the notation y_{jk}, we are not having to specify any given subject i, since it is understood that there is only a **single subject per cell**. This notation is not necessarily standard across authors, but rather is simply how we choose to designate the difference between the replicated and nonreplicated designs.

As we will see, models such as this in which there is but a single observation per cell, referred to as **nonreplicated** designs, present some interesting challenges in terms of analysis. Because there is only a single observation per cell, it becomes impossible to derive ε_{jk}, the error term, because there is no within-cell variation to speak of. Hence, in these $n = 1$ per cell designs, **the error term and the interaction term are confounded**, which, as we will see, has implications for how F-ratios are constructed.

6.3.2 Additive Randomized Block Design

If we either assume or otherwise conclude (e.g., through **Tukey's test for nonadditivity**, as we will discuss later) there to be no factor x block interaction, then we can simplify and rewrite the model of (6.1) as

$$y_{ijk} = \mu + \alpha_j + \beta_k + \varepsilon_{ijk}$$

Notice that now we are no longer modeling an $(\alpha\beta)_{jk}$ interaction. What are the consequences of either including or not including a factor x block interaction in the model? This depends primarily on two things: whether treatment and block are considered fixed or random, and the extent of replication within cells. Consider the data in Table 6.2 and the accompanying cell layout of Table 6.3.

At first glance, we note differences among sample means for factor A (21.33 versus 30.67 versus 60.33). We note as well the differences among sample means between blocks in the right-hand margin of the layout. Even if we regard block as fixed, we are usually not interested in mean differences. We

TABLE 6.2 Fictional Data on Y, A, and Block

Y	A	Block
11	1	1
12	1	2
14	1	3
16	1	4
51	1	5
24	1	6
27	2	1
29	2	2
31	2	3
26	2	4
36	2	5
35	2	6
57	3	1
45	3	2
65	3	3
95	3	4
54	3	5
46	3	6

TABLE 6.3 Cell Layout of Data on Y, A, and Block

Block	Factor A			Means
	1	2	3	
1	11	27	57	31.67
2	12	29	45	28.67
3	14	31	65	36.67
4	16	26	95	45.67
5	51	36	54	47.00
6	24	35	46	35.00
Means	21.33	30.67	60.33	37.44

are most interested in simply removing block from the error term of the overall ANOVA. And certainly, when block is random, as we have already argued that it should be designated as such in most contexts, mean differences are definitely not of interest. Likewise, we are usually more interested in simply extracting it from the error term of the model as to provide a more sensitive test for factor A.

For pedagogical purposes, we run the ANOVA model specifying both **factor** and **block** as fixed effects (where f.block in what follows designates block as a factor in R):

```
> block.data <- read.table("blocking.txt", header = T)
> attach(block.data)
> f.a <- factor(a)
> f.block <- factor(block)
> fit.block <- aov (y ~ f.a*f.block)
> anova(fit.block)
```

```
Analysis of Variance Table

Response: y
              Df Sum Sq Mean Sq F value Pr(>F)
f.a            2 4976.4 2488.22
f.block        5  827.8  165.56
f.a:f.block   10 2136.2  213.62
Residuals      0    0.0
```

Notice that R was unable to generate F-ratios for any of the effects (i.e., blank spaces under F value). Also note that it was unable to compute a sum of squares for error. Why did this occur? It occurred because our design is one **without replication** per cell. Recall that within each factor x block cell combination, there exists only a **single observation**. Because of the fact that each cell has $n = 1$, we are unable to generate an error term separate from the interaction term, and according to fixed effects theory of previous chapters, both fixed effects in the model should be tested against MS error. Since we cannot generate an MS error term, we cannot test either effect, the effect due to factor A or that due to block.

Suppose now we decided not to test the factor A by block interaction, but still kept factor and block fixed:

```
> fit.additive <- aov (y ~ f.a + f.block)
> anova(fit.additive)
```

```
Analysis of Variance Table

Response: y
              Df Sum Sq Mean Sq F value   Pr(>F)
f.a            2 4976.4 2488.22  11.648 0.002444 **
f.block        5  827.8  165.56   0.775 0.589224
Residuals     10 2136.2  213.62
```

We note that when we do not test the interaction, R is able to generate F-tests for both factor and block. Why is this so? Because without a specified interaction term, this source of variance is relegated to **MS residual**. Notice that the mean squares of 213.62 of the previous interaction term in the full factorial model is now the mean squares for "error" (residual) in the model without interaction.

As the interaction term, its expected mean squares in the nonadditive case was that of interaction variance and error. However, that term has now become simply residual. And in line with expected mean squares for the fixed effects model, both fixed effects factor and block are tested against MS residual. It should no longer be called MS error since variability within cell is impossible in this case. As Casella (2008, p. 45) notes, "Within error, sometimes called 'pure' error, is very different from a 'residual.' As the name implies, a residual is something that is left over. In statistics, the residual is left over from the model fit." We will revisit this distinction in our study of structural equation models later in the book.

What was the point of these two analyses? On a purely statistical level, it was a simple demonstration that single cell designs do best (at an arithmetical level, not necessarily theoretical, see Casella (2008)) without the testing of an interaction term. On a more substantive and scientific level, however, it served as a demonstration that the decisions you make in how you set about testing your model will have a serious impact on the outcome of the model, even to the point of whether terms in the model are even **testable**. We summarize with the following:

In an **unreplicated** randomized block design which includes an interaction term, if both **factor** and **block** are specified as fixed effects, it is impossible to test either effect since we are unable to generate an error term distinct from the interaction term. If the design did contain more than a single observation per cell, we would be able to test each fixed effect against MS error because we would be able to distinguish an error term over and above the interaction term (Hays, 1994). Hence, from a scientific perspective, if you are wanting to test effects of interest, you must ensure your data meet specifications required of the model.

The above principle generalizes to virtually any model you choose to test and is not a sole property of the randomized block design. **A model can only test effects if you have supplied it with enough information to do so**. If you do not have enough information to fit your model, it does not necessarily mean the untested effects would not have existed had you supplied the requisite data to test them. **Good models are correctly specified and are based on enough data to test effects of theoretical interest. Plan and design your experiments carefully.**

6.4 TUKEY TEST FOR NONADDITIVITY

We noted that in the nonreplicated situation, it was impossible to test for the presence of an interaction effect, and if block were designated as fixed, we used MS residual to test our effects. It may seem impossible then to be able to tell if an interaction exists in data for which there is only a single observation per cell. Now, had we been able to **replicate per cell** in the block design, then we could obtain a true error term estimate due to having more than a single observation per cell, and be able to provide a test for the interaction effect. But it would seem at first glance in the nonreplicated situation, testing for an interaction effect would be impossible. However, not all hope is lost. There does exist a method for testing the presence of **nonadditivity** (i.e., the presence of a factor by block interaction, in this case). This test is known as the **Tukey test for nonadditivity** (1949). We do not demonstrate the test here, though the interested reader is encouraged to consult the R package `additivityTests` (Simeckova, Rusch, and Simecek, 2014) for details on how to run the Tukey test and other similar tests. If such a test does not indicate the presence of an interaction effect, then one may assume it safe to pool interaction variance into MS residual. If such a test does not rule out the possibility of an interaction, then pooling, in general, would be ill-advised, unless of course on a theoretical level you did not hypothesize an interaction effect to begin with. As always, theory should be guiding your work, and if not, then you should be noting that you're engaging in exploratory pursuits rather than claiming each "finding" as a theoretically driven result.

6.5 ASSUMPTIONS FOR THE COVARIANCE MATRIX

Recall that in a between-subjects design, it was assumed that the covariance between treatment conditions was equal to zero and that errors ε_i had a covariance equal to zero:

$$\text{cov}\left(\varepsilon_{ij}, \varepsilon_{i'j'}\right) = 0$$

In **between-subjects designs**, we had no good reason to suspect that treatment conditions would be **correlated**. For instance, in our melatonin example, we had no cause to suspect that subjects in the 1 mg group would be correlated to subjects in the 3 mg group. They were independently sampled subjects, randomly assigned to conditions, and hence assuming a covariance of zero between independent groups seemed reasonable. Even if you post-hoc correlated the groups and found a small correlation (incidentally, all data are correlated to some degree in the sample, you will likely find one), it would still not imply that the **prior expectation of zero correlation** between groups was violated. The key point is that in a classic between-subjects design, a data point in one group has no reason for being "related" to a data point in another group, assuming the experiment was conducted properly and subjects randomly assigned to conditions.

In the randomized block design (and as we will see later, repeated measures design as well), as a result of subjects being **matched** across treatments, there is now a reasonable expectation that measurement occasions will have a covariance between them unequal to zero. **The fact that we expect this covariance to be unequal to zero requires us to consider the randomized block model in a different light than the classic between-subjects analysis of variance model**. Instead of simply specifying an assumption about variances, we now need to also incorporate an assumption about **covariances** between treatment levels (or **measurement occasions**, in the case of repeated measures). That is, we need to also model the **covariance structure** that may be inherent in our data instead of simply assuming it to be zero between treatments. As Hays (1994) put it:

> The key point is that the identity or the matching of subjects not only introduces a dependency among the resulting scores under different treatments but also creates dependencies among the treatment populations, as reflected in the covariances. (p. 572)

That the covariance between treatments is expectantly unequal to zero will pose some challenges for how we construe our analysis. But we can minimize or simplify this challenge by at least assuming that the **pairwise covariance** across treatment populations is a **constant** value. Intuitively, such an assumption would seem to make things "easier" than if we had to hypothesize a different covariance value for each treatment-to-treatment pair. If we define the covariance as $\rho\sigma_T^2$, where ρ is the population correlation coefficient that is presumed constant between treatment populations (Kirk, 1995) and σ_T^2 is the variance in any of the treated populations (i.e., the variance of any treated group, which is assumed constant from group to group), then we can define the **common covariance** between observations sampled from any two population pairings y_{i1} and y_{i2} to be

$$\text{cov}\left(y_{i1}, y_{i2}\right) = \rho\sigma_T^2$$

We can see that ρ in $\rho\sigma_T^2$ is, in effect, drawing on a proportion of the variance. The extent to which $\rho \to 1.0$ is the extent to which $\rho\sigma_T^2 = \sigma_T^2$, since $(1)\sigma_T^2 = \sigma_T^2$.

The assumption of constant values for σ_T^2 and constant values for $\rho\sigma_T^2$ for each treatment is known as the **compound symmetry assumption** (Hays, 1994, p. 573). In a randomized block or repeated measures analysis, the assumption of compound symmetry is a **sufficient** condition for carrying on with the analysis, though it is not a **necessary** condition. What this means is that even if compound symmetry is not achieved, so long as a "lesser" assumption (typically easier to achieve) known as **sphericity** is satisfied, then the analysis can proceed without adjustment.

Though sphericity can be defined in terms of orthogonal contrasts (e.g., see Lane, 2016), it can also be related to a different condition, known as **homogeneity of treatment difference variances**. This states that for any two treatment levels, difference scores $y_{i1} - y_{i2}$ will have identical variances. The assumption of homogeneity of treatment difference variances asserts that these variances will be equal across pairs of within-subject treatment levels. Satisfying this assumption has been shown to be equal to satisfying the more complex definition of the assumption of sphericity (see Huynh and Feldt (1970) and Rouanet and Lépine (1970) for details). Hence, historically it has been the assumption of sphericity, not compound symmetry, that is the one under evaluation in randomized block or repeated measures models. Further details on the sphericity assumption can be found in Kirk (1995, pp. 274–279) who provides a thorough treatment involving matrices.

Cutting to the chase, when sphericity is violated, the ensuing F-statistic from the ANOVA may not be distributed as F from the theoretical density of F. The classic test, though not necessarily the most powerful (e.g., see Cornell et al., 1992) used to test the null hypothesis of a spherical matrix is that given by Mauchly (1940). **Mauchly's test** has been shown to be somewhat problematic (Stevens, 2009), and the **Greenhouse–Geisser** adjustment to degrees of freedom is often recommended regardless of the outcome of Mauchly's test.

Many researchers have evaluated the effect of violations of sphericity (e.g., see Box, 1954; Geisser and Greenhouse, 1958; Huynh and Feldt, 1970). If sphericity is violated, one can compute an epsilon value, ε, to assess the degree to which the covariance matrix departs from the ideal form under the null hypothesis (Kirk, 1995). See Howell (2002, p. 487) for the computation of ε.

Some features of ε include the following:

- When the assumption of a spherical matrix is perfectly met, ε will equal 1.0.
- The extent to which the assumption of sphericity is not met, ε will decrease from 1.0.
- The minimum value of ε is $1/(J - 1)$, the lower bound on ε, where J is the number of levels of the within-subjects factor.

When the assumption of sphericity is not tenable, the **Greenhouse–Geisser conservative F-test** (Geisser and Greenhouse, 1958) is recommended by most authors (though other adjustments such as the **Huynh–Feldt** exist, to be discussed in our software examples later). The Greenhouse–Geisser **adjusts degrees of freedom downward**, making it more difficult to reject the null hypothesis than it would be if sphericity were not violated (i.e., the test "punishes" for not attaining sphericity). We will discuss and interpret the G–G adjustment when we perform a repeated measures analysis using software later in this chapter.

6.6 INTRACLASS CORRELATION

Recall the population intraclass correlation of the previous chapter in our discussion of random effects and mixed models, of which an estimate of ρ was given by:

$$\hat{\rho} = \frac{\hat{\sigma}_A^2}{\hat{\sigma}_A^2 + \hat{\sigma}_e^2} = \frac{\hat{\sigma}_A^2}{\hat{\sigma}_y^2}$$

In our discussion of random effects models, the intraclass correlation was based on the fact that within any treatment population there existed two sources of variance, σ_A^2 and σ_e^2, such that the total variation, σ_y^2 could be regarded as a sum of these two components, $\sigma_y^2 = \sigma_A^2 + \sigma_e^2$. We will now use the intraclass correlation to demonstrate the influence of **pairwise treatment covariance** in the randomized block or repeated measures analysis of variance, where now σ_T^2 replaces σ_y^2.

From our definition of ρ, some algebra reveals that

$$\rho = \frac{\sigma_A^2}{\sigma_T^2}$$

$$\rho\sigma_T^2 = \sigma_A^2$$

which also implies that

$$\sigma_e^2 = \sigma_T^2(1-\rho) \tag{6.3}$$

which in words means that error variance is equal to a proportion of the variance σ_T^2. Equation (6.3) is the case since the balance of whatever ρ is not accounting for in $\rho\sigma_T^2 = \sigma_A^2$ is $1-\rho$, which yields the right-hand side. And since whatever is not accounted for is error, this quantity of $\sigma_T^2(1-\rho)$ is thus defined as such, σ_e^2. This form of σ_e^2 will be useful in our consideration of the expected mean squares for repeated measures models. By (6.3), we can now assess the effect that ρ has on our estimate of σ_e^2. For instance, for $\rho = 0$, it follows that

$$\sigma_e^2 = \sigma_T^2(1-\rho)$$
$$= \sigma_T^2(1-0)$$
$$= \sigma_T^2$$

That is, under the condition that the intraclass correlation is equal to 0, the factor (which is **blocks** in a **block design**, and **subjects** in a **repeated measures design**) is accounting for no variance, and the total variance σ_T^2 within any treated group is made up of simply σ_e^2. When ρ is equal to 1, on the other hand, then

$$\sigma_e^2 = \sigma_T^2(1-\rho)$$
$$= \sigma_T^2(1-1)$$
$$= 0$$

Under this scenario, the factor (again, which is blocks in a block design, and subjects in a repeated measures design) is accounting for 100% of the variance, and hence σ_e^2 is equal to 0. This would imply that block (or subject) is accounting for all the variance. As noted by Hays (1994, p. 574), "a large correlation ρ implies a large MS value, and hence a large proportion of variance accounted for by subject differences … a large correlation ρ actually lowers the proportion of MS treatments or (MS A) that is caused by error variability."

Of course, neither of these two extremes (i.e., $\rho = 0$ or $\rho = 1$) will usually dominate in practice, and ρ will often be a value somewhere between the range of 0 and 1. The point of considering the upper and lower limits of σ_e^2 under maximum and minimum values for ρ is simply to reveal how influential ρ is in either increasing or decreasing σ_e^2. That is, **the intraclass correlation has an impact on the size of σ_e^2**. This should not be surprising, however, since the inclusion of any factor in an ANOVA, if the factor is worthwhile, should typically help to reduce (or account for) error variance. In the current case, block or subject is the factor we are considering, and what the above shows is that if the blocking or subject factor is important to the design (i.e., if repeated measures are effective from a statistical point of view), then it should result in a **reduction in error variance**, which is precisely what an experimenter desires, that of having a way to reduce unexplained variation as to isolate and emphasize treatment differences for the factor of interest. For a classic (and excellent) discussion of intraclass correlation, see none other than Fisher (1925).

6.7 REPEATED MEASURES MODELS: A SPECIAL CASE OF RANDOMIZED BLOCK DESIGNS

A generalization of the randomized block design is the **repeated measures model**, often called **within-subjects** or **longitudinal** models. Recall that in a randomized block design, each block consisted of subjects homogeneous as possible on one or more nuisance factors. This was done to make subjects "alike" as much as possible within each block. What if we wanted to take "likeness" to its absolute extreme? Instead of blocking on subjects **similar** to one another, what if we blocked on the **same** subject? Repeated measures models take the idea of blocking to the limit, where now each block consists of the **same** subject. As was true for the block design, the hope is that the similarity of a subject's responses under testing conditions can be exploited and the covariance between testing conditions can be removed to boost sensitivity and power of the ensuing *F*-test.

Before we discuss repeated measures models further, it should be emphasized that since you have already been exposed to randomized block designs, you are already "familiar" with repeated measures models. The only difference is the criteria used to form the blocks. The skill required to understand statistical modeling is in part disentangling the jargon used in different fields. Some writers, for example, Kirk (1995), discuss the randomized block design at length and make only minor mention of repeated measures. Other writers present repeated measures models without hardly any mention of the underlying randomized block theory. This is fine too since longitudinal data dominates many fields, and data analyzed on the same subjects over time have their own peculiarities that may not be present in purely blocking designs. The approach followed in this book is that if one understands the randomized block design, one has their foot in the door of even the most complex of analysis of variance models, which include the repeated measures model as a special case. Indeed, a course or book of the sort "**Randomized Block Designs and Their Special Cases**" would not be an unreasonable title for an all-inclusive analysis of variance text, since so many ANOVA models can be subsumed under that title. What unites them all is covariance and correlation as well as fixed and random factors. For instance, **time series analysis** is another type of modeling technique that assumes a covariance between measurements at different time points (Upton and Cook, 2002), often going by the name of **autocorrelation**. These models are useful for analyzing such things as seasonal variation and other components that help explain variance in the given sequence of measurements. These models are beyond the scope of this book, though we make a brief note about them in Chapter 7. The interested reader is encouraged to consult Chatfield (2019) for more details.

6.8 INDEPENDENT VERSUS PAIRED-SAMPLES *T*-TEST

We develop the idea of repeated measures by building on familiar tests previously learned, those of the **independent samples** and **paired samples *t*-tests**. Recall the independent samples *t*-test:

$$t = \frac{\bar{y}_1 - \bar{y}_2}{\sqrt{\frac{s_1^2}{n_1} + \frac{s_2^2}{n_2}}}$$

Recall that the denominator $\sqrt{\frac{s_1^2}{n_1} + \frac{s_2^2}{n_2}}$ is called the **estimated standard error of the difference between means**. All else equal, the lower the standard error of the difference, the greater the resulting t, because any observed difference in means $\bar{y}_1 - \bar{y}_2$ in the numerator compared to a number in the denominator that gets increasingly smaller has the effect of making the numerator look quite large and impressive.

For example, imagine for a given experiment that the difference in means were equal to $\bar{y}_1 - \bar{y}_2 = 5$. Suppose the standard error of the difference were equal to 10. Then the value of the resulting t would be $5/10 = 0.5$. Now, contrast this to the situation where the value of the standard error of the difference was instead equal to 1. The value of the resulting t would be $5/1 = 5$. Notice that in the second scenario, where we have a much smaller standard error of the difference, this is equated with a much larger t-statistic (and hence, yielding more strength against the null hypothesis). **Anything that makes the standard error smaller in any statistical test serves to boost power, because all else equal, it gives us a bigger statistic**. As Casella (2008) remarks in his discussion of a wide variety of models, "It's all about the denominator!" (p. 5). Indeed, from a **statistical** point of view, it is. From a **scientific** point of view, we are much more concerned with the numerator. Being mindful of this distinction can facilitate understanding a majority of significance tests.

Now, recall the paired-samples t-test:

$$t = \frac{\bar{y}_1 - \bar{y}_2}{\sqrt{\frac{s_1^2}{n_1} + \frac{s_2^2}{n_2} - 2\text{cov}(\bar{y}_1, \bar{y}_2)}}$$

where the variance of the difference was equal to, in the denominator,

$$\frac{s_1^2}{n_1} + \frac{s_2^2}{n_2} - 2\text{cov}(\bar{y}_1, \bar{y}_2)$$

The subtraction of $2\text{cov}(\bar{y}_1, \bar{y}_2)$ in the denominator served to lower the variance of the difference. The extent to which pairs of observations had a covariance unequal to 0 was the extent to which the paired-samples t-test provided a more sensitive test (i.e., a more **powerful** test) relative to the independent-samples t-test. The matching of subjects exploited the covariance among columns.

When we consider the wider and more elaborate repeated measures model, the concept is analogous to the paired samples t-test used in matched-pairs designs. The essential idea is to exploit the correlational structure between measurement conditions so that we may use this information to extract variation from the error term in our ensuing F-ratios.

6.9 THE SUBJECT FACTOR: FIXED OR RANDOM EFFECT?

In the **randomized block design**, as mentioned, we usually regarded block as a random factor, since in most circumstances we were not only interested in the **particular** blocks sampled but were interested instead in generalizing to the population of blocks of which our selected sample was simply a randomly chosen subset. Our argument was that in **most** contexts, block should be regarded as a random factor, though ultimately this decision should be made by the investigator, assuming adequate information (i.e., sample size per cell) to estimate the given model.

In **repeated measures models**, it stands to reason that our new blocking factor (i.e., subjects) again be designated as a random effect. It is easy to understand why this should be so. If you have sampled, say, 10 subjects in a repeated measures design, are you actually interested in **these particular**

subjects? Usually not. What you are most often interested in is in generalizing to the population of subjects of which your chosen subjects is merely a **sample**. This fits the precise definition of a random effect, and hence in a repeated measures design, the blocking factor **subject** will under most circumstances be designated as random.

Indeed, in most statistical models, whether by the name of fixed, random, or mixed, "subject" or "observation" is implicitly regarded as a random effect. Why is this so? In the hypothetical sample of 10 subjects, if we were truly only interested in **these particular 10 subjects**, then subject would no longer be considered a random effect. It would be considered a fixed effect, since in hypothetical replications of the experiment, **the same 10 subjects would be used again**. Further, the same "level" (not necessarily the same measured value) of error $\varepsilon_1, \varepsilon_2, \varepsilon_3$, etc., would show up in accordance with the same "level" of subjects s_1, s_2, s_3, etc. However, beyond this correspondence, level of error could still reasonably be expected to be sampled at random from a population of levels and hence fluctuate from sample to sample regardless of whether subject is fixed or random. By its very nature then, ε_i is virtually guaranteed to be a random effect because one can argue that its "level" is determined by more than simply the level of subject chosen. Theoretically, its true levels from experiment to experiment may not even be "knowable" if they are truly governed by a random process.

6.10 MODEL FOR ONE-WAY REPEATED MEASURES DESIGN

Recall that in the randomized block design, we distinguished between **nonadditive** and **additive** models. The nonadditive model contained a factor x block interaction. The additive model did not. For the repeated measures model, we are usually not interested in modeling a factor x subject interaction, and so our repeated measures model will usually be an additive one of the kind we are already familiar:

$$y_{jk} = \mu + \alpha_j + \beta_k + \varepsilon_{jk}$$

where α_j is a fixed effect and β_k is the random effect associated with each subject. Aside from not having much interest in modeling the interaction, there is an additional reason for not including the interaction effect. Recall the issues that presented themselves in the randomized block design when each cell contained a single observation. In such situations, we could not distinguish an error term distinct from the interaction term due to the fact that each cell contained only a single observation. In the repeated measures model, since we will also have a single observation per cell, the same issue that was present in the randomized block design exists in the repeated measures model. That is, **MS error in an unreplicated (i.e., $n = 1$ per cell) repeated measures design cannot be distinguished from the factor x subject interaction term**. Hence, by not specifically modeling $(\alpha\beta)_{jk}$, we "free up" this term to serve as the **residual term** for the factor in our model that we **are** interested in testing.

6.10.1 Expected Mean Squares for Repeated Measures Models

Just as we did for ANOVA models of previous chapters, the designation of fixed versus random effects helps inform us on how to generate suitable F-ratios to test effects of interest. To learn of the appropriate denominators, we need to consider the expected mean squares for the various sources of variation. In a one-way repeated measures model, we will have three sources of variation: (1) **subjects**, (2) **treatments**, and (3) **residual**. As we will see, the residual term here will usually result from not testing the subject x treatment interaction effect, and hence as we demonstrated earlier in the chapter with the blocking design, the interaction term can be pooled into the residual so that the main effect(s) of interest have a suitable denominator for generating the ensuing F-ratio.

We do not detail the derivation of the expected mean squares here. The interested reader can find details in Hays (1994) and Kirk (1995). It suffices, for now, to know **why** we are wanting to compute expected mean squares and how to interpret them in generating F-ratios. The expected mean squares for the additive repeated measures model turn out to be:

$$E(\text{MS subjects}) = \sigma_e^2 + K\sigma_{\text{subjects}}^2$$

$$E(\text{MS treatment}) = \sigma_e^2 + \frac{J\sum_j \alpha_j^2}{K-1}$$

$$E(\text{MS residual}) = \sigma_e^2 = \sigma_T^2(1-\rho)$$

where J is the number of rows for subjects and K is the number of columns for treatment (i.e., the treatment on which the repeated measurements are being taken). It is easy to see from the expected mean squares that for a test of treatments, the correct denominator is MS residual, since when $\alpha_j^2 = 0$, we have

$$E(\text{MS treatment}) = \sigma_e^2 + \frac{J\sum_j 0_j^2}{K-1}$$

$$= \sigma_e^2$$

for which σ_e^2 is the expectation for MS residual alluded to just above, which recall, is conflated with the subject x treatment interaction. That is, arithmetically, **the residual is the subject x treatment interaction term**.

6.11   ANALYSIS USING R: ONE-WAY REPEATED MEASURES: LEARNING AS A FUNCTION OF TRIAL

To demonstrate a simple, one-way repeated measures analysis of variance, consider the data in Table 6.4 (also briefly featured in Chapter 4) where rats were tested three times to measure the elapsed time it took to press a lever in an operant conditioning chamber. The response variable is the time (measured in minutes) it took for the rats to learn the lever press response. We would expect that if learning is taking place, the time it takes to press the level should generally **decrease** across trials.

TABLE 6.4   Learning as a Function of Trial (Hypothetical Data)

Rat	Trial 1	2	3	Rat Means
1	10.0	8.2	5.3	7.83
2	12.1	11.2	9.1	10.80
3	9.2	8.1	4.6	7.30
4	11.6	10.5	8.1	10.07
5	8.3	7.6	5.5	7.13
6	10.5	9.5	8.1	9.37
Trial Means	$M = 10.28$	$M = 9.18$	$M = 6.78$	

We identify our object in R as `learn` and request R to read headers from the data:

```
> learn <- read.table("rat.txt", header = T)
> library(car)
> some(learn)

  rat trial time
1   1     1 10.0
2   1     2  8.2
3   1     3  5.3
```

Note that the data are in so-called **long format**, with each record for rat having a single row (i.e., there are three rows for each rat representing the three different measurement occasions). When the same data are analyzed in SPSS, we will require the data to be in **wide format**, where each variable header will represent a given trial.

Before running any inferential tests, we generate a `qplot` to get a first glimpse of the data:

```
> attach(learn)
> library(ggplot2)
> qplot(trial, time)
```

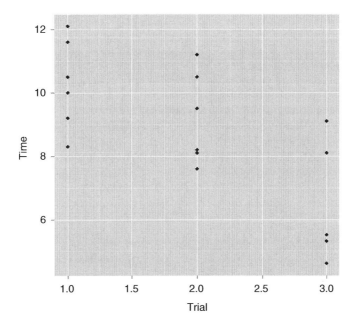

It is evident from the plot that response time generally decreases over trials. To prepare the data for analysis, we first generate a factor variable from trial and from rat:

```
> f.trial = factor(trial)
> f.rat = factor(rat)
```

We first conceptualize, for demonstration only, the model as one of a randomized block design:

```
> rat.block <- aov(time ~ f.trial + Error(block/f.trial), data = learn)
```

Note that in the above model formula, (block/f.trial) communicates the fact that **trials are nested within block**. Of course, our block factor is actually **rat**. Hence, in the spirit of repeated measures, we write out the model as:

```
> rat.block <- aov(time ~ f.trial + Error(f.rat/f.trial), data = learn)
```

What we have just specified is a repeated measures analysis where trial measurements are nested within rat. The point of the competing model formulations is simply to reveal the analogy of a block design versus a repeated measures design so you may observe their similarity. In the latter, our blocks are **subjects** (rats in this case). When we run the analysis, we obtain:

```
> summary(rat.block)
Error: f.rat
          Df Sum Sq Mean Sq F value Pr(>F)
Residuals  5  35.62   7.124

Error: f.rat:f.trial
          Df Sum Sq Mean Sq F value   Pr(>F)
f.trial    2  38.44  19.220   72.62 1.11e-06 ***
Residuals 10   2.65   0.265
---
Signif. codes:  0 '***' 0.001 '**' 0.01 '*' 0.05 ',' 0,1 ' ' 1
```

We note a statistically significant effect for trial ($p = 1.11$e–06).

We contrast this with a slightly different situation. Consider the data in Table 6.5, in which we have added a second factor to the design (i.e., **treatment**). Perhaps some rats were treated with a drug before the trials began, and in addition to response time, we were interested in estimating the effect of such a treatment.

TABLE 6.5 Learning as a Function of Trial and Treatment (Hypothetical Data)

Treatment	Rat	Trial 1	Trial 2	Trial 3	Rat Means
Yes	1	10.0	8.2	5.3	7.83
No	2	12.1	11.2	9.1	10.80
Yes	3	9.2	8.1	4.6	7.30
No	4	11.6	10.5	8.1	10.07
Yes	5	8.3	7.6	5.5	7.13
No	6	10.5	9.5	8.1	9.37
	Trial Means	$M = 10.28$	$M = 9.18$	$M = 6.78$	

We run this analysis as follows in R:

```
> f.treat <- factor(treat)
> rat.two.way <- aov(time ~ f.trial*f.treat + Error(f.rat/f.trial),
data = learn)
```

Note carefully again how we specified the error term. The statement `Error(f.rat/f.trial)` communicates that **measurements on trial are nested within rat**. Recall that this does not literally mean that **trials** are nested within rat, since it is clear from the data layout that each rat is receiving the same trials. To say that trials (or any other factor) are nested within rat would imply that some rats got, say, trials 1, 2, and 3, while others received say, trials 4, 5, and 6. Of course, this would not make sense in a repeated measures such as this, but it is still worth mentioning so that the distinction between **measurements being nested within rats versus trials being nested within rats** is conceptually clear. In the current situation, they amount to the same thing, yet this still serves as a good example to help understand the nature of nesting. In general, when you come across nesting in research papers, pay close attention to what exactly is being nested, as at times it may prove challenging to disentangle.

We obtain a summary of the fitted model:

```
> summary(rat.two.way)

Error: f.rat
            Df Sum Sq Mean Sq F value  Pr(>F)
f.treat      1  31.73   31.73   32.68 0.00463 **
Residuals    4   3.88    0.97
---
Signif. codes:  0 '***' 0.001 '**' 0.01 '*' 0.05 '.' 0.1 ' ' 1

Error: f.rat:f.trial
                Df Sum Sq Mean Sq F value   Pr(>F)
f.trial          2  38.44  19.220  91.403 3.09e-06 ***
f.trial:f.treat  2   0.96   0.482   2.293    0.163
Residuals        8   1.68   0.210
---
```

The effect for interaction `f.trial:f.treat` is not statistically significant ($p = 0.163$), however there is a main effect for trial ($p = 3.09e-06$). **Mauchly's test of sphericity** can be obtained in R via `mauchly.test()`, though not demonstrated here (we demonstrate the test shortly in SPSS).

A **nonparametric** alternative to the one-way randomized block design is the **Friedman rank sum test**, easily computed in R:

```
> friedman <- friedman.test(time ~ f.trial | f.rat)
> friedman

        Friedman rank sum test

data:  time and trial and rat
Friedman chi-squared = 12, df = 2, p-value = 0.002479
```

We note that though the observed p-value of 0.002 is sufficient to reject the null hypothesis, it is a larger p-value than in the equivalent parametric test. For an explanation and demonstration of the test, see Howell (2002, pp. 720–722).

6.12 ANALYSIS USING SPSS: ONE-WAY REPEATED MEASURES: LEARNING AS A FUNCTION OF TRIAL

We now demonstrate the one-way repeated measures on the same data analyzed previously in R. Entered into SPSS, the learn data appear as follows:

	trial_1	trial_2	trial_3
1	10.00	8.20	5.30
2	12.10	11.20	9.10
3	9.20	8.10	4.60
4	11.60	10.50	8.10
5	8.30	7.60	5.50
6	10.50	9.50	8.10

Notice that each column in the data represents a trial, and each of the rows represents a subject (rat in this case). Entering the data is easy enough, but should you require further direction, see Denis (2019) for a visual depiction of the data view for entering the data into SPSS, as well as a demonstration of the syntax below in GUI format.

We request the analysis via the following syntax:

```
GLM trial_1 trial_2 trial_3 [requests a general linear model with levels 1 through 3 of trial]
/WSFACTOR=trial 3 Polynomial [requests a polynomial contrast for the within factor]
/METHOD=SSTYPE(3)
/PRINT=ETASQ
/CRITERIA=ALPHA(.05)
/WSDESIGN=trial. [requests a model that includes the within-subjects factor "trial" (which is
```
all the current model can produce anyway since it's the only factor)]

Multivariate Tests[a]						
Effect		Value	F	Hypothesis df	Error df	Sig.	Partial Eta-Squared
trial	Pillai's trace	0.942	32.251[b]	2.000	4.000	0.003	0.942
	Wilks' lambda	0.058	32.251[b]	2.000	4.000	0.003	0.942
	Hotelling's trace	16.126	32.251[b]	2.000	4.000	0.003	0.942
	Roy's largest root	16.126	32.251[b]	2.000	4.000	0.003	0.942

[a]Design: Intercept within-subjects design: trial.
[b]Exact statistic.

The above are **multivariate tests of significance**, which can be interpreted either in conjunction with or in replacement of, the univariate tests. A multivariate model, discussed much more extensively in Chapter 11, features more than a single response variable. In our current model, instead of conceiving of trial (1 versus 2 versus 3) as a predictor of a single response, we can instead visualize it as three different response variables (but related through their difference scores across trials 1, 2, 3, hence they will be transformed into 2 variables (Stevens, 2009)). The fact that we now have three response variables (before transformation) instead of just the one makes the model **multivariate**, which is why SPSS also presents us with such multivariate tests of significance. In the language of MANOVA, we are analyzing a **linear combination** of responses (i.e., on trials 1, 2, 3). Many authors have contributed to the analysis of longitudinal data through a MANOVA approach (e.g., see Potthoff and Roy, 1964, for an example of an early paper).

For our applied purposes, it is enough to be familiar with the conventional rule that one interprets the multivariate tests or the Greenhouse–Geisser correction if one has evidence that sphericity has been violated. However, since tests of sphericity such as Mauchly's can be problematic in their own right (e.g., sensitive to sample size), MANOVA or the Greenhouse–Geisser correction is often recommended regardless of the results of Mauchly's test of sphericity (Howell, 2002; Stevens, 2009).

All four multivariate tests suggest to reject the null hypothesis ($p < 0.001$). For a description of these multivariate tests, refer to Chapter 11 or to Johnson and Wichern (2007, p. 336). We do not detail them here as matrix concepts cannot be avoided in their explanation, which are better left postponed to later in the book.

Mauchly's Test of Sphericity[a]								
Measure: MEASURE_1								
Within-Subjects Effect	Mauchly's W	Approx. Chi-Square	df	Sig.	Epsilon[b]			
					Greenhouse–Geisser	Huynh–Feldt	Lower bound	
trial	0.276	5.146	2	0.076	0.580	0.646	0.500	

Tests the null hypothesis that the error covariance matrix of the orthonormalized transformed dependent variables is proportional to an identity matrix.

[a]Design: Intercept within-subjects design: trial.

[b]May be used to adjust the degrees of freedom for the averaged tests of significance. Corrected tests are displayed in the tests of within-subjects effects table.

Mauchly's test of sphericity evaluates the null hypothesis that, as noted by SPSS, **the error covariance matrix of the orthonormalized transformed dependent variables is proportional to an identity matrix** (see Appendix for details on identity matrices). A statistically significant result for Mauchly's ($p < 0.05$ or similar) suggests the assumption of sphericity to be violated. For the test on our data, we do not reject the null hypothesis ($p = 0.076$).

The output for Mauchly's also reports **epsilon** values. Recall that these are values indicating the extent to which one should correct the degrees of freedom associated with the univariate test results in order to account for a violation of sphericity. We discuss these adjustments now in the context of the univariate effects.

Tests of Within-Subjects Effects						
Measure: MEASURE_1						
Source		Type III Sum of Squares	df	Mean Square	F	Sig.	Partial Eta-Squared
trial	Sphericity assumed	38.440	2	19.220	72.620	0.000	0.936
	Greenhouse–Geisser	38.440	1.160	33.131	72.620	0.000	0.936

(Continued)

(Continued)

Tests of Within-Subjects Effects

Measure: MEASURE_1

Source		Type III Sum of Squares	df	Mean Square	F	Sig.	Partial Eta-Squared
Error (trial)	Huynh–Feldt	38.440	1.292	29.750	72.620	0.000	0.936
	Lower hound	38.440	1.000	38.440	72.620	0.000	0.936
	Sphericity assumed	2.647	10	0.265			
	Greenhouse–Geisser	2.647	5.801	0.456			
	Huynh–Feldt	2.647	6.461	0.410			
	Lower bound	2.647	5.000	0.529			

The first correction on degrees of freedom in the SPSS output is the **Greenhouse–Geisser**. Notice that the degrees of freedom for it are 1.160 and 5.801 (for error). These numbers were obtained by using the correction factor **epsilon** listed under Greenhouse–Geisser in the report of Mauchly's test of sphericity. That value is equal to 0.580. This means to take 0.580 of the original degrees of freedom (for both numerator and denominator), and use this as our new "corrected" degrees of freedom. When we take 0.580 of 2, we get 1.16, which are the degrees of freedom given for the numerator of the Greenhouse–Geisser. When we take 0.580 of 10, we get 5.801, which are the degrees of freedom for the denominator of the Greenhouse–Geisser. The F-test for the G–G (i.e., Greenhouse–Geisser) correction is evaluated on 1.16 and 5.801 degrees of freedom instead of the original 2 and 10. Note that the F-statistic produced for G–G is the same as that produced when sphericity is assumed. The difference is only on the degrees of freedom on which the obtained F is evaluated. When we evaluate on 1.160 and 5.801, we note the p-value is **greater** than what it is for when sphericity is assumed (if you double-click on the p-values in the SPSS output, you will get the representative decimal places). It makes sense that the p-value should rise, since we are evaluating on less (and hence, more conservative) degrees of freedom. That is, G–G is issuing a "penalty" of sorts on degrees of freedom to account for the violation of sphericity.

The second correction provided by SPSS is the **Huynh–Feldt**. This time, we take 0.646 (i.e., epsilon value under Huynh–Feldt in Mauchly's test of sphericity) of the original degrees of freedom. This amounts to $0.646(2) = 1.29$ and $0.646(10) = 6.46$.

Finally, the third option for using a correction factor is the **Lower Bound** provided by SPSS. It is computed as $1/(J-1)$, equal to $1/(3-1) = 0.50$ for our data (i.e., corresponding to degrees of freedom 1 and 5 for our data). Recall "J" here is the number of levels of the within-subjects factor (in our case, 3, because there are three trials). This correction represents the most strict and conservative adjustment on the degrees of freedom.

6.12.1 Which Results Should Be Interpreted?

We have explored five different options for interpreting the F-test in a repeated measures analysis: univariate results with sphericity assumed, MANOVA, Greenhouse–Geisser, Huynh–Feldt, and the Lower bound correction. Which to use, and when? The literature in this area is not conclusive, although a general "workable" recommendation, primarily due to Girden (1992), is that when epsilon values are greater than 0.75, the Huynh–Feldt correction should be used. When epsilon values are less than 0.75, the Greenhouse–Geisser correction should probably be interpreted. And if nothing is known about sphericity, or one suspects that Mauchly's test cannot be interpreted accurately due to small or large sample sizes or questionable distributional assumptions, **Greenhouse–Geisser is still usually the best option**.

A practical recommendation, for most cases, is to report Greenhouse–Geisser and multivariate results. If one desired a less univariate conservative correction, Huynh–Feldt can be reported, keeping in mind that relatively small differences in p-values should not lead to disparate **scientific conclusions** regardless of the correction used (see Chapter 2, Section 2.28). For instance, if Greenhouse–Geisser yielded a p-value of 0.07 while Huynh–Feldt yielded 0.04, since neither of these should be used **exclusively** as a scientific indicator of the existence of a phenomenon from your experiment or study (i.e., recall you should be also simultaneously interpreting effect size), such small differences in p-values for correction factors turns out to be much more a **statistical** issue than it is a **scientific** one (i.e., do not lose sleep over which test to report given slightly different p-values). As a researcher, you presumably conducted the repeated measures in the hope of finding a scientific effect. In this spirit, use the correction factors as a **guide** to determining whether something occurred in the study, and interpret p-values only as a means to this end. Given all we know about the behavior of p-values, a p-value of 0.05 versus 0.04, for example, should not be causing you to make a drastically different **scientific decision** in each case.

SPSS next provides us with the between-subjects effects:

	Tests of Between-Subjects Effects				
	Measure: MEASURE_1				
	Transformed Variable: Average				
Source	Type III Sum of Squares	df	Mean Square	F	Sig.
Intercept	1378.125	1	1378.125	193.457	0.000
Error	35.618	5	7.124		

What is "Error" in the output above? What SPSS is calling **error** is actually the effect of "subjects" (or **rats**, for our data). Indeed, in the R analysis featured earlier, R specifically designated the error as that due to rat:

```
Error: f.rat
          Df Sum Sq Mean Sq F value Pr(>F)
Residuals  5  35.62   7.124
```

Notice the degrees of freedom for error are equal to 5, which is equal to the number of subjects (6) minus 1. When we introduce a second between-subjects factor, as we will do shortly by including treatment (i.e., recall some rats received treatment, some did not), we will further partition this SS due to subjects. Hence, both SPSS and R are telling us that this is essentially the **subjects variability that is available for further partition** upon introducing another between-subjects factor into the design.

There is no test for the subjects effect because we are not able to produce an error term distinct from the subjects by trial interaction term used to test the within-subjects effect. We are usually not interested in testing the effect of subject anyway, since it is usually considered nothing more than a nuisance factor (Hays, 1994). One would expect subjects to differ from one another, which is a good thing, since we are able to partial this variance out of the error term. That SPSS is not providing us a test for subject is not a problem.

Within-subjects contrasts can also be generated, though not shown here. For example, a **simple contrast** would generate comparisons of trial 1 versus trial 3 and trial 2 versus trial 3, obtained through:

```
GLM trial_1 trial_2 trial_3
  /WSFACTOR=trial 3 Simple
```

Other contrasts can also be performed. Post-hocs can also be obtained in SPSS for the trial factor by:

```
GLM trial_1 trial_2 trial_3
  /WSFACTOR=trial 3 Polynomial
  /METHOD=SSTYPE(3)
  /EMMEANS=TABLES(trial) COMPARE ADJ(BONFERRONI)
```

6.13   SPSS TWO-WAY REPEATED MEASURES ANALYSIS OF VARIANCE MIXED DESIGN: ONE BETWEEN FACTOR, ONE WITHIN FACTOR

Having demonstrated the analysis of repeated measures data for the one-way model, we now demonstrate an analysis of a two-way model in SPSS. We refer to the two-factor layout earlier cited in Table 6.5, where, recall, in addition to being assessed over trials, some of the rats were given a medical treatment hypothesized to promote efficiency at learning the task (treatment = "yes" in Table 6.5).

Recall this is now a 2×3 repeated measures ANOVA that contains both a **between factor** and a **within factor**. Such a design is often referred to as a **mixed design**. The term "mixed design" here is used to indicate the presence of a mix of between-subjects and within-subjects factors. It is not equivalent in meaning to the term **mixed model** that we have been discussing. However, since subject, as already discussed, is usually considered to be a random factor, the mixed design is more often than not analyzed as a mixed model (still, you should not equate mixed design with mixed model, as it will mislead you in the study of more advanced models).

When entered into SPSS, the data file appears as follows:

trial_1	trial_2	trial_3	treat
10.00	8.20	5.30	1.00
12.10	11.20	9.10	0.00
9.20	8.10	4.60	1.00
11.60	10.50	8.10	0.00
8.30	7.60	5.50	1.00
10.50	9.50	8.10	0.00

Notice that in entering the data into SPSS, as before, each level of the repeated measure has a unique column (**trial_1, trial_2, trial_3**). The levels of the between-subjects factor are represented by a single column (**1 = yes treatment, 0 = no treatment**) to denote the grouping effect. Again, if you require a visual depiction of how the data are entered into SPSS along with window outtakes, see Denis (2019).

We request the repeated measures ANOVA:

GLM trial_1 trial_2 trial_3 BY treat [requests a general linear model with three dependent variables trial_1 through trial_3 and a single independent variable "treat"]
    /WSFACTOR=trial 3 Polynomial
    /METHOD=SSTYPE(3)
    /PRINT=ETASQ
    /CRITERIA=ALPHA(.05)
    /WSDESIGN=trial [specifies the within-subjects factor]
    /DESIGN=treat.[specifies the between-subjects factor]

SPSS first provides us with multivariate tests of significance:

Effect		Value	F	Hypothesis df	Error df	Sig.	Partial Eta-Squared
Multivariate Tests[a]							
trial	Pillai's trace	0.963	38.569[b]	2.000	3.000	0.007	0.963
	Wilks' lambda	0.037	38.569[b]	2.000	3.000	0.007	0.963
	Hotelling's trace	25.713	38.569[b]	2.000	3.000	0.007	0.963
	Roy's largest root	25.713	38.569[b]	2.000	3.000	0.007	0.963
trial * treat	Pillai's trace	0.427	1.117[b]	2.000	3.000	0.434	0.427
	Wilks' lambda	0.573	1.117[b]	2.000	3.000	0.434	0.427
	Hotelling's trace	0.745	1.117[b]	2.000	3.000	0.434	0.427
	Roy's largest root	0.745	1.117[b]	2.000	3.000	0.434	0.427

[a]Design: Intercept + treat within-subjects design: trial.

[b]Exact statistic.

The multivariate tests all suggest the presence of a main effect for trial. Evidence for an interaction effect is not supported across all multivariate tests ($p = 0.434$).

Next, we are given the findings of Mauchly's test:

Within-Subjects Effect	Mauchly's W	Approx. Chi-Square	df	Sig.	Epsilon[b]		
Mauchly's Test of Sphericity[a]							
Measure: MEASURE_1							
					Greenhouse–Geisser	Huynh–Feldt	Lower bound
trial	0.392	2.811	2	0.245	0.622	0.991	0.500

Tests the null hypothesis that the error covariance matrix of the orthonormalized transformed dependent variables is proportional to an identity matrix.

[a] Design: Intercept + treat within-subjects design: trial

[b] May be used to adjust the degrees of freedom for the averaged tests of significance. Corrected tests are displayed in the tests of within-subjects effects table.

Mauchly's test of sphericity is not statistically significant ($p = 0.245$), and hence if we were to trust this test, it would suggest that we do not have a violation of the sphericity assumption. However, based on our previous recommendation, we will nonetheless interpret both the multivariate tests and Greenhouse–Geisser F-test when drawing conclusions regarding the within-subject effect, along with the interaction involving the within-subjects factor.

Next are provided the tests for the within-subjects effects:

		Type III Sum of Squares	df	Mean Square	F	Sig.	Partial Eta-Squared
		Tests of Within-Subjects Effects					
		Measure: MEASURE_1					
trial	Sphericity assumed	38.440	2	19.220	91.403	0.000	0.958
	Greenhouse–Geisser	38.440	1.244	30.909	91.403	0.000	0.958
	Huynh–Feldt	38.440	1.982	19.399	91.403	0.000	0.958
	Lower bound	38.440	1.000	38.440	91.403	0.001	0.958
trial * treat	Sphericity assumed	0.964	2	0.482	2.293	0.163	0.364
	Greenhouse–Geisser	0.964	1.244	0.775	2.293	0.194	0.364
	Huynh–Feldt	0.964	1.982	0.487	2.293	0.164	0.364
	Lower bound	0.964	1.000	0.964	2.293	0.205	0.364
Error (trial)	Sphericity assumed	1.682	8	0.210			
	Greenhouse–Geisser	1.682	4.975	0.338			
	Huynh–Feldt	1.682	7.926	0.212			
	Lower bound	1.682	4.000	0.421			

All univariate tests suggest the presence of a main effect ($p = 0.000$) while the trial by treat interaction term still has a relatively large p-value (i.e., $p = 0.194$ for the Greenhouse–Geisser). A plot of means (insert after /METHOD) reveals (/PLOT=PROFILE(trial*treat)):

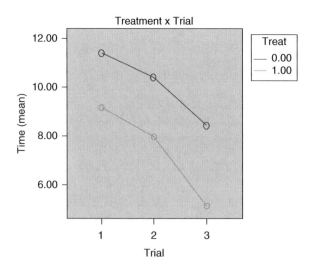

We can see that essentially, a similar "story" of mean differences between treatments is being told across trials, though at the third trial, it would appear that the decrease in time for the treated group is somewhat greater than for the nontreated group. But with such a small sample size, we likely did not have the power to detect such an effect. In this small sample, as evidenced by partial eta-squared, approximately 36% of the variance in the dependent variable is accounted for by the interaction term.

Next are the between-subjects effects:

Source	Type III Sum of Squares	df	Mean Square	F	Sig.	Partial Eta-Squared
Intercept	1378.125	1	1378.125	1419.122	0.000	0.997
treat	31.734	1	31.734	32.678	0.005	0.891
Error	3.884	4	0.971			

Tests of Between-Subjects Effects

Measure: MEASURE_1

Transformed Variable: Average

Notice that we have a statistically significant effect for treatment ($p = 0.005$). This is the difference in treatment means resulting from collapsing across levels of the repeated measure.

6.13.1   Another Look at the Between-Subjects Factor

For pedagogical purposes, we now run an ordinary between-subjects ANOVA, testing the null hypothesis that population means on treatment are equal. Our only factor in the model is treat, hence, we run the model as time BY treat:

```
UNIANOVA time BY treat
  /METHOD=SSTYPE(3)
  /INTERCEPT=INCLUDE
  /CRITERIA=ALPHA(0.05)
  /DESIGN=treat.
```

Tests of Between-Subjects Effects

Dependent Variable: Time

Source	Type III Sum of Squares	df	Mean Square	F	Sig.
Corrected model	31.734[a]	1	31.734	11.290	0.004
Intercept	1378.125	1	1378.125	490.315	0.000
treat	31.734	1	31.734	11.290	0.004
Error	44.971	16	2.811		
Total	1454.830	18			
Corrected total	76.705	17			

[a]R-squared = 0.414 (adjusted R-squared = 0.377).

The above computation of the sums of squares for treatment is the same as seen in the between-subjects effects in the repeated measures ANOVA (i.e., SS = 31.734). However, notice the error term, it is **not** equal to 0.971 as it is in the repeated measures output. Why not? The above analysis investigates the effect of treatment, and what remains is relegated to the error term (i.e., effect due to trial and subject). As discussed in this chapter, this is one advantage of performing repeated measures—**it allows you to remove variability due to subject that would otherwise make its way into the error term**. This is analogous to the simpler case of the paired samples t-test in which the covariance between treatments or testing conditions is removed from the error term. **In paired t-tests, randomized blocks, or repeated measures, all three methods remove variability due to block or subject that would otherwise make its way into the error term, and consequently, often provide a more powerful test of effects of interest**.

Contrasts, post-hocs, and simple effects can also be calculated on this data. For simple effects, one can compare trials at each level of treatment by (output not shown) the following:

```
/emmeans = tables(trial*treat) compare (trial) adj (Bonferroni)
```

6.14   CHAPTER SUMMARY AND HIGHLIGHTS

- In classical **between-subjects designs**, subjects or objects are randomly assigned to a condition on the independent variable with the hope that nuisance factors more or less balance out across groups.

- The goal of a **randomized block design** is to attempt to reduce the **error term** in the analysis of variance by administering levels of the independent variable across homogeneous subsets of individuals that are relatively alike. In this way, randomized block designs attempt to capture within-group homogeneity and model it out of the error term so that the factor(s) of interest in the design can be tested with greater sensitivity and power.

- The **randomized block design** can be conceptualized as an extension of the simpler **matched samples design**.

- **Randomized block designs** can be analyzed as **fixed effects**, **random effects**, or **mixed models**. Most often, because it makes most sense to consider block as a random effect, the mixed model is appropriate (assuming at least one other factor is fixed).

- A **nonadditive** model is one that includes an interaction term. An **additive** one is one that does not.

- In designs where there is a single observation per cell (the so-called **nonreplicated** design), it becomes impossible to generate an error term separate from the interaction term. This is an important consideration both in planning a scientific investigation as it is in building and interpreting a statistical model.

- The **Tukey test for nonadditivity**, as well as other so-called additivity tests, may be used for testing the presence of an interaction effect in $n = 1$ per cell designs.

- In both randomized block designs and repeated measures, because **measurements are nested within block/subject**, the expectation of zero covariance between treatments is no longer reasonable. We must instead make assumptions about the **correlational structure** between treatments. **Compound symmetry** and **sphericity** are common assumptions made for these models.

- There are many **adjustments to degrees of freedom** available if the assumption of sphericity is violated or is suspect. Of these, the **Greenhouse–Geisser conservative F-test** is often recommended.

- The **intraclass correlation** ρ is useful in demonstrating the influence of pairwise treatment covariance in randomized block or repeated measures analysis of variance. When $\rho = 0$, σ_e^2 is equal to treatment variance alone. When $\rho = 1$, $\sigma_e^2 = 0$.

- A **repeated measures model** can be conceptualized as a special case of a randomized block design in which subjects are the blocks.

- **Repeated measures** can be conceptualized as an extension of the **paired samples t-test**.

- **Randomized block designs** and **repeated measures models** can be analyzed in both R and SPSS.

REVIEW EXERCISES

6.1.   What distinguishes a **between-subjects design** from a **repeated measures** (or "within-subjects" design)? Explain how these two designs are different from one another.

6.2.   Define a **randomized block design**. What is the general purpose of such a design?

6.3.   Explain how **subjects** are nested within blocks in a randomized block design.

6.4.   Discuss how a **randomized block design** can be conceptualized as an extension of the matched-pairs design.

6.5.   Under what situations is a **block** best considered a **fixed** or **random** factor? Explain.

6.6.   Distinguish between the **additive** and **nonadditive** randomized block designs.

6.7.   In a randomized block design **where $n = 1$ per cell**, discuss the problems with designating both effects as fixed effects and why a test of these effects is not possible under the nonadditive model.

6.8.   Briefly explain the purpose of **Tukey's test for nondditivity**.

6.9.   Discuss how the **intraclass correlation** can be used to demonstrate the influence of pairwise treatment covariance in a randomized block or repeated measures ANOVA.

6.10.   Explain why **repeated measures ANOVA** is best considered a special case of the randomized block design.

6.11.   Consider the data in Table 6.6. Nitrogen in blood plasma was recorded in six rats across 360 days.

TABLE 6.6   Nitrogen in Blood Plasma

	Age	25	37	50	60	80	100	130	180	360
Rat	1	0.83	0.98	1.07	1.09	0.97	1.14	1.22	1.20	1.16
	2	0.77	0.84	1.01	1.03	1.08	1.04	1.07	1.19	1.29
	3	0.88	0.99	1.06	1.06	1.16	1.00	1.09	1.33	1.25
	4	0.94	0.87	0.96	1.08	1.11	1.08	1.15	1.21	1.43
	5	0.89	0.90	0.88	0.94	1.03	0.89	1.14	1.20	1.20
	6	0.83	0.82	1.01	1.01	1.17	1.03	1.19	1.07	1.06
	Means	0.86	0.90	1.00	1.04	1.09	1.03	1.14	1.20	1.23

(a) Perform a one-way repeated measures ANOVA in R.

(b) Perform the same one-way repeated measures ANOVA in SPSS.

(c) Do you have evidence to doubt the assumption of **sphericity**? Why or why not? Any issues with obtaining a test?

(d) Does interpretation of the **Greenhouse–Geisser correction** provide a different conclusion than when sphericity is assumed?

(e) Estimate the trend of blood plasma from day 25 to day 360. What **polynomial** best accounts for the trend?

7

LINEAR REGRESSION

By this method, a kind of equilibrium is established among the errors which, since it prevents the extremes from dominating, is appropriate for revealing the state of the system which most nearly approaches the truth.

(Legendre, 1805, pp. 72–73)

I found it hard at first to catch the full significance of the entries in the table, which had curious relations that were very interesting to investigate. They came out distinctly when I "smoothed" the entries by writing at each intersection of a horizontal column with a vertical one, the sum of the entries in the four adjacent squares, and using these to work upon. I then noticed that lines drawn through entries of the same value formed a series of concentric and similar ellipses.

(Galton, 1886, pp. 254–255)

Suppose a biologist would like to be able to predict the heights of offspring once they are grown adults. For a randomly chosen adult offspring, what is a good guess at its height? A reasonable guess might be the population mean of all adult offspring, especially if it was desired to minimize the signed error in prediction. However, guessing the mean would likely still result in imprecise predictions, and on the whole, result in much **error in prediction**. Knowing that parental height is correlated to offspring height, the biologist seeks a statistical method to exploit this correlation to reduce his error in predicting offspring height. The statistical method that will be of use to the biologist is **simple linear regression**.

Simple linear regression is a statistical method useful for making predictions about a continuous response variable based on knowledge of a second variable, usually also continuous, though categorical variables can also be modeled via dummy-coded regressors. The designation **simple** linear regression denotes the fact that the regression model features only a **single** explanatory variable. Models with two or more explanatory variables will be discussed in Chapters 8 and 9.

Applied Univariate, Bivariate, and Multivariate Statistics: Understanding Statistics for Social and Natural Scientists, With Applications in SPSS and R, Second Edition. Daniel J. Denis.
© 2021 John Wiley & Sons, Inc. Published 2021 by John Wiley & Sons, Inc.
Companion Website: www.wiley.com/go/denis/appliedstatistics2e

More than simply making predictions, regression seeks to predict values on the response variable such that the average error in prediction is **less** than what would be the case had the explanatory variable not been used as a predictor. What this means statistically is that there must be a correlation between the response and explanatory variable for linear regression to be effective. Otherwise, in the absence of such a correlation, predictions would be generally no more accurate than if the explanatory variable were not used at all.

Draper and Smith (1998) is a classic resource on regression analysis that also features topics on weighted least-squares, ridge regression, nonlinear estimation, and robust regression. Fox (2016) is a definitive thorough treatment of regression and related models, which includes generalized linear models. Fox also provides a rather in-depth study of diagnostics for linear models and also includes chapters on the geometry of such models. Cohen et al. (2003) is also a classic resource on applied regression with a focus toward the behavioral sciences. Pedhazur (1997) provides a thorough treatment targeted toward behavioral scientists. Neter et al. (1996) feature wide coverage of linear models in general. Wright and London (2009) is a useful resource for fitting regression models in R.

7.1 BRIEF HISTORY OF REGRESSION

Regression analysis has a very deep history. The techniques of **correlation** and **regression**, as applied to empirical observations, are generally attributed to **Francis Galton** (1822–1911), an English Victorian who made countless contributions to science in fields such as anthropology, geography, psychology, and statistics (Figure 7.1). For a discussion of Galton, see Fancher and Rutherford (2011). For a read of some of Galton's original works in the area of statistics, see Galton (1886, 1888).

Several historians, however (e.g., Hald, 1998), have noted that the mathematics of correlation and regression predated Galton by many years. Adrien Marie Legendre (1752–1883) is generally credited with the development of primitive **least-squares theory**, the exact method later employed by Galton in

FIGURE 7.1 Francis Galton. Innovator of correlation and regression.

analyzing empirical observations (Stigler, 1986). Legendre published his method in 1805 in "**Nouvelles méthodes pour la détermination des orbites des comètes**" which included a section on "**Sur la méthode des moindres quarrés**" (which in English translates to "On the method of least-squares.").

In addition to Legendre, correlational theory was likely developed in one form or another by other pioneers as well (see Denis (2001) and Walker (1929) for details). Among perhaps the most significant of these is Auguste Bravais (1811–1863), a professor of astronomy and physics, who wrote a paper in 1846 titled "**Analyse mathématique sur les probabilités des erreurs de situation d'un point**," which translated means "Mathematical analysis on the probability of errors of a point." Karl Pearson (1920) credits Bravais with having discussed the theorems of correlation in this paper. He essentially discovered what Galton would later call the **regression line** by an investigation of elliptical areas, but is thought to have not fully realized it. As Walker (1929) notes, Bravais could not make the "leap" required for a full-fledged discovery of correlation and regression. Depicted in Bravais' work was the geometrical ellipse, which within it, for all purposes, was the regression line later discovered by Galton:

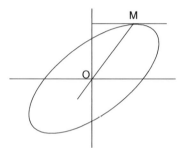

As Pearson (1920) noted regarding Bravais' geometrical analysis:

> He gets the line [i.e., "OM"] which corresponds to Galton's regression-line [sic]. But this is not a result of observing x and y and determining their association, but of the fact that x and y are functions of certain independent and directly observed quantities. (p. 32)

If we compare Bravais' work to that of Galton's 50 years later, the similarity is apparent (see Figure 7.2). Indeed, as noted by Friendly, Monette, and Fox (2013):

> It is not stretching the point too far to say that a large part of modern statistical methods descend from these visual insights: correlation and regression [Pearson (1896)], the bivariate normal distribution, and principal components [Pearson (1901), Hotelling (1933)] all trace their ancestry to Galton's geometrical diagram. (p. 2)

Galton's correlational diagram related the heights of mid-parents (the average adjusted height (female heights were multiplied by a constant) of the mother and father) with their adult children. The numbers in the table correspond to the numbers of mid-parent to adult children height combinations. And though Galton's correlational diagram is somewhat more complex than we shall detail in this chapter, one can appreciate the general similarity between his work and that of Bravais. Most significantly, both men obtained the **correlational ellipse**. In the case of Galton's work, the ellipse represented an **empirical reality first**, whereas in the case of Bravais, it appeared to be mostly a theoretical deduction. The distinction between the two discoveries is why Pearson (1920) referred to Galton's correlation as that of "**organical association**."

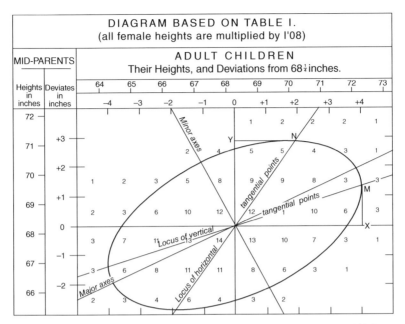

FIGURE 7.2   Galton's 1886 correlational surface. Source: Galton (1886).

7.2   REGRESSION ANALYSIS AND SCIENCE: EXPERIMENTAL VERSUS CORRELATIONAL DISTINCTIONS

Oftentimes researchers associate the use of analysis of variance models with experimental data and the use of correlational and regression techniques with nonexperimental data. The reason for this is largely historical rather than technical. There is nothing "experimental" or "nonexperimental" about a statistical technique, any more than there is anything experimental or nonexperimental about your pocket calculator. Either analysis of variance or regression can be used with either type of data, and often are. **Whether a scientific study is experimental or not has everything to do with the research design, not the statistical model used to analyze findings**.

The reason for the popular distinction is likely because analysis of variance designs arose in the context of experimental studies, whereas regression analysis, predating ANOVA by about 30–40 years (depending on when you consider the "origin" of regression to be), had its origins in the context of nonexperimental, correlational investigations. As we will see, one might view ANOVA as a **subcategory** of regression analysis, one for which the partitioning of variability is made much simpler for models with categorical predictors than with continuous ones. Indeed, some have argued that had high-speed computing machines been available during the advent of regression analysis, Fisher's analysis of variance (beginning in the 1920s), as a distinct technique, may not have come into existence at all, but rather may have forever been naturally subsumed under the wider regression model. Fisher's genius was in providing researchers with a useful and convenient (and marketable) statistical methodology for partitioning variability, originally in agricultural and biological settings. Mathematically, however, the two statistical methods, that of ANOVA and regression, overlap a great deal. This is why often in rather deep studies of regression analysis, analysis of variance models are presented as special cases of regression models rather than as distinct models in their own right (Fox, 1997).

7.3   A MOTIVATING EXAMPLE: CAN OFFSPRING HEIGHT BE PREDICTED?

To help motivate our discussion of regression analysis, we consider the original data collected by Galton in 1886 on the heights of parent and their grown offspring (we surveyed this data somewhat in Chapter 2). Some of Galton's data appears below (there are 928 cases in total):

```
> library(HistData)
> library(car)
> some(Galton)

    parent child
32    65.5  63.2
56    69.5  64.2
84    67.5  64.2
```

The question we (as did Galton) would like to ask of this data is the following:

Is child height able to be predicted by knowledge of parent height?

If the answer to the above question is **yes**, then we would expect there to be a relationship between these two variables. A plot and imposed regression line suggests there to be a somewhat **linear relationship**. That is, the data points, which are a subset of the Cartesian product, hint at a polynomial of degree 1 as perhaps the best functional rule for accounting for the scatter:

```
> attach(Galton)
> plot(parent, child, main = "Scatterplot of Child and Parent Heights")
```

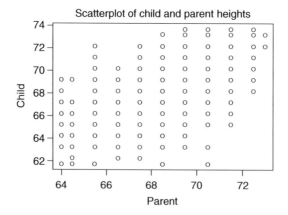

We conduct the linear regression using R's lm hypothesizing child as a function of parent:

```
> reg.model <- lm(child ~ parent)
> reg.model
```

```
Call:
lm(formula = child ~ parent)

Coefficients:
(Intercept)        parent
   23.9415        0.6463
```

We can fit a least-squares regression line using `abline` (a red one, see Figure 7.3):

```
> abline(reg.model)
> abline(reg.model, col = "red")
```

Referring to the coefficient estimates obtained in the `lm` output of 23.94 for intercept and 0.64 for slope, the raw-score regression equation representing the line of the best fit in Figure 7.3 is given by

$$y_i' = a + bx_i + e_i$$
$$= 23.9415 + 0.6463(x_i) + e_i$$

Using the estimated regression equation, we could obtain a predicted value y_i' for a given value x_i. For instance, what is the predicted height of offspring for a parent height of $x_i = 68$? The predicted height is computed:

$$y_i' = a + bx_i + e_i$$
$$= 23.9415 + 0.6463(x_i) + e_i$$
$$= 23.9415 + 0.6463(68)$$
$$= 67.89$$

That is, for a parent height of 68, the predicted child height is 67.89.

Informally, note how the least-squares regression was fit. It was fit in such a way that it provided the **best fit** to the data swarm. How this "best fit" idea is operationalized and defined is an idea we will

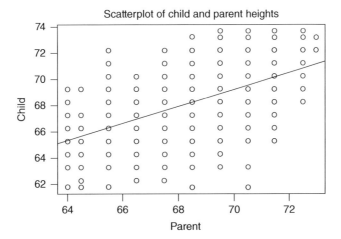

FIGURE 7.3　Regression line of child on parent.

unpack shortly. For now, it is enough to intuit that the regression line is fit in the sense of **minimizing** or **maximizing** some function of the data. As we will soon see, the OLS ("ordinary least-squares") line is that which **minimizes the sum of squared (vertical) errors** around the line.

7.4 THEORY OF REGRESSION ANALYSIS: A DEEPER LOOK

As an introduction to the theory of regression, as is true when learning any new statistical method, it is first helpful to recall where we have already been. Recall the one-way fixed-effects analysis of variance model of previous chapters:

$$y_{ij} = \mu + \alpha_j + \varepsilon_{ij}$$

Recall that the purpose in coming up with a model equation was to be able to theorize, on a quantitative level, **how the data were generated**. We theorized that any randomly sampled observation from the population in a random group, that is, y_{ij} (individual i in group j) was a function of three components:

- μ, which was an overall grand mean of the population
- α_j, which was the effect of being in one population versus another (i.e., the corresponding sample effect was $a_j = \bar{y}_j - \bar{y}.$)
- ε_{ij}, which was random error that we assumed had a mean of 0 and finite variance σ_e^2

The first part of the model, that of $\mu + \alpha_j$, was the **systematic** portion of the model, while the last component of the model ε_{ij}, represented, in a sense, our failure to account for individual differences unruly. In other words, we may have been able to make reasonable predictions of y_{ij} via $\mu + \alpha_j$, but in the end, we had to concede that our predictions might still deviate from expectation. Recall that the expectation for each group j was the mean for that group, $E(\bar{y}_j) = \mu_j$, which, if the null hypothesis held, then $E(\bar{y}_j) = \mu$ across all groups. To account for our errors in prediction, we introduced the error term in the model ε_{ij}, and assumed it to behave in a random **unsystematic** fashion.

In regression, we put forth another model that serves as a "theory" for how data were generated. Though at first glance the model may appear different than that studied in ANOVA, as we learn more about the model and its variations, we will learn that it is not that different from the analysis of variance setup. Though at first their similarities can be difficult to grasp, our study of regression, especially the multiple regression models of the following chapter, will help in revealing their likeness.

The simple linear regression model is given by

$$y_i = \alpha + \beta x_i + \varepsilon_i \tag{7.1}$$

where,

- y_i is an observed value of the dependent or "response" variable.
- α is the population intercept and is **fixed** for the given population, meaning that for a given population we are modeling, we are assuming it to have only a **single** intercept term.
- β is the population slope parameter, and like α, is also fixed for the given population, which as was true for α, implies that we are assuming the given population to have only a single β term. We do not cover the case of **random intercepts** and **random slopes** in this book. For a discussion, see Raudenbush and Bryk (2002).

- ε_i is the error associated with predictions of y_i, and unlike α or β, is not fixed, but random (just as in ANOVA models).

Let's compare side-by-side for a moment the ANOVA model to the regression model:

$$y_{ij} = \mu + \alpha_j + \varepsilon_{ij} \text{ versus } y_i = \alpha + \beta x_i + \varepsilon_i$$

We note the following similarities:

- In both models, we are wanting to predict a randomly sampled observation. In the ANOVA model, these observations are subscripted by ij to denote individual i in group j. In the regression model, observations are subscripted only by i. This is because in a simple regression model such as this, there are no "groupings" on the predictor variable. Or, if you wish, the actual groupings are infinitely small "categories" of the continuously-natured predictor variable, which have a limiting probability equal to 0 as the "slices" become smaller and smaller.
- The first term in the ANOVA model is μ, representing the overall **grand mean**. In the regression model, the first term is α, which represents the **intercept** for the regression line. Recall that the intercept of a line is where the line meets the ordinate axis. Though μ and α are different "things," by centering our predictor x_i in the regression model, we can transform α to represent the predicted value of y_i at the mean of x_i rather than when $x_i = 0$ as in the uncentered case. The point to emphasize right now is that both μ and α can be said to represent "starting points" to the model before the actual "exciting" part of the model takes place (which is included in α_j in the ANOVA and β in the regression). Both μ and α can be conceptually interpreted as all that is "common" to observations in the given data for ANOVA in terms of an overall mean, and the starting point of the line of best fit in regression.
- The second term in the ANOVA model is the population effect, α_j, while the second term in the regression model is the slope parameter, β. In each model, this is where the "action" is. Why is this so? Consider the case where population effects α_j are all equal to zero in the ANOVA model and the slope effect β is equal to zero in the regression model. What would this imply? Under this circumstance, the expectation for each model, if we allowed for error, would be $E(y_{ij}) = \mu + \varepsilon_{ij}$ for the ANOVA model and $E(y_i) = \alpha + \varepsilon_i$ in the regression model. Notice the similarity between these two expectations. In each case, where the treatment effect or slope "effect" is equal to 0, our best prediction is that of the population mean in ANOVA and the population intercept in regression. Incidentally, do not confuse α_j in the ANOVA model (i.e., population effects) with α in the regression model, as they are not the same thing.
- The last term in each model is ε_i and represents **deviation from expectation**. That is, in both models, ε_i represents that which is unaccounted for or unexplained by the systematic portion of the model. When we work with sample data, we typically refer to e_i instead of ε_i.

Once the intercept and slope have been estimated by respective estimators a and b, one enters a value for x_i to obtain the predicted value for y_i. We will designate the predicted value of y_i by the notation, y'_i (i.e., y_i "prime"). As mentioned, for the regression model under discussion, the values for x_i are assumed to be fixed rather than random quantities. That is, their individual values are assumed to be selected in advance by the researcher, rather than being sampled at random as one would have with random regressors. However, as noted by Fox (2016), much of the theory for fixed effects transfers over to random effects in the more realistic case where values of the predictor are considered a random sample from a wider population (e.g., such as one would have in an observational study). See Fox (2016, p. 228) for required assumptions.

The constants α and β are traditionally estimated by **ordinary least-squares** (**OLS**), though other estimation procedures are also available (e.g., **maximum likelihood**, **weighted least-squares**, etc.). As we will discuss when we lay out the assumptions of the regression model later, the expectation of y_i, $E(y_i)$, is equal to $\alpha + \beta x_i$ and the expectation of ε_i, $E(\varepsilon_i)$, is equal to zero. We also assume that the expectation of y_i given any chosen value of x_i, that is the **conditional expectation**, is equal to $\alpha + \beta x_i$. That is, $E(y_i/x_i) = \alpha + \beta x_i$.

We see then that both the analysis of variance model and the regression model share very similar characteristics in terms of their model equations. Which is the more **general model**? As we will continue to learn, the regression model is "king" of the two, since ANOVA can be subsumed under the wider regression model by a relatively simple reparameterization.

7.5   MULTILEVEL YEARNINGS

For readers with at least some familiarity with **hierarchical** or **multilevel** regression modeling, you may have experienced the temptation to "free" α and β in (7.1) thereby allowing them to be **random** rather than **fixed** effects. Indeed, one advantage of extending the fixed linear regression model to one with random effects is this ability to estimate **variance components** associated with these parameters (as one ordinarily does in virtually all models with ε_i, since recall ε_i is a random effect (Fox, 2016)) to learn how much variance in the response variable can be accounted for by such parameters. This idea of freeing parameters and thereby conceiving obtained sample statistics to be a random sample of a wider set of possible parameter values is analogous to how we conceptualized random effects and multilevel models in the previous chapters on ANOVA. The only essential difference is the nature of the parameters. For details on fitting multilevel regression models in R, see Gelman and Hill (2007).

7.6   THE LEAST-SQUARES LINE

Consider the depiction of the least-squares line in Figure 7.4.

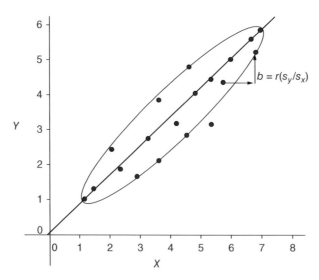

FIGURE 7.4   Linear regression of Y on X, where s_y and s_x are the standard deviations of Y and X, respectively; $b = r(s_y/s_x)$ is the slope of Y on X.

About Figure 7.4:

- s_y is the sample standard deviation of values of the response variable.
- s_x is the sample standard deviation of values of the predictor variable.
- $b = r(s_y/s_x)$ is the regression slope for y_i predicted from x_i, computed in this case as the correlation coefficient r multiplied by the ratio of standard deviations s_y to s_x. Under the condition that $s_y = s_x$, as such would occur if both variables were **standardized** to have variances each equal to 1, b becomes simply a function of r for the single predictor case, since

$$b = r(s_y/s_x) = r(1) = r$$

Such coefficients computed on standardized x_i and y_i are referred to as **standardized regression coefficients** β, or "**Betas**."

The regression line featured in Figure 7.4 can be conceptualized as a sort of **floating mean**, in that it traces the conditional distribution of y_i given a particular value of x_i. The extent to which the linear model accounts well for the data is the extent to which points fall perfectly along the regression line. Though beyond the scope of this chapter, much insight into how regression (and other statistical methods) works can be gleamed via a deeper understanding of ellipses (of the kind depicted in Figure 7.4). See Friendly, Monette, and Fox (2013) for an exceptional treatment and discussion.

7.7 MAKING PREDICTIONS WITHOUT REGRESSION

It is often taught that the purpose of regression analysis is to make predictions. However, are we not able to make predictions without regression? Of course we are. For instance, if a meteorologist wanted to predict tomorrow's temperature, could she not do it without using predictive weather models? Of course she could. Assuming she was not concerned with minimizing some function of the errors in making her predictions, she could predict **any** temperature she chooses, perhaps even "ball-parking" it based on her memory of last year's daily temperatures over the course of the year. However, her **accuracy** in prediction might not be very good. Yes, drawing informally on her memory of last year's temperatures might be better than if she simply drew temperatures "out of a hat" and completely at random, but her accuracy in prediction would likely still be quite poor.

This is where regression analysis comes in. Regression analysis helps us **improve our overall accuracy in making predictions**. Oftentimes, we implicitly guess **average** values when predicting, but regression tells us we can usually do better than that, especially if we have other variables correlated to the variable we are seeking to make predictions about.

To illustrate, consider again Figure 7.3, the plot of Galton's data, but now altered to denote virtually zero correlation between parents and their grown-up children (Figure 7.5).

Suppose now, under this circumstance, Galton wished to make as accurate of predictions of child height as possible. Which value should he guess? To keep the sum of squared errors $\sum_{i=1}^{n} \varepsilon_i^2$ to a minimum, it stands that he should guess the **mean** child height. Why? Because under the condition of zero correlation (or no predictor), **the mean guarantees that the average squared deviation will be smallest when it is used as the predicted value** (Hays, 1994, p. 188). The mean child height is equal to 68.1, and thus his prediction for any given child, would be 68.1.

The Galton example here emphasizes the fact that if there is no correlation between x_i and y_i, then our best "line" of prediction, so to speak, assuming our goal is to minimize squared errors in predicting,

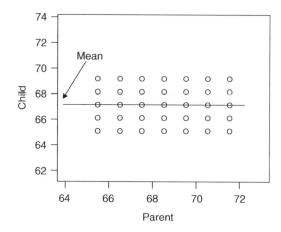

FIGURE 7.5 Galton data adjusted to show no correlation.

would be the **horizontal line** representing the mean. In such a case, our error in prediction would be equal to the standard deviation of y_i. **This horizontal line is, in actuality, the regression line of y_i on x_i under the situation of either no predictor, or equivalently, under the condition of zero linear correlation between y_i and x_i.** Understanding this idea is a powerful "first step" to understanding what regression analysis is all about.

Now, consider the situation where there is a linear relationship between x_i and y_i, as depicted in Galton's original data (Figure 7.6). We will **tilt** the regression line upward so that it becomes our new **floating mean** from which we will make predictions. With this tilted line in place, we will never predict a **single** value y_i for all values of x_i as we did for the horizontal "regression line." Rather, with the tilted line in place (the one computed based on knowledge of the correlation between x_i and y_i), our predictions will be values that fall on the line **conditional** upon our selection of values for x_i (where x_i is subscripted here to emphasize that we are selecting a given value of the variable for input into the regression equation).

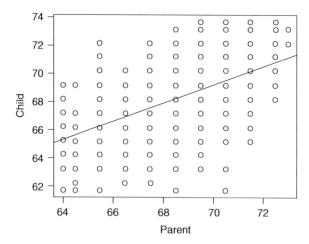

FIGURE 7.6 Linear regression of child height on parent height.

We can summarize the primary features of regression in the following statement:

In the absence of correlation, the regression line that minimizes the sum of squared error is the horizontal line corresponding to the arithmetic mean. In the presence of correlation, the regression line that minimizes the sum of squared error is the "tilted" line (titled up or down depending on the sign of the correlation) corresponding to the "new" regression line about which we can make predictions while keeping the average sum of squared errors of prediction to a minimum value compared to any other place where we could fit the line.

7.8   MORE ABOUT ε_i

We said that the expectation, or **mean** of the random variable ε_i is equal to 0. But what does this imply, exactly? It implies that on average, our predictions will be correct and without error. That is, on average, it will be true that $y_i' = y_i$ (i.e., $E(y_i') = E(y_i)$ (Hays, 1994, p. 601)). Recall that though the arithmetic mean is a good summary statistic for a sample or population of data, it is not always a useful summary statistic, at least in a practical sense, for any **single observation** from that set of data, **especially if the set of data exhibits much variability**. For example, that the mean home price in your city is $250,000 does little to describe your home value of $200,000. Simply because the expectation of ε_i is equal to 0 does not definitively tell us what happens for any subset of values "on the road" to $E(\varepsilon_i) = 0$. Likewise in regression, for any given value of x_i, it is certainly reasonable that our model may generate an **error in prediction**, which we denote, for our sample data by:

$$e_i = y_i - (a + bx_i)$$

It seems natural that in fitting a function to a bivariate plot of data that we should want to **minimize**, on average, errors in prediction. But like with any arithmetic mean, we cannot guarantee that for any given value of x_i we will not still make a relatively substantial error. What we can guarantee is that in estimating parameters using ordinary least-squares, **on average**, and given that our regression assumptions are more or less satisfied (to be discussed shortly), **our errors of prediction will be smaller than anywhere else we may have fit the line**. This is the essence of ordinary least-squares regression. It does not guarantee we will make **few** errors in prediction or even precise predictions each time, no better than the mean house price of your city guarantees a precise prediction of **your** home value. What it does guarantee, however, is that **on average, squared errors will be minimized**.

7.9   MODEL ASSUMPTIONS FOR LINEAR REGRESSION

As was the case in analysis of variance models, we likewise need to impose a set of assumptions on our regression model before we can use the model to make inferences. We list and briefly discuss these assumptions below that are typically held in ordinary least-squares (OLS) regression. Some of these are properties of the regression model per se, while others are specifically required for OLS estimation. We do not distinguish between the two, and for convenience, discuss them collectively in a single discussion. Other types of estimation do not necessarily require all of these assumptions. We state the assumptions for linear regression:

- **Linearity in Parameters**. We assume that the relationship between y_i and x_i or $x_1, x_2, \ldots x_k$ (in the case of multiple predictors, see Chapter 8) is linear in the unknown parameters of α and β (or again $\beta_1, \beta_2, \ldots \beta_k$ in the case of multiple predictors). It is important to emphasize that linearity in the parameters does not mean we cannot have higher-order powers of x_i. For instance,

$$y_i = \alpha + \beta_1 x_i + \beta_2 x_i^2 + \varepsilon_i$$

still exhibits linearity in parameters since α, β_1, and β_2 are still raised to the first power. In contrast,

$$y_i = \alpha + \beta_1 x_i + \beta_2^2 x_i + \varepsilon_i$$

is not linear, since β_2 is raised to a higher power than 1 (i.e., in this case, it is raised to the 2nd power).

We also assume that the expected value of y_i (i.e., "$E(y_i)$"), is equal to $\alpha + \beta x_i$, where $\alpha + \beta x_i$ is sometimes referred to as the **model function** since it contains the systematic predictive element of the model, whereas the random component is in the error term, ε_i. The expectation of the error term, however, is equal to 0 (i.e., $E(\varepsilon_i) = 0$), and so

$$E(y_i) = E(\alpha + \beta x_i + E(\varepsilon_i))$$
$$= \alpha + \beta x_i + 0$$
$$= \alpha + \beta x_i$$

We also assume that the expectation of y_i given any chosen value of x_i, is equal to $E(\alpha + \beta x_i + \varepsilon_i)$ (i.e., $E(y_i/x_i) = E(\alpha + \beta x_i + \varepsilon_i)$) (Fox, 2016, p. 107), but again, since $E(\varepsilon_i) = 0$, this reduces to

$$E(y_i/x_i) = \alpha + \beta x_i + 0$$
$$= \alpha + \beta x_i$$

- **Normal and Identically Distributed Errors**. That is, $\varepsilon_i \sim \text{NID}(0, \sigma_\varepsilon^2)$, which says that errors are distributed with a mean of zero and a finite variance, $\sigma^2 > 0$ (i.e., it has a positive variance which is measurable). For each conditional distribution of y_i, (i.e., y_i/x_i), normality should hold (Fox, 2016, p. 107). We will see how we can visually and informally test this and other assumptions through an examination of **residuals** toward the end of the chapter.

- **Homoscedasticity**. For each population denoted by values of the variable x_i, the variances of these populations on y_i are equal (i.e., $\sigma_{(\varepsilon/x_i)}^2 = \sigma_\varepsilon^2$). If distributions are not homoscedastic, then a problem of **heteroscedasticity** is said to exist. Heteroscedasticity (sometimes written as "heteroskedasticity") essentially means **unequal variances**. Equivalently, we may state this assumption as one of the variance of errors being constant across values of x_i and not conditional upon it. If the variance of the errors is conditional upon the value of x_i, then this will show up in residual analysis plots as a potential "fan shape" or other irregularity, as we will see when we discuss residual analyses. In the case of a single predictor, one can simply inspect the original scatterplot to see if distributions of y_i have the approximate same variance for each x_i, but in the case of multiple regression, we shall require residual analyses proper to verify this assumption due to the high dimensionality of the data.

- **Independence of Errors**. The errors ε_i both within conditional distributions of y_i/x_i and between conditional distributions of y_i/x_i are independent. Practically, what this means is that no single observation in the set of data is dependent (in a probabilistic way) on any other observation. That is, it is assumed that observations are more or less sampled independently (Fox, 2016, p. 108). The assumption is typically satisfied by the **method of data collection**, although residual plots, as we will see, may also be used to somewhat evaluate, in an imperfect sense at least, the tenability of the assumption.

- **Absence of Influential or Outlying Data**. We assume that our data does not contain observations that will influence the regression solution to such an extent that it is no longer feasible to believe that the fitted regression line is a suitable model for the obtained data. Though outliers should generally never be removed from a sample unless there is good **substantive** (in addition to **statistical**) reason for doing so, the linear regression model assumes that there are no extreme observations that would otherwise contribute to gross misspecification of the fitted model.

7.9.1 Model Specification

There are two additional general assumptions we should make, or at minimum consider, when fitting a simple linear regression model to empirical data. The first is that **we have specified the correct model**. The assumption essentially implies that we have chosen predictors that are thought to account for variance in the measured response variable (or "variables" in the case of the wider multivariate multiple regression model), and that we have not left any "important" variables out of the regression model. At its extreme, we would like to assume that all sources of variation accountable for explaining y_i have been incorporated into the model. Of course, this is fanciful and unrealistic thinking, since whatever predictors we choose are likely to be only **some** of the many that may account for variance in y_i. This is one reason for generally preferring experimental designs over nonexperimental ones, since the process of randomization helps to ensure (but in no way **guarantees**) that the innumerable nuisance factors, either observable or latent, are distributed somewhat **evenly** across treatment levels. Regardless, however, of whether you are working in the context of an experimental or nonexperimental design, if you are aware of additional variables that account for significant sources of variance in your dependent variable **and you do not include such variables in your model**, then the model can be said to be **incorrectly specified**. When a model is incorrectly specified, not only is it substantively less meaningful, but **parameter estimates are likely to also be biased** (see Draper and Smith, 1998, pp. 235–242).

As an example, suppose we are studying the predictive ability of depression scores based on a measure of anxiety. We know *a priori*, however, that socioeconomic status (SES) is also an important predictor of depression rates. Then simply regressing depression on anxiety would constitute a misspecified model. Further, as we will discuss in the following chapter on multiple regression, other predictors included in the model can have a significant impact on the interpretation of statistical outcomes and the estimation of parameters. In brief then, we may summarize with the following:

> **A properly specified model is one in which you are identifying and accounting for, at minimum, the correct "already-known" sources of variation in y_i, to the extent that you are knowledgeable or able to. If you are testing a model for which you are aware that important predictors are being left out, and could have a significant impact on the model under consideration, then your model is misspecified.**

It does not take long to realize that, on an idealistic level, **all models are misspecified to some degree**, especially if we stretch our definition to include all **possible** sources of variation, both known and **unknown**. Even models emanating from experimental designs can be a struggle to specify well. We gain solace and comfort in George Box's wise words, in that **all models are wrong, some are useful**. Beyond that, we must do the best we can.

7.9.2 Measurement Error

Speaking of unrealistic or otherwise unattainable assumptions, a final assumption implicitly made in linear regression is that there is no **measurement error** in predictors. As Fox (1997) notes:

> The regression model accommodates measurement error in the **dependent** variable, because measurement error can be conceptualized as a component of the general error term ε, but the independent variables in regression analysis are assumed to be measured without error. (p. 130)

The consequences of measurement error in a given predictor is generally to **attenuate** (i.e., lower) the regression coefficient, and in the context of multiple predictors (see Chapter 8), to diminish the utility of the predictor as a statistical control (see Fox, 1997, pp. 130–132). The assumption of no measurement error is an unrealistic one for the most part, and the degree to which it is violated will depend to some extent on the sophistication and accuracy of measurement instruments used. For example, if one is measuring reaction time, one can probably do so with a relatively small amount of measurement error. If one is measuring IQ, on the other hand, the risk of measurement error will likely be of greater concern, unless of course your operational definition of IQ implies it can be measured simply and quite precisely. The bottom line is that if you have "sloppy" measurement, the regression equation will not somehow "cure it." Rather, as mentioned, it will have an influence on the computed regression coefficients. Measurement error exists in virtually all measurement, and it is the careful scientist who guards against this by developing quality tools (whether in the physical or social sciences) to access the phenomena for which he or she is assigning a number to according to a set of rules (i.e., the process of measurement).

7.10   ESTIMATION OF MODEL PARAMETERS IN REGRESSION

From a purely technical standpoint, the problem of linear regression boils down to estimating model parameters subject to particular constraints. In the model equation $y_i = \alpha + \beta x_i + \varepsilon_i$, we wish to estimate parameters α and β such that they are estimated in such a way that conforms to the overall purpose of building the model. What constraint or condition is appropriate? As mentioned, on both technical and commonsense substantive grounds, it seems appropriate to choose a and b, which are estimators of α and β respectively, such that the sum of squared errors

$$\sum_{i=1}^{n} e_i^2 = \sum_{i=1}^{n} \left[y_i - (a + bx_i) \right]^2$$

is kept to a **minimum** value, which means the line we are fitting to the data, **the least-squares regression line**, guarantees that we are fitting the line that, overall and "on average," has the least amount of prediction error compared to any other line we could fit to the data.

Methods of estimation in linear regression include the aforementioned ordinary least-squares, but also **maximum likelihood**, **weighted least-squares**, among others. Under the assumptions of the linear model, least-squares estimation equates to being the maximum-likelihood estimator (Fox, 2016). Weighted least-squares is suitable for situations where the variance of the response variable varies over the range of predictor values. That is, when we have **nonconstant error variance**. Instead of minimizing $\sum_{i=1}^{n} \varepsilon_i^2$ as is true of least-squares, WLS seeks to minimize $\sum_{i=1}^{n} (w_i) \varepsilon_i^2$, where w_i is some weight applied to the errors ε_i^2. Reciprocals of the variance can be used as weights, although other weights can also be applied. For details on fitting weighted least-squares models, see Fox (2016) and Venables and Ripley (2002).

We now briefly discuss some of the principles behind ordinary least-squares estimation and briefly review how to obtain the least-squares estimators. Though our primary goal is the **application of linear regression**, a brief overview of how the least-squares estimators are obtained can prove fruitful for an understanding of estimation in general.

7.10.1   Ordinary Least-Squares (OLS)

Recall the regression model of (7.1) for a single predictor:

$$y_i = \alpha + \beta x_i + \varepsilon_i$$

where α, β are parameters to be estimated from sample data. To be able to fit the least-squares line, we require good **estimators** (i.e., statistics based on the sample data) for these parameters, analogous to how we required good estimators for other parameters such as the population mean, μ (which recall turned out to be the sample mean, $\bar{y}$).

The least-squares estimators are obtained by first taking **partial derivatives** of $\sum_{i=1}^{n} \varepsilon_i^2$ with respect to both α and β. Recall what it means to take a partial derivative. It means to **differentiate the function with respect to one variable while holding the other variable(s) in the equation constant**. In the case of least-squares, we wish to:

- differentiate $\sum_{i=1}^{n} \varepsilon_i^2$ with respect to α while holding β constant

- differentiate $\sum_{i=1}^{n} \varepsilon_i^2$ with respect to β while holding α constant

First, with respect to α, in terms of sample quantities, we have (Draper and Smith, 1998, p. 23):

$$\frac{\partial \sum_{i=1}^{n} e_i^2}{\partial a} = -2 \sum_{i=1}^{n} (y_i - a - bx_i) \tag{7.2}$$

Then, with respect to β,

$$\frac{\partial \sum_{i=1}^{n} e_i^2}{\partial b} = -2 \sum_{i=1}^{n} x_i (y_i - a - bx_i) \tag{7.3}$$

Solutions to (7.2) and (7.3) generate the ensuing **normal equations**:

$$\sum_{i=1}^{n} y_i - na - b \sum_{i=1}^{n} x_i = 0$$

$$\sum_{i=1}^{n} x_i y_i - a \sum_{i=1}^{n} x_i - b \sum_{i=1}^{n} x_i^2 = 0$$

From these equations, we obtain the least-squares coefficients:

$$a = \bar{y} - b\bar{x}$$

as an estimator of α, and

$$b = \frac{\sum (x_i - \bar{x})(y_i - \bar{y})}{\sum (x_i - \bar{x})^2}$$

as an estimator of β.

In words, the formula for b requires us to sum the **cross-products** of x_i and y_i, and then divide by the **sum of squares** for x_i. The estimate for b can also be computed by

$$b = \frac{\text{cov}_{xy}}{s_x^2}$$

since when expanded,

$$b = \frac{\text{cov}_{xy}}{s_x^2} = \frac{\frac{\sum_{i=1}^{n}(x_i-\bar{x})(y_i-\bar{y})}{n-1}}{\frac{\sum_{i=1}^{n}(x_i-\bar{x})(x_i-\bar{x})}{n-1}} = \frac{\sum_{i=1}^{n}(x_i-\bar{x})(y_i-\bar{y})}{n-1} \cdot \frac{n-1}{\sum_{i=1}^{n}(x_i-\bar{x})(x_i-\bar{x})} = \frac{\sum_{i=1}^{n}(x_i-\bar{x})(y_i-\bar{y})}{\sum_{i=1}^{n}(x_i-\bar{x})(x_i-\bar{x})} = \frac{\sum(x_i-\bar{x})(y_i-\bar{y})}{\sum(x_i-\bar{x})^2}$$

Least-squares estimators are **unbiased**. That is, $E(a) = \alpha$ and $E(b) = \beta$, which in words, means the expectation of their sample quantities is equal to the parameters they seek to estimate. Furthermore, as noted by Fox (1997):

> If the errors are independently distributed with zero expectation and constant variance, then the least-squares estimator b is the most efficient linear unbiased estimator of β. That is, of all unbiased estimators that are linear functions of the observations, the least-squares estimator has the smallest sampling variance and, hence, the smallest mean-squared error. (p. 217)

This result is generally known as the **Gauss–Markov theorem**. Consult Fox (1997) for a discussion, as well as Hastie, Tibshirani, and Friedman (2009).

Recall that ordinary least-squares is often mistaken to be the **only** method by which parameters in regression can be estimated, since it is usually the preferred choice for estimating parameters and is most often taught in introductory regression courses. Indeed, historically, OLS ("ordinary least-squares") has become more or less synonymous with regression (see Stigler, 1986 for details). However, as mentioned already, even if you should never require alternative methods of estimation in your practice of statistical regression, you should be aware that there are several other methods of estimation available, which include maximum likelihood, weighted least-squares, etc. Nonparametric approaches to regression also exist and may prove useful from time to time. Regression analysis is much more than one particular way of estimating parameters, which would explain why entire books and courses are devoted to the subject.

7.11 NULL HYPOTHESES FOR REGRESSION

Obtaining estimates of α and β is one thing, but testing null hypotheses about their population values is quite another. We are most interested in testing the null hypothesis $H_0 : \beta = 0$ against the alternative hypothesis $H_1 : \beta \neq 0$, where β is the population regression coefficient. As with all significance tests, we require the requisite estimate of the **standard error**. Recall why we require standard errors. Even if our sample data yield a sample slope of say, $b = 0.7$, this in no way, on its own, suggests that the null hypothesis is false and that $\beta \neq 0$ in the population from which the sample data were drawn, analogous to how in the analysis of variance a mean difference in our sample did not by itself imply a mean difference in the population.

What is needed is an estimate of how much sampling variability exists from sample to sample if we were to draw repeatedly from a population for which the null hypothesis were "true." That is, if we sampled repeatedly from a population such that $\beta = 0$, how much variation around expectation (i.e., 0) would we experience in our sampling? If the estimate of $b = 0.7$, for instance, was sampled from a population (and its corresponding sampling distribution) in which there is expected much variation in slopes, then 0.7 may not be regarded as that **unlikely** under the null hypothesis. However, if $b = 0.7$ were sampled from a population for which there is exceedingly little variation in slopes, then 0.7 may very well suggest to us that $\beta = 0$ is not true and that $\beta \neq 0$ is a more accurate reality. We need an estimate of variability of our sample statistic, b. In other words, we need to know the **standard error of the slope**.

The variance of b is given by

$$s_b^2 = \frac{\dfrac{\sum_{i=1}^{n}\left(y_i - y_i'\right)^2}{n-2}}{\sum_{i=1}^{n}\left(x_i - \bar{x}\right)^2}$$

where,

- s_b^2 is the variance of the slope estimator.

- $\dfrac{\sum_{i=1}^{n}\left(y_i - y_i'\right)^2}{n-2}$ is the variance of residuals, or **variance of the estimate**, or again, **MS residual**. Its square root is the **standard error of the estimate** (which is the standard deviation of residuals, but with $n-2$ in the denominator rather than $n-1$ for simple linear regression).

- $\sum_{i=1}^{n}\left(x_i - \bar{x}\right)^2$ is the sum of squares for the predictor.

We can appreciate why s_b^2 is the way it is. We are taking a ratio of error variance (i.e., numerator) relative to a measure of variability in our predictor (i.e., denominator). The extent to which average prediction error is large relative to variability in the predictor is the extent to which b will be estimated **imprecisely**, meaning that there is expected to be much fluctuation from sample to sample. On the other hand, if average prediction error is small relative to variability in the predictor, b will be estimated more precisely, which implies less fluctuation in b from sample to sample.

Computing the standard error from the variance of b is straightforward. As usual, we simply take the square root:

$$s_b = \sqrt{\frac{\dfrac{\sum_{i=1}^{n}\left(y_i - y_i'\right)^2}{n-2}}{\sum_{i=1}^{n}\left(x_i - \bar{x}\right)^2}}$$

Though in most contexts (other than perhaps in multilevel models), we are not especially concerned with estimating the intercept parameter, in some cases, we like to center predictors so that the intercept value corresponds to a value on the response variable at the mean of the predictor. Either way, an inferential test on α may sometimes be of interest. Also, since the intercept plays a much more significant role in advanced modeling techniques such as mixed models and the aforementioned multilevel model, understanding how to obtain a significance test is worthwhile.

As we did for b, we require an estimate of the variance of a. The variance of a turns out to be

$$s_a^2 = \left(\frac{\sum_{i=1}^{n} (y_i - y_i')^2}{n-2} \right) \left(\frac{\sum_{i=1}^{n} x_i^2}{n \sum_{i=1}^{n} (x_i - \bar{x})^2} \right)$$

where,

- s_a^2 is the variance of the intercept estimator.
- $\dfrac{\sum_{i=1}^{n} (y_i - y_i')^2}{n-2}$ is, as before, the variance of residuals (i.e., MS residual).

To get the standard error of s_a^2, we again take the square root:

$$s_a = \sqrt{ \left(\frac{\sum_{i=1}^{n} (y_i - y_i')^2}{n-2} \right) \left(\frac{\sum_{i=1}^{n} x_i^2}{n \sum_{i=1}^{n} (x_i - \bar{x})^2} \right) }$$

Having now obtained sampling variances (and their corresponding standard errors) for the slope and intercept parameters, we are now in a position to test null hypotheses on these parameters. We consider these tests next.

7.12 SIGNIFICANCE TESTS AND CONFIDENCE INTERVALS FOR MODEL PARAMETERS

Recall that we said that if predictor x_i does not afford additional predictive power over and above simply predicting the mean of y_i, then it implied a horizontal slope (i.e., a horizontal "regression line"), which also implied that we have no basis for rejecting $\beta = 0$. When we obtain a sample estimate of β, say, $b = 0.7$, we ask the question:

Does $b = 0.7$ deviate enough from expectation that we can reject the null hypothesis $\beta = 0$ and infer $\beta \neq 0$?

Now that we have measures of sampling variability for b (i.e., the variance and standard error we discussed earlier), we can now test the observed deviation $b - \beta$ relative to expectation under H_0. That is, we can test the **statistical significance** of b against a value of the parameter β under the null hypothesis using a t-statistic:

$$t = \frac{(b - \beta_{\text{null}})}{s_b}$$

where t is distributed on $n - 2$ degrees of freedom and β_{null} represents some value under the null hypothesis, usually equal to 0. The logic of the t-test is clear: we are comparing an **observed deviation** $b - \beta_{\text{null}}$ to a **deviation we would expect**, s_b, under the null hypothesis, or equivalently, under repeated sampling of b statistics from a population for which β_{null} is true.

Likewise, the statistical significance of a can be assessed by a t-statistic:

$$t = \frac{(a - \alpha_{\text{null}})}{s_a}$$

where t is again distributed on $n - 2$ degrees of freedom. The logic of the test is analogous to that for b. We are comparing an observed deviation of the kind $a - \alpha_{\text{null}}$ to an average deviation s_a we would expect under the null. As was true for β_{null}, with some programming, we are free to specify α_{null} as a value different from zero if we really so desired, and incorporate this into our test, but in the absence of any particular reason to do so, the default test value will be zero (remember, null hypotheses can always be rejected if they are unreasonable or unrealistic to begin with; you must choose the null wisely so the significance test makes substantive sense for your given research problem).

A $100(1 - \alpha)$ confidence interval for b can be constructed as follows:

$$b \pm t_{(\alpha/2)} s_b$$

where b is the sample estimate of β, $t_{(\alpha/2)}$ is the two-tailed critical value for the $100(1 - \alpha)$ confidence level on $n - 2$ degrees of freedom, α is the significance level for the level of confidence, for which the confidence interval divides it by 2 (i.e., $\alpha/2$) to make it "two-sided," and s_b is the estimated standard error of the slope.

Likewise, a $100(1 - \alpha)$ confidence interval for a can be constructed:

$$a \pm t_{(\alpha/2)} s_a$$

where a is the sample estimate of α.

We will see significance tests and confidence intervals "in action" when we consider software applications shortly. Tests for comparing slopes from two different samples are also available as are tests that two correlation coefficients are equal to some number, usually zero (see Howell, 2002, p. 276). An alternative to estimating confidence intervals analytically is to employ a **bootstrap procedure**. For an example of how the bootstrap can be used in this regard, see Crawley (2013, pp. 478–481). The **jack-knife** procedure, which has been shown to be an approximation to the bootstrap, is also an alternative strategy for obtaining confidence intervals, though as noted by Fox (1997), may not perform as well when compared to the bootstrap.

7.13 OTHER FORMULATIONS OF THE REGRESSION MODEL

In addition to specifying the regression model as we did in (7.1), we can express the model via other configurations. Not only does doing so constitute an interesting algebraic exercise, but also formulating the regression model in different formats helps us better understand just what regression is **doing** in terms of its mechanics. That is, greater insight into regression can be "experienced" by representing the model in a variety of algebraically equivalent formats.

For instance, we have already seen that the equation for computing the intercept term is given by $a = \bar{y} - b\bar{x}$. When we substitute this into the model equation of (7.1), for any sample of observations, we obtain:

$$y_i = a + bx_i + e_i$$
$$= (\bar{y} - b\bar{x}) + bx_i + e_i$$

Removing the parentheses, we have

$$y_i = \bar{y} - b\bar{x} + bx_i + e_i$$

Notice that b is common to the terms $b\bar{x}$ and bx_i, which means we can factor b out and get

$$y_i = \bar{y} - b\bar{x} + bx_i + e_i$$
$$= \bar{y} + b(x_i - \bar{x}) + e_i \tag{7.4}$$

What is the advantage of the formulation in (7.4)? It emphasizes the fact that y_i is a function first of its mean, $\bar{y}$, which is the expected value of y_i under the circumstance of zero correlation between y_i and x_i, **adjusted** by the extent to which the term "x_i" changes the prediction, over and above ε_i (though recall, $E(\varepsilon_i) = 0$, so this is not a concern, it is not a **systematic** component). This idea of $\bar{y}$ being "adjusted" is a powerful way to understand regression. If our best prediction of y_i given no information is the mean of y_i, then when we do have more information in the form of x_i, we adjust our prediction line accordingly to reflect this influence. The sample estimator b tells us the degree and direction for which we should be making such an adjustment to our original horizontal line.

We can use the formulation (7.4) to show that when x_i is equal to $\bar{x}$, the best prediction for y_i is indeed $\bar{y}$, in the sense of minimizing sum of squared errors, $\sum e_i^2$.

$$y_i = \bar{y} + b(x_i - \bar{x}) + e_i$$
$$= \bar{y} + b(\bar{x} - \bar{x}) + e_i$$
$$= \bar{y} + b(0) + e_i$$
$$= \bar{y} + e_i$$

That is, when x_i is equal to $\bar{x}$, then $\bar{x} - \bar{x}$, and so $b(0)$ is equal to 0, leaving us with only $y_i = \bar{y} + e_i$.

7.14 THE REGRESSION MODEL IN MATRICES: ALLOWING FOR MORE COMPLEX MULTIVARIABLE MODELS

The simple algebraic model formulations thus far employed, though sufficient for simple linear regression, will not be for discussing the multiple regression model of the following chapter. The "vehicle" for multiple regression and multivariable methods, in general, is that of **vectors** and **matrices**.

In this section, we briefly introduce and detail the simple linear regression model in matrix form so that when we arrive at multiple regression, we will be in a position to extend on this simple formulation for multiple predictors, and in some cases, multiple response variables as well (analogous to the case of **multivariate analysis of variance**).

In a simple regression model, we can write each component of the model as its own **vector**. The expectation for the response variable y_i, that is, $E(y_i)$, is now written as an expectation of a vector of responses, $E(\mathbf{y})$, which is a matrix containing n rows and a single column. We write $E(\mathbf{y})$ as

$$E(\mathbf{y}) = \begin{bmatrix} E(y_1) \\ E(y_2) \\ E(y_3) \\ . \\ . \\ . \\ E(y_n) \end{bmatrix} \tag{7.5}$$

In (7.5), we are now simply denoting each value of the response vector as a single column, of which we take the expectation on each y_i.

Since in the simple linear regression model the expectation of a randomly chosen value for the response is $E(y_i) = \alpha + \beta x_i$, we can express the vector $E(\mathbf{y})$ as

$$E(\mathbf{y}) = \begin{bmatrix} E(y_1) \\ E(y_2) \\ E(y_3) \\ . \\ . \\ . \\ E(y_n) \end{bmatrix} = \begin{bmatrix} \alpha + \beta x_{i=1} \\ \alpha + \beta x_{i=2} \\ \alpha + \beta x_{i=3} \\ . \\ . \\ . \\ \alpha + \beta x_{i=n} \end{bmatrix}$$

where $\alpha + \beta x_{i=1}$, $\alpha + \beta x_{i=2}$, $\ldots \alpha + \beta x_{i=n}$ represents the **systematic portion** of the model applied to each observation y_1 through y_n in the data. It stands that the error term will also have its own $n \times 1$ (i.e., n rows and 1 column) vector:

$$\boldsymbol{\varepsilon} = \begin{bmatrix} \varepsilon_{i=1} \\ \varepsilon_{i=2} \\ \varepsilon_{i=3} \\ . \\ . \\ . \\ \varepsilon_{i=n} \end{bmatrix}$$

When we put all the pieces together, the full expression of the simple linear regression model is given by

$$\begin{bmatrix} E(y_{i=1}) \\ E(y_{i=2}) \\ E(y_{i=3}) \\ . \\ . \\ . \\ E(y_{i=n}) \end{bmatrix} = \begin{bmatrix} \alpha + \beta x_{i=1} \\ \alpha + \beta x_{i=2} \\ \alpha + \beta x_{i=3} \\ . \\ . \\ . \\ \alpha + \beta x_{i=n} \end{bmatrix} + \begin{bmatrix} \varepsilon_{i=1} \\ \varepsilon_{i=2} \\ \varepsilon_{i=3} \\ . \\ . \\ . \\ \varepsilon_{i=n} \end{bmatrix}$$

LINEAR REGRESSION

More compactly, and adjusted slightly, we may write the model as:

$$
\begin{aligned}
\mathbf{y} &= [E(\mathbf{y})] + \boldsymbol{\varepsilon} \\
&= [\boldsymbol{\alpha} + \boldsymbol{\beta}'\mathbf{X}] + \boldsymbol{\varepsilon}
\end{aligned}
\tag{7.6}
$$

where, all we did to get from $[E(\mathbf{y})] + \boldsymbol{\varepsilon}$ to $[\boldsymbol{\alpha} + \boldsymbol{\beta}'\mathbf{X}] + \boldsymbol{\varepsilon}$ was to recognize that $E(\mathbf{y}) = \boldsymbol{\alpha} + \boldsymbol{\beta}'\mathbf{X}$.

Note that in this formulation, we are grouping the intercepts α (which once estimated, are a constant for all observations) into their own vector $\boldsymbol{\alpha}$.

The formulation (7.6) of the regression model is fine, but it becomes awkward when one considers such things as **multilevel** models in which the intercept term (and potentially the slope) is a random variable and hence not fixed. That is, in such models, often researchers wish to free the intercept term so that each particular **individual** or specified **subgroup** in the data has their own intercept. Consequently, it would be useful to format the regression model such that the inclusion of the intercept term α for each y_i is designated as a potentially **unique** quantity. In the formulation of the regression model in (7.6), all intercepts are assumed to be a constant for all observations on the response.

We can accomplish this more easily in our notation by re-expressing $\boldsymbol{\alpha} + \boldsymbol{\beta}'\mathbf{X}$ into two components, one component representing the values for x_i and the other component the parameters of our model. We need two vectors to accomplish this, being sure to appropriately index the intercept parameter with a column of 1's:

$$
\mathbf{X} = \begin{bmatrix} 1 & x_{i=1} \\ 1 & x_{i=2} \\ 1 & x_{i=3} \\ . & . \\ . & . \\ . & . \\ 1 & x_{i=n} \end{bmatrix} \quad \boldsymbol{\beta} = \begin{bmatrix} \alpha \\ \beta \end{bmatrix}
$$

The components of the full simple linear regression model can thus now be written as

$$
\mathbf{y} = \begin{bmatrix} y_{i=1} \\ y_{i=2} \\ y_{i=3} \\ . \\ . \\ . \\ y_{i=n} \end{bmatrix} \quad \mathbf{X} = \begin{bmatrix} 1 & x_{i=1} \\ 1 & x_{i=2} \\ 1 & x_{i=3} \\ . & . \\ . & . \\ . & . \\ 1 & x_{i=n} \end{bmatrix} \quad \boldsymbol{\beta} = \begin{bmatrix} \alpha \\ \beta \end{bmatrix} \quad \boldsymbol{\varepsilon} = \begin{bmatrix} \varepsilon_1 \\ \varepsilon_2 \\ \varepsilon_3 \\ . \\ . \\ . \\ \varepsilon_n \end{bmatrix}
$$

or more compactly as:

$$
\mathbf{y} = \mathbf{X}\boldsymbol{\beta} + \boldsymbol{\varepsilon}
\tag{7.7}
$$

To summarize,

- $\mathbf{y}$ is a vector of responses on a single response variable in the case of simple or multiple regression. In the case of **multivariate regression**, this vector could include several response variables (see our discussion of MANOVA in Chapter 11 for an introduction to the multivariate landscape, though in the context of ANOVA models). In this book, we typically designate such a vector with potentially several responses as **Y** (whereas the smaller case bold vector **y** has only a single column of scores on a given response variable).

- **X** is generally known as the **model or design matrix**; in regression analysis, it typically contains values on one or more predictors; in ANOVA models, it can be adapted to represent group membership in the form of indicator variables such as "0" or "1" to denote classification on a categorical predictor.
- **β** is a matrix of regression coefficients, including the constant α.
- **ε** is a vector of errors; this vector is typically a single column in most models because no matter how complex our regression model, only a single error exists for any predicted value of the response. In multilevel models, one could parameterize error terms for each level of the hierarchy or grouping structure, though we do not consider such possibilities here.

7.15 ORDINARY LEAST-SQUARES IN MATRICES

Having surveyed the use of matrices in the representation of a regression model, it follows that we should be able to express the least-squares solutions and estimators in matrices as well. That is, all that was done on scalars in simple regression should have their corresponding matrix counterparts. Our formulation to follow mirrors that of the formulation of least-squares using scalar quantities. The only difference is that now, instead of formulating the model in terms of scalars with a **single** predictor, we generalize the least-squares solutions in **matrices** to better prepare for more complex modeling to come, including multivariate modeling and structural equation models.

Formally, the method of least-squares and of minimizing $\varepsilon'\varepsilon$ (squared error) is analogous to finding an orthogonal projection of observed data in **y** onto a new space that is said to be **spanned** by the predictor variable(s), where the span is the set of matrices (or vectors) that is expressible as **linear combinations**. One may think of a projection in this case as merely the estimated regression line, of which the information contained in predictor(s) is used in "projecting" points onto this line (or surface, in the case of multiple regression).

In both (7.1) and (7.7), the systematic portion of the model is contained in $\alpha + \beta x_i$ and $\mathbf{X}\boldsymbol{\beta}$, respectively. The error ε_i in (7.1), recall, was equal to $\varepsilon_i = y_i - y_i'$. The equivalent for expressing a difference between an observed and predicted value (i.e., $y_i - y_i'$) using matrices is thus

$$\boldsymbol{\varepsilon} = (\mathbf{y} - \mathbf{X}\boldsymbol{\beta})$$

The matrix equivalent of minimizing the sum of squared errors $\sum_{i=1}^{n} \varepsilon_i^2$ in scalar quantities becomes

$$\begin{aligned}\boldsymbol{\varepsilon}'\boldsymbol{\varepsilon} &= (\mathbf{y} - \mathbf{X}\boldsymbol{\beta})'(\mathbf{y} - \mathbf{X}\boldsymbol{\beta}) \\ &= \mathbf{y}'\mathbf{y} - \boldsymbol{\beta}'\mathbf{X}'\mathbf{y} - \mathbf{y}'\mathbf{X}\boldsymbol{\beta} + \boldsymbol{\beta}'\mathbf{X}'\mathbf{X}\boldsymbol{\beta} \\ &= \mathbf{y}'\mathbf{y} - 2\boldsymbol{\beta}'\mathbf{X}'\mathbf{y} + \boldsymbol{\beta}'\mathbf{X}'\mathbf{X}\boldsymbol{\beta}\end{aligned}$$

where $(\boldsymbol{\beta}'\mathbf{X}'\mathbf{y})' = \mathbf{y}'\mathbf{X}\boldsymbol{\beta}$ yield the same result (Draper and Smith, 1998, p. 135).

We can clearly see that the expression $\varepsilon'\varepsilon$ is nothing more than the equivalent of ε_i^2 in scalar algebra, since $(\mathbf{y} - \mathbf{X}\boldsymbol{\beta})'(\mathbf{y} - \mathbf{X}\boldsymbol{\beta})$ essentially translates to $(\mathbf{y} - \mathbf{X}\boldsymbol{\beta})$ "squared."

After taking the relevant partial derivatives as we did in the simpler regression situation where we were not using matrices, the **normal equations** using matrices end up being

$$(\mathbf{X'X})\,\mathbf{b} = \mathbf{X'y}$$

We can now solve for **b**. The analog in scalar algebra would be to divide the left and right-hand sides by $\mathbf{X'X}$ as to isolate **b**. That is, if we were to express the above formulation **naively** in matrix terms, pretending for a moment that we were still doing scalar algebra, solving for **b** would look like this:

$$(\mathbf{X'X})\,\mathbf{b} = \mathbf{X'y}$$
$$\mathbf{b} = \frac{\mathbf{X'y}}{\mathbf{X'X}}$$

In matrix algebra, however, as reviewed in the Appendix, division such as just performed is not permitted. Rather, to "divide" in this case, we multiply by the inverse:

$$\mathbf{b} = (\mathbf{X'X})^{-1}\mathbf{X'y}$$

Hence, the solution for obtaining the least-squares estimators is $\mathbf{b} = (\mathbf{X'X})^{-1}\mathbf{X'y}$. We note that should $\mathbf{X'X}$ not be invertible, solutions for **b** cannot be obtained using this set of equations since if $(\mathbf{X'X})^{-1}$ cannot be computed, we end up with

$$\mathbf{b} = (\mathbf{X'X})^{-1}\mathbf{X'y}$$
$$= (\emptyset)\mathbf{X'y}$$
$$= \emptyset$$

where $\emptyset$, in this case, simply represents a quantity that cannot be calculated. If we cannot solve $(\mathbf{X'X})^{-1}$ (since it does not exist, (Draper and Smith, 1998, p. 136)), we cannot solve for **b** in the given model.

Again, what was the purpose of demonstrating the matrix development in simple regression if using "ordinary" algebra would have sufficed? That is, we certainly did not require matrices to conceptualize the simple linear regression model. However, when we get to multiple regression models, those containing several predictors, or multivariate models, those containing several dependent variables, the employment of matrices is not simply an equivalent way of conceptualizing these models. It is a **requirement**. Matrices allow for the generalization from simple to complex statistical models, and hence our brief survey above was simply a preparation for this further work and for when the complexity we encounter will not be reducible to simpler configurations. It should be noted that understanding the matrices underlying regression (and other) models does not necessarily provide one with greater insight into the **nature of regression**, but it does allow one to be able to understand how statistical models, even complex ones, are configured in the most general of cases.

7.16   ANALYSIS OF VARIANCE FOR REGRESSION

When we run a simple linear regression, in addition to parameter estimates, we obtain an **analysis of variance** table. At first sight, students familiar with ANOVA, but newcomers to regression are often somewhat surprised to see an ANOVA table in regression output. What is an ANOVA table doing in a regression chapter? However, remember that ANOVA and regression are both versions of the **general**

linear model, and at their base, seek to accomplish very much the same thing, that of partitioning the variance of a response variable into **explained** and **unexplained** components. The analysis of variance **method** or **procedure** was that featured in earlier chapters. However, we come across the analysis of variance **partition** in both ANOVA **and** regression models. Further, as will become clearer the more you study these models, you will see that though model specifications and parameterizations are different in each one, at their respective cores, both accomplish a remarkably similar analytical goal.

Table 7.1 shows the partition of the sums of squares in a simple linear regression.

We unpack each of the terms in the table:

- $\sum_{i=1}^{n} (y_i' - \bar{y})^2$ is the sum of squares due to regression.

- $\sum_{i=1}^{n} (y_i - y_i')^2$ is the sum of squares due to error (often called **SS residual** in this context).

- $\sum_{i=1}^{n} (y_i - \bar{y})^2$ is the sum of squares total for the entire data.

- Degrees of freedom for regression are k, where k is the number of predictors.
- Degrees of freedom for error (residual) are $n - k - 1$, where n is the number of data points.
- Degrees of freedom for total are $n - 1$, where again n is the number of data points.
- Mean squares are computed by taking the relevant sums of squares and dividing by respective degrees of freedom, analogous to what is done in ANOVA models.
- F-ratio is computed by MS regression/MS residual and is evaluated on k and $n - k - 1$ degrees of freedom.

The identity **SS total = SS regression + SS residual** in regression is conceptually (though not computationally) analogous to the identity **SS total = SS between + SS within** in the analysis of variance. Hence in regression, the partition is:

$$\sum_{i=1}^{n} (y_i - \bar{y})^2 = \sum_{i=1}^{n} (y_i' - \bar{y})^2 + \sum_{i=1}^{n} (y_i - y_i')^2$$

The identity tells us that we can break down the total variation in a data set into two parts, the sums of squares due to regression and the sums of squares due to residual. Naturally, as a researcher, your

TABLE 7.1 Analysis of Variance for Linear Regression

Source	Sum of Squares	df	Mean Square	F
Regression	$\sum_{i=1}^{n} (y_i' - \bar{y})^2$	k	$\dfrac{\sum_{i=1}^{n} (y_i' - \bar{y})^2}{k}$	$\dfrac{\sum_{i=1}^{n} (y_i' - \bar{y})^2/k}{\sum_{i=1}^{n} (y_i - y_i')^2/(n-k-1)}$
Residual	$\sum_{i=1}^{n} (y_i - y_i')^2$	$n - k - 1$	$\dfrac{\sum_{i=1}^{n} (y_i - y_i')^2}{n-k-1}$	
Total	$\sum_{i=1}^{n} (y_i - \bar{y})^2$	$n - 1$	$\dfrac{\sum_{i=1}^{n} (y_i - \bar{y})^2}{n-1}$	

TABLE 7.2 **Analysis of Variance Summary Table in Matrices**

Source	Sum of Squares	df	Mean Square	F
Regression	$\mathbf{b'X'y} - (1/n)\mathbf{y'Jy}$	$p - 1$	$\dfrac{\mathbf{b'X'y} - (1/n)\mathbf{y'Jy}}{p-1}$	$\dfrac{\mathbf{b'X'y} - (1/n)\mathbf{y'Jy}/(p-1)}{\mathbf{y'y} - \mathbf{b'X'y}/(n-p)}$
Error	$\mathbf{y'y} - \mathbf{b'X'y}$	$n - p$	$\dfrac{\mathbf{y'y} - \mathbf{b'X'y}}{n-p}$	
Total	$\mathbf{y'y} - (1/n)\mathbf{y'Jy}$	$n - 1$	$\dfrac{\mathbf{y'y} - (1/n)\mathbf{y'Jy}}{n-1}$	

Source: Adapted from Neter et al. (1996, p. 229).

hope is that the sums of squares due to regression is much larger than the sums of squares due to residual. To the extent that SS regression is small relative to SS residual is the extent to which the model is not very effective at accounting for observed data (e.g., there is far too much scatter around the estimated regression line for SS regression to be dominating SS residual).

Having laid out some of theory of regression in matrices, we can also represent the ANOVA summary table in matrix form (see Table 7.2). Because we are using matrices, such a summary table is applicable to either simple or multiple regression.

We do not discuss or describe Table 7.2 at any length. Given the matrix formulation of the regression model discussed earlier, you should be able to recognize familiar elements in Table 7.2. The purpose of showing it is mostly to reveal how ANOVA for regression can be generalized to matrices for more complex cases than that of a single predictor. For instance, the sums of squares for regression in the simple bivariate case are equal to $\sum_{i=1}^{n} \left(y_i' - \bar{y}\right)^2$ whereas the equivalent matrix formulation is that of $\mathbf{b'X'y} - \left(\frac{1}{n}\right)\mathbf{y'Jy}$, where $\mathbf{b}$ is a vector of regression coefficients, $\mathbf{X}$ is a matrix of predictors, $\mathbf{y}$ is a vector of response variables, n is the sample size, and p are the number of estimated parameters (e.g., intercept and slope in the case of simple regression). The matrix $\mathbf{J}$ is defined as a square matrix of 1's.

The $\mathbf{H}$ matrix is the so-called **hat matrix** defined as $\mathbf{H} = \mathbf{X}(\mathbf{X'X})^{-1}\mathbf{X'}$, where $\mathbf{H}$ is used in estimating the fitted values of $\mathbf{y}$ by $\mathbf{y'} = \mathbf{Hy}$. The sums of squares for regression, $\mathbf{b'X'y} - \left(\frac{1}{n}\right)\mathbf{y'Jy}$ in Table 7.2 can be expressed as $\mathbf{y'}\left[\mathbf{H} - \left(\frac{1}{n}\right)\mathbf{J}\right]\mathbf{y}$, which, as can be seen, explicitly uses $\mathbf{H}$. As we will discuss shortly, the hat matrix is often used in defining various regression diagnostics. The sums of squares for error $\mathbf{y'y} - \mathbf{b'X'y}$ can likewise be expressed in terms of the hat matrix by $\mathbf{y'}(\mathbf{I} - \mathbf{H})\mathbf{y}$.

To reiterate, in a simple linear regression problem, one would not use nor need to refer to such matrix formulations as in Table 7.2. The advantage of presenting these results now, however, is that it readily prepares the reader to handle larger more complex models as one would see in a more "general" **multivariable** context (i.e., it "unpacks" the SS in more detail), as well as initiates him or her to the use of matrices in the specification of linear models in general. **It is essential that if you plan on specializing in statistical modeling to any advanced degree, that you become familiar with matrices and how linear models are depicted using them.** When multivariate techniques such as the **multivariate analysis of variance (MANOVA)** and **factor analysis** are considered later, the requisite employment of matrix formulations should not catch the reader by surprise. All simpler model formulations which do not explicitly require matrix operations can be considered "special cases" of the wider matrix framework. Hence, the initial "complexity" of matrix operations actually, in the end, makes things a whole lot simpler because it widens the landscape. It may be said that perspective in and understanding of whatever craft one studies is facilitated by aspiring to a comprehension of the most **global** and **universal** principles, of which all others are special cases. **Always strive to see the "bigger picture" so that smaller pictures are interpreted from that vantage point.**

To demonstrate the use of matrix operations in computing least-squares solutions, we will consider a simple example using R shortly in which all computations are done "manually" (i.e., by matrix computations). Familiarity with this example will help greatly "demystify" matrix calculations in regression and will open the door for the reader to much more complex modeling employing matrices.

7.17 MEASURES OF MODEL FIT FOR REGRESSION: HOW WELL DOES THE LINEAR EQUATION FIT?

Fitting a regression model to data is relatively easy. It is essentially and simply the imposition of a functional polynomial form on what is usually, relatively speaking, a **messy data surface**, especially if one's data arise from social processes (i.e., lots of variability). Determining how well the model actually fits is where our interests really lie. Any data can accommodate a regression line, meaning it is possible to fit a regression line to any quality of data. **Good models, however, accommodate that data well**. We must define more precisely a measure of "wellness" or "goodness" of fit.

The most popular measure of model fit in a simple regression setting is the **coefficient of determination**, r^2. In a multiple regression context, the statistic is usually referred to as R^2 and denotes the **coefficient of multiple determination**. Regardless of whether the model is simple or multiple, we compute these by the ratio of sums of squares of regression to sums of squares total. The resulting coefficient will range from 0 to 1 (coefficients of determination cannot typically be negative under normal parameterizations of the regression model), with low values indicating poor fit and increasingly larger values indicative of a model that fits increasingly well. We can define R^2 as:

$$R^2 = \frac{\sum_{i=1}^{n} \left(y_i' - \bar{y}\right)^2}{\sum_{i=1}^{n} \left(y_i - \bar{y}\right)^2}$$

We interpret R^2 to be **the proportion of variance in the response variable accounted for by knowledge of the predictor variable(s)**. Hence, if $R^2 = 0.70$, the interpretation is that 70% of the variance in the response variable is accounted for or "explained" by knowledge of the set of predictors (even if that set consists of only a single variable). R^2 is also simultaneously **the squared correlation between observed values on y_i and predicted values y_i' based on the regression**.

But what does R^2 tell us **exactly**? In the case of zero linear correlation between x_i and y_i, R^2 is equal to 0. Why is this so? This is the case because under the condition of zero linear correlation, our "best" prediction (in the sense of minimizing sum of squared errors) is that of the mean, $\bar{y}$ (Hays, 1994). That is, y_i' must equal $\bar{y}$, which means the numerator of R^2 will equal 0. That is,

$$R^2 = \frac{\sum_{i=1}^{n} \left(y_i' - \bar{y}\right)^2}{\sum_{i=1}^{n} \left(y_i - \bar{y}\right)^2} = \frac{\sum_{i=1}^{n} \left(\bar{y} - \bar{y}\right)^2}{\sum_{i=1}^{n} \left(y_i - \bar{y}\right)^2} = \frac{0}{\sum_{i=1}^{n} \left(y_i - \bar{y}\right)^2} = 0$$

Notice that since the numerator of the second term is now $\sum_{i=1}^{n} \left(\bar{y} - \bar{y}\right)^2$, R^2 can be nothing else other than 0 regardless of how much the denominator $\sum_{i=1}^{n} \left(y_i - \bar{y}\right)^2$ is greater than zero. That is, even if there is a lot of variability in one's data (i.e., $\sum_{i=1}^{n} (y_i - \bar{y})^2$ may be large in the particular data set you are considering), one still obtains an R^2 equal to 0 since $\sum_{i=1}^{n} (\bar{y} - \bar{y})^2 = 0$.

7.18   ADJUSTED R^2

A related statistic to R^2 is **adjusted** R^2 given by (Draper and Smith, 1998, p. 140):

$$R_{\text{adj}}^2 = 1 - \left(1 - R^2\right) \left(\frac{n-1}{n-p}\right)$$

where n, as before, is the number of observations in the sample and p is the number of parameters being estimated, including the intercept term, α. The logic behind R_{adj}^2 is to make an adjustment for the number of parameters being fit in the model relative to their additive "value" in the regression. This adjustment may be meaningful for a given model because R^2 can typically only increase given the inclusion of additional predictors. Hence, a model that fits "everything under the sun" may at first glance appear impressive, yet due to a lack of parsimony, may be less so relative to a competing model that contains lesser predictors but explains similar amounts of variance. As noted by Heumann and Shalabh (2016), because of this "correction," R_{adj}^2 can be used to compare the fit of models with differing numbers of covariates. Not all statisticians find this use helpful, however, and recommend it only as a "gross indicator" when comparing models (see Draper and Smith, 1998, p. 140). If one attempts to do the same (i.e., model comparison) with R^2 rather than R_{adj}^2, one should inquire more into each model that is being compared to learn of the number of covariates utilized in each. As sample size increases, the expected difference between R^2 and R_{adj}^2 goes to zero. For smaller sample sizes, however, differences between the two statistics can be fairly substantial. Even if you are to report R^2 in your manuscript, a glance over at the behavior of R_{adj}^2 is always a good idea.

7.19   WHAT "EXPLAINED VARIANCE" MEANS AND MORE IMPORTANTLY, WHAT IT DOES NOT MEAN

We have said that R^2 and R_{adj}^2 both measure the extent to which one or more predictor variables "explain" variance in a response variable. But what does this mean, exactly? To understand what it means, let us first consider what it does **not** necessarily mean, arbitrated by the context in which it is used. **It does not necessarily mean that if you change values of the predictor(s), this will lead to a change in values of the response.** R^2 presumes to know nothing whatsoever about your **research design**. It can be computed on nonexperimental data just as it can on experimental data, but beyond that, the exact interpretation of "variance explained" is completely up to the researcher to disentangle given the particular research context, the degree of experimental controls, and other **research-related** (not statistically-related) matters. For instance, consider the following statement:

Melatonin ingestion explains 30% of the variance in sleep duration.

The above statement does not alone tell us the true **strength of the evidence** we might have for the effectiveness of melatonin as a sleep aid. Were subjects randomly assigned to dose conditions? If so, then we can interpret the 30% figure in the context of a true **experimental design**. If subjects were not randomly assigned and we simply correlated the amount of drug subjects took with their sleeping behavior, our strength of evidence, even if still at 30% variance explained, is not nearly as **strong**. Simply put, there could be an inordinate number of **covariates** linked to melatonin ingestion on such a nonexperimental study, that it is impossible, on methodological grounds, to associate the number 30% with any kind of strong, directional claim. If an experiment had been performed, then presumably an effort was made to exclude or control the infinite number of nuisance factors that may have played a role, and hence the 30% figure would mean something more on a **scientific** level.

The point of this discussion is to emphasize that both experimental evidence and nonexperimental evidence can generate impressive R^2. Even made up fictitious data can generate moderate to large R^2. However, **not all R^2 should be considered equal in scientific credibility or worth**. Experimental evidence, due to the process of random assignment and the attempt to balance out nuisance factors, is typically always more credible.

Furthermore, just as is true for evaluating the size of such measures as Cohen's d, R^2 values should never be evaluated in a "vacuum" without some reference knowledge of the given research area. Is 30% variance explained for melatonin on sleep latency a strong effect? Perhaps in an **absolute** sense explaining 30% may be impressive in its own right. However, if a competing medication explains 70% of variance in sleep (with no side effects and at reasonable cost), then suddenly the 30% figure is not quite as impressive. When it comes to interpreting effect sizes, one should only rarely rely on the **absolute** size of the effect. Ideally, one should rather have enough **prior knowledge of the research area** to know whether the effect before them is impressive or not. Seasoned researchers do not need any arbitrary guidelines or "rules of thumb" to tell them whether an effect is impressive or not. Rather, they are well aware of current and past findings in the area and are in that sense prepared to properly interpret results from their study. This highlights the importance of always conducting a thorough prior review of the research area before attempting to interpret new findings.

At the same time, explaining 20% of the variance yet speaking as though "X explains Y" without highlighting that most of the phenomenon remains **unexplained**, is unethical or misunderstood science, and is seen everyday in both conference presentations and in publications. In this case, X explains "some" of the variance in Y, yet to speak of it as if it simply "explains Y" (especially when communicating to the media) without **de-emphasizing the phrase**, is severely misguided. Of course, it is a much "sexier" account of reality than concluding "Most of the variance in Y remains unexplained, and what we found is quite minimal. We essentially know quite little thus far about the phenomenon." A discussion of **ethics in science** is beyond the scope of this book. Simply be aware that **many researchers absolutely love to exaggerate their findings way (way) beyond what the data actually say** (it is called "career-building" and has little to do with good science), and no branch of science (whether physics, psychology, medicine) is immune to this. Science is, fundamentally, a **social discipline**, which means it is not immune to these factors. Complete "objectivity" in science is a fruitful goal, but also, and sadly, too often a myth.

7.20 VALUES FIT BY REGRESSION

Recall that a purpose in performing a regression is to obtain a vector (i.e., a "column") of fitted values that are conditioned on the observed value of the predictor x_i entered into the model. For the case of simple linear regression, we can express the **fitted** or **predicted values** as a function of our model equation:

$$y_i' = a + bx_i$$

For larger models cast in matrices, predicted values for **y** are computed as

$$\mathbf{y}' = \mathbf{Xb}$$

It stands that if a given observation is perfectly predicted, that is, $y_i' = y_i$, then the observed residual is equal to

$$e_i = y_i - y_i' = 0$$

or

$$\varepsilon = \mathbf{y} - \mathbf{y}' = \mathbf{0}$$

in matrices.

It should be emphasized that in the formulation of the regression model (7.1), $y_i = \alpha + \beta x_i + \varepsilon_i$, ε_i is considered to be the actual **true error** in terms of what would exist had we actual model parameters and the population (instead of merely sample estimators of those parameters as we have in the case of sample data), whereas e_i is an **observed error** (or "residual") in an empirical application of the regression model. If the assumptions that are made about ε_i are correct, then the observations for e_i should be at least somewhat representative of their behavior; however, a true error is still not the same as a residual and they should not be equated. This distinction is further elaborated on in Chapter 15 when we survey the structural equation model.

7.21   LEAST-SQUARES REGRESSION IN R: USING MATRIX OPERATIONS

One of the joys of learning and performing matrix algebra in R is that one can reproduce virtually any relatively elementary statistical analysis by the simple, and perhaps at times complex, manipulation and construction of basic matrices.

As an example, we once again consider the problem of least-squares, but this time instead of invoking a canned "routine" in software, we demonstrate how the solution can be obtained by simple matrix construction and operations. Such a demonstration provides some insight into what the actual regression procedure is "doing" when working behind the scenes in software, and also helps the user better understand possible computer error messages if and when they arise.

Readers unfamiliar with general matrix operations, and who choose to study this section, are encouraged to consult the Appendix where these concepts are reviewed. For the following demonstration, we assume some familiarity with such operations as matrix multiplication, and move rather quickly to computing least-squares estimates using a hypothetical and easy example. The purpose of doing it the "long way" is simply to demonstrate the matrix computations and the principles involved. Rarely, if ever, baring a software power outage, will you need to calculate regression the long way with the matrices we are about to compute, given that virtually all the regressions you will perform will be via software.

To generate some hypothetical data, suppose we took a measurement of one's quantitative and verbal abilities on some standardized test and wished to predict one's quantitative score based on knowledge of that person's verbal score. Hence, quantitative (or "Q") is the response variable. Verbal (or "V") is the predictor. For each scale, a score of 0 represents minimal ability (it cannot represent **no** ability, it is not measurable on a ratio scale) while a score of 10 represents maximal ability. The data are given in Table 7.3.

TABLE 7.3 **Quantitative and Verbal Scores on Nine Subjects (Hypothetical Data)**

Subject	Quantitative	Verbal
1	5	2
2	2	1
3	6	3
4	9	7
5	8	9
6	7	8
7	9	8
8	10	10
9	10	9

Our first step is to create the respective vectors for Q and V:

```
> Q <- c(5, 2, 6, 9, 8, 7, 9, 10, 10)
> V <- c(2, 1, 3, 7, 9, 8, 8, 10, 9)
```

We next obtain a plot of the data:

```
> plot(V, Q)
```

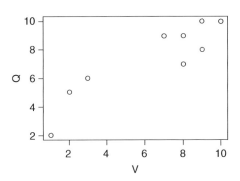

Though the sample size is very small and the number of data points are sparse, we can see, in general, that a somewhat **linear relationship** appears to exist between Q and V (try imposing a line on the bivariate data to confirm the general linear trend).

Recall that we need to compute $\mathbf{X'X}\ \mathbf{b} = \mathbf{X'y}$. To get the matrix $\mathbf{X}$ needed to compute $\mathbf{X'X}\ \mathbf{b} = \mathbf{X'y}$, we generate the first vector by the name of $\mathbf{I}$ (for "intercept") then bind both $\mathbf{I}$ and $\mathbf{V}$ into $\mathbf{X}$. Since everyone in our data gets the intercept, the vector $\mathbf{I}$ will simply be a list of 1's:

```
> I <- c(1, 1, 1, 1, 1, 1, 1, 1, 1)
> V <- c(2, 1, 3, 7, 9, 8, 8, 10, 9)
> Q <- c(5, 2, 6, 9, 8, 7, 9, 10, 10)
> X <- cbind (I, V)
> X
```

```
        I   V
 [1,]   1   2
 [2,]   1   1
 [3,]   1   3
 [4,]   1   7
 [5,]   1   9
 [6,]   1   8
 [7,]   1   8
 [8,]   1  10
 [9,]   1   9
```

We could have also used I <- rep(1,9) to generate **I** (try it). We can now solve for $\mathbf{X}'\mathbf{X}$ quite easily by premultiplying **X** by the transpose of **X** (t(X)):

```
> XTX <- t(X)%*%X
> XTX
    I   V
I   9  57
V  57 453
```

Likewise, we solve for $\mathbf{X}'\mathbf{y}$, only this time using **Q**:

```
> XTY <- t(X)%*%Q
> XTY
   [,1]
I    66
V   483
```

Making the appropriate substitutions, our equation thus far reads:

$$\mathbf{X}'\mathbf{X}\,\mathbf{b} = \mathbf{X}'\mathbf{y}$$

$$\begin{bmatrix} 9 & 57 \\ 57 & 453 \end{bmatrix}\mathbf{b} = \begin{bmatrix} 66 \\ 483 \end{bmatrix}$$

Our next step is to solve for **b**:

$$\mathbf{b} = (\mathbf{X}'\mathbf{X})^{-1}\mathbf{X}'\mathbf{y}$$

$$\mathbf{b} = \begin{bmatrix} 9 & 57 \\ 57 & 453 \end{bmatrix}^{-1}\begin{bmatrix} 66 \\ 483 \end{bmatrix}$$

The inverse of $\mathbf{X}'\mathbf{X}$ is computed by solve in R:

```
> XTX.I <- solve(XTX)
> XTX.I

            I           V
I   0.54710145 -0.06884058
V  -0.06884058  0.01086957
```

Now that we have the inverse, $(\mathbf{X}'\mathbf{X})^{-1}$, the solution for **b** is thus:

```
> XTX.I %*% XTY
        [,1]
I 2.8586957
V 0.7065217
```

Hence, the estimate of the intercept term is 2.859 while the estimate of the slope parameter is 0.707. For this data, we conclude that **for a one-unit increase in verbal ability, one can expect, on average, an approximately 0.71 unit increase in quantitative ability**.

7.22   LINEAR REGRESSION USING R

We now perform the equivalent regression using R's lm function that will also allow us to obtain a measure of model fit, which we did not bother with in our manual computations:

```
> reg.fit <- lm(Q~V)
> reg.fit

Call:
lm(formula = Q ~ V)

Coefficients:
(Intercept)              V
     2.8587         0.7065
```

We request a more complete picture of the regression output via the summary function:

```
> summary(reg.fit)

Call:
lm(formula = Q ~ V)

Residuals:
    Min      1Q  Median      3Q     Max
-1.5652 -1.2174  0.4891  0.7826  1.1957

Coefficients:
            Estimate Std. Error t value Pr(>|t|)
(Intercept)   2.8587     0.8874   3.221 0.014626 *
V             0.7065     0.1251   5.648 0.000776 ***
---
Signif. codes:  0 '***' 0.001 '**' 0.01 '*' 0.05 '.' 0.1 ' ' 1

Residual standard error: 1.2 on 7 degrees of freedom
Multiple R-squared: 0.8201,    Adjusted R-squared: 0.7944
F-statistic:  31.9 on 1 and 7 DF,  p-value: 0.0007758
```

Notice that R's output for the linear model requested by `summary(reg.fit)`, though containing much of the same information as SPSS (as we will soon see), is organized much more succinctly than in SPSS. R also includes quartile information about residuals before presenting the main results of the regression. Some general features of the output include the following:

- **Multiple R-squared** for the model is equal to 0.82, with an **adjusted R-squared** of 0.79. Though as discussed, magnitude of effect sizes should never be appraised or otherwise evaluated in a vacuum (i.e., they are usually only considered "large" or "small" in comparison to other effect sizes in the area of investigation), it might be safe to assume in this case that an R^2 of 0.82 likely indicates a rather large effect. Indeed, if we could predict 82% of the variance in quantitative scores based on knowledge of verbal scores, we are definitely "on to something" in a scientific sense. Recall that "relative to the research area" simply means to position the obtained effect in contrast to other results in the particular area. **An R^2 of say, 0.20, is still small in the big scheme of things**. It may be **relatively large** in a given research area, but it still does not permit us to say a "lot" of variance is being explained. **Humility when interpreting effect sizes is a must for serious scientists**.

- The regression model, evaluated on 1 and 7 degrees of freedom, yields an F-statistic of 31.9 with an associated p-value of 0.00078, which is statistically significant at even a very conservative significance level. There is a single degree of freedom for numerator ($k = 1$) and 7 degrees of freedom for denominator because there is a total of nine cases (i.e., $n - k - 1 = 9 - 1 - 1 = 7$).

- The regression coefficient for V, equal to 0.7065, yields a p-value of 0.00078, which is statistically significant when evaluated against conventional significance levels (e.g., 0.01, 0.05).

- Since there is only a single predictor, both the inferential question regarding both the model and that of the predictor yield identical p-values (i.e., of 0.000776). In the presence of multiple predictors (i.e., as one would have in multiple regression), the p-values for predictors will usually not be identical to the model p-value. In multiple regression models, statistical significance of individual predictors is not equivalent to the statistical significance of the model taken as a whole for reasons that will be better understood when we survey multiple regression.

Recall that in every regression, there is a partition of variance of **total variability into SS reg + SS residual**. This is known as the **analysis of variance for regression** featured earlier in this chapter. To view the ANOVA table from the regression, we request:

```
> anova(reg.fit)

Analysis of Variance Table

Response: Q
          Df Sum Sq Mean Sq F value    Pr(>F)
V          1 45.924  45.924  31.904 0.0007758 ***
Residuals  7 10.076   1.439
---
Signif. codes:  0 '***' 0.001 '**' 0.01 '*' 0.05 '.' 0.1 ' ' 1
```

We can surmise directly from the output how multiple R^2 was computed, that of the ratio 45.924 to $(45.924 + 10.076) = 45.924/56 = 0.82$. We can also request confidence intervals for model parameters with `confint`:

```
> confint(reg.fit)

                2.5 %    97.5 %
(Intercept) 0.7602728 4.957118
V           0.4107442 1.002299
```

The above is a 95% confidence interval for *V*. It is interpreted to mean that **with 95% confidence, the true regression slope likely lay between the lower limit of 0.41 and the upper limit of 1.00**. Note that this interval is symmetrical about the actual sample value of 0.7065 that was estimated. Recall as well that when interpreting confidence intervals, it is the **sample** on which the interval is computed that is the random component. The parameter we are seeking to estimate is assumed to be a **fixed** value.

7.23 REGRESSION DIAGNOSTICS: A CHECK ON MODEL ASSUMPTIONS

Recall that whenever one fits a model to data, and more importantly, **interprets** that model, one is doing so under the assumption that the model assumptions originally postulated are more or less satisfied. Even though your regression analysis may boast statistical significance, this result is only as good as the "goodness" of the model assumptions that underlie the analysis. If your assumptions are not at least **tentatively** reasonable, then the ensuing inferential statistics and *p*-values may not be accurate. For instance, you may underestimate or overestimate the probability of a type I error.

There is a whole field of expertise in the area of so-called "**regression diagnostics**," which comprise a host of statistical indicators and tools used for the specific purpose of revealing, both through numerical summaries and graphics, potential problems with a fitted regression model. These include ways in which the model might be improved by either altering the functional polynomial form or implementing data transformations of empirical variables (e.g., taking square roots or logarithms, for instance). Following Fox (2016), who has written thoroughly on this topic with tremendous precision and accuracy, diagnostics can be generally divided into three very broad categories:

- **Unusual and influential data:** This includes the attempt to detect data that does not conform to our model.
- **Nonlinearity, nonconstant error variance, and nonnormality:** These are methods used to detect deviations from the model assumptions on which our regression model is based.
- **Collinearity**, where one or more predictors in the model are highly correlated: These are meaningful only in regression models with multiple predictors (i.e., multiple regression models) and can be evaluated using such measures as the **variance inflation factor** or **tolerance**, to be discussed in the following chapter.

We must warn the reader that our treatment of regression diagnostics is very brief and incomplete compared to more thorough treatments on the subject. We literally only scratch the surface so that the reader may gain at least some rudimentary understanding of how model assumptions may be evaluated using residuals and other techniques. This is not to suggest that diagnostics are not important, especially those relating specifically to verifying model assumptions. To the contrary, **evaluating the tenability of model assumptions is important**. For instance, if the assumption of linearity is more or less violated, then fitting an alternative polynomial (other than a line) to one's data may be called for. Likewise, if two variables are highly collinear, this may also pose consequences for one's regression model. And though the detection of unusual and influential observations is also important, we regard it as less so largely because of the view generally advanced in this book concerning outliers. Recall that though comparing models with and without designated outliers or other otherwise influential points is good practice, we do not generally recommend the **deletion** of such data points **unless there is a very good scientific rationale for doing so**. Detecting an observation that is influential to the model is one thing. Excluding that observation from the **science** you have undertaken is quite another.

To reiterate, **never casually and automatically delete an observation simply because it is distant from the others**. Such constitutes dishonest data analysis, unethical science, and if you are to be that careless about your empirical observations, it is questionable whether you should be analyzing data at all. If the data point was sampled correctly, and you have no **substantive** reason to delete it, it should remain in your data and be a part of your conclusions. If you delete it for the sake of making your model "fit better," you are no longer practicing science, but rather are demonstrating that you know how to make a sample of data points conform neatly to a model, which is usually of little use or interest to anyone on a **scientific** level. Anyone can make a model fit well through such means. A scientist's model fits well because of the knowledge and insight that went into hypothesizing, data collection, and eventual analysis of that data.

If in doubt, a practical solution would be to run the analysis **with the outlier**, then run it **without**, and present both results to your audience (whether your audience be readers of your publication, and/or conference attendees). Of course, if you do find a serious substantive anomaly with the given case (e.g., the participant was sleeping during the memory task, or someone in your sample had a learning disability which may have influenced your measurement), then by all means, consider deleting the case, but record your deletion in your manuscript write-up, and inform the reader exactly **why** the case was deleted.

Another possibility is to run a model in which outliers have a lesser influence such as **robust regression** or plotting a "resistant" regression line (see Venables and Ripley, 2002, pp. 156–163). Never simply **ignore** data, however, even data that you "dislike." **Science is about uncovering empirical truths, not getting caught up in the aesthetics of model-fitting**. Too often researchers "massage" their data so much through both deletion of points and replacement of missing values that the true empirical nature of their investigation becomes suspect. In this age of advanced computing techniques and the facility of fast data analysis, the temptation to delete "offending" data may exist. However, the data analysis should never be placed above the science. If you collected the data properly, and your model does not fit well, it is most likely the **model that is in error**, **not the data**.

7.23.1 Understanding How Outliers Influence a Regression Model

To help appreciate the effect of extreme observations on a regression model, one need only draw a simple analogy to an empirical distribution of a sample on which a mean is computed. To demonstrate, consider the following distribution of numbers:

$$2 \quad 5 \quad 7 \quad 8 \quad 9$$

The mean of these numbers is 6.2. If we added another data point, 30, our new distribution becomes

$$2 \quad 5 \quad 7 \quad 8 \quad 9 \quad 30$$

The mean of this new distribution is 10.2. By simply adding one rather **extreme data point** relative to the others, our arithmetic mean has shifted in magnitude from 6.2 to 10.2.

Extreme data points have a similar effect on a regression model. Recall that least-squares regression guarantees to fit the line such that the sum of squared errors is kept to a minimum value. In both cases, that of the simple arithmetic mean and regression line, extreme data points demand a **shift toward the center of gravity of the distribution**. This was easily demonstrated for the mean with a shift from 6.2 to 10.2. In regression, we can demonstrate the effect as follows. Consider the regression line of y_i on x_i in Figure 7.7.

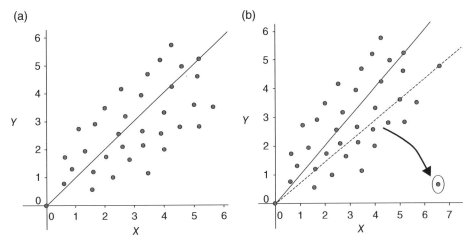

FIGURE 7.7 Least-squares regression line of Y on X (a). Dotted line is the new least-squares regression line (b) after adding outlier. Arrow shows the "pulling down" of the regression line to accommodate the gravitational center of the "floating mean."

We see the effect that the bivariate outlier has on the least-squares regression line. Since the regression line can be conceived as a "floating mean," it is being pulled down in effort to "accommodate" all data points (the effect is somewhat exaggerated in the right-hand figure to motivate our discussion). And though the regression line (Figure 7.7b) is still keeping the sum of squared errors to a minimum, **this minimum value is larger because of the outlier**. The regression line, just as is true of the arithmetic mean, must find a way to "appease" all data points, even those that are quite extreme. In this way, means and regression lines can be said to be quite "democratic," in that analogous to political affiliation, even those holding extreme positions, whether on the far right or far left, are incorporated into the gravitational center of the political climate, and "good" politicians must find a way to appease them. The reader should be forewarned that outlying observations can at times **greatly influence the regression solution**. For a good example of this effect, see Dalgaard (2008, p. 221). If in doubt, fit the model **with**, then **without** the given data point(s) if for no other reason than to get a clearer picture of its influence. Then, on a **scientific basis** (as opposed to statistical), you can make the decision as to whether to include or exclude the observation from your data.

7.23.2 Examining Outliers and Residuals

In an attempt to evaluate outliers and other assumptions in a regression model, we can investigate **residuals**. Recall that the residual e_i is the difference between an observed score y_i and a predicted score y_i', that is, $y_i - y_i'$.

A **standardized residual** is simply a raw score residual e_i divided by an estimate of its standard deviation:

$$e_{\text{st}} = \frac{e_i}{s_{e_i}}$$

where e_i is the given raw-score residual and s_{e_i} is an estimate of its standard deviation. However, as noted by Fox (2016, p. 272), $s_{e_i}^2$ (i.e., the square of s_{e_i}) is not a satisfactory estimate of the variance of ε_i (the error term). A better estimate of the variance of errors ε_i is $s_{e_i}^2 (1 - h_{ij})$, where h_{ij} represents a value along the principal diagonal of the "hat matrix" briefly discussed earlier. Incorporating this estimate of the variance of errors into our formula for the standardized residual gives us

$$e_{st} = \frac{e_i}{s_{e_i}\sqrt{(1 - h_{ij})}}$$

It is possible to define other residuals as well. The **studentized residual** is the same as e_{st} only that s_{e_i} is based on all observations except that for the given residual. The logic is to estimate deviations of the kind $y_i - y'_i$, but where the given observation (on which we are computing the residual) is not involved in the computation of fitted values. See Fox (2016, p. 272) for further details.

7.23.2.1 Errors Versus Residuals

7.23.2.1 Errors Versus Residuals   Recall the distinction between an **error** and a **residual**. The error, ε_i can be said to be the difference between the observed value from the true value (i.e., $E(y_i)$), whereas the residual can be conceptualized as the difference between the observed value and that predicted by the model. Since we can never actually obtain "true" values, **we cannot compute errors directly**. We are relegated to working with residuals which essentially represent that source of variance which is left over from the fitted model, that unaccounted for by the model equation. We will revisit this distinction between an error versus a residual in Chapter 15 when we discuss structural equation modeling, as it is in such latent variable modeling where the distinction becomes even more important and relevant to appreciate.

7.23.2.2 Residual Plots

7.23.2.2 Residual Plots   **Residual plots** are useful because they allow us to evaluate whether at least some of the model assumptions we stated before the fitting of our model are actually **valid** in some sense. Keep in mind that in any statistical model, no assumption will typically ever be **perfectly** satisfied. The goal, however, through residual analyses is to learn whether or not **most assumptions are at least tenable**. In some cases, it may be exceedingly difficult to even know whether some assumptions are satisfied or not, even through residual analyses. However, residual plots are definitely helpful in attempting a verification of them. Though many plots and variations thereof are possible (see Fox, 2016 for more plots), we focus our attention on a single plot that can usually suffice as a **quick diagnostic** in terms of evaluating the behavior of residuals.

Consider the plot of residuals against predicted values in Figure 7.8. The absence of a relationship (whether linear or otherwise) between residuals and predicted values is ideal. A violation of this assumption could, at least in theory, take the form of the plot on the right side of Figure 7.8 in which a positive linear relationship is evident (Johnson and Wichern, 2007, p. 382). However, as noted by Fox

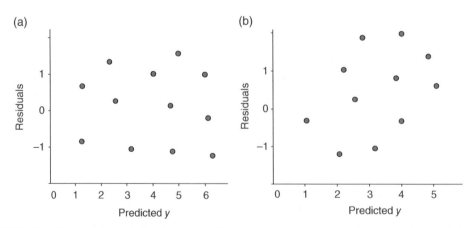

FIGURE 7.8   Absence of correlation between residuals and predicted values (a). Linear relationship between residuals and predicted values (b).

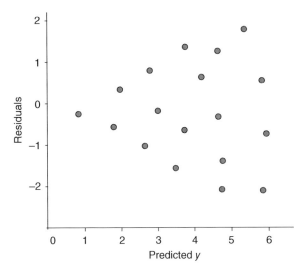

FIGURE 7.9   Evidence of heteroscedasticity of errors.

(2016), since the least-squares fit guarantees that the correlation between the fitted values and residuals is equal to 0, a linear form that dominates the plot (and is not simply present in a subset of the plot) would usually suggest numerical calculations are not correct and/or the intercept has been omitted from the model (Johnson and Wichern, 2007). Indeed, any kind of systematic pattern in the plot (e.g., a curvilinear trend) may also suggest the need for more terms in the model. Overall, and in general, the important point to remember is that **any lack of independence of the residuals from predicted values (whether linear, curvilinear, etc., in any sub-area of the plot) may suggest that revisions or double-checks on the model are required**.

Residual plots can also be used to identify potential problems with the **variance of errors**. Distributions of residuals (or their standardized counterparts) should be relatively evenly distributed across values of the predicted value or predictor(s) (in the case of multiple regression) as is the case in Figure 7.8a. A violation of this assumption might look something like that shown in Figure 7.9, where it is clear that the distribution of residuals is not constant across predicted values.

Though not demonstrated here, problems of **heteroscedasticity** (i.e., unequal variances) can be potentially remedied through power transformations or through a weighted least-squares solution instead of traditional OLS. For details, see Fox (2016) or Draper and Smith (1998).

7.23.2.3   *Time Series Models*   Serial correlation, also generally known as **autocorrelation**, can sometimes exist in residuals, especially in time series models which feature the behavior of a response variable over time. The **Durbin–Watson test** is useful in detecting such patterns (see Neter et al. 1996, p. 504 for details). As an example of a time series, consider data from Arbuthnot's analysis of the ratio of males to female births in London from 1629 to 1710. Recall that Arbuthnot used this analysis to argue that more males were being christened than females over the course of this period. As an example of some of his data, consider the time series for males:

```
> library(car)
> library(HistData)
> attach(Arbuthnot)
> scatterplot(Year, Males)
> library(ggplot2)
> qplot(Year, Males, data = Arbuthnot, geom = "line")
```

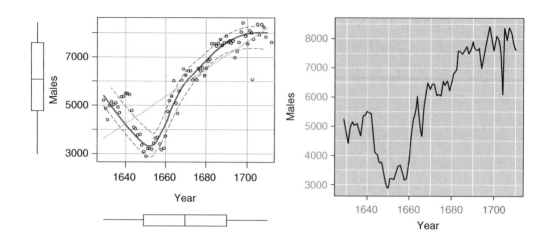

We can see that after a gradual dip from 1629 to about 1660, christenings (i.e., "baptisms") for males increased quite rapidly. A goal of time series analysis is to study the frequency of an event over a specified period of time as to make inferences about trend. Of interest as well is to estimate the extent to which events in previous time periods are related to events in later time periods. The so-called **time lag** allows one to forecast future data based on data from previous periods. Often in time series, residuals (as proxies for errors) are **autocorrelated**, which may imply, for instance, that residuals from recent periods are more related to current residuals than residuals from more distant past periods. Such indicators are used in models as predicting stock market share prices over time. Though because all models are based on data we actually **have** and not data we are **going to have**, even for seasoned time series specialists, making money from the stock market can still be quite challenging! A time series cannot tell you if a competitor will release a new (and better) product in the near future thereby causing your stock price to potentially drop. Only **substantive knowledge** of the research domain (in this case, economics and business, and perhaps insider trading) can do that. **Statistical modeling on its own is not a panacea in any field**.

Times series models are a huge topic and other than our brief commentary are beyond the scope of this book. Many books have been written on the topic. Our purpose in mentioning them at all was simply to highlight circumstances that may feature highly correlated residuals. Such models have often proved useful in modeling **econometric**, **geographical**, **financial**, and **astronomical** data, among other uses. For details on both fitting and decomposing time series in R, consult Crawley (2013, pp. 785–808) or Venables and Ripley (2002, pp. 387–418). Teetor (2011) also provides a useful chapter on how to extract a variety of indicators for time series, such as computing percentage change over time, moving averages, and testing for autocorrelation.

7.23.3 Detecting Outliers

Outliers that fall relatively near the mean or center of the predictor(s) are usually not considered to be that influential on the final regression solution. Outliers that are relatively far from the mean of predictor(s) and somewhat distant from the regression line or plane (i.e., high or strong **leverage** data points) have greater influence on the final regression solution. As noted by Fox (2016), "The combination of high leverage with a regression outlier therefore produces substantial **influence** on the regression coefficients" (p. 267).

Hence, **influence** can be defined as the **product of leverage and discrepancy**, where discrepancy is a measure of how "outlying" an observation is relative to the fitted regression model. Influential observations are those that are both outliers away from the mean of the predictor(s) and have a rather strong impact on the regression solution because they are also relatively distant from the regression line (or plane, in the case of multiple regression). For example, in Figure 7.7 discussed earlier, the outlier in question would likely exhibit a relatively strong influence on the regression solution.

To compute a measure of leverage, recall once more that predicted values in regression can be defined as a function of hat values in

$$\mathbf{y}' = \left[\mathbf{X}(\mathbf{X}'\mathbf{X})^{-1}\mathbf{X}'\right]\mathbf{y}$$
$$= \mathbf{Hy}$$

where $\mathbf{X}(\mathbf{X}'\mathbf{X})^{-1}\mathbf{X}'$ is the so-called "hat matrix." **Hat values provide a measure of leverage** (Izenman, 2008). Values that exceed twice to three times their average, where their average is computed as $\bar{h} = (k + 1)/n$ (where k is the number of predictors and n is sample size) are usually considered to be noteworthy (Fox, 2016, p. 270). For our Q-V linear regression, we can compute hat values in R by:

```
> hatvalues(reg.fit)
        1         2         3         4         5         6         7         8
0.3152174 0.4202899 0.2318841 0.1159420 0.1884058 0.1413043 0.1413043 0.2572464
        9
0.1884058
```

The mean hat value for our data is 0.22 (i.e., $\bar{h} = (k + 1)/n = (1 + 1)/9$). Hence, values exceeding $2\bar{h}$, or $2(0.22) = 0.44$, might be of concern. For our small data set, we have no such extreme values, though the hat value for the second observation (i.e., 0.42) is fairly close to meeting this criterion.

As a measure of influence, **dfbeta** ("difference in beta values") and their standardized counterparts, "**dfbetas**" (note that the standardized versions are simply designated as plural) can be computed. A dfbeta is defined as **the difference between a regression coefficient computed with the given observation included and then without**. That is, the dfbeta is defined as

$$d_{ij} = b_j - b_{j(-i)}$$

where d_{ij} is the dfbeta for the given observation, b_j is the regression coefficient(s) computed with the observation i included in the model, and $b_{j(-i)}$ is the regression coefficient(s) computed with the observation i not included in the model (Fox, 2016, p. 276). Dfbeta values capture influence by deleting the given data point from the estimated regression coefficient and recomputing it to assess how different it would be with that observation removed. In this way, it is measuring the degree of influence for the given observation on the regression solution. For more details, see Fox (2016, Chapter 11).

Cook's D statistic provides a useful index combining information about discrepancy and leverage:

$$D_i = \left(\frac{e_i'^2}{k + 1}\right) \cdot \left(\frac{h_i}{1 - h_i}\right)$$

where $e_i'^2$ is the **squared standardized residual** for the ith observation and h_i is the given hat value for the ith observation. Cook values are a general measure of **multivariate** distance. Relatively large values suggest the given data point may exert a rather strong influence on the estimated regression coefficients. To get Cook d values in R, we compute:

```
> cooks.distance(reg.fit)
            1                2                3                4                5                6
0.1238382524 1.0642718447 0.1425192487 0.0736656900 0.1472491909 0.1519522308
            7                8                9
0.0159258458 0.0009376835 0.0608529820
```

A relatively large value occurs for the second case (value of 1.06), and hence may be one worth looking into as a potentially **statistically** (yet recall, perhaps not **scientifically**) problematic observation. We could also obtain Cook values along with a host of other indicators (e.g., dfbeta) using `influence.measures(reg.fit)`. One can also compute what are known as **partial regression plots** to visualize the joint influence of observations on a given model. For details, see Fox (1997).

7.23.4   Normality of Residuals

Recall that another of the assumptions of the regression model is that errors follow a normal distribution with constant variance. That is, $\varepsilon_i \sim \text{NID}(0, \sigma^2)$. A failure to meet this assumption can weaken the efficiency with which parameters are estimated.

We can generate Q–Q plots which allow us to informally evaluate the normality of residuals assumption. The Q–Q plot should reveal an approximate linear trend between theoretical quantiles and residuals:

```
> library(car)
> qqPlot(reg.fit)
```

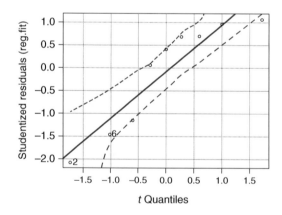

This particular plot features that of **studentized residuals** against t quantiles. The dashed lines are a point-wise confidence envelope (Fox, 2016, p. 298) that R includes automatically when generating the plot. For such a small data set, the assumption of normality can be quite challenging (and more or less hopeless) to assess.

7.24   REGRESSION IN SPSS: PREDICTING QUANTITATIVE FROM VERBAL

We now demonstrate linear regression on the Q–V data in SPSS. Our results will, not surprisingly, parallel those obtained earlier in R. We nonetheless provide a rather thorough explanation of findings to ensure that the reader has the opportunity to master the interpretation of essential statistics learned in this chapter, as well as how they are reported in SPSS, since "regression in SPSS" will likely be a common strategy for him or her in a multitude of research contexts.

First, we generate a plot of Q on V:

```
GRAPH
  /SCATTERPLOT(BIVAR)=V WITH Q
  /MISSING=LISTWISE.
```

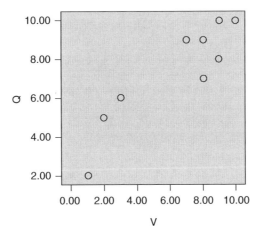

As can be seen from the plot (and as we noted in our previous analysis using R), the relationship between Q and V appears to be more or less linear. We proceed with the regression analysis, requesting only the most essential output:

```
REGRESSION
  /MISSING LISTWISE
  /STATISTICS COEFF OUTS R ANOVA
  /NOORIGIN
  /DEPENDENT Q
  /METHOD=ENTER V.
```

	Variables Entered/Removed[a]		
Model	Variables Entered	Variables Removed	Method
1	V[b]		Enter

[a]Dependent variable: Q.
[b]All requested variables entered.

The first piece of output above simply states which variables have been entered into the equation, and which have been left out. Since we are performing a **simple** linear regression, there will always only be a single variable entered, and **no variables removed**. In a multiple regression, the topic of the following chapter, this report will prove useful as a quick verification that the predictors we intended to include in the model are indeed included, while the ones we wished to not include have been left out. Under "**Method**," we see the word "**Enter**." This simply means that we are conducting a **full entry regression** (also known as **simultaneous regression**) in which all predictors are entered into the model at the same time, and hence all parameters (in our case, α and β) are estimated simultaneously. Such output will again prove more useful to us when we analyze multiple regression models, where we are able to choose one of several techniques for performing the multiple regression (e.g., forward regression, stepwise regression, etc.).

Model Summary				
Model	R	R-Square	Adjusted R-Square	Std. Error of the Estimate
1	0.906	0.820	0.794	1.19977

The **model summary** contains the essential information regarding how well our regression model fit the observed data. We see that R is equal to 0.906. Since this is a model with a single predictor, R is equal to the Pearson product-moment correlation coefficient for Q and V.

R-square is equal to $(0.906)^2$, which yields 0.820. Recall that this expresses the proportion of variance in Q that is "explained" or "accounted for" by knowledge of V. For our data, this means that approximately 82.0% of the variance in Q can be accounted for by knowledge of V. Recall the substantive issues with interpreting this number, the statistic itself in no way suggests verbal ability "leads to" or "causes" quantitative ability. It simply means that if you know one's verbal ability, you can reasonably well predict their quantitative ability. When interpreting any statistics, be very cautious about "over-interpreting" what they are able to tell you. Evidence for any type of causation or stronger statements on the evidence must come from your **research design**, not the statistics used to model the data.

Next, we come to the ANOVA summary table:

ANOVA					
Model	Sum of Squares	df	Mean Squares	F	Sig.
Regression	45.924	1	45.924	31.904	0.001
Residual	10.076	7	1.439		
Total	56.000	8			

Recall that the **regression sum of squares** tells us how much variability is accounted for by the regression model based on the fitting of the least-squares line. The **residual sum of squares** tells us how much variability is **unaccounted** for by the regression model. The total variability is the sum of both regression and residual variability, due to the identity that total variability is a sum of these two components (i.e., regression + residual). Recall that the extent to which the regression sum of squares is large relative to the residual sum of squares is the extent to which more variability than not is accounted for by our model. Note that we also get R^2 from the summary table by taking the ratio $45.92/56.00 = 0.82$, which agrees with R^2 reported earlier (and as we computed in R). The value of adjusted-R^2, as expected, is less than R^2.

Recall that the mean squares are computed just as they are in ANOVA-type models, that is, by taking the relevant sum of squares and dividing by the corresponding degrees of freedom. Mean squares regression is computed by dividing SS regression by df for regression, which for our data is 45.92/1 = 45.92. Mean squares residual is computed by dividing SS residual by df for residual, which for our data is 10.08/7 = 1.44.

The F-ratio for the regression model is computed by taking the ratio of MS regression to MS residual, that is, 45.924/1.439 = 31.91. The test of significance reveals that the probability of obtaining an F statistic as the one we have obtained or more extreme from an F distribution on 1 and 7 degrees of freedom is very small with an associated p-value equal to 0.001. Hence, we can reject the null hypothesis that R^2 in the population is equal to zero, and conclude the statistical alternative hypothesis that it is not equal to zero. That is, we conclude $H_1 : R^2 \neq 0$.

Next in our output are provided the regression model coefficients. Included are significance tests for the intercept (which SPSS calls "Constant"), as well as for predictor variable V.

Coefficients					
	Unstandardized Coefficients		Standardized Coefficients Beta	t	Sig.
	B	Std. Error			
Constant	2.859	0.887		3.221	0,015
V	0.707	0.125	0.906	5.648	0.001

Our estimated regression equation is thus:

$$y_i' = 2.859 + 0.707(x_i)$$

Recall the correct way to interpret the coefficient for b: **For a one-unit increase in V, on average, we expect Q to increase by 0.707 units**.

Recall why we need to include the statement "on average" or at minimum, "expect." Though our regression line is a functional relation (it is a perfect polynomial of degree 1), the data on which we are fitting the perfect functional form is less than perfect. It is rather messy and contains much variation. So in interpreting the coefficient 0.707, it would be technically **incorrect** to say that **a one-unit increase in V is associated with a 0.707 unit increase in Q**. We need "on average" or "expected" (or both) to denote the fact that we are dealing with a **statistical** model rather than a functional or otherwise purely **deterministic** one that assumes no error in prediction. If we are speaking only of the regression line itself, then avoiding the words "expect" and "on average" would be fine. However, since we are applying the line to data, and interpreting the regression line in this context, including these words provides a more accurate interpretation. This is a good example of what differentiates **mathematics** (e.g., precalculus, where lines are drawn) and **statistics** (where lines are fit, and estimation takes precedence). These are not trivial distinctions and should be appreciated and recognized by the reader.

The constant is equal to 2.859, which means that the least-squares regression line touches the ordinate axis at that particular value. It can be interpreted as the predicted value for Q when $V = 0$, as easily demonstrated:

$$y_i' = 2.859 + 0.707(x_i)$$
$$= 2.859 + 0.707(0)$$
$$= 2.859$$

Though in this circumstance the interpretation of the intercept as a predicted value is somewhat mean-ingful since it is presumably theoretically possible to get a value of zero on our scale of verbal ability (perhaps not on **actual** verbal ability itself, as a construct, however), recall that in many situations the **substantive** (as opposed to strictly mathematical) interpretation of the intercept term is ambiguous at best. For example, if the predictor was that of weight and the response was that of height, then it would indeed be nonsensical to conclude that when one weighs zero pounds, one's predicted height is y_i'. In such cases, the intercept should not be interpreted without further adjustment to the model. The prin-ciple here is that **though the intercept is usually estimated for all linear models, it is only as sub-stantively interpretable as it makes good research, scientific, or even common sense to interpret**. Having said this, there are ways of making the intercept more interpretable if it is not already readily interpretable. One common method is to **mean-center** predictors such that one obtains a predicted value for y_i' when x_i is equal to its mean, instead of 0. We survey this possibility in Chapter 9 when we consider the case of interaction regressors.

Next are the estimated standard errors associated with both the intercept and the slope parameters. Recall that the standard errors, in this case, provide us with a measure of how much we should expect the given estimated coefficient to vary under the assumption of the null hypothesis. The standard error is the standard deviation of the corresponding sampling distribution of the statistic, which in this case, the statistic is that of the **sample regression slope**. Consider the standard error for the intercept, equal to 0.887. What this means is that even if the null hypothesis is true (i.e., that $\alpha = 0$), we would expect repeated samplings of the sample intercept to vary on average by 0.887 units. Recall that this variation from expectation usually goes by the name of **sampling error** or **chance**. The question we want to ask is as follows:

How large is our obtained value of the intercept relative to how much we should expect it to vary across theoretical repeated samplings?

If the value of the intercept is large relative to its standard error, it gives us reason to believe the true intercept in the population (from which these data were presumably drawn) is not equal to zero. This forms the logic of the t-test that follows (on the right-hand side of the output), in which the obtained intercept term is compared to its standard error by means of a ratio. For the intercept, the t-ratio is equal to $2.859/0.887 = 3.22$. Evaluated for statistical significance on $n - k - 1$ degrees of freedom (which in our case is equal to 7), we have evidence to reject the null hypothesis that the population intercept is equal to 0, since the obtained p-value is relatively small ($p = 0.015$). In other words, we have evidence to infer $H_1 : \alpha \neq 0$, which is a statement that the intercept is not equal to zero in the population from which these data were drawn.

We interpret the standard error for the slope analogously. The null hypothesis is that the population regression slope is equal to zero, $H_0 : b^* = 0$, where b^* is the population slope parameter (we use b^* in this case to represent the population parameter instead of β, which we reserve for the standardized coef-ficient, Beta, in our current example). The statistical alternative hypothesis is $H_1 : b^* \neq 0$. If we were to repeatedly sample slope statistics from this given population, how much should we expect them to vary from sample to sample? The answer lies in the standard error, equal to 0.125. Since the obtained slope is relatively large compared to its standard error, as confirmed by the t-ratio, we reject the null hypothesis and infer $H_1 : b^* \neq 0$. The probability of obtaining such sample slopes under H_0 is relatively small ($p = 0.001$). In other words, the deviation from 0 that we are witnessing in our particular sample is likely not simply a byproduct of sampling error. It could be, but it **probably** is not. Does the slope of 0.707 rep-resent a **substantively meaningful slope**? We cannot answer this question based on the significance test alone, and any measure of importance ascribed to the size of effect, represented by R^2, recall, must be tempered by the size of effects in the given literature for the phenomenon under investigation, or at minimum, knowledge of the researcher. Remember that **effect sizes** should only rarely be interpreted in a vacuum, and rough guidelines should be used only when no other reference points are available. And if the effect is small, you should not speak of it as if it is large or that it simply "exists" without mentioning how small it is. An R^2 of 0.30 may be **relatively** large in relation to your field, but it is still **absolutely** quite small. Hence, concluding that "X predicts Y" as a blanket statement without contextualizing how weak the effect is would be entirely misleading, regardless of whether you rejected or did not reject the null hypothesis. Recall that in most cases, simply because you rejected the null does not on its own necessarily imply anything **scientifically** (as opposed to **statistically**) meaningful has occurred. What it does mean is that you witnessed a relatively unlikely statistic, which on a scientific level, may be quite trivial. **Size of effect and whether experimental controls, etc. were implemented are much more relevant to the scientist.**

Next in the output is the value for β (Beta), which recall is the **standardized regression coefficient**. This is the slope coefficient that is generated when Q and V are both standardized to have a mean of 0 and standard deviation equal to 1 (i.e., transformed into z-scores). The value for β for our data is 0.906. The interpretation for β is as follows:

> **For a one-standard deviation increase in V, on average, we expect Q to increase by 0.906 of a standard deviation.**

Notice as well that 0.906 for this model is actually equal to coefficient r, the bivariate correlation between V and Q. This is as a result of the model having only a single predictor. In multiple regression, where we have several predictors and are required to interpret **partial regression coefficients**, this relationship will, of course, no longer hold.

7.25 POWER ANALYSIS FOR LINEAR REGRESSION IN R

The function `pwr.f2.test` in the `pwr` package (Champely, 2020) can be used to estimate power for linear regression models. We once again require the transformation of f^2 into R^2, originally featured in our discussion of power estimation in the context of ANOVA models (see Chapter 3):

$$f^2 = \frac{R^2}{1 - R^2}$$

To demonstrate the estimation of power, suppose a researcher hypothesized or expected an R^2 of 0.40 for a simple linear regression model. Suppose also the degrees of freedom for the model were 1 (for numerator) and 7 (for denominator), which implies there to be a total of **nine observations** (i.e., $n - k - 1 = 9 - 1 - 1 = 7$). The value for f^2 is therefore equal to 0.66 (i.e., $0.40/(1 - 0.40)$). We enter these parameters, specifying `power = NULL` so that power is estimated:

```
> library(pwr)
> pwr.f2.test(u = 1, v = 7, f2 = .66, sig.level = .05, power = NULL)

     Multiple regression power calculation
```

```
        u = 1
        v = 7
       f2 = 0.66
sig.level = 0.05
    power = 0.5552861
```

The estimated power for this regression is approximately 0.55. That is, the probability of rejecting the null hypothesis in such a model given that the null is actually false is a bit higher than that of getting a head or tail on a flip of a fair coin. Even with a relatively large effect size (0.40), we would require a greater sample than nine to achieve respectable power levels (e.g., power equal to 0.90 or higher).

Power and sample size for regression can also be quite easily estimated using G*Power. We select: Tests -> Correlation and regression -> Linear multiple regression: Fixed model, R^2 deviation from zero. We enter the same parameters as we did in R: $f^2 = 0.66$, $\alpha = 0.05$, power = 0.55, and number of predictors = 1:

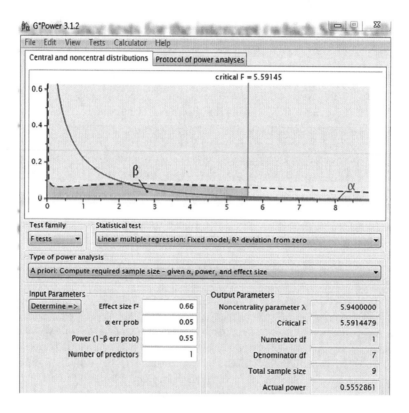

We see that the total sample size of nine matches that estimated in R. How much of a sample size would we require to achieve 0.90 power? We keep all other numbers the same, but adjust our power level to 0.90:

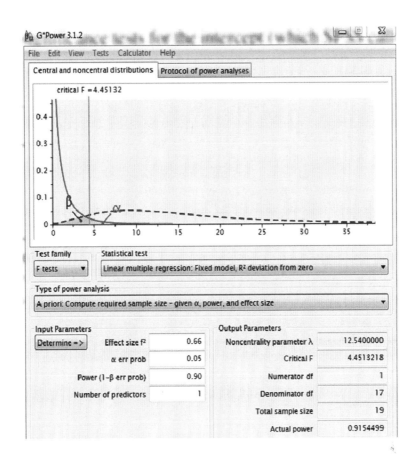

We can see that to reach power of 0.90 or so, requires a total sample size of approximately 19 subjects.

7.26 CHAPTER SUMMARY AND HIGHLIGHTS

- Linear regression is a statistical method that capitalizes on the correlation between two variables to aid in or improve upon the prediction of a response variable based on knowledge of a predictor variable. The goal of regression analysis is not simply to make predictions, as we can make predictions without using regression. Rather, the goal is to on average, **minimize error in prediction**.
- **Simple linear regression** features the use of a single variable in predicting a continuous response. **Multiple regression** models feature multiple predictor variables predicting a continuous response.
- The **history of regression analysis** has its roots with Legendre, Bravais, and Galton, among others. The distinguishing feature of Galton's discovery is that he "saw" regression in real, empirical data.

- Regression models can be used to model both **experimental** and **nonexperimental** data. The statistical tool of regression is no more experimental or nonexperimental than is the arithmetic of ANOVA models.

- By comparing the regression model to the analysis of variance models of previous chapters, one can more easily assimilate and understand the technique of regression. Whereas ANOVA models contain **population effects** (estimated by **sample effects**), regression models contain **slope parameters**. Both models include ε_i, error terms. Since error terms are not observable and hence cannot be known, we usually study **residuals** as estimates of the errors, which are differences between **predicted** and **observed values** on the response variable.

- The reader familiar with **multilevel models** can easily note how the regression model of the current chapter can be extended to allow the **intercept** and **slope** terms to vary, thereby designating them as **random effects** arising from a probability distribution.

- The **assumptions of regression analysis** include **linearity in parameters**, **normality of errors**, **homoscedasticity**, **independence of errors**, **absence of influential or outlying data**, and that the **correct model has been specified**. The classic regression model also assumes that predictors are free from **measurement error**. Most assumptions in regression on real data are rarely if ever definitively "true," but it generally suffices if we can verify that we do not have evidence that they are grossly violated. Through residual analyses and plots, many of the assumptions can be examined, though often imperfectly. "Assumption-checking" in regression or any other statistical model is by no means an "exact science."

- The **estimation of parameters** in linear regression usually takes the form of **ordinary least-squares**, though maximum-likelihood estimators (as well as others) are also available (e.g., weighted least-squares).

- The **null hypothesis** for a regression model usually takes the form $H_0 : b* = 0$ and is tested against an alternative hypothesis of the form $H_1 : b* \neq 0$.

- **Alternative formulations** of the regression model exist, such as $y_i = y - bx + bx_i + e_i$. One advantage of conceptualizing the regression model in different forms is that it allows one to appreciate the role played by various inputs to the model.

- The regression model can be constructed using **matrices**, which is especially useful for multiple (and multivariate) regression models. For simple linear regression, scalar algebra is usually sufficient for describing and working with the model.

- R^2, the "**coefficient of determination**," is the typical measure of model fit for the regression model, having a minimum value of 0 and an upper maximum value of 1. R^2 provides us with the proportion of variance explained in the response variable by knowledge of the predictor variable(s).

- **Adjusted** R^2 is a useful alternative (or supplementary statistic) to R^2, and is adjusted to incorporate the number of parameters being fit in the model and the extent to which they serve to increase R^2.

- Regression analysis can be performed in R through **matrix operations**. Understanding such matrix operations is useful, but pragmatically, regressions are typically run through "canned" routines such as lm in R or REGRESSION in SPSS.

- The area of **regression diagnostics** is a field unto itself. Its purpose is to help identify **outlying** or **influential data**, as well as to aid in verifying assumptions of the regression model. Diagnostics can generally be divided into three categories: unusual and influential data, nonlinearity, nonconstant error variance and nonnormality, and collinearity.

- **Power analyses** and estimating sample size can be conducted either in R or G*Power, as well as with several other on-line power estimation programs.

REVIEW EXERCISES

7.1. Describe the overall goal and **purpose of regression analysis**.

7.2. Why can it be said that the phrase "**The purpose of regression is to make predictions**" is not a complete description of what regression is about? Why is such a description necessarily incomplete?

7.3. Discuss how **analysis of variance** is no more "experimental" of a statistical method than is regression analysis.

7.4. Compare the classic one-way **ANOVA model** with the **simple linear regression** model of this chapter. How are they similar? Different?

7.5. Discuss the nature of the **least-squares** regression line. What is its objective, and how does it go about accomplishing this goal?

7.6. A colleague says to you, "By fitting the least-squares regression line to data, one is guaranteed to achieve a **small sum of squared errors term**." Comment on your colleague's statement, and correct it if appropriate.

7.7. Discuss how **averages can be misleading**, and how this lesson can be extended to cautions regarding regression analysis. Can you provide a simple demonstration to illustrate that you understand the concept?

7.8. Distinguish between the **standard error of a statistic** and the **standard error of the estimate** as employed in regression.

7.9. Distinguish between the **standard error of the estimate** and its square.

7.10. List and briefly discuss the **assumptions of a regression analysis** estimated by ordinary least-squares.

7.11. Discuss what **linearity in the parameters** means by comparing and contrasting the model $y_i = \alpha + \beta_1 x_i + \beta_2 x_i^2 + \varepsilon_i$ with the model $y_i = \alpha + \beta_1 x_i + \beta_2^2 x_i + \varepsilon_i$.

7.12. Discuss what is meant by a **model function**.

7.13. Discuss the difference between a **mathematical** model and a **statistical** one, with reference to the model function and error term.

7.14. Why is the assumption of a correct **model specification** virtually never met in practice, but is nonetheless an assumption that should be carefully considered when formulating regression models?

7.15. Discuss why the assumption of no **measurement error** in variables is usually unrealistic with virtually any data, especially those arising from the social and behavioral sciences.

7.16. What does it mean to **estimate a parameter** in regression analysis?

7.17. Explain the role of **differentiation** as it relates to obtaining the least-squares solutions in linear regression. Why are derivatives necessary?

7.18. Under what condition(s) are the **least-squares estimators** of α and β also **maximum-likelihood estimators**?

7.19. Give a research example where $H_0 : \beta = 0$ may not be regarded as a suitable **null hypothesis** for the particular problem at hand.

7.20. For $s_b > 0$, under what condition(s) would $t = 0$ in $t = \dfrac{(b - \beta_{null})}{s_b}$?

7.21. Referring to the following form of the **regression equation**,

$$y_i = \bar{y} + b(x_i - \bar{x}) + e_i$$

what is the predicted value for y_i under the condition that $x_i = \bar{x}$? Why does this make good sense given what you know about regression analysis?

7.22. Why is expressing the regression model in **matrices** typically not essential for simple linear regression, but necessary for multiple linear regression and more complex models?

7.23. Describe each component of the classic **linear regression model** $\mathbf{y} = \mathbf{X}\boldsymbol{\beta} + \boldsymbol{\varepsilon}$.

7.24. Define what is meant by a **residual** in regression analysis. What is the difference between it and an **error**?

7.25. Discuss why solving for $\mathbf{b}$ in $\mathbf{X}'\mathbf{X}\,\mathbf{b} = \mathbf{X}'\mathbf{y}$ is problematic if $\mathbf{X}'\mathbf{X}$ is not **invertible**.

7.26. Describe and explain each of the equations in Table 7.1. (**Analysis of Variance for Linear Regression.**)

7.27. Distinguish between R^2 and R^2_{adj}. How are they similar? Different? What is the potential benefit to interpreting R^2_{adj} over R^2?

7.28. Describe the overall purpose of **regression diagnostics**, what they are for, and how they should be used. How might the assumptions of **normality of residuals** and **homoscedasticity** be checked?

7.29. Consider data by Snedecor in **Calculation and Interpretation of Analysis of Variance and Covariance** (1934). These data consist of 14 freshman mathematics classes (Table 7.4) with mean ability scores and final grades recorded for each class.

TABLE 7.4   Mean Ability Scores and Final Grades for 14 Mathematics Classes

Class	Number in Class	Mean Ability Score	Mean Final Grade
1	17	5.00	4.12
2	20	2.55	2.55
3	20	1.95	2.45
4	19	2.84	3.11
5	17	2.18	2.47
6	21	2.33	2.24
7	18	2.94	2.94
S	18	4.17	4.22
9	20	3.05	2.90
10	20	3.20	3.25
11	21	3.33	3.14
12	21	2.33	2.05
13	19	0.95	1.47
14	20	3.00	2.50

Source: Based on Snedecor (1934).

(a) Run a regression analysis in R using `lm` denoting **mean final grade** as the response variable and **mean ability score** as the predictor.

(b) Assess the model fit in the analysis run in part (a). Is it a well-fitting model? Why?

(c) Using graphical displays and plots, evaluate whether the assumptions of **normality of residuals** and **independence of errors** are at least tentatively satisfied. Why might it be difficult to confirm that the assumption of independence of errors is not violated for these data?

Further Discussion and Activities

7.30. As briefly discussed in the introduction to this chapter, there are many who contributed to the invention of regression analysis as we know it today. Most associate **Francis Galton** with the history of regression, though as mentioned, a character by the name of **Auguste Bravais** was also influential in discovering the bivariate normal surface. Read Denis (2001) and comment on the historical evidence that exists **for** and **against** claims that Bravais is more relevant to the origins of correlation and regression than is Galton.

8

MULTIPLE LINEAR REGRESSION

> If there is any good reason to fear disturbances of results by other variables than the one with which we are immediately concerned, the proper method to be employed is, it seems to me, that of 'multiple correlation.' This method enables us to deal with facility with three variables, and if need be with more, and to form coefficients of correlation between any two of the variables while eliminating the effects of variations in the third. Such 'net coefficients' will probably play an important part in future statistical researchers.
>
> (Yule, 1896, p. 615)

Whereas the simple linear regression model of Chapter 7 featured a single explanatory variable, the more general **multiple regression** model is able to accommodate several predictors, given by

$$y_i = \alpha + \beta_1 x_1 + \beta_2 x_2 + \ldots + \beta_k x_k + \varepsilon_i \tag{8.1}$$

where, as was the case of simple regression, y_i is an observed value of the response variable and α is the population intercept. Note that instead of only a single population coefficient β, (8.1) now contains terms β_1, β_2, β_k, where β_1 is the **partial regression slope** parameter for predictor x_1, β_2 is the partial regression slope parameter for predictor x_2, and β_k is the partial regression slope parameter for predictor x_k. As before, ε_i is the error associated with predictions of y_i. Parameters $\alpha, \beta_1, \beta_2, \ldots \beta_k$ are also typically estimated by ordinary least-squares (OLS). The expectation for y_i is now $\alpha + \beta_1 x_1 + \beta_2 x_2 + \ldots + \beta_k x_k$, which also implies $E(\varepsilon_i) = 0$. Finally, we also assume:

$$E(y_i / x_1, x_2, \ldots, x_k) = \alpha + \beta_1 x_1 + \beta_2 x_2 + \ldots + \beta_k x_k \tag{8.2}$$

Analogous to simple regression, the systematic portion of the model featured in (8.2) is sometimes called the **model function**.

Applied Univariate, Bivariate, and Multivariate Statistics: Understanding Statistics for Social and Natural Scientists, With Applications in SPSS and R, Second Edition. Daniel J. Denis.
© 2021 John Wiley & Sons, Inc. Published 2021 by John Wiley & Sons, Inc.
Companion Website: www.wiley.com/go/denis/appliedstatistics2e

The origins of multiple regression lay with Yule (1871–1951) and Pearson (1857–1936) in late nineteenth-century Britain. One of the earliest complete multiple regressions was published by Yule (1899) in an article titled "**An investigation into the causes of changes in pauperism in England, chiefly during the last two intercensal decades**" published in **Journal of the Royal Statistical Society**. The paper featured a thorough analysis of the then debated predictors of pauperism in England, which followed Yule's "tour de force" (Stigler, 1986) work of 1897 in which he laid out much of the theory of correlation. The social and political factors that motivated the use of multiple regression are discussed elsewhere (e.g., see Denis and Docherty (2007)). A largely unknown figure in the history of regression, Charles Stewart Loch (then secretary of the **Charity Organization Society** in London), in responding to Charles Booth's study **The Aged Poor in England and Wales**, also figured somewhat prominently in the social (as opposed to technical) uprise of multiple regression.[1]

Before surveying the theory of multiple regression, a review of partial and semipartial correlation is pedagogically useful.

8.1   THEORY OF PARTIAL CORRELATION

Multiple regression has its roots in the theory of partial correlation. Recall that the Pearson coefficient of correlation is a measure of the **linear relationship** between two variables. The coefficient of **partial correlation** is a measure of the linear relationship that still exists when the linear influences of one or more variables are removed. In a sense, partial correlation attempts to provide an **after the fact** estimate of what the bivariate correlation **might have been** had we been able to control for the aforementioned linear influences. **The partial correlation coefficient, however, is not an antidote for the absence of experimental controls**.

Partial correlations may be obtained in more than a single way. A partial correlation between variables x_1 and x_2 controlling for z_i is obtained, the "long way," as follows:

- Regress x_1 on z_i and obtain a column of residuals.
- Regress x_2 on z_i and obtain a column of residuals.
- Correlate the residuals from x_1 on z_i to those from x_2 on z_i.

The logic of the partial correlation is to first account for the predictive power of z_i in both cases, then correlate what is "left over." This remainder is contained in the residuals. Hence, we see that partial correlation is actually a **correlation of residuals**.

An easier way for computing partial correlations, though perhaps less pedagogical, is through the following:

$$r_{12.3} = \frac{r_{12} - (r_{13})(r_{23})}{\sqrt{\left(1 - r_{13}^2\right)\left(1 - r_{23}^2\right)}} \tag{8.3}$$

where $r_{12.3}$ is the correlation between variables 1 and 2 after removing the variability due to variable 3. So long as its limitations are appreciated, we can interpret the partial correlation as the correlation between variables 1 and 2 after **controlling** for variable 3. As already mentioned, however, nothing

---

[1] The interested reader is encouraged to consult Stigler (1986), Desrosières (1998), Denis and Docherty (2007) for details on the social and political forces that helped motivate the use of multiple regression in the poverty debate. See also **Further Discussion and Activities** of the current chapter. This is not to say that multiple regression would not have arose without such a rich social history. It is only to emphasize the fact that statistical techniques rarely, if ever, come into mainstream use without some **purpose**, often social or political in nature.

is actually being "controlled" when computing partial correlation. We are simply partialling out variability, nothing more. We will have much more to say on this matter later in this chapter when we survey multiple regression in its entirety. The concept of "statistical control" is a key feature across the majority of statistical models, not just multiple regression.

Some features of (8.3) are worth noting. For one, notice that the numerator starts with r_{12}, which is the actual correlation coefficient we want to obtain. We then subtract out the product $(r_{13})(r_{23})$, which has the effect of removing the linear influence of variable 3 on both variables 1 and 2. If variable 3 has no linear influence, then it stands that $(r_{13})(r_{23})$ will equal 0, which then the numerator becomes $r_{12} - 0 = r_{12}$.

We can demonstrate the computation of partial correlations using the package `corpcor` (Schäfer et al., 2014) in R. Our demonstration uses data from Hotelling (1936) which we will discuss more extensively in our upcoming discussion of canonical correlation in Chapter 12. For our purposes here, we wish simply to demonstrate the computation of partial correlations. Hotelling's matrix on four variables, which we name `cancor.matrix`, is the following:

```
> library(corpcor)
> cancor.matrix

        [,1]     [,2]     [,3]     [,4]
[1,]  1.0000   0.6328   0.2412  0.0586
[2,]  0.6328   1.0000  -0.0553  0.0655
[3,]  0.2412  -0.0553   1.0000  0.4248
[4,]  0.0586   0.0655   0.4248  1.0000
```

We now generate all partial correlations in the matrix:

```
> library(corpcor)
> cor2pcor(cancor.matrix)

             [,1]         [,2]         [,3]          [,4]
[1,]   1.0000000   0.6758534   0.3852133   -0.1558443
[2,]   0.6758534   1.0000000  -0.3229349    0.1770043
[3,]   0.3852133  -0.3229349   1.0000000    0.4520026
[4,]  -0.1558443   0.1770043   0.4520026    1.0000000
```

Be sure to note the correct interpretation of the coefficients in the matrix. For example, the partial correlation between variable 1 and 2 is 0.676 (row 1, column 2). Consistent with our definition, this is the correlation between 1 and 2 after removing linear influences of both variables 3 and 4. In this case, two variables were partialled out of the relationship of interest. There is no limitation to partialling out several more variables had we also wished to remove their influence from the relationship. For instance, theoretically, in a five-variable problem, one could compute the partial correlation between variables 1 and 2 removing the linear influences of variables 3, 4, and 5.

8.2   SEMIPARTIAL CORRELATIONS

When we computed a partial correlation, both variables x_1 and x_2 were adjusted to remove the linear regression on variable z_i. As we have seen, the partial correlation $r_{12.3}$ is actually the correlation between the two adjusted variables, $x_{(1.3)}$ and $x_{(2.3)}$. That is, both variables have been adjusted for the 3rd variable, which, as mentioned, ends up being a correlation of residuals.

Considering now the **part** or **semipartial** correlation, we will still want to remove the part of x_2 due to z_i to form the new variable $x_{(2.3)}$, but we do not want to adjust variable x_1 at all. That is, in computing the semipartial correlation, we want the correlation between the **unadjusted** variable x_1 and the **adjusted** variable $x_{(2.3)}$. This correlation can be symbolized as $r_{1(2.3)}$.

Computing the semipartial correlation between x_1 and x_2, where only x_2 is adjusted for z_i, we follow these steps:

- Regress x_2 on z_i and obtain a column of residuals.
- Correlate the residuals from x_2 on z_i to the unadjusted values of x_1.

Notice carefully that in the computation of the semipartial correlation, only x_2 is adjusted, while x_1 is left unadjusted.

We can compute the semipartial correlation between x_1 and x_2 in which only variable x_2 has been adjusted for z_i:

$$r_{1(2.3)} = \frac{r_{12} - r_{13}r_{23}}{\sqrt{1 - r_{23}^2}} \tag{8.4}$$

Again, as was true for partial correlation, it is evident from (8.4) that if the linear influence of variable 3 is nonexistent, then the numerator reduces to simply r_{12}, since $r_{12} - 0 = r_{12}$. The point of noting this fact is to emphasize that $r_{1(2.3)}$ starts off with r_{12}, and then "adjusts" it accordingly through the subtraction of $r_{13}r_{23}$.

One can use the `ppcor` package (Kim, 2012) in R to compute semipartial (and partial, for that matter) correlations in R, though we do not demonstrate their computation here. Both partial and semipartial correlations are also easily obtainable via SPSS.

8.3 MULTIPLE REGRESSION

Having surveyed partial and semipartial correlation, we can now build up the multiple regression model from these first principles, where the idea of partialling out variability due to other variables will again be featured.

Recall that in simple linear regression, the least-squares normal equations were given by

$$a = \bar{y} - b\bar{x}$$

and

$$b = \frac{\sum(x_i - \bar{x})(y_i - \bar{y})}{\sum(x_i - \bar{x})^2}$$

Recall we could also compute b by

$$b = \frac{\text{cov}_{xy}}{s_x^2}$$

These solutions guaranteed that the line fit to the sample data would be the **best fit line** in the sense of minimizing the sum of squared errors, $\sum_{i=1}^{n} \varepsilon_i^2$, the so-called least-squares regression line.

In multiple regression, we again seek estimators for our model equation that guarantee a "best fit" in the least-squares sense and from which we can draw inferences regarding parameters $\alpha, \beta_1, \beta_2, \ldots \beta_k$.

Parameters $\beta_1, \beta_2, \ldots \beta_k$ we will call by the name of **partial regression coefficients**. The raw partial regression coefficient of y_i on x_1 holding z_i constant is interpreted as follows:

For a one-unit increase in x_1, we expect, on average, y_i to increase (or decrease, depending on the sign) by b_1 units, controlling for z_i.

Going from the raw, or **unstandardized** partial regression coefficients to the **standardized** ones is straightforward, since they are simply linear transformations of one another. A bit of algebra shows that given b_1, β_1, the standardized coefficient, is easily obtained:

$$\beta_1 \frac{s_y}{s_{x_1}} = b_1$$

$$\beta_1 = \frac{b_1}{\frac{s_y}{s_{x_1}}} = b_1 \cdot \frac{s_{x_1}}{s_y}$$

where s_y, as before, denotes the standard deviation for y_i, and s_{x_1} denotes the standard deviation for variable x_1. That is, the standardized partial regression coefficient is computed by multiplying the raw partial regression coefficient by the ratio of standard deviations of x_1 to y_i. The interpretation of the coefficient is:

For a one-standard deviation increase in x_1, we expect, on average, y_i to increase (or decrease, depending on the sign) by β_1 standard deviations, controlling for z_i.

A key point to remember is that in multiple regression, whether for raw or standardized coefficients, these coefficients are always only interpretable **relative to the model in which they are estimated, and never independently of the tested model**. This is the very nature of multivariable relationships, to incorporate dependencies of variables in the context of other variables and not independent of them. **Be sure you are aware of the distinction between interpreting a simple linear regression and a multiple linear regression as it pertains to interpreting regression coefficients. Partial coefficients should never be interpreted as though they were computed in a simple linear regression context.**

8.4 SOME PERSPECTIVE ON REGRESSION COEFFICIENTS: "EXPERIMENTAL COEFFICIENTS"?

Partial regression coefficients, regardless of the complexity of the model in which they are interpreted, estimate how much a response variable will change, on average, given a 1-unit increase in the predictor while holding all other variables in the model "constant." It is important to realize that slight changes in regression coefficients, given the addition or subtraction of additional predictors, do not **necessarily** carry with them any importance, nor should such small deviations be overanalyzed by the researcher.

For example, should a partial regression coefficient change from 0.72 to 0.70 given the inclusion of an additional predictor in a regression model, this change, though **numerically** noteworthy, does not necessarily equate to being **scientifically** meaningful, nor should investigators in most cases busy themselves with trying to explain such small changes. Regression weights are not "experimental" coefficients. Adding a new variable to a regression model and observing the change in coefficients currently in the model is not an experimental exercise of manipulation and control. Rather, it is simply an exercise in **variance partitioning**. Unless one is working in the context of a controlled experiment or is otherwise dealing with extremely sensitive material as objects of measurement, such slight changes usually cannot be attributed to an underlying substantive mechanism or process, nor should researchers in most cases look for one. This is true especially in cases where there is significant measurement error in variables employed in the regression.

The author has noted that in some circles, researchers exercise a ritual in which correlational variables are added then subtracted from a model and even slight changes in partial regression coefficients are then subjected to critical analysis in hopes of giving them substantive meaning, almost akin to believing they are manipulating an independent variable and observing the effect on a measured response. True, for those who wish to test mediational hypotheses, and can muster a philosophical basis for doing so (see Section 8.15), huge drops in coefficients may be both statistically and substantively meaningful. However, small coefficient changes on what is in many cases error-prone, and very sample-specific data, usually has little if any meaning. On the other hand, if measurements are precise and virtually error-free, then even small changes in coefficients may be noteworthy. The key point is to not "over-theorize" on the outcomes of variance partitioning, which in the end, is all that ANOVA or regression (or virtually any statistical model) can promise to accomplish. The scientific claims should usually and properly arise from one's research design.

8.5 MULTIPLE REGRESSION MODEL IN MATRICES

The essential statistical theory for the regression model was surveyed in Chapter 7 for the case of a single predictor variable. It is a simple matter to extend on that model to obtain the multiple regression model of the current chapter. Recall the expectation for the simple linear regression model:

$$
E(\mathbf{y}) = \begin{bmatrix} E(y_1) \\ E(y_2) \\ E(y_3) \\ . \\ . \\ . \\ E(y_n) \end{bmatrix} = \begin{bmatrix} \alpha + \beta x_{i=1} \\ \alpha + \beta x_{i=2} \\ \alpha + \beta x_{i=3} \\ . \\ . \\ . \\ \alpha + \beta x_{i=n} \end{bmatrix}
$$

where $E(\mathbf{y})$ was the expectation of the vector of responses on $y_1, y_2, \ldots y_n$, $E(y_1) \ldots E(y_n)$ was the expectation of each response $y_1, y_2, \ldots y_n$, and $\alpha + \beta x_{i=1} \ldots \alpha + \beta x_{i=n}$ constituted the so-called **model function** which was the **systematic** portion of the model. It was the expectation of each response.

The expectation for the multiple linear regression model can be easily extended to incorporate additional predictors:

$$
E(\mathbf{y}) = \begin{bmatrix} E(y_1) \\ E(y_2) \\ E(y_3) \\ . \\ . \\ . \\ E(y_n) \end{bmatrix} = \begin{bmatrix} \alpha + \beta_1 x_{1(i=1)} + \beta_2 x_{2(i=1)} + \ldots + \beta_k x_{k(i=1)} \\ \alpha + \beta_1 x_{1(i=2)} + \beta_2 x_{2(i=2)} + \ldots + \beta_k x_{k(i=2)} \\ \alpha + \beta_1 x_{1(i=3)} + \beta_2 x_{2(i=3)} + \ldots + \beta_k x_{k(i=3)} \\ . \qquad . \qquad . \\ . \qquad . \qquad . \\ . \qquad . \qquad . \\ \alpha + \beta_1 x_{1(i=n)} + \beta_2 x_{2(i=n)} + \ldots + \beta_k x_{k(i=n)} \end{bmatrix}
$$

where, instead of the expectation $E(y_1) \ldots E(y_n)$ being equal to $\alpha + \beta x_{i=1} \ldots \alpha + \beta x_{i=n}$ as in the case of the simple linear regression model, the expectation is now equal to

$$\alpha + \beta_1 x_{1(i=1)} + \beta_2 x_{2(i=1)} + \ldots + \beta_k x_{k(i=1)} \ldots \alpha + \beta_1 x_{1(i=n)} + \beta_2 x_{2(i=n)} + \ldots + \beta_k x_{k(i=n)}$$

for observations $y_1 \ldots y_n$. Associated with each $y_1 \ldots y_n$ is still $\varepsilon_1 \ldots \varepsilon_n$, which once more is given by a single column vector:

$$\boldsymbol{\varepsilon} = \begin{bmatrix} \varepsilon_{i=1} \\ \varepsilon_{i=2} \\ \varepsilon_{i=3} \\ . \\ . \\ . \\ \varepsilon_{i=n} \end{bmatrix}$$

8.6   ESTIMATION OF PARAMETERS

Recall that in the simple linear regression model constants α and β were chosen so that $\sum_{i=1}^{n} \varepsilon_i^2$ is minimized. In a multiple regression model, because there is more than a single predictor variable, we choose scalars $\alpha, \beta_1, \beta_2, \beta_3, \ldots \beta_k$ so that $\sum_{i=1}^{n} \varepsilon_i^2$ is again kept at a minimum. Analogous to the simple regression case, estimators for $\alpha, \beta_1, \beta_2, \beta_3, \ldots \beta_k$ are obtained by again taking **partial derivatives** of $\sum_{i=1}^{n} \varepsilon_i^2$ with respect to each of $\alpha, \beta_1, \beta_2, \beta_3, \ldots, \beta_k$ instead of simply α and β as in the case of simple linear regression. We do not detail the derivation here, as on an applied level, it is not enlightening. For details on the estimation of model parameters in a multiple regression, see DeGroot and Schervish (2002, p. 648).

8.7   CONCEPTUALIZING MULTIPLE R

There are various algebraically equivalent ways of both conceptualizing and computing multiple R, and its square, the **coefficient of multiple determination**. R^2 can be defined as the ratio of SS regression to SS total,

$$\frac{\sum_{i=1}^{n} \left(y_i' - \bar{y} \right)^2}{\sum_{i=1}^{n} \left(y_i - \bar{y} \right)^2}$$

or as the squared Pearson product-moment correlation between observed values y_i and predicted values y_i' on the response variable y_i:

$$r\left(y_i, y_i' \right)^2 \tag{8.5}$$

We also noted in Chapter 7 how R^2 could be computed via matrices. All of these ways of conceptualizing and computing R^2 are applicable to the multiple regression case, the only difference is that now y_i' is a function of more "information" (i.e., predictor variables) than in the case of simple regression. Regardless of the number of predictors, however, there is still only one way to compute the bivariate correlation between observed and predicted values, and hence (8.5) will still apply even for the case of multiple predictors.

8.8   INTERPRETING REGRESSION COEFFICIENTS: CORRELATED VERSUS UNCORRELATED PREDICTORS

Partial regression coefficients in a multiple regression model should not be interpreted as if they were **zero-order regression coefficients** in a simple regression model. The reason for this is that correlation between predictors makes the zero-order interpretation incorrect. When predictors are correlated, which is virtually the case in all samples, $R_{y \cdot x_1, x_2, \ldots x_k}$ is a more complex function of zero-order correlations, and can be written as the square root of a **weighted sum** of the relevant $\beta_1, \beta_2, \beta_k$ weights (Hays, 1994, pp. 697–698):

$$R_{y \cdot x_1, x_2, \ldots x_k} = \sqrt{\beta_1 r_{y \cdot 1} + \beta_2 r_{y \cdot 2} + \ldots + \beta_k r_{y \cdot k}} \tag{8.6}$$

Equation (8.6) tells us what $R_{y \cdot x_1, x_2, \ldots x_k}$ actually is. **It is a linear combination of zero-order correlation coefficients, each weighted by variables' respective standardized regression coefficients**. When we square $R_{y \cdot x_1, x_2, \ldots x_k}$, we get the aforementioned coefficient of multiple determination.

8.9   ANDERSON'S IRIS DATA: PREDICTING SEPAL LENGTH FROM PETAL LENGTH AND PETAL WIDTH

We demonstrate a simple example of multiple regression on Fisher's iris data where a single response variable is hypothesized as a function of two predictors. These data were first made available by Anderson (1935), and hence the iris data has come to be also known also as Anderson's iris data. The data consist of a total of 150 observations on three species of iris, 50 on iris **setosa**, 50 on iris **virginica**, and 50 on iris **versicolor**. The length and width of both sepals and petals were recorded. The data are of historical significance and have been used in countless papers as a model demonstration of numerous statistical methods. We request some of the iris data in R:

```
> attach(iris)
> library(car)
> some(iris)
```

	Sepal.Length	Sepal.Width	Petal.Length	Petal.Width	Species
3	4.7	3.2	1.3	0.2	setosa
12	4.8	3.4	1.6	0.2	setosa
14	4.3	3.0	1.1	0.1	setosa
33	5.2	4.1	1.5	0.1	setosa
37	5.5	3.5	1.3	0.2	setosa
86	6.0	3.4	4.5	1.6	versicolor
116	6.4	3.2	5.3	2.3	virginica

122	5.6	2.8	4.9	2.0	virginica
129	6.4	2.8	5.6	2.1	virginica
148	6.5	3.0	5.2	2.0	virginica

For this analysis, we concern ourselves only with predicting sepal length from knowledge of petal length and petal width.

We would first like to get a picture of the data. We can generate a 3D scatterplot in R using the `scatterplot3d` package (Ligges and Mächler, 2003). For example, we obtain a plot of sepal length by petal length by petal width:

```
> library(scatterplot3d)
> scatterplot3d(Sepal.Length, Petal.Length, Petal.Width)
```

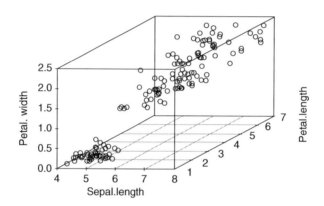

The above plot is useful if for nothing more than to gain an appreciation of the complexities associated with analyzing and visualizing multivariate data even in only three dimensions. An at least approximate linear relationship among the three variables appears to be present. The actual initial screening of multivariate data usually takes the form of **univariate histograms** or **stem-and-leaf plots**, as well as **bivariate scatterplots**. These plots are useful in the early detection of potential outliers and as a quick check on whether the hope for "multivariate linearity" (i.e., linear relationships holding other variables constant) is at least tenable. Even if linearity holds for y_i on x_1 and y_i on x_2, this does not necessarily imply **multivariable** linearity. **Marginal relationships** (e.g., y_i on x_1) are not equivalent to **partial relationships** (e.g., y_i on x_1 **controlling** for x_2), and hence bivariate plots can typically only help rule out problematic nonlinear data in two dimensions. They also cannot in any way guarantee that our empirical observations are multivariate normal. Generally, satisfying assumptions in lower dimensions does not guarantee that these same assumptions will hold in higher dimensions. In verifying assumptions in multiple dimensions, the analysis of residuals from the fitted model is a must.

We carry on now with the regression analysis, designating the response variable to be `Sepal.Length` and predictor variables `Petal.Length` and `Petal.Width`:

```
> reg.fit.iris <- lm(Sepal.Length ~ Petal.Length + Petal.Width)
> summary(reg.fit.iris)
```

```
Coefficients:
            Estimate Std. Error t value Pr(>|t|)
(Intercept)  4.19058    0.09705  43.181  < 2e-16 ***
Petal.Length 0.54178    0.06928   7.820 9.41e-13 ***
Petal.Width -0.31955    0.16045  -1.992   0.0483 *
---
Signif. codes:  0 '***' 0.001 '**' 0.01 '*' 0.05 '.' 0.1 ' ' 1

Residual standard error: 0.4031 on 147 degrees of freedom
Multiple R-squared:  0.7663,     Adjusted R-squared:  0.7631
F-statistic:   241 on 2 and 147 DF,  p-value: < 2.2e-16
```

Some general features of the output:

- The estimated model equation is $y_i' = 4.19 + 0.54(x_1) - 0.32(x_2)$.
- The overall model yields an F-statistic of 241 on 2 and 147 degrees of freedom and is statistically significant yielding a p-value of 2.2e−16. Degrees of freedom for numerator are equal to 2 since there are two predictors, whereas degrees of freedom for denominator are equal to 147 (computed as $n - k - 1 = 150 - 2 - 1 = 147$).
- Both petal length and petal width are statistically significant predictors of sepal length, with p-values of 9.41e−13 and 0.0483, respectively.
- For a 1 unit increase in petal length, we can expect, on average, sepal length to increase by 0.54 units, when petal width is held constant (or equivalently, given the **model** currently under test).
- For a 1 unit increase in petal width, we can expect, on average, sepal length to decrease by 0.32 units (i.e., negative coefficient), when petal length is held constant (or again equivalently, given the **model** currently under test).
- The value of multiple R^2 is equal to 0.7663, indicating that the model as a whole accounts for nearly 77% of the variance in sepal length.
- The value for adjusted R^2 is equal to 0.7631, slightly smaller than R^2, but still indicative of the overall effect.
- The standard error of the estimate (residual standard error) is equal to 0.4031. This is the standard deviation of residuals around the fitted regression line, but with $n - k - 1 = 150 - 2 - 1 = 147$ in the denominator (Fox, 2016, p. 87). It is an estimate of the standard deviation of errors (James et al., 2013, pp. 68–69). The variance of the estimate (i.e., MS residual) is the square of 0.4031, equal to $(0.4031)^2 = 0.1625$. Notice that the variance of residuals is quite small, suggesting a relatively good fit of the model to the empirical data (which is also confirmed by a relatively large R^2).

We can obtain the predicted (or **fitted**) values for our model, where we print only the first five:

```
> fitted(reg.fit.iris)

4.885160     4.885160     4.830983     4.939338     4.885160
```

What are these fitted values? The first fitted value, that of 4.885, is the predicted value for the first observation in our sample data. The first observation has values

```
    Sepal.Length Sepal.Width Petal.Length Petal.Width  Species
            5.1         3.5          1.4         0.2   setosa
```

For demonstration, we can compute the predicted value using our model equation for the first observation to match that generated by R:

$$
\begin{aligned}
y_i' &= 4.19 + 0.54(x_1) - 0.32(x_2) \\
&= 4.19 + 0.54(1.4) - 0.32(0.2) \\
&= 4.19 + 0.756 - 0.064 \\
&= 4.88
\end{aligned}
$$

We could likewise generate the remainder of the fitted values using the regression equation. How well did the model predict this first observation? To evaluate how well it did, we compute the residual (after rounding 4.88 up to 4.9):

$$
\begin{aligned}
r_i &= y_i - y_i' \\
&= 5.1 - 4.9 \\
&= 0.2
\end{aligned}
$$

We can see that for this observation, we are in slight **error**. The model did not predict this observation **perfectly**. We can obtain residuals for all observations by computing (we print only the first five):

```
> residuals(reg.fit.iris)

   0.214839668   0.014839668  -0.130982616  -0.339338047   0.114839668
```

We note that the residual for the first observation, 0.21, matches that which we computed manually.

How much variability is there in these residuals? A model with large residuals, overall, would suggest the model does not fit very well. A model with small residuals, would, on the other hand, imply a well-fitting model. Recall that simply summing the residuals will not answer our question, since **the sum of residuals will always equal 0**. Of course, the solution, as always, is to sum the **squared residuals**. We compute a standard deviation of residuals:

```
> sd(residuals(reg.fit.iris))
[1] 0.4003412
```

The standard deviation of residuals is equal to 0.4003. Note however that this standard deviation of residuals is slightly different from the standard error of the estimate featured earlier (which recall was an estimate of the true standard deviation), equal to 0.4031. For practical purposes, they will be virtually the same (as in the current case) especially for large n, but the standard error of the estimate recall uses $n - k - 1$ in its denominator, not n or $n - 1$ as in the typical standard deviation. Still, it is convenient to simply refer to the standard error of the estimate as the **standard deviation of residuals**, so long as you keep in mind the loss of greater degrees of freedom in the denominator due to estimation. The above standard deviation of 0.4003 will still give you a reasonable estimate, it is just not the most accurate one, since, as mentioned, the degrees of freedom will be slightly off.

Recall that one way of conceptualizing multiple R is that it is the bivariate correlation between observed values of y_i and predicted values y_i'. We can easily compute this:

```
> pred <- fitted(reg.fit.iris)
> cor(Sepal.Length, pred)
```

```
[1] 0.8753635
```

Notice that, within slight rounding error, the value of 0.875 is the value of the square root of multiple R^2 reported by R (0.766), as it should be.

We can also obtain confidence intervals for the estimated regression coefficients by confint in R:

```
> confint(reg.fit.iris, level = 0.95)
                  2.5 %        97.5 %
(Intercept)    3.9987971   4.382367716
Petal.Length   0.4048602   0.678694143
Petal.Width   -0.6366424  -0.002458754
```

With 95% confidence, the interval from 0.40 to 0.68 likely covers the true population partial regression coefficient for petal length. Similarly, with 95% confidence, the interval from −0.64 to −0.002 likely covers the true population partial regression coefficient for petal width. We could request R to compute 99% intervals as follows:

```
> confint(reg.fit.iris, level = 0.99)
                  0.5 %       99.5 %
(Intercept)    3.9373230   4.44384187
Petal.Length   0.3609733   0.72258101
Petal.Width   -0.7382818   0.09918069
```

Our interpretation of these is analogous to the 95% intervals. That is, with 99% confidence, the interval 0.36 to 0.72 likely covers the true population partial regression coefficient for petal length. With 99% confidence, the interval −0.74 to 0.099 likely covers the true population partial regression coefficient for petal width.

Be sure to note that the limits for the 99% confidence intervals are wider than those for the 95% confidence intervals. Recall from our review of confidence intervals in Chapter 2 that more certainty (in terms of adjusting the confidence limits) entails a widening of the interval, not a narrowing. An easy way to remember this principle is to recall what, in theory, a 100% interval would look like. An interval with ranges −∞ to +∞ would have to theoretically exist for 100% of samples drawn (on which intervals are computed) to capture the true parameter. Conversely, a 0% interval is equivalent to a point estimate, for which the probability of it equaling the parameter has a limiting value of 0. This is precisely why we often prefer interval estimates over point estimates. They give us a bit more flexibility for covering the true parameter and provide a sense of precision in the given estimate.

8.10 FITTING OTHER FUNCTIONAL FORMS: A BRIEF LOOK AT POLYNOMIAL REGRESSION

We have up to now assumed a linear model, but there is nothing preventing us from trying out other polynomials to see which best fits the iris data. The general name for this is **polynomial regression**, of which linear regression can be considered a special case. Recall that linear forms are but one type of **function**. Others include **quadratic**, **cubic**, **quartic**, **quintic**, and so on, each representing terms raised to a different exponent. Polynomial regression is not necessarily synonymous with **nonlinear regression**. A polynomial regression model is still **linear in the parameters**. For instance, consider the following model in which a quadratic term is included:

$$y_i = \alpha + \beta_1 x_1 + \beta_2 x_2^2 + \varepsilon_i$$

The model is still a **linear** model, even though x_2^2 (a quadratic term) is included. What makes it still linear is that the parameters β_1 and β_2 are still raised to the first exponent. Hence, though polynomial regression is useful for capturing nonlinear trends in data, it is still a linear model.

As an example of polynomial regression, we could have tried the following **quadratic model** for the iris data, where the term I(Petal.Width^2) is the new squared term added to the model:

```
> quad.fit.iris <- lm(Sepal.Length ~ Petal.Length + Petal.Width + I
(Petal.Width^2))
> summary(quad.fit.iris)

Coefficients:
                  Estimate Std. Error t value Pr(>|t|)
(Intercept)        4.26600    0.09283  45.955  < 2e-16 ***
Petal.Length       0.71892    0.07621   9.433  < 2e-16 ***
Petal.Width       -1.52224    0.30775  -4.946 2.05e-06 ***
I(Petal.Width^2)   0.34795    0.07759   4.484 1.47e-05 ***
---
Signif. codes:  0 '***' 0.001 '**' 0.01 '*' 0.05 '.' 0.1 ' ' 1

Residual standard error: 0.3792 on 146 degrees of freedom
Multiple R-squared:  0.7946,    Adjusted R-squared:  0.7903
F-statistic: 188.2 on 3 and 146 DF,  p-value: < 2.2e-16
```

We note that fitting a quadratic term for petal width appears to be worthwhile, as it is statistically significant and the overall model fit has improved somewhat. The AIC for this model is 140.69 (AIC (quad.fit.iris)) while the AIC for the model without the quadratic term was 158.05. Incidentally, a convenient way to test different competing models in R is to use the update function (request >?update in R to learn more or consult Venables and Ripley (2002)). **Regression splines** are a technique that can be used in place or in conjunction with polynomial regression, which provide a flexible (and often preferable) way of dealing with nonlinearity while keeping the degree of the polynomial fixed. Essentially, the technique works by fitting lines in regions of the predictor space, for which these spaces are defined by what are called "knots." For details, see James et al. (2013).

8.11   MEASURES OF COLLINEARITY IN REGRESSION: VARIANCE INFLATION FACTOR AND TOLERANCE

In addition to the typical model assumptions one must make for simple least-squares regression, for multiple regression, we must also make the assumption that the **rank of the data matrix is equal to the number of columns in the data matrix**. **Rank** of a matrix is the number of **linearly independent rows** or **columns** of that matrix (see Appendix). Such independence translates more substantively in an applied setting to the assumption of a lack of **multicollinearity** among predictors. If one predictor is an exact linear combination of another predictor, then the data matrix is not of **full rank** since one or more columns are linearly dependent. As multicollinearity increases to the point of linear dependence, the matrix product $\mathbf{X'X}$ in $\mathbf{b} = (\mathbf{X'X})^{-1}\mathbf{X'y}$ is singular (Fox, 2016), which means the least-squares equations cannot generate a unique solution. That is, perfect collinearity results in least-squares coefficients that are not unique (Fox, 2016, p. 342). However, even less than perfect collinearity will cause serious difficulty. As summarized by Fox (2016):

When the regressors in a linear model are perfectly collinear, the least-squares coefficients are not unique. Strong, but less-than-perfect collinearity substantially increases the sampling variances of the least-squares coefficients and can render them useless as estimators. (p. 344)

To help diagnose problems with collinearity, both the **variance inflation factor** (VIF) and its reciprocal, **tolerance**, have been proposed. To understand VIF, consider first how we may write the variance for a given i*th* partial regression coefficient (Cohen et al., 2003, p. 423):

$$s_{b_i}^2 = \frac{s_y^2}{s_{x_i}^2} \left(\frac{1 - R_{y.12...k}^2}{n - k - 1} \right) \left(\frac{1}{1 - R_{i.12...(i)...k}^2} \right) \tag{8.7}$$

where $R_{y.12...k}^2$ is the variance explained by the hypothesized model (i.e., the regression that is being run based on all predictors), and $R_{i.12...(i)...k}^2$ is the variance explained by the model in which the given predictor (x_i) is being regressed on the remaining predictors in the model. For instance, if we are calculating the variance for x_1, $R_{i.12...(i)...k}^2$ is computed by taking predictors x_2 and x_3 and using them simultaneously to predict x_1.

The variance inflation factor is the last term in (8.7):

$$\text{VIF} = \left(\frac{1}{1 - R_{i.12...(i)...k}^2} \right)$$

Since the denominator of VIF is $1 - R_{i.12...(i)...k}^2$, where recall $R_{i.12...(i)...k}^2$ represents the regression of the predictor of interest (i.e., for which VIF is being computed) on the remaining predictors, what this means is that **the extent to which the given predictor is highly correlated with the remaining predictors is the extent to which VIF will be large**. That is, as $R_{i.12...(i)...k}^2$ approaches 1.0, VIF will be increasingly large. For instance, suppose $R_{i.12...(i)...k}^2$ were equal to 0.90, which is, in most contexts, an impressive coefficient of determination. This means that the predictor on which we are computing VIF is highly correlated with the remaining predictors in the model. The computation of VIF would be:

$$\text{VIF} = \left(\frac{1}{1 - R_{i.12...(i)...k}^2} \right) = \left(\frac{1}{1 - 0.90} \right) = 10$$

The VIF of 10 in this case suggests that the variance of the regression coefficient will be quite "inflated." This implies that the standard error, which recall is simply the square root of $s_{b_i}^2$, will likewise be large. A **large standard error** suggests that the given parameter (in our case, β_i) is not being estimated **precisely**. That is, in theoretical samplings of the given partial regression statistic from a population in which the null hypothesis is true (i.e., population $\beta = 0$), a large standard error indicates that we can expect quite a bit of sampling fluctuation in the infinite number of samples we theoretically collect. The variance inflation factor then is just what the name suggests: **it is a factor by which the variance of the given partial regression coefficient increases due to the given variable's extent of correlation with the other predictors in the model**.

The minimum VIF can be is 1.0. A VIF of 1.0 can only occur when $R_{i.12...(i)...k}^2$ is equal to 0, which implies that the given predictor has **zero** linear relationship with other predictors in the model. With $R_{i.12...(i)...k}^2$ equal to 0, we find VIF to have no influence on the estimation of $s_{b_i}^2$:

$$s_{b_i}^2 = \frac{s_y^2}{s_{x_i}^2}\left(\frac{1 - R_{y.12...k}^2}{n - k - 1}\right)\left(\frac{1}{1 - R_{i.12...(i)...k}^2}\right)$$

$$= \frac{s_y^2}{s_{x_i}^2}\left(\frac{1 - R_{y.12...k}^2}{n - k - 1}\right)\left(\frac{1}{1 - 0}\right)$$

$$= \frac{s_y^2}{s_{x_i}^2}\left(\frac{1 - R_{y.12...k}^2}{n - k - 1}\right)$$

Such also reveals that VIF values cannot be less than 1.0, since 1.0 represents the ideal situation of no correlation with other predictors. Also implied is that VIF cannot be negative.

Tolerance is simply the reciprocal of VIF and is thus computed as

$$\text{Tol} = 1/\text{VIF}$$

Whereas large values of VIF are undesirable, large tolerances are preferable to smaller ones. It stands as well that the maximum value of tolerance must be 1.0. Cohen et al. (2002) suggest VIF values of 10 or more to be of potential concern. Using a strict cutoff here, however, is probably not the best strategy. Increasingly larger values for VIF, in addition to inflating variance, are also of potential concern for they may indicate a **substantive** issue with your model in the sense that scientific parsimony is likely not being achieved. Given a relatively large VIF statistic, it may be worth re-hypothesizing your model by possibly dropping the given predictor that is largely a function of other predictors in the model. One can easily compute vif in R (`vif(model)`).

Another measure that may be used to detect the presence of multicollinearity is known as the **condition index**, though not discussed here. See Lattin, Carroll, and Green (2003) for details.

8.12   R-SQUARED AS A FUNCTION OF PARTIAL AND SEMIPARTIAL CORRELATIONS: THE STEPPING STONES TO FORWARD AND STEPWISE REGRESSION

To set the stage for the consideration of model-building procedures such as forward and stepwise regression to be discussed shortly, it is imperative to get a sense of how R^2 can be decomposed into partial and semipartial correlations. It can be shown (Hays, 1994, p. 713) that R^2 can be written as a function of the following product of partial correlations:

$$1 - R_{y.12}^2 = \left(1 - r_{y1}^2\right)\left(1 - r_{y2.1}^2\right) \tag{8.8}$$

where, as before, $R_{y.12}^2$ is the variance explained in y_i by x_1 and x_2, r_{y1}^2 is the variance explained in y_i by x_1, and $r_{y2.1}^2$ is the variance explained in y_i by x_2, after partialling out x_1. By rearranging (8.8) slightly, we get

$$1 - R_{y.12}^2 = \left(1 - r_{y1}^2\right)\left(1 - r_{y2.1}^2\right)$$

$$R_{y.12}^2 = 1 - \left[\left(1 - r_{y1}^2\right)\left(1 - r_{y2.1}^2\right)\right]$$

That is, $R^2_{y\cdot12}$ can be expressed as a function of 1 minus the product of partial correlation "variance unexplained" terms (i.e., $\left(1-r^2_{y1}\right)\left(1-r^2_{y2\cdot1}\right)$). The essential point to grasp here is that the coefficient of multiple correlation can be decomposed into **partial** correlations.

Similarly, for the three-variable case, we write $R^2_{y\cdot123}$ as

$$R^2_{y\cdot123} = 1 - \left[\left(1-r^2_{y1}\right)\left(1-r^2_{y2\cdot1}\right)\left(1-r^2_{y3\cdot12}\right)\right]$$

and for the four-variable case,

$$R^2_{y\cdot1234} = 1 - \left[\left(1-r^2_{y1}\right)\left(1-r^2_{y2\cdot1}\right)\left(1-r^2_{y3\cdot12}\right)\left(1-r^2_{y4\cdot123}\right)\right]$$

We can also decompose R^2 as a function of **semipartial** correlations. Again, for the case of two predictors, we can define $R^2_{y\cdot12}$ as

$$R^2_{y\cdot12} = r^2_{y1} + r^2_{y(2\cdot1)} \tag{8.9}$$

where r^2_{y1} is, as before, simply the proportion of variance explained from regressing y_i on x_1. The quantity $r^2_{y(2\cdot1)}$ is the proportion of variance explained from regressing y_i on x_2, of which the influence of x_1 is removed from x_2. That is, $r^2_{y(2\cdot1)}$ is the **squared semipartial correlation** between y_i and x_2. Be sure to note the difference in notation between $r^2_{y2\cdot1}$ and $r^2_{y(2\cdot1)}$. The first is the partial correlation, the second is the semipartial correlation. As discussed earlier, each tells us something different.

Equation (8.9) is especially relevant because understanding it is the "gateway" to understanding **stepwise regression**. The equation tells us that the proportion of explained variance in y_i based on a regression of y_i on x_1 and x_2 is a function of y_i on x_1, plus the **additional contribution** of x_2 after "controlling" (which recall, really means "partialling out") for x_1. The quantity $r^2_{y(2\cdot1)}$ is the additional variance explained **over and above** that already contributed by x_1. Hence, we see that **the increment in variance explained is described by the squared semipartial correlation**. This is the basis on which **forward** and **stepwise regression** procedures operate, which we now consider as we turn to model-building strategies.

8.13 MODEL-BUILDING STRATEGIES: SIMULTANEOUS, HIERARCHICAL, FORWARD, STEPWISE

Model selection in statistics is a difficult problem, and arriving at the "best" model is not a simple task, largely because how one defines "best" varies depending on what criteria are used in the definition. What is more, predictor selection is not simply a **statistical problem**, but also a **scientific** one as well. And though statisticians busy themselves with developing criteria such as **AIC, R^2, adjusted-R^2**, and others to aid in model selection, the fundamental problem of choosing a "best" model actually begins long before the software routine is run. It begins with the scientist choosing an initial pool of what can be called **candidate predictors** that will even stand the chance at being chosen into the model in the first place.

To unpack this idea a bit further, consider the following figure, adapted from Denis (2020) (Figure 8.1).

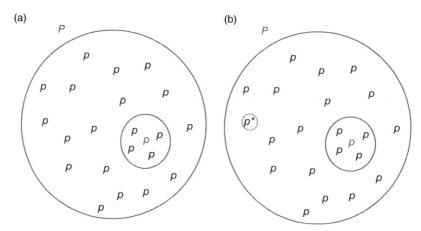

FIGURE 8.1 Model selection unofficially begins with the selection of p candidate predictors (smaller circle) from a wider set of P predictors (a). The problem of omitted variables, where p^* represents an important but omitted variable, potentially unknown to the scientist, or unknowable based on the current status of the scientific discipline (b). The effect of $p*$ on the candidate predictors in building a statistical model could be substantial. Source: Denis (2020). Reproduced with permission from John Wiley & Sons.

The figure emphasizes the fact that model selection begins with the scientist or researcher making initial selections of variables into the pool of candidate predictors, as indicated in the smaller circle in part (a). These are the predictors the scientist deems worthwhile as a subset from which to then attempt to construct a statistical model. It needs to be recognized, however, that if one or more candidate predictors are omitted from this initial selection, as depicted in part (b), then whatever model is built using the candidate predictors will necessarily result in a **faulty model**, due to the already-discussed fact that all models are context-dependent. Recall what this means. In any statistical model, magnitudes and signs of coefficient estimates are dependent on which other variables also appear in the model. As we will see later, this forms the very bedrock of estimating **mediation models**, for instance, as the inclusion of one predictor in the model may offset the coefficient(s) of another or several other predictors. **Whenever you test a model, you are always testing that model in a given context of other variables, even if other variables are potentially omitted, in which case, the "context" is a missingness of influential predictors that should have been included in the model.**

The problem of **omitted variables** is a huge one (Marais and Wecker, 1994), and the philosophical problems that accompany model selection are significant and much deeper than we have space for here (see Kieseppä, 2001, for an insightful discussion). Even without considering the problem of omitted variables and assuming we have successfully chosen the ideal set of candidate predictors, the number of total models possible for a problem of only 20 predictors is equal to $2^p = 2^{20} = 1,048,576$, and for 30 predictors, 1,073,741,824. That is, this is when we fit a separate model for each combination of the predictors (James et al., 2013). Hence, the number of potential models to select from is exceedingly large even for problems of still relatively few predictors. This is all to say, then, that model selection in statistics and science is not an easy problem that features an easy solution. What is more, even if a model is selected which maximizes criteria on statistical grounds, as will be discussed and expanded on later, this may not equate to maximizing **scientific usability** or **practicality**. For now, however, we concern ourselves only with statistical criteria for selecting a "best" model and postpone a more pragmatic discussion of model selection after surveying a few options.

Several approaches to predictor selection have been proposed for building regression models. Of these, the most popular are **simultaneous entry**, **hierarchical**, and **forward** or **backward** regression.

Toward the conclusion of the chapter, we will also survey **lasso regression**, which can also be employed as a model-selection procedure. It needs to be recognized and appreciated from the outset, however, that each of these model-building approaches come with their own set of benefits and drawbacks, and it behooves the researcher to be familiar with them as to simply be knowledgeable about the tool they are using. There is, unfortunately, no panacea when it comes to model selection, and if a scientist is unfamiliar with the approach they are using, it can easily lead to erroneous scientific conclusions about the substantive matter subjected to the given algorithm.

8.13.1 Simultaneous, Hierarchical, Forward

The most straightforward way to build a regression model is to estimate all parameters in the model at the same time without proposing any kind of hierarchical structure or order of entry. Given a set of linear equations,

$$y_1 = \alpha + \beta_1 x_{11} + \beta_2 x_{12} + \ \ldots + \beta_k x_{1k} + \varepsilon_1$$
$$y_2 = \alpha + \beta_1 x_{21} + \beta_2 x_{22} + \ldots + \beta_k x_{2k} + \varepsilon_2$$
$$.$$
$$.$$
$$y_n = \alpha + \beta_1 x_{n1} + \beta_2 x_{n2} + \ldots + \beta_k x_{nk} + \varepsilon_n$$

so-called **simultaneous regression** seeks to solve for parameters $\alpha, \beta_1, \beta_2, \ldots \beta_k$ for k predictors on n observations. For instance, for a model having three predictors, x_1, x_2, x_3, the system of linear equations would be given by:

$$y_1 = \alpha + \beta_1 x_{11} + \beta_2 x_{12} + \beta_3 x_{13} + \varepsilon_1$$
$$y_2 = \alpha + \beta_1 x_{21} + \beta_2 x_{22} + \beta_3 x_{23} + \varepsilon_2$$
$$.$$
$$.$$
$$y_n = \alpha + \beta_1 x_{n1} + \beta_2 x_{n2} + \beta_3 x_{n3} + \varepsilon_n$$

Simultaneous regression is the "default" way to estimate parameters in a regression model, and in most cases, at least for reasonably theory-driven models, it is the preferable strategy.

A **hierarchical** (or **sequential**) regression features entering predictors in a **preconceived order of entry**, presumably based on theory, as opposed to entering all predictors simultaneously. "Hierarchical" implies the building of an **order** or the execution of a **systematic plan** for model-building. "Hierarchical" in this sense is not equivalent to **hierarchical linear modeling** discussed earlier in the book.

It is very important to recognize that **hierarchical regression is also neither equivalent to stepwise regression** nor other selection methods such as forward regression or backward elimination. These latter methods use selection order based on the sequential statistical significance of predictors rather than theory to determine entry of predictors into the regression model. They are entirely different methods than the theory-driven hierarchical regression approach.

As an example, suppose a researcher would like to predict depression based on one's anxiety but knows from prior research that socioeconomic status, or SES, is a predictor relevant to the model. Given this, the researcher might enter SES on the first "step" of the model, then add anxiety at the second step. The result would reveal how much anxiety predicts depression **over and above SES**.

It needs to be noted that in procedures such as **forward**, **backward**, and **stepwise**, different criteria may be used and the exact details of the given algorithm may differ depending on the particular package or program. So while what we describe below provides a general idea of how most of these function, do not assume that if you look up a package or software program and it reads "forward" or "stepwise," etc., that it guarantees that it works exactly as described in the following. You may assume it functions in a similar manner, but it behooves you to study its manual to learn of precisely how it proceeds, how it defines entry and removal of variables, and so on. Different algorithms may maximize different criteria, so it is important that you familiarize yourself with the procedure before running it (and most importantly, interpreting it).

We first consider the selection procedure known as **forward regression**. In this procedure, once a predictor is selected into the model, it cannot be removed. Other predictors may be added at future steps, but predictors already in the model remain in the model. As we will see, this is different from stepwise regression in which we can specify entrance criteria for both adding and removing predictors at each step.

The following is the "logic" of how forward selection generally proceeds. It is imperative that we detail these steps rather thoroughly so that you have a solid grasp of how selection procedures work before you use (and interpret) them:

- **Step 1**—The predictor with the largest squared correlation with the response is entered into the model. Since this is the first step of the selection procedure, entering the predictor with the largest squared correlation is equivalent to entering the predictor with the largest **squared semipartial correlation** as well. It may seem trivial at this point to bring up the idea of semipartial correlation at step 1 of the procedure, but we do so because at subsequent steps, the criterion for entrance into the regression equation will be the squared semipartial correlation (or equivalently, the amount of variance contributed by the new predictor over and above variables already entered into the equation). This process equates to adding the predictor variable with the largest F-statistic that is also statistically significant (Izenman, 2008).

- **Step 2**—The predictor with the **largest squared semipartial correlation** with the response is selected. That is, the predictor with the largest correlation with y_i **after being adjusted for the first predictor** is entered if it meets entrance criteria in terms of preset statistical significance for entry, what SPSS refers to as "PIN" (probability of entry, or "in") criteria. Be sure to note that even when this new predictor is entered at step 2, the predictor entered at step 1 remains in the equation, even if its new semipartial correlation with y_i is now less than what it was at step 1. This is the nature of the forward selection procedure; **forward selection does not reevaluate already-entered predictors into the model after adding new variables**. It only adds predictors to the model (assuming these predictors meet entrance criteria). In the stepwise procedure, to be discussed shortly, in addition to entrance criteria being specified for new variables, **removal criteria** are also specified at each stage of the variable-selection procedure.

- **Step 3**—The predictor with the largest squared semipartial correlation with the response is selected. That is, the predictor with the largest correlation with the response after being adjusted for **both of the first predictors** is entered. Be sure to note that the entrance of this variable is conditional upon its relationship with the previously entered variables at steps 1 and 2. Hence, for a variable to be entered at step 3, the algorithm asks the question: **"Which among available variables currently not entered into the regression equation contribute most to variance explained in y_i given that variables entered at steps 1 and 2 remain in the model?"** Translated into statistical language, what this question boils down to is selecting the variable among those still available in the pool that has the largest statistically significant squared semipartial correlation with y_i.

- **Steps 4, 5, 6, ...** proceed in analogous fashion to previous steps, the number of steps ultimately determined by how many variables in the pool we have available for entrance into the model and that meet entrance (PIN) criteria.

We can summarize the general rule for how forward regression operates:

Forward regression, at each step of the selection procedure from step 1 through subsequent steps, chooses the predictor variable with the greatest squared semipartial correlation with the response variable for entry into the regression equation. The given predictor will be entered if it also satisfies entrance criteria (significance level, PIN) specified in advance by the researcher. Once a variable is included in the model, it cannot be removed regardless of whether its "contribution" to the model decreases given the inclusion of new predictors.

In a similar spirit as forward selection, **backward elimination** begins with all predictors entered into the model and then at each subsequent step removes the predictor with the smallest squared semi-partial correlation that meets removal criteria (Hays, 1994). This equates to removing that predictor that carries with it a nonstatistically significant contribution to the model in the presence of predictors already included in the model. The algorithm ceases when there are no further variables that meet statistical criteria for removal.

8.13.2   Stepwise Regression

Stepwise regression operates in a similar fashion as forward selection in that it selects predictors into the model that have the highest semipartial correlation with the response, or equivalently, yields the largest F value that is statistically significant. However, at each step of the procedure, **predictors already entered into the model are re-evaluated for their contribution in the presence of the newly entered predictor(s)**. Hence, in addition to having to specify a PIN value, the user also needs to specify a POUT ("probability out") value, which is the p-value criteria that designates removal of the given predictor. Hence, stepwise regression can be conceptualized as a mixture of sorts between the approaches of forward and backward elimination.

For example, we might set PIN at 0.05 and POUT at 0.10 for each step of the procedure. What this would mean is that a variable that meets PIN criteria is entered into the model and variables already in the model are simultaneously evaluated for POUT criteria, **the least significant of which is removed from the model**. It should be noted that POUT must be set at a value greater than PIN, otherwise, the stepwise routine might engage in a cyclical simultaneous acceptance and rejection of the same predictor.

You might think of forward selection as very "loyal" to predictors. Once you are in, you stay in. Stepwise regression is not very loyal. Once you are in, you are in until another predictor entered at a future step diminishes your stock value, then you are out. A further caution about stepwise regression is that significance levels typically do not represent true error rates by the very manner in which predictors are entered into the model. As noted by Draper and Smith (1998, pp. 342–343), though worthy of concern, this issue alone should not prevent you from using the procedure. That being said, stepwise methods, and to some extent variable-selection methods in general, though they are useful in many contexts have been found to **bias parameter estimates** and impose a degree of doubt on the inferential process. The degree of bias introduced can vary from model to model but can be as much as in the order of 1–2 standard errors (Izenman, 2008). Some are very critical of stepwise methods. For instance, Hastie et al. (2009) conclude:

> Other more traditional [software] packages base the selection on F-statistics, adding "significant" terms, and dropping "non-significant" terms. These are out of fashion, since they do not take proper account of the multiple testing issues ... the standard errors are not valid, since they do not account for the search process. (p. 60)

But what does this mean pragmatically? It means it is probably best to employ stepwise programs as a **guide to model selection**, and not place undo weight or importance on observed p-values. Now, that does not imply that if a p-value for a predictor is equal to 0.90, for instance, that the predictor is likely not all that valuable, but it does suggest that p-values coming out just above or below PIN or POUT criteria may still be worthwhile, and may be coming out smaller or larger simply because of instability in the standard error associated with the coefficient. But we have discussed this matter before, and that is, when it comes to scientific selection over and above statistical selection, a certain degree of common sense, flexibility, and judgment is necessary. **Do not get "lost" in p-values. Use them as a guide, and interpret them within a wider scientific context.**

8.13.3   Selection Procedures in R

There are a variety of packages in R that perform similar tasks as the stepwise procedure just reviewed. For instance, the `bestglm` package (McLeod and Xu, 2020) uses selection criteria such as AIC and BIC to inform the user on the best model. The `leaps` package in R (Lumley, 2020) can also be used for search procedures.

8.13.4   Which Regression Procedure Should Be Used? Concluding Comments and Recommendations Regarding Model-Building

It is important to realize that there is no golden rule regarding which model-building procedure one should use in any given context. When idealized statistical methods meet the harsh realities of the real world of applied empirical research, it becomes clear that the statistical criteria by which a regression model is chosen is only a small part of the input required to make intelligent decisions regarding selection criteria. **The final decision regarding any model, regardless of model selection procedure chosen, will be whether it is useful or theoretically meaningful to the researcher in the accomplishment of a wider endeavor.**

As an example to emphasize this concept of **utility**, imagine you were to enter five predictors into a model and run a stepwise regression on these candidate predictors. Suppose predictors x_1, x_3, and x_5 were selected into the final model. Is this then the "best" model? Yes, in the stepwise sense of how predictors were chosen to maximize model R^2, it is. However, the model may only be "best" in the statistical but not **substantive** sense. In the real empirical world of research, predictors are not abstract variables. They are **real** and correspond to actual things we are modeling. In this sense, the decision-making process regarding which model is best can hardly be fully relegated to a statistical algorithm of selection and removal criteria. Instead, you, the researcher, must have the final input into the model, presumably because you are the one most familiar with the variables you are modeling. You know them, or at least **should** know them, very well. **You must guide your own work.** Do not rely on software or mathematical optimization to completely guide your decision-making process. **Researchers run the machines. The machines should not be running researchers.**

To illustrate the point, suppose the model choosing predictors x_1, x_3, and x_5 accounted for 35% of the variance in the response. Suppose that a competing model with predictors x_1, x_3, and x_4 also accounted for 30%. Which model is better? This is a very difficult question to answer unless we first know something more about what these variables actually **are**, what they are supposed to represent, how they were measured, etc., and more importantly, **what the model we are building is actually for**. On a practical level, if x_5 were a very difficult and expensive piece of information to collect from subjects, but x_4 were a much easier (and cheaper) item to collect, then in this sense, x_4 may very well be the "better" predictor pragmatically when compared to x_5, especially if the reduction in variance explained is worth the cost of not having to collect x_5. **Maximizing utility is not the same as maximizing expected value.**

Following an example found in Denis (2020), if a graduate school committee basing student entrance selection criteria finds undergraduate GPA, number of research publications, and quality of letters of recommendation to be all strong predictors statistically, this may be only the beginning, not end, of finalizing the model, as measurement and psychometric issues may still remain inherent in these predictors. That is, if letters of recommendation are seen by committee members to be only vaguely associated with the student, and more a function of the letter writer (i.e., maybe the letter writer drafts "glowing" letters for everyone), then though it may turn out to be a strong predictor, it may not be prioritized in final model selection. Likewise, if committee members see second authorship on research publications as simply academic supervisors attaching their students' names to their research publications to impress deans, with the student providing only minimal if any contribution, then the statistical criteria may not matter much. Pragmatically speaking, letters of recommendation may be seen as a less important predictor when compared to undergraduate GPA. Always mind psychometric issues (e.g., such as validity and reliability) that may be lurking behind the scenes. Statistical "fit" criteria are not enough.

The major point is this:

Good models have to be evaluated in a wider context than statistical criteria alone, and results of statistical modeling should always be interpreted in the wider framework of decision-making for which there may be numerous inputs to the decision that lay outside of the results of statistical modeling. Statistics are meant to inform our decisions, not make them for us. Judgment must always trump protocol when it is most needed.

8.14 POWER ANALYSIS FOR MULTIPLE REGRESSION

We can use R to estimate power for multiple regression models just as easily as for simple regression models. Once again, we use `pwr.f2.test` in the `pwr` package (Champely, 2020). As an example, for a model with two predictors with a combined R^2 of 0.50 (f2 = 1, that is, $R^2/(1 - R^2)$), assuming a significance level of 0.05, on 20 subjects (i.e., $v = n - k - 1 = 20 - 2 - 1 = 17$), we estimate power to be:

```
> library(pwr)
> pwr.f2.test(u = 2, v = 17, f2 = 1, sig.level = .05, power = NULL)

     Multiple regression power calculation

              u = 2
              v = 17
             f2 = 1
      sig.level = 0.05
          power = 0.9630578
```

Hence, given these parameters, the probability of rejecting the null hypothesis given that it is false is equal to 0.96. Power for multiple regression can also quite easily be estimated using G*Power, although we do not demonstrate such estimation here.

8.15 INTRODUCTION TO STATISTICAL MEDIATION: CONCEPTS AND CONTROVERSY

We consider now a brief demonstration of statistical mediation along with a somewhat critical commentary regarding potential issues that may arise whenever mediation models are fit to data. Since path

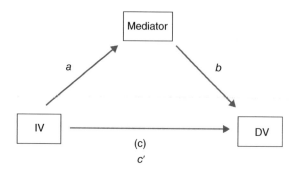

FIGURE 8.2   Classic single-variable mediation model.

analysis and structural equation models (discussed in Chapter 15) often feature mediational models, the warnings and caveats discussed here are equally applicable in the context of those models as well.

We discuss only the single "classic" mediation model in this chapter, usually attributed to Baron and Kenny (1986). Mediation is an extensive topic for which models can feature more than a single mediator. Our survey is only meant to provide a glimpse into what mediation models are about and comment on the nature of mediational hypotheses. For a thorough introduction to mediation, see MacKinnon (2008).

In the classic mediation model, an independent variable is hypothesized to predict a dependent variable **through** a mediator. A diagram for the single mediator model is given in Figure 8.2.

We define the following:

- IV—the independent variable (or predictor) hypothesized to predict the dependent variable DV (or response).
- MEDIATOR—the hypothesized mediator in which it is typically believed that the independent variable acts on the dependent variable through (in some sense) the hypothesized mediator; at minimum, it is hypothesized that the mediator "explains" the relationship between IV and DV.
- a is the estimated regression coefficient for the IV on the mediator.
- b is the estimated regression coefficient for the mediator on the DV.
- c is the estimated regression coefficient for the IV on the DV.
- c' is the estimated regression coefficient for IV predicting DV when the mediator is included in the model. According to the **mediational hypothesis**, when the mediator is introduced over and above the IV, the change from c to c' is evidence of **complete mediation** if $c' = 0$. If $c' \neq 0$, but still **decreases** substantially from c (i.e., $c' < c$), then **partial mediation** is said to exist.

In the single mediation model, we can distinguish different types of effects (MacKinnon, 2008):

- The path from IV to DV, that is, c, unadjusted by the mediator, is known as the **total effect model**.
- The **indirect effect** is estimated by the product of the coefficients a and b. The indirect effect can also be computed as $c - c'$. According to mediation theory, the product ab (or the difference $c - c'$) is also known as the **mediated effect**.

Suppose for some data we computed paths a, b, and c to be 0.80, 0.70, and 0.40, respectively. The estimate of the indirect effect is thus computed as $ab = 0.56$. Since it can be shown that path c can be **decomposed** into $c = c' + ab$, we can immediately know the extent to which c has changed as a result of including the mediator:

$$c = c' + ab$$
$$c' = c - ab$$
$$= 0.40 - 0.56$$
$$= -0.16$$

That is, our original path c has dropped to $c' = -0.16$ as a result of including the mediator.

To construct **confidence intervals** and **significance tests**, we must, of course, first obtain an estimate of the standard error for the statistic we are computing. Since we are interested in a confidence interval for $c - c'$ (or equivalently, ab), we need to construct the requisite standard errors. The estimated standard errors based on ab are known as **product of coefficient standard errors**, while the standard errors based on $c - c'$ are known as **difference in coefficients** standard errors (see MacKinnon, 2008, p. 51). Sobel (1982) proposed standard errors for the mediational effect, but according to mediation specialists, other methods are often preferable. See MacKinnon (2008) and Hayes and Scharkow (2013) for further details comparing a variety of methods for testing mediation. MacKinnon et al. (2002) is also a useful resource and should be consulted.

8.15.1 Statistical Versus True Mediation: Some Philosophical Pitfalls in the Interpretation of Mediation Analysis

Having demonstrated a simple example of mediation, some strong cautionary caveats are warranted. Applied researchers are sometimes wary of getting too "philosophical" about their research and claim to simply want to "look at the data." However, as a user of mediational analyses, or virtually any other statistical technique in research, you must be able to defend some rather obvious philosophical issues that present themselves when conducting and interpreting such analyses. Otherwise, your research will have no legs to stand on regardless of its quantitative sophistication. **In general, if you cannot convince the critical philosopher of your evidentiary claims, it usually suggests a methodological problem**. This is as true for the physical scientist as it is for the medical or social scientist. If you cannot make a strong methodological claim to demonstrate effectiveness of your design or intervention, then you have little in the way of science, but much in the way of hypothetical conjecture.

For instance, with the outbreak of COVID-19 in 2020, one "convinces" the critical consumer of the effectiveness of a treatment if it improves health or saves lives. One cannot simply believe the treatment works independent of philosophically demonstrating its effectiveness, which is usually done in these situations via **experiment**, which is about as philosophically rigorous a tool for demonstrating evidence that is known to humankind (though not without its own weaknesses and shortcomings—no method of investigation is perfect). Indeed, this is one of the great critiques of psychoanalytic theory, in that it is very difficult to subject to test. Does that mean it is "wrong?" Not at all. But being unable to subject its matter to test makes it less scientific, even if using it helps provide useful narratives for people's lives (which of course, is a good thing). A theory may be absolutely true in the end, but if it is not amenable to scientific test, then it is generally not part of what is regarded as "good science." In psychology, for instance, this is one reason why Skinner's behaviorism took flight in the early 1900s. Though it was fraught with its own difficulties, it was more scientifically "testable" than was psychoanalytic theory, for instance. In the case of COVID-19, the philosophical issues are a bit easier, in that a person's death is a strong indicator that the hypothesized treatment isn't having the desired effect. In this sense, theories in medicine and the like are much easier to "falsify," and though someone can entertain beliefs about a treatment, the data, in the end, will justify whether that belief is warranted. **Real, true science is very unforgiving**, and it is never the subject matter that deems it credible (there is good and bad science everywhere, from physics to psychology to medicine), it is the **method of inquiry** (e.g., the scientific method is quite credible) that is most important in the establishment of evidence. When you are doing good science, you are much more often "wrong" than you are "right," and if nature isn't constantly "slapping you in the face" with its data, guiding you on revising your theory and proving you incorrect, then you are probably doing something wrong. Titanic was believed "unsinkable." Nature had other ideas.

First and foremost, there is nothing "wrong" with performing statistical mediation. It is a rather elegant applied statistical procedure, and one that is quite popular in the social and behavioral sciences. **There is something seriously wrong however with drawing conclusions from a mediation analysis that are not warranted from the context in which you are applying the technique**. What does it mean to say one variable "*mediates*" the relationship between two others? Such language seems to imply some sort of **physical** process, or at minimum, an "action" of sorts. However, based on a test of statistical mediation alone, **no such processes can necessarily be inferred**. Whether there is a physical or even **directional** process of any kind must be concluded **separate** from the statistical test.[2]

As noted by MacKinnon (2008), the origins (or at minimum, an early documented example) of mediation analysis in psychology were with Woodworth in 1928 in which a stimulus–organism–response (S–O–R) model was hypothesized:

> Woodworth (1928) outlined a stimulus-organism-response (S-O-R) model for explaining how the organism mediates the relationship between the stimulus and response by postulating different mediating mechanisms operating in the organism. Mediating mechanisms are what determines how an organism responds to a stimulus. For example, a stimulus may trigger a memory mechanism that identifies the stimulus as a threat that leads to an avoidance response, or a stimulus may trigger an attraction process that leads to a physiological response such as pupil dilation and an approach response. (p. 2)

MacKinnon's characterization of an early mediation model is useful here, because it provides the **scientific context** in which mediation analysis arose. Note the key word "trigger" in the above quote. This denotes a physical, or at minimum, **directional event** of one variable onto another. In the S–O–R model, such an **assumption** of physical or directional causation was reasonable due to the nature of the scientific material under investigation. That is, mediation occurred in the S–O–R model because there **was real evidence that true mediation actually happens**. The process was not simply one of **statistical** mediation.

However, when a researcher takes variables x_i, y_i, and z_i, throws them into a regression program, and draws the conclusion that z_i mediates the relationship between x_i and y_i, what does this mean, exactly? In truth, we have no idea what it means until we know more about what x_i, y_i, and z_i actually **are** and whether or not a mediational hypothesis is actually plausible for these variables. Does a student's self-esteem mediate the relationship between grade-point average and probability of being accepted into graduate school? You can obtain all the statistical mediation evidence you like, but until you can actually convince someone that a **real mediational process** is occurring, you simply have no evidence for **true mediation**. What you have is evidence for **statistical mediation**, which usually is not that meaningful if you cannot use the statistical model to describe a real process. Too often in the research literature, mediational hypotheses are advanced simply because of evidence for even slight statistical mediation. Caution needs to be exercised in properly evaluating the research context in which the model is fit, so that one can then make an intelligent appraisal as to whether it is realistic or not to conclude a true mediational process. **A statistical test alone usually tells you little, in a direct sense, about any kind of physical or substantive process that may be present. Physical, concrete, or directional inferences have little to do with statistics and everything to do with principles of sound research design.**

Our brief discussion only begins to survey this relatively deep issue. The interpretative issue is much more salient in structural equation models where **causality** hypotheses are too often advanced without contextual evidence for any causal processes whatsoever. As will be discussed in Chapter 15, this difficulty, in part, can be traced back to the historical origins of path analysis.

---

[2] A similar discussion can be had about so-called **suppressor** variables. For details, see Howell (2002, pp. 557–558) and MacKinnon, Krull, and Lockwood (2000).

8.16 BRIEF SURVEY OF RIDGE AND LASSO REGRESSION: PENALIZED REGRESSION MODELS AND THE CONCEPT OF SHRINKAGE

We close this chapter with a brief survey of ridge and lasso regression, and refer the reader to additional follow-up resources should they wish to venture into these areas and consequently fit these models using software. These are relatively advanced procedures, even if at first glance they can appear deceptively straightforward. A deeper study of them however is required before one can appreciate or even fully understand details of their output.

So, what is so special about these methods? These techniques extend on the ordinary multiple regression model in that they implement what are known as **penalty functions** on regression parameters as a way of dealing with potential instability of OLS estimates due to factors such as **collinearity** among variables. **Ridge regression** (Hoerl and Kennard, 1970) sought to essentially get a handle on potential instability in OLS estimators through a slight modification when solving for least-squares coefficients, which recall were given by

$$\mathbf{b} = (\mathbf{X'X})^{-1}\mathbf{X'y}$$

Recall from our discussion of regression in Chapter 7 that due to the **Gauss–Markov theorem**, least-squares estimators are **unbiased** and thus generally preferable over other linear estimators (Hastie et al., 2009). Statistically speaking, unbiased estimators have much to recommend them. However, in some situations, it may be worth purposely introducing (or "allowing") bias into the estimator if, in some circumstances, it may lead to a lower mean squared error than if the OLS estimator were used. This brings us to the motivation for ridge regression. **Ridge regression sacrifices a degree of bias for a potential improved reduction in error**, known generally as the "bias-variance trade-off" (Hastie et al., 2009), which features the trading off of bias for a reduction in variance. In other words, we jettison the idea of an unbiased estimator, use a biased one, but with the hope that it helps us reduce variance more than would the unbiased estimator.

The technical theory behind ridge regression is that it adds a small constant quantity to the diagonal entries of $\mathbf{X'X}$ above in $(\mathbf{X'X})^{-1}$, which serves as a way to "track" the instability in estimation caused by such things as collinearity among variables (Izenman, 2008). Hence, in this respect, ridge regression may be useful as one possible way of dealing with collinearity or other causes of instability. When the value added to the diagonal of $\mathbf{X'X}$ is equal to 0, then ridge regression simply reduces to ordinary OLS estimation. Otherwise, ridge regression generates different estimators than OLS. This addition of a constant to ease instability of estimation translates to minimizing a slightly different quantity than in traditional least-squares regression. Recall that in OLS regression the sum of squared errors $\sum_{i=1}^{n} \varepsilon_i^2$ was minimized, which when "unpacked" yielded

$$\sum_{i=1}^{n} \varepsilon_i^2 = \sum_{i=1}^{n} \left(Y_i - \beta_0 - \sum_{j=1}^{k} \beta_k X_k \right)^2$$

In ridge regression, the following quantity is instead minimized,

$$\sum_{i=1}^{n} \varepsilon_i^2 = \sum_{i=1}^{n} \left(Y_i - \beta_0 - \sum_{j=1}^{k} \beta_k X_k \right)^2 + \lambda \sum_{j=1}^{k} \beta_k^2$$

where λ is known as a **tuning parameter** that serves to regulate the degree of penalty imposed on the regression estimates. For any value of $\lambda \geq 0$ then, the trailing term $\lambda \sum_{j=1}^{k} \beta_k^2$ will also be greater than zero and is known as the **shrinkage penalty** term. It stands as well that for increasing values of λ, the degree of penalty likewise will increase.

But why would we want to impose a penalty term on a regression? As mentioned, the primary motive for doing so is to deal with potential instability in OLS estimates, and it turns out that this translates to "shrinking" one or more coefficients **toward values of zero**. What this translates into substantively then is that ridge regression can be said to identify those predictor variables that are, in a sense, the most "important" in the regression, and it is hoped that less important variables will have their coefficients "shrink" toward zero. A key point to note, however, is that **ridge regression retains all predictors in the model**, and thus even though it shrinks coefficients toward zero, it is not considered a variable or model selection procedure. That is, all predictors remain in the model, since even the shrunken coefficients will still typically have coefficients greater than zero in magnitude.

A derivative, and thus related procedure to ridge regression is that of **lasso regression**, which, like ridge regression, employs the tuning parameter λ, however does so by minimizing a slightly different quantity than in ridge regression. Recall that whereas the trailing term in ridge was $\lambda \sum_{j=1}^{k} \beta_k^2$, that is, it involved the sum of **squared** regression coefficients, in lasso regression, the trailing term is instead, $\lambda \sum_{j=1}^{k} |\beta_k|$. That is, lasso regression features the **absolute value** of coefficients in the sum instead of squared values as in ridge. Hence, what lasso regression is minimizing then, is

$$\sum_{i=1}^{n} \varepsilon_i^2 = \sum_{i-1}^{n} \left(Y_i - \beta_0 - \sum_{j=1}^{k} \beta_k X_k \right)^2 + \lambda \sum_{j=1}^{k} |\beta_k|$$

As was the case for ridge regression, when λ, the tuning parameter, is equal to 0, the term minimized by lasso regression reduces to that minimized by OLS, since $0 \sum_{j=1}^{k} |\beta_k|$ will always equal 0 regardless of the sum of absolute values for β_k. And, as was true for ridge, for values of $\lambda \geq 0$, the influence of the penalty term likewise increases. In mathematical parlance, ridge regression and lasso regression differ by their norms, where $|\beta_k|$ is known as the **L1-norm**, and β_k^2, the **L2-norm**. But as mentioned, each is modulated by the size of the tuning parameter, λ.

Why might one prefer lasso regression over ridge? Recall whereas ridge will not perform variable selection, lasso regression, on the other hand, **does drive coefficients all the way to zero**, and hence can be considered in its own right to be a variable-selection procedure. That is, when performing lasso regression, one may decide to retain predictors with coefficients greater than zero, and drop those equal to zero. Ridge regression, on the other hand, will retain all predictors, and hence is considered less useful for variable selection. However, this does not imply that a user of ridge cannot drop predictors manually that are headed toward zero, only that because it does not drive them exactly to zero, solutions to ridge may be somewhat more substantively difficult to interpret relative to the lasso.

This is as far as we take our brief survey of ridge and lasso regression in closing out this chapter. Despite the simplicity of our short overview, as mentioned, these techniques are actually quite complex and have quite intricate relationships with **principal components analysis**. To understand them in sufficient detail requires more study than we have space for, and their software implementation is likewise much more detailed than running classical OLS regression models, for instance. Performing ridge and lasso regression often involve simultaneously running what are known as **grid searches** to help select optimal values of the relevant tuning parameter that minimizes mean squared error, as well as splitting up data into a **train** and **test set** in helping determine the value of this important parameter through a variety of techniques of **cross-validation**. For full, detailed, and excellent treatments of ridge and lasso regression, including implementation in R software, see James et al. (2013), and for an even deeper and more technical treatment that cuts no corners, the reader is strongly encouraged to consult Hastie et al. (2009) or Izenman (2008). In the spirit of history, for a good introduction to ridge regression as originally conceived, Hoerl and Kennard (1970) should be consulted.

8.17 CHAPTER SUMMARY AND HIGHLIGHTS

- Whereas **linear regression** dealt with regressing a single response variable on a single predictor variable, **multiple linear regression** models are useful for regressing a single response variable on two or more predictors variables.
- The **multiple linear regression** model is given by $y_i = \alpha + \beta_1 x_1 + \beta_2 x_2 + \ldots + \beta_k x_k + \varepsilon_i$ where k is the number of predictors, and β_1 through β_k are the population partial regression parameters, usually estimated by ordinary least-squares.
- In developing the multiple regression model, it is useful, pedagogically, to first consider coefficients of **partial** and **semipartial correlation**.
- The **partial correlation** between variables x_1 and x_2 controlling for z_i is found by first regressing x_1 on z_i to obtain a column of residuals, then regressing x_2 on z_i to obtain a second column of residuals, then obtaining the Pearson correlation coefficient between these two columns of residuals.
- The **semipartial correlation** between x_1 and x_2, where only x_2 is adjusted for z_i, is computed by regressing x_2 on z_i to obtain a column of residuals, then computing the Pearson correlation coefficient between these residuals and unadjusted values of x_1.
- The **multiple regression model** can also be given in matrices, allowing one to see the components of the model more clearly and also aiding in generalizing the model to more complex cases (e.g., multivariate multiple regression).
- **Unstandardized** or **standardized partial regression coefficients** cannot be interpreted as coefficients are in a simple linear regression. That is, partial regression coefficients are not the same as the slope in a simple linear regression model (i.e., based on bivariate data).
- The idea of **statistical control** has nothing to do with **experimental control**. When variables are "controlled" in a multiple regression, it simply implies a partialling of variability.
- The **variance inflation factor** (VIF) and **tolerance** are common measures used to evaluate **multicollinearity** among predictors in a multiple regression model. Tolerance is the reciprocal of VIF. Elevated levels of VIF (or, equally, low values of tolerance) may indicate a problem with collinearity.
- R^2 can be written as a function of **partial correlations**. It can also be written as a function of **semipartial correlations**. This fact forms the analytical basis for model-building strategies such as forward and stepwise regression.
- **Hierarchical regression**, in which predictors are entered in a preconceived order by the researcher, is not the same as **stepwise regression**.
- **Sample size** for multiple regression can be estimated relatively easily in R and is equally as easy to compute using G*Power.

- The distinction between **statistical mediation** versus **true mediation** is an important one. Regardless of whether one finds statistical support for mediation, true mediation can ordinarily only be justified through resort to the research context or paradigm.
- **Ridge** and **lasso regression** are relatively advanced extensions of the original multiple regression model. These models invoke a **penalty term** in minimizing the sum of squared errors, and seek to shrink one or more coefficients toward (**ridge**), or to zero (**lasso**), as a way of reducing potential instability of OLS estimates, as well as aiding in predictor selection.

REVIEW EXERCISES

8.1. Explain why **multicollinearity** in regression can be considered both a statistical and a substantive concern.

8.2. Discuss how the **variance inflation factor** (VIF) can be instrumental in influencing the size of the standard error for a partial regression coefficient, β_i.

8.3. What is **tolerance**, and how is it defined? All else equal, would you prefer to see a high tolerance value or a low tolerance value for a given predictor? Why?

8.4. Justify why **tolerance** for any given predictor must range between 0.0 and 1.0.

8.5. Discuss what it means, in a very general sense, to solve a system of linear equations such as the following:

$$y_1 = \alpha + \beta_1 x_{11} + \beta_2 x_{12} + \ldots + \beta_k x_{1k} + \varepsilon_1$$
$$y_2 = \alpha + \beta_1 x_{21} + \beta_2 x_{22} + \ldots + \beta_k x_{2k} + \varepsilon_2$$
$$.$$
$$.$$
$$y_n = \alpha + \beta_1 x_{n1} + \beta_2 x_{n2} + \ldots + \beta_k x_{nk} + \varepsilon_n$$

8.6. Discuss the procedure of **hierarchical regression**, and come up with one example in which a researcher may be especially interested in performing this type of regression.

8.7. Verify that $R^2_{y\cdot 12}$ can be expressed as a function of **partial correlations**.

8.8. Verify that $R^2_{y\cdot 12}$ can be expressed as a function of **semipartial correlations**.

8.9. Discuss the difference between $r^2_{y2\cdot 1}$ and $r^2_{y(2\cdot 1)}$ and why this distinction is important. What role does each play in defining $R^2_{y\cdot 12}$?

8.10. Explain, in detail, the following equation, and comment on why it is important:

$$R^2_{y\cdot 1234} = 1 - \left[\left(1 - r^2_{y1}\right)\left(1 - r^2_{y2\cdot 1}\right)\left(1 - r^2_{y3\cdot 12}\right)\left(1 - r^2_{y4\cdot 123}\right) \right]$$

8.11. Describe, in detail, the "logic" of **forward regression**. Imagine you were explaining the procedure to a colleague.

8.12. For a **forward regression**, what significance level would you suggest setting PIN at? What kinds of things should this decision depend on?

8.13. Why does the **semipartial correlation** figure so prominently in the discussion of forward, step-wise, and backward elimination regressions?

8.14. How would you recommend a researcher set **PIN** and **POUT** in the typical **stepwise regression**? Why would you recommend this?

8.15. Conceive of a research example in which **stepwise regression** would be the preferred method of regression over **simultaneous**, **forward**, or **backward elimination**.

8.16. Consider the following data on variables where x and y are continuous and z binary.

```
> x <- c(0, 5, 8, 3, 9, 10, 15, 4, 8, 2)
> y <- c(9, 7, 4, 8, 2, 6, 5, 4, 8, 9)
> z <- c(0, 0, 0, 0, 0, 1, 1, 1, 1, 1)
```

Perform a **multiple linear regression** in which y is the response variable and x, z are predictors. Evaluate both effects for x and z. What would you conclude for the effect for x? Why? Next, run a **simple linear regression** with only x in the model. Did your conclusion for x change from what it was when z was included in the model? Why or why not? Briefly explain.

Further Discussion and Activities

8.17. Applied statistical methods are often taught with little regard to any of the historical or political and social influences that may have been instrumental in promoting the techniques. One prime example is the history of regression. Gaining an appreciation of how and why a statistical method came into prominence is very useful since it helps one contextualize statistical methods in a wider social framework, instead of seeing the method as a mere computational algorithm. Read Denis and Docherty (2007). Briefly summarize how the advent and rise of multiple regression, though traditionally associated with the likes of Karl Pearson and George Udny Yule, can also be said to be a product of the sociopolitical debate between Charles Booth and Charles Stewart Loch.

9

INTERACTIONS IN MULTIPLE LINEAR REGRESSION

In this chapter, we briefly survey the analysis of interactions in the context of multiple regression. Analogous, but not identical to the ANOVA context, situations arise in regression where a researcher hypothesizes that a given predictor is useful in predicting a response, but that its predictive power is not **constant** across the range of a second predictor. The concept of an interaction in multiple regression parallels that of an interaction in ANOVA, and although different in computation, both analyses essentially accomplish something similar. In ANOVA, to break apart an interaction, we study its **simple effects**. In regression, we break apart interactions by **simple slopes**. Recall that a simple main effect in ANOVA is the mean difference on one factor at a particular level of a second factor. A simple slope in regression is defined as the slope of y_i on x_i at a particular value of a second predictor z_i. So-called "moderation regression" has become quite popular in the social and natural sciences. For thorough accounts that extend on the topics discussed in this chapter, including estimation of simple slopes, see Aiken and West (1991) and Cohen et al. (2003).

As an example of where an interaction in a regression model may be relevant and of interest, suppose a research psychologist would like to predict **treatment success** (y_i) (measured on some continuously scaled questionnaire purported to evaluate overall outcome of treatment) based on length of therapy (x_i) and also hypothesizes that this regression will be contingent on a client's age. Perhaps the researcher believes that treatment success will be better predicted by length of treatment for clients who are **young** compared to **older** clients. Consider the hypothetical plots in Figure 9.1.

We note in Figure 9.1 that the slope for **age = young** is somewhat steeper than the slope for **age = old**. It should be noted as well that for our demonstration, only two age "groups" have been selected. We of course could have conceived age as having an infinite number of potential values and therefore plotted the **regression lines at each value**. Perhaps the actual relationship would be as in Figure 9.2, where the slope of treatment success on length of treatment decreases as age increases.

Applied Univariate, Bivariate, and Multivariate Statistics: Understanding Statistics for Social and Natural Scientists, With Applications in SPSS and R, Second Edition. Daniel J. Denis.
© 2021 John Wiley & Sons, Inc. Published 2021 by John Wiley & Sons, Inc.
Companion Website: www.wiley.com/go/denis/appliedstatistics2e

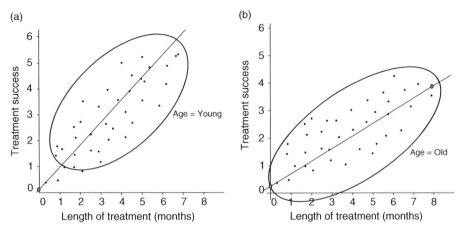

FIGURE 9.1 Hypothesized slope of treatment success on length of treatment for young (a). Hypothesized slope of treatment success on length of treatment for old (b).

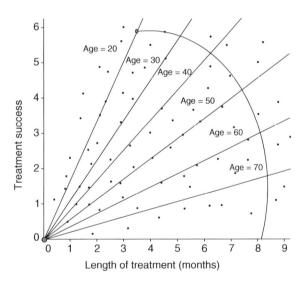

FIGURE 9.2 Length of treatment and age interact in predicting treatment success. As age increases from 20 to 70, the slope of treatment success on length of treatment decreases.

9.1 THE ADDITIVE REGRESSION MODEL WITH TWO PREDICTORS

A regression model with two predictors x_i and z_i is given by:

$$y_i = \alpha + \beta_1 x_i + \beta_2 z_i + \varepsilon_i \tag{9.1}$$

Recall from our discussion of ANOVA that the model in (9.1) is considered an **additive** model because no terms (i.e., x_i, z_i) are crossed (multiplied). When we specify an **interaction model**, we cross terms x_i and z_i to produce a **product term** $x_i z_i$. We can specify an interaction term for $x_i z_i$ (i.e., z_i multiplied by x_i) by adding this term to the model:

$$y_i = \alpha + \beta_1 x_i + \beta_2 z_i + \beta_3 x_i z_i + \varepsilon_i \qquad (9.2)$$

The term $x_i z_i$ is the product term in (9.2) that represents the crossing of x_i with z_i. This is the **interaction term**.

9.2 WHY THE INTERACTION IS THE PRODUCT TERM $x_i z_i$: DRAWING AN ANALOGY TO FACTORIAL ANOVA

You may ask why we are multiplying two variables to get the interaction term. That is, why is the interaction term in (9.2) defined as a product? To understand why this is so, I ask you to draw on your knowledge of factorial analysis of variance for a useful, even if inexact, analogy. Recall that in ANOVA, we defined a **cell effect** as the mean for the given cell minus the grand mean of all the data (or, of the mean of all the means for a balanced design). Recall the typical factorial ANOVA table of Chapter 4, reproduced in Table 9.1.

Recall that the cell means were generated by "crossing" factor 1 with factor 2. That is, $\bar{y}_{jk}$ for each cell can be defined as the **intersection** of the given factor level for each variable (e.g., level 1 of factor 1 with level 1 of factor 2). In Table 9.1 is featured only a 2×3 design. Imagine now if we increased factor levels on each variable to a much larger number, say 30 on factor 1 and 20 on factor 2. When generating cell effects, we would thus have $30 \times 20 = 600$ "cells" in our design. Imagine now we increase the number of levels on factor 1 to 300 and on factor 2 to 100. We would now have 30,000 "cells" in our design. But, each cell would still contain **unique information** (e.g., a mean and variance) in the given crossing. If you continue expanding the number of levels for each factor, you will eventually arrive at a state of approximate **continuity** for each factor. That is, **each factor will have an infinite number of "levels."** And even with this infinite number, we are still theoretically interested in what is contained within each combination. We keep putting "levels" in quotes, because when we are working with continuous variables, we seldom think of values of the variable as "levels" at all. But in terms of drawing the analogy between factorial ANOVA and a product term in regression, it is helpful to temporarily equate the two concepts.

The point of this discussion is to emphasize that crossing (or "multiplying") predictors in multiple regression accomplishes a conceptually similar result as crossing factors in ANOVA. We are interested in the **joint relationship** or **intersection** of where the values of the variable (i.e., factor or predictor) meet up. This "cell," which is obvious in ANOVA, is much less so in regression because the "cell" contains (or is "able" to contain, the exact nature of the "cells" will depend on the data) but only one score if both predictors are continuous. Hence, this is a useful conceptual analogy as to why it makes good sense to obtain a product term to represent the interaction $x_i z_i$ in regression. Figuratively at least, we have been doing it all along in our ANOVA models.

TABLE 9.1 Cell Means Layout for 2×3 Factorial Analysis of Variance

	Factor 2		
Factor 1	Level 1	Level 2	Level 3
Level 1	$\bar{y}_{jk}$	$\bar{y}_{jk}$	$\bar{y}_{jk}$
Level 2	$\bar{y}_{jk}$	$\bar{y}_{jk}$	$\bar{y}_{jk}$

9.3 A MOTIVATING EXAMPLE OF INTERACTION IN REGRESSION: CROSSING A CONTINUOUS PREDICTOR WITH A DICHOTOMOUS PREDICTOR

Consider some hypothetical data where the response variable is **final grade** in a statistics course for a given student. The predictor variables of interest are **study time** devoted to that course, measured in hours, and whether or not a student was seated at the front (class = 1) or rear (class = 0) of the **class**. We are interested in learning whether study time and class seating over the duration of the short course is predictive of final grade.

Our data are as follows:

```
> grades
   final study class
1     85   1.0     0
2     74   1.2     0
3     62   1.8     0
4     78   1.3     0
5     61   1.5     0
6     96   2.1     1
7     74   1.5     1
8     64   1.8     1
9     42   1.1     1
10    69   1.3     1
```

Before conducting any inferential tests, we can attempt to estimate via graphical methods whether study and class interact in predicting final. We plot the data and reveal group membership by class, where circles and squares represent the different class seating (squares represent class = 0, circles represent class = 1):

```
> plot(study, final, pch = as.integer(class))
```

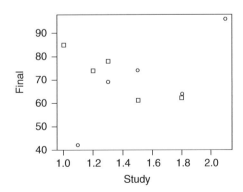

We fit the regression line to the entire data:

```
> model <- lm(final ~ study + class)
> abline(model)
```

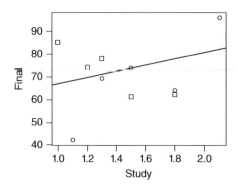

We can see that the **common regression line** suggests a **positive slope**. However, what if we draw regression lines separately for each group? Would they each conform to the **slope** of the common regression line? That is, would the slopes of each group be the same? Should they not conform in this way, this may suggest an **interaction** between study and class in the sample. When we run the regression including this product term **study × class**, we get:

```
> model.int <- lm(final ~ study*class)
> summary(model.int)

Coefficients:
            Estimate Std. Error t value Pr(>|t|)
(Intercept)   113.31      22.09   5.131  0.00216 **
study         -30.38      15.92  -1.908  0.10503
class        -107.25      29.49  -3.637  0.01088 *
study:class    70.72      20.07   3.524  0.01246 *
---
Signif. codes:  0 '***' 0.001 '**' 0.01 '*' 0.05 '.' 0.1 ' ' 1

Residual standard error: 9.711 on 6 degrees of freedom
Multiple R-squared:  0.7114,    Adjusted R-squared:  0.567
F-statistic: 4.929 on 3 and 6 DF,  p-value: 0.04653
```

Notice that the interaction term `study:class` is statistically significant ($p = 0.01246$). It is suggesting that `study` is predictive of `final`, but this prediction may depend on whether you are seated at the front or rear of the class.

We generate an **interaction plot** using R's scatterplot function to better visualize the interaction:

```
> library(car)
> scatterplot(final ~ study | class, data = grades)
```

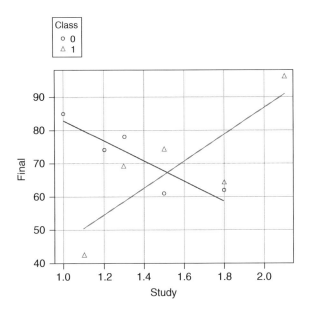

Clearly, the plot reveals evidence of a study × class interaction, where class = 0 is indicated by circles and class = 1 is indicated by triangles. The simple slopes suggest that an increase in study time is related to an increase in final grade for those seated at the front of the class. If you are seated at the back of the class, however, the plot reveals a negative relationship between study time and grade (i.e., the circles in the plot). Perhaps those students seated at the back of the class are misunderstanding or not recording information correctly off the board, and hence as they increase their study of "wrong" information, they do progressively worse on their evaluations.

We can easily duplicate this analysis in SPSS. Our corresponding data file is the following:

	final	study	class	study_class
1	85.00	1.00	0.00	0.00
2	74.00	1.20	0.00	0.00
3	62.00	1.80	0.00	0.00
4	78.00	1.30	0.00	0.00
5	61.00	1.50	0.00	0.00
6	96.00	2.10	1.00	2.10
7	74.00	1.50	1.00	1.50
8	64.00	1.80	1.00	1.80
9	42.00	1.10	1.00	1.10
10	69.00	1.30	1.00	1.30

The `study_class` column corresponds to the product term, which we created by multiplying `class` by `study` coding the following:

```
COMPUTE study_class=study*class.
EXECUTE.
```

We now run the regression by entering all terms:

```
REGRESSION
  /MISSING LISTWISE
  /STATISTICS COEFF OUTS R ANOVA
  /CRITERIA=PIN(.05) POUT(.10)
  /NOORIGIN
  /DEPENDENT final
  /METHOD=ENTER study class study_class.
```

Coefficients[a]

Model		Unstandardized Coefficients		Standardized Coefficients	t	Sig.
		B	Std. Error	Beta		
1	(Constant)	113.312	22.086		5.131	0.002
	study	−30.376	15.923	−0.721	−1.908	0.105
	class	−107.255	29.492	−3.830	−3.637	0.011
	study_class	70.724	20.069	4.139	3.524	0.012

[a]Dependent variable: final.

We see above the SPSS output mirrors that generated in R earlier. Recall as well that SPSS will report the model fit statistics separate from the coefficients, whereas R reported them all in one output. The corresponding model fit and ANOVA table are the following:

Model Summary

Model	R	R Square	Adjusted R Square	Std. Error of the Estimate
1	0.843[a]	0.711	0.567	9.71143

[a]Predictors: (Constant), study_class, study, class.

ANOVA[a]

Model		Sum of Squares	df	Mean Square	F	Sig.
1	Regression	1394.629	3	464.876	4.929	0.047[b]
	Residual	565.871	6	94.312		
	Total	1960.500	9			

[a]Dependent variable: final.

[b]Predictors: (Constant), study_class, study, class.

9.4   ANALYSIS OF COVARIANCE

The models we have just surveyed in which a dichotomous or polytomous predictor is crossed with a continuous predictor may be likened to **analysis of covariance** models. Though the ANCOVA model is parameterized differently than the dummy-regression models we have just surveyed, the model will nonetheless provide us with the same fit (Fox, 2016). However, in the ANCOVA model, aside from evaluating what is generally known as the **homogeneity of regression** assumption that assumes an absence of an interaction effect between the continuous predictor and the dichotomous or polytomous predictor, **the interaction effect is typically not modeled**. Just as ANOVA can be conceptualized as a subcategory of the wider regression model, so too can ANCOVA be considered a special case of regression analysis in which one or more continuous predictors are partialled out of mean differences of a dichotomous or polytomous independent variable on some continuous response variable. One can also compute **adjusted means** that take into consideration the removal of the covariate from the relationship of interest (see Hays, 1994, p. 823).

We can demonstrate a simple ANCOVA on the IQ data using SPSS, where **verbal** is the dependent variable, **group** is the independent variable of interest, and **quant** is the designated covariate. We show only the first 10 cases below (group actually has three levels to it, though only the first level "0" appears below due to showing only the first 10 cases):

	verbal	quant	analytic	group
1	56.00	56.00	59.00	0.00
2	59.00	42.00	54.00	0.00
3	62.00	43.00	52.00	0.00
4	74.00	35.00	46.00	0.00
5	63.00	39.00	49.00	0.00
6	68.00	50.00	36.00	0.00
7	54.00	54.00	29.00	0.00
8	56.00	52.00	57.00	0.00
9	51.00	46.00	65.00	0.00
10	49.00	39.00	61.00	0.00

```
UNIANOVA verbal BY group WITH quant
  /METHOD=SSTYPE(3)
  /INTERCEPT=INCLUDE
  /CRITERIA=ALPHA(0.05)
  /DESIGN=quant group.
```

In the above, `UNIANOVA verbal BY group WITH quant` designates verbal as the dependent variable, group as the independent variable, and quant as the covariate. The output of the ANCOVA now follows:

Tests of Between-Subjects Effects

Dependent Variable: verbal

Source	Type III Sum of Squares	df	Mean Square	F	Sig.
Corrected model	3683.268[a]	3	1227.756	26.641	0.000
Intercept	1710.893	1	1710.893	37.125	0.000
quant	10.402	1	10.402	0.226	0.639
group	495.963	2	247.981	5.381	0.011
Error	1198.198	26	46.085		
Total	164,168.000	30			
Corrected total	4881.467	29			

[a]R squared = 0.755 (adjusted R squared = 0.726).

We can see from the output that the independent variable of group is statistically significant ($p = 0.011$). Recall that it is hoped by including the covariate, we experience a **reduction in MS error**, thus making the test more sensitive (Montgomery, 2005).

For our data, however, including the covariate actually had the effect of increasing the p-value (if you try the ANOVA without the covariate, you will see that the p-value for group drops). The reader should be aware that sometimes this can occur in ANCOVA, especially in cases of low statistical power. Other issues when conducting ANOVA include the **type of sums of squares** that are used to estimate effects. Though we do not detail the issue here, **type I SS** are sometimes preferred over SPSS's traditional **type III**. For details, see Warner (2013) who provides a very useful discussion of using type I versus type III sums of squares (we have used type III as default above). While some authors recommend type I SS, others (such as Tabachnick and Fidell (2007)) use the more popular type III SS. A discussion of the differences between sums of squares for ANOVA is beyond the scope of this book, and in most cases, type III sums of squares suffice.

While ANCOVA makes the typical assumptions of classic ANOVA, such as normality and homogeneity of variances, as mentioned, contrary to seeking out interactions in regression as we have done in this chapter, we actually wish for the **absence of an interaction** between the independent variable and covariate in ANCOVA. What this translates to is that at each level of the independent variable, the regression of the dependent variable on to the covariate should be **linear and approximately the same in each level**. This assumption is generally referred to as **homogeneity of regression slopes** and can be easily tested in SPSS by specifying a custom model where the interaction term is a designated effect in the model:

```
UNIANOVA verbal BY group WITH quant
  /METHOD=SSTYPE(3)
  /INTERCEPT=INCLUDE
  /CRITERIA=ALPHA(0.05)
  /DESIGN=group quant group*quant.
```

Notice above that the /DESIGN statement now includes the product term group*quant. When we run this model, we obtain:

Tests of Between-Subjects Effects

Dependent Variable: verbal

Source	Type III Sum of Squares	df	Mean Square	F	Sig.
Corrected model	3886.994[a]	5	777.399	18.761	0.000
Intercept	1057.475	1	1057.475	25.520	0.000
group	73.396	2	36.698	0.886	0.426
quant	14.975	1	14.975	0.361	0.553
group * quant	203.726	2	101.863	2.458	0.107
Error	994.473	24	41.436		
Total	164,168.000	30			
Corrected total	4881.467	29			

[a]R squared = 0.796 (adjusted R squared = 0.754).

As we can see, the group*quant interaction effect is not statistically significant ($p = 0.107$), and, potential power issues aside (i.e., if you plotted the data you would likely discover an interaction in the sample), we will assume for demonstration purposes the absence of an interaction effect. In R, an ANCOVA can be conducted a few different ways, one common way is to simply use the aov function as in aov(dv ~ iv + cov), where "cov" is the continuous covariate.

9.4.1 Is ANCOVA "Controlling" for Anything?

The language often used in describing ANCOVA models is that of "controlling for" the continuous covariate so that it does not unduly influence the relationship between the predictor and response we are interested in. However, as noted in our discussion of multiple regression models in Chapter 8, the phrase "controlling for" means nothing more than a partialling out of variance unless one is actually implementing controls through experimental design. One may go as far as to say ANCOVA tells us what the group means (or cells, in the case of factorial designs) **might have been** (i.e., through the computation of **adjusted means**) had we been able to control for the covariate, but unless we **did** control for the covariate **for real** (such as in an experimental design), such statements should still be interpreted in the realm of **statistical variation** rather than having anything to do with true experimental control. **ANCOVA, or any other statistical method, will never tell you what would have been had you conducted a real experiment with your correlational data**. At most, they will suggest to you what **might** have been. Some authors also maintain that ANCOVA should not be associated with even pseudo-control at all (Miller and Chapman, 2001), and should only be employed for increasing power for detecting the effects in the model of interest. For more information on ANCOVA and its relation to regression, see Fox (1997, pp. 192–195). Hays (1994) also gives a good account of ANCOVA as an extension of ANOVA rather than as a side note to regression models. Howell (2002, pp. 603–654) provides a good overview of how analysis of covariance can be conceptualized under the wider general linear model.

9.5   CONTINUOUS MODERATORS

When the moderator can at least theoretically take on an **infinite** number of values, or at least practically can take on enough values that we may deem it to have enough categories to be able to consider it continuous, then it is called a **continuous moderator**. The concept of a continuous moderator parallels that of both dichotomous and polytomous moderators.

An interaction between a continuous x_i and a continuous z_i would mean that the slope of y_i on x_i differs depending on the "level" chosen for variable z_i. Analogously, we could also say that the slope of y_i on z_i differs depending on the "level" chosen for variable x_i. Of course, for a truly continuous variable, it has no real "levels" analogous to a dichotomous or polytomous moderator. Recall that true continuity (see Appendix) implies that any values are possible on the infinitely dense real line. When we step down from the ideal of theory and into the world of research, however, we quickly come to realize the limitation that for us to actually work with a variable, we must somehow reduce it down to being a **categorical** one of sorts, even if we consider it to have infinitely many of these categories. Recall that continuity does not truly exist on real variables that we model, even if we do proceed as though it does in many of our analytical approaches. For examples and demonstrations of interactions with continuous moderators and the computation of simple slopes, see Aiken and West (1991).

9.6   SUMMING UP THE IDEA OF INTERACTIONS IN REGRESSION

Interactions in regression may at first glance appear rather daunting. However, if you use your knowledge of factorial analysis of variance as a springboard to understanding them, you will quickly see the similarities between the two interaction models. Let us again review the situation of y_i on x_i. In an ANOVA-type model, x_i is categorical. In a regression-type model, x_i is (typically) continuous. In the ANOVA model, we are interested in mean differences on y_i across categories of x_i. In the regression model, we are interested not in mean differences on y_i, but rather in the **slope** of y_i on x_i. Now, suppose we introduce the variable z_i as a moderator in each case. Here is a summary of how the interpretation differs based on the model:

Analysis of Variance—If z_i is a moderator of y_i on x_i, then this implies that mean differences of y_i for categories of x_i differ depending on the level chosen for z_i.

Regression—If z_i is a moderator of y_i on x_i, then this implies that the slope of y_i on x_i differs depending on the "level" chosen for z_i.

Depending on whether z_i is dichotomous, polytomous, or continuous, interpretation of the variable is slightly different, but the essential role of z_i as a moderator in each case is the same.

9.7   DO MODERATORS REALLY "MODERATE" ANYTHING?

9.7.1   Some Philosophical Considerations

Having reviewed some of the theory of interactions in regression, we have called these "third variables" by the name of "moderators" only because that is how they are commonly referred to in the literature. However, I personally have never liked the name moderator for the reason that I believe it unduly implies a **physical** action of some sort that has nothing to do whatsoever with variance partitioning, which, after all, is all we are accomplishing in **any** statistical model. Just as a caveat had to be issued when discussing mediation, one likewise needs to be advanced in a discussion of moderation.

To say that z_i **moderates** the relationship between y_i and x_i seems to imply a physical model such that z_i is somehow "regulating" the y_i on x_i relation, analogous to how a moderator might oversee a negotiation in a business deal. And if you have substantive variables such that this is actually the case, then yes, z_i can and should be called a moderator. **However, simply because one is calling interactions in regression by the name of moderated regression does not give one philosophical license to ascribe any powers to z_i that it did not have before you conducted the regression.** To do so would imply more of a **functional** role, as opposed to a **statistical** role, where "function" in the sense used here implies a physical act or contribution (such as the function of a gas engine in a car to move a vehicle forward). For instance, consider the following conclusion, the methodological kind (not necessarily substantive, I am pulling the example out of a hat) of which is often proclaimed as evidence in the social science literature:

> **Our research and statistical analyses suggest that self-esteem moderates the relationship between stress and propensity to engage in violence.**

What does such a conclusion mean, really? Statistically, we know what it suggests, and so long as we associate the idea of moderation with the discovery of statistical interactions, then all is well and good. However, if we get a bit too "in love with our theory," we may begin to actually believe that self-esteem **impacts** the relationship between stress and violence. Does it? What evidence do we have to say that self-esteem impacts anything? We usually have scant evidence of this on an experimental level. But in speaking of it as a moderator, it seems to imply a directional **causal** force of some kind. Statistically, however, all we have discovered is an interaction, a **statistical** relationship.

The key point to remember is to never ascribe powers to empirical variables unless you have generated their outcomes in such a way that such powers can then be substantiated by the statistical analysis. Statistical analysis can hardly ever be considered justification alone for the existence of a phenomenon, regardless of the field to which it is applied.

9.8 INTERPRETING MODEL COEFFICIENTS IN THE CONTEXT OF MODERATORS

There is a big difference between interpreting regression coefficients in a model that contains an interaction term versus a model that contains only "main effect" terms, the so-called additive model. Coefficients in a main-effects-only model estimate something different than coefficients in a model that includes a product term.

When we test the main-effects-only model, we interpret β_1 as reflecting the expected change in y_i given a one unit change in x_i across z_i, or, equally, over all values of z_i. That is, the interpretation of β_1 assumes we are **generalizing** or **averaging** over values of z_i. Likewise, we interpret β_2 as the expected change in y_i given a one unit change in z_i across x_i, or, again, generalizing or averaging over all values of x_i. This interpretation of "averaging over" only holds true when we do not have an interaction term in the model.

When we include the product term, $x_i z_i$, β_1 now reflects the expected change in y_i given a one-unit change in x_i when $z_i = 0$, and β_2 reflects the expected change in y_i given a one unit change in z_i when $x_i = 0$. In essence, when we interpret the "main effects" in a nonadditive multiple regression model (i.e., one with a product term), we are actually interpreting **simple slopes** for values of 0 on the moderating variable. As an example, if β_1 was equal to 2.0 in the nonadditive model, we would say that the expected (or average) change in y_i for a one unit change in x_i is 2.0, when $z_i = 0$.

To summarize:

- When estimating a model that contains only main effect terms, regression coefficients estimate "general" relationships averaging across the levels of the other predictor.

- When estimating a model that contains a product term, regression coefficients for "main effects" estimate **conditional relationships** focused on a specific value of the other predictor. That value of the other predictor (moderator) is equal to 0.
- It is very important to not interpret main effects in an interactive model as you would in a purely additive one.

9.9 MEAN-CENTERING PREDICTORS: IMPROVING THE INTERPRETABILITY OF SIMPLE SLOPES

We have said that when we have a product term in a multiple regression, the partial regression coefficients are interpreted differently than they would be if we did not have the product term included in the model. Referring once again to the product term model,

$$y_i = \alpha + \beta_1 x_i + \beta_2 z_i + \beta_3 x_i z_i + \varepsilon_i$$

the partial regression coefficient β_1 is interpreted as the expected change in y_i for a one unit change in x_i, when $z_i = 0$. For β_2, we interpret as the expected change in y_i for a one unit change in z_i when $x_i = 0$.

Theoretically, the above is sound enough and is algebraically correct. However, one practical difficulty arises when one considers the plausibility of letting $z_i = 0$ in real empirical data. For instance, does it make sense to calculate a simple slope for when one weighs 0 pounds? This remains a difficulty whether we have an additive model or a nonadditive one. Technically, the numbers are correct, since when $z_i = 0$, we surely can interpret the model **mathematically**. We may not however be able to do so **substantively**. For instance, in the product-term regression model, if both $x_i = 0$ and $z_i = 0$, then we have

$$y_i' = a + b_1(0) + b_2(0) + b_3(0)(0)$$
$$- \bar{u}$$

That is, the predicted value for y_i is the intercept term. What would we like this intercept to represent? If zero on both scales is not interpretable, then perhaps we can linearly transform x_i and z_i so that when we interpret the predicted value for y_i at $x_i = 0$ and $z_i = 0$, these zero values actually represent a quantity that is both more realistic, and more importantly, of more interest than an actual **true** zero value (i.e., with regard to the scale, not necessarily the **thing** that is being measured; see Chapter 2 for a discussion of measurement scales).

The way to **mean center** a predictor is to subtract the mean of that predictor from each value. We can use the `QuantPsyc` package (Fletcher, 2012) in R to easily perform mean-centering. For instance, consider the grades data featured earlier in Section 9.3. Suppose we wish to mean center study:

```
> library(QuantPsyc)
> mc.study <- meanCenter(study)
> mc.study
 [1] -0.46 -0.26  0.34 -0.16  0.04  0.64  0.04  0.34 -0.36 -0.16
```

We can now incorporate the mean-centered predictor into our regression:

```
> model.cent <- lm(final ~ mc.study*class)
> summary(model.cent)
```

```
Coefficients:
                Estimate Std. Error t value Pr(>|t|)
(Intercept)       68.962      4.626  14.908 5.73e-06 ***
mc.study         -30.376     15.923  -1.908   0.1050
class             -3.997      6.462  -0.619   0.5589
mc.study:class    70.724     20.069   3.524   0.0125 *
```

The estimated intercept value of 68.96 still represents the predicted mean grade when study time is equal to 0, but now, unlike in the regression where study was not centered, "0" does not really mean "0." Because we have centered study, "0" is equivalent to the **mean** study time, and so the number 68.962 is equal to the expected mean grade on the final at a mean amount of study time for those in class 0. Hence, the correct interpretation of the intercept of 68.96 is now:

The predicted mean final grade when one studies an average amount of time in class = 0 is equal to 68.96.

We can again visualize the interaction by generating scatterplots for each class, this time noting that mc.study is a centered variable along the abscissa:

```
> library(car)
> scatterplot(final ~ mc.study | class, data = grades)
```

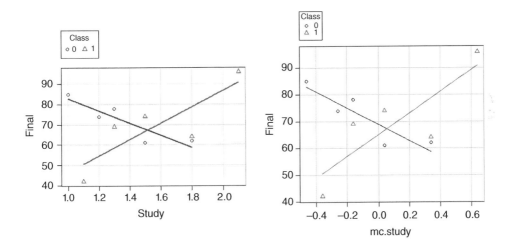

We can see from the above (right plot) that 0.0 on the mc.study variable now represents the mean study time, which turns out to be equal to 1.46 when not centered. For reference, we reproduced the original plot with uncentered data (left) plot. Notice both plots communicate the same information, only that the rightmost plot now features a "zero" that is meaningful, since it represents the mean study time. The zero point on the left plot represents the predicted final grade for someone having **literally studied zero hours**, which greatly extrapolates on the data and is far off to the left of the plot (e.g., try drawing a line beyond the plot to the point study = 0 for class = 0, and you will get the intercept value we obtained in the original regression, of 113.31. Hence, it makes much more sense in this case to center study, thus assigning it a meaningful zero point.

Product terms in a multiple regression can also in some cases cause problems of multicollinearity. Recall that we also know from our discussion of multiple regression that multicollinearity among predictors in a regression model can pose a serious problem, both substantively and technically. On a technical level, if one predictor is a linear combination of another predictor, the data matrix is of less than **full rank**, which implies that the determinant of the matrix will equal 0 (i.e., it will be **singular**), which further implies that the matrix will not be **invertible**. This will cause serious problems for whatever regression program you are using, and you will not be able to obtain a solution, or at minimum, for less severe cases, estimated regression coefficients will not be very stable. And recall that on the substantive side, we know that multicollinearity is a problem because it suggests that two or more predictors are accounting for the same variance in the response variable, and if for no other reason than parsimony (e.g., **Ockham's razor**), having two highly correlated predictors in a multiple regression is not in any way ideal. The usual course of action is to simply delete one of them, being sure to retain the one that is most substantively meaningful.

Centering predictors before producing product terms can also sometimes aid in reducing collinearity. For details on this issue, see Aiken and West (1991). Jaccard and Turrisi (2003) should also be consulted. It should be noted however that not all researchers and statisticians recommend centering in all cases. For a more in-depth and critical discussion of centering that distinguishes between **nonessential** versus **essential collinearity**, consult Dalal and Zickar (2011). Iacobucci et al. (2016) also provide a critical appraisal and should be consulted.

9.10   MULTILEVEL REGRESSION: ANOTHER SPECIAL CASE OF THE MIXED MODEL

Recall our discussion from Chapter 5 in which we introduced the multilevel model as a special case of the mixed model. There we discussed these models in the context of ANOVA-type models. In a regression style model, we could likewise test a model in which α is random but β is fixed. What this implies is that α terms now vary and have a probability distribution associated with them. We could also test a model in which slope β is random, while intercepts α are fixed. Or, we could test a model in which **both intercepts and slopes are random**. Allowing these parameters to be random is especially relevant in a multilevel context where a **clustering effect** is apparent. For example, consider Figure 9.3 reproduced from Demidenko (2004) where the relationship between sales and price is considered.

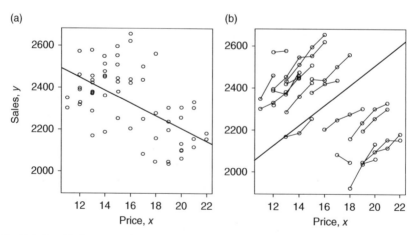

FIGURE 9.3   Relationship between sales and price using OLS (a) versus mixed modeling via cluster commodity (b). Source: Demidenko (2004). Reproduced with permission from John Wiley & Sons, Inc.

In the classical least-squares regression fit to these data (Figure 9.3a), the relationship between sales and price is negative. However, when type of commodity is taken into account, where each commodity represents a "cluster," we get a very different conclusion (Figure 9.3b). Of course, the clustering effect alone does not absolutely necessitate using a mixed modeling approach. A lesser analysis would have one fitting a different least-squares line to each commodity and keeping effects fixed. However, for clustered data, especially in situations where designating terms as random makes good sense, the mixed modeling approach is usually the preferred choice, and has numerous advantages such as facilities for dealing with missing data and more sophisticated estimation algorithms. **Linear mixed models** are well suited for taking into account dependencies in data, whether those dependencies arise from a clustering or other nesting effect.

9.11 CHAPTER SUMMARY AND HIGHLIGHTS

- **Interaction effects** can be tested in multiple regression by generating **product terms** between predictors.
- **Simple slopes** in multiple regression are analogous to **simple effects** in the analysis of variance. Both are effects conditional upon the value of another predictor (independent variable).
- The **additive** model is one in which there is **no interaction** term specified. The **nonadditive** model is one in which a **product term** is specified.
- Analogous to cell effects in ANOVA, a cell effect for a **product term**, conceptually, can be considered to be the joint occurrence of a score on each variable making up the product term. For continuous data, there are theoretically an infinite number of "cells" that are contained in the product term. This is a powerful way to conceptualize product terms in regression.
- **Moderators** may be dichotomous, polytomous, or continuous.
- The interpretation of "**main effect**" coefficients in a model containing a product term is not analogous to the interpretation of coefficients in a model not containing product terms. When a product term is present, the effect of one variable is implicitly evaluated at a value of 0 on the other variable. That is, main effects are interpreted as **simple slopes** where the value of the moderator is equal to 0.
- **Mean-centering** predictors can aid in interpretability and can also help in some cases to reduce **collinearity** between the predictor and the product term.

REVIEW EXERCISES

9.1. Discuss the similarities and differences between hypothesizing an **interaction** in **ANOVA** versus hypothesizing one in a **regression** context.

9.2. Conceive a substantive example where a **moderation analysis** in regression would be useful.

9.3. Referring to the following model, discuss how you would parameterize z_i differently for a **continuous moderator** than for a **dichotomous** or **polytomous** one:

$$y_i = \alpha + \beta_1 x_i + \beta_2 z_i + \beta_3 x_i z_i + \varepsilon_i$$

9.4. Consider an example where the slope of y_i on x_i is not **linear** across levels of the moderator z_i.

9.5. Discuss and explore the issue of whether or not moderators actually "moderate" anything. What are some of the **philosophical pitfalls** to using such words as "moderation" in the context of what are otherwise interactions?

9.6. Explain how coefficients in a **nonadditive** model should be interpreted differently than coefficients in an **additive** one.

9.7. Discuss the purpose of **mean-centering** predictors in a regression model.

9.8. Consider the following data where we wish to hypothesize that raise is a function of learning and privileges. Test the model `lm(raises ~ learning*privileges)` and provide a full summary of its findings, especially that of interpreting the product term generated by crossing two continuous variables.

```
> attach(attitude)
> attitude
    rating complaints privileges learning raises critical advance
1      43         51         30       39     61       92      45
2      63         64         51       54     63       73      47
3      71         70         68       69     76       86      48
4      61         63         45       47     54       84      35
5      81         78         56       66     71       83      47
6      43         55         49       44     54       49      34
7      58         67         42       56     66       68      35
8      71         75         50       55     70       66      41
9      72         82         72       67     71       83      31
10     67         61         45       47     62       80      41
```

10

LOGISTIC REGRESSION AND THE GENERALIZED LINEAR MODEL

> Linear models customarily embody both systematic and random (error) components, with the errors usually assumed to have normal distributions. The associated analytic technique is least-squares theory … Techniques developed for non-normal data include probit analysis, where a binomial variate has a parameter related to an assumed underlying tolerance distribution, and contingency tables, where the distribution is multinomial and the systematic part of the model usually multiplicative. In both these examples there is a linear aspect to the model.
>
> (Nelder and Wedderburn,1972, p. 370)

The class of models surveyed up to this point in the book have generally been of two types, analysis of variance models (e.g., fixed effects, random effects, and mixed models), and linear regression models (e.g., simple linear regression and multiple linear regression). In all cases, we have made many model assumptions (such as normality and independence of errors), but none more relevant to least-squares than assuming the relationship between the response variable and the explanatory variables is **linear** in form. That is, up to now, we have assumed **linearity in the parameters**. For instance, recall the multiple regression model of Chapter 8:

$$y_i = \alpha + \beta_1 x_1 + \beta_2 x_2 + \cdots + \beta_k x_k + \varepsilon_i$$

in which there were k predictors. Recall that the constants α and β_1 through β_k represented the intercept and partial regression parameters, to be estimated from sample data, and ε_i was the random error associated with each prediction of y_i. When we say we are making the assumption of **linearity in the parameters**, what we mean is that the exponent on each estimated parameter α, β_1, etc., is understood or implied to be equal to 1. This is what defines the model as **linear**. The model

Applied Univariate, Bivariate, and Multivariate Statistics: Understanding Statistics for Social and Natural Scientists, With Applications in SPSS and R, Second Edition. Daniel J. Denis.
© 2021 John Wiley & Sons, Inc. Published 2021 by John Wiley & Sons, Inc.
Companion Website: www.wiley.com/go/denis/appliedstatistics2e

$$y_i = \alpha + \beta_1 x_1 + \beta_2 x_2^2 + \varepsilon_i$$

is also linear in the parameters since parameters in the model are still raised to the exponent 1. Simply because $\beta_2 x_2^2$ contains the term x_2^2 does not in itself make the model **nonlinear**. What would make the model nonlinear is if β_2 were squared, for instance, as in

$$y_i = \alpha + \beta_1 x_1 + \beta_2^2 x_2 + \varepsilon_i \tag{10.1}$$

The model in (10.1) is properly considered **nonlinear** because it is **nonlinear in the parameter** β_2.

Up to now in the book, we have yet to consider models of the type in (10.1) where parameters are raised to any other exponent than 1. Indeed, there is a good reason for our emphasis on linear models. The **general linear model** is easily the most popular and relevant of models in statistical analysis, and many scientific phenomena can be modeled relatively precisely under the assumption of linearity.

There are times, however, when linear models are definitely **not** appropriate. Such situations include, but are not exclusive to, circumstances where a **nonlinear** relationship between the response variable and predictor variable is hypothesized or expected. Nonlinear relationships might be hypothesized for at least a couple reasons:

- The actual empirical relationship between the response and predictor variable is thought to be nonlinear in form. For example, the classic Yerkes and Dodson (1908)[1] inverted U curve (i.e., inverted parabola) is one famous example of a nonlinear relationship, specifically the relationship between performance and arousal.

- The empirical relationship between the response and predictor variable is nonlinear as a result of how the response variable is operationalized or defined. Recall that one assumption for linear models is that the response variable is conditionally normally distributed with independent errors. To have any chance of satisfying this assumption, an essential requirement is that the response variable be, at least in a practical sense, **continuous**. If the response variable is not measured on a continuous scale, then assuming normality can become quite difficult or even impossible. For instance, if the response variable is a Bernoulli variable (i.e., a binary-coded variable), then it is impossible to assume it to be normally distributed. In situations such as this, where the response is binary or even multinomial, the relationship between the response and predictor variable is generally poorly described by linearity by simply a consequence of how the response variable is **defined** and **measured**. "Pass versus fail" is, by nature, a binary response, as is "survive versus perish." In data where nonlinearity is clearly present because of how the response variable is operationalized, we may require a model other than the linear model to fit to such data.

A useful distinction to make when referring to nonlinear models, in general, is that between models that are **intrinsically linear** versus those that are not (see Neter et al., 1996, pp. 534–535). Nonlinear models that can be linearized through a transformation are usually considered to be intrinsically linear. For instance, the exponential response function is considered to be an intrinsically linear model since if we take the log of the function we get a linear function. However, as Neter et al. (1996) note, that a nonlinear response function is intrinsically linear does not necessarily mean that linear regression is still suitable, since even after the transformation, the linearization may generate an error term that is **not** normally distributed with constant variance, which recall, is an assumption required of least-squares estimation.

---

[1] An on-line version can be accessed on **Classics in the History of Psychology** website: http://psychclassics.yorku.ca/

In this chapter, we treat models such as the **logistic** and **Poisson** models as **special cases** of generalized linear models, where generalized linear models may in turn be considered special cases of the wider nonlinear framework. However, we do not treat **nonlinear estimation** in any detail such as that provided by functions as `nls` (nonlinear least squares) in R. Such a topic is beyond the scope of this book. For details on how to estimate nonlinear models in R, see Crawley (2013, Chapter 20).

We begin the chapter with a brief general discussion of nonlinear and generalized linear models, then spend the rest of the chapter discussing one very specific and popular case of a nonlinear model, that of **logistic regression**. To understand logistic regression, one first requires a familiarity with exponential and logarithmic functions. These concepts are also reviewed.

The classic resource for generalized linear models is that of McCullagh and Nelder (1989). Fox (2008b, Chapter 15) is also an excellent and readable overview.

10.1 NONLINEAR MODELS

A general form for nonlinear regression models can be given by:

$$\mathbf{y} = \mathbf{X}\boldsymbol{\gamma} + \boldsymbol{\varepsilon}_i \tag{10.2}$$

where $\mathbf{y}$ denotes a vector of observations, $\mathbf{X}$ is the model matrix, $\boldsymbol{\gamma}$ is the parameter vector, and $\boldsymbol{\varepsilon}_i$ is the error associated with each observation in $\mathbf{y}$, typically assumed to be independent and normally distributed (Neter et al., 1996). Note that the model in (10.2) is identical to the classic regression model of (7.7) of Chapter 7, $\mathbf{y} = \mathbf{X}\boldsymbol{\beta} + \boldsymbol{\varepsilon}$, only that now we are replacing $\boldsymbol{\beta}$ with $\boldsymbol{\gamma}$ to denote the nonlinearity. In (10.2), we are simply using a different symbol to represent a different model (i.e., one that is nonlinear). Otherwise, the two model statements are quite similar.

One common nonlinear model is the **exponential regression** model given by:

$$\mathbf{y} = \gamma_0 \exp(\gamma_1 x_i) + \boldsymbol{\varepsilon}_i$$

where γ_0 and γ_1 are parameters, x_i are fixed values for the explanatory variable, and $\boldsymbol{\varepsilon}_i$ are independent normally distributed errors. Analogous to linear regression in which the least-squares criterion assured us of the minimization of the sum of squared errors,

$$\sum_{i=1}^{n} \varepsilon_i^2 = \sum_{i=1}^{n} [y_i - (\alpha + \beta x_i)]^2$$

likewise, in nonlinear regression, we seek to minimize the sum of squared errors (Neter et al., 1996)

$$\sum_{i=1}^{n} \varepsilon_i^2 = \sum_{i=1}^{n} [y_i - \gamma_0 \exp(\gamma_1 x_i)]^2$$

As another example of a nonlinear model, consider a hypothetical relationship between hours of therapy and number of suicide attempts in Figure 10.1. We can see that the relationship between suicide attempts and hours of therapy is nonlinear. As hours of therapy increase, the number of suicide attempts decreases.

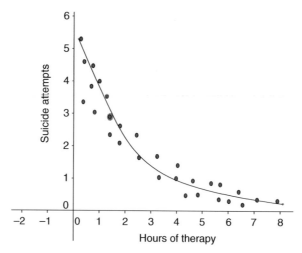

FIGURE 10.1   Hypothetical nonlinear relationship between suicide attempts and hours of therapy.

10.2   GENERALIZED LINEAR MODELS

An important class of models was proposed by Nelder and Wedderburn (1972) to incorporate not only the classic linear model but also situations where nonlinearity is present. As we have already noted and will see further, the general linear model can be considered a special case of the wider class of **generalized** linear models, which includes models that allow for noncontinuous, binary and multinomial responses, as well as responses which are in the form of **counts**. Generalized linear models utilize what is known as a **link function** to essentially transform what is a nonlinear model into one that is **linear**.

For example, the logistic regression model, the topic of most of this chapter, is a nonlinear model. However, through the appropriate link function that transforms the nonlinear response into one that is **linear**, we are able to interpret the model almost analogous to how we would interpret a "naturally occurring" linear model. Of course, there will be exceptions and specific details about the transformation that we will need to tend to, but the essence of the generalized linear model is basically to **make a nonlinear model linear through the relevant link function**. This is done for the purpose of aiding the statistical analysis and facilitating the interpretation of estimated coefficients. For instance, in the case of a binary response coded 0 and 1, not only will the relevant link function linearly transform the response variable, but it will also free up the range on the response variable so that it can assume values beyond simply 0 and 1. In fact, as we will see, the appropriate link function for the binary response will transform the variable into one that is continuous with a range $-\infty$ to $+\infty$, that is, the entire range of the real line. As noted by Fox (2016):

> Beyond the desire to select a link function that renders the regression of Y on the X's linear, a promising link will remove restrictions on the range of the expected response ... This is not to say that the choice of link function is entirely determined by the range of the response variable, just that the link should behave reasonably in relation to the range of the response. (p. 420)

Depending on how the response variable is defined (e.g., binary, multinomial, and count), there is a specific link function appropriate to the given form. This specific link function is generally known as the **natural link function** for the given family of distributions (e.g., Gaussian, Binomial, Poisson) and is often referred to as the **canonical link function** for the given family.

We now survey the logic and conceptualization of the generalized linear model more closely and develop the concepts of linear predictors and link functions more in depth. This will pave our way to a consideration of a popular generalized linear model featured in this chapter, the logistic model.

10.2.1   The Logic of the Generalized Linear Model: How the Link Function Transforms Nonlinear Response Variables

The regression model $y_i = \alpha + \beta x_i + \varepsilon_i$ surveyed up to now can be said to have two "parts" or "components" to it. These components are as follows:

- A **systematic** component, equal to $\alpha + \beta x_i$. It is called a systematic component because this is the **predictive** part of the model. It is the part for which so long as we estimate α and β intelligently (i.e., via a good estimation procedure such as ordinary least-squares), we can make a good prediction of y_i, while on average, keeping the sum of squared error of prediction to a minimum. In developing the generalized linear model, it is helpful if we identify $\alpha + \beta x_i$ as a **linear predictor**, denoted simply by η_i. That is, $\eta_i = \alpha + \beta x_i$.
- A **random** or **stochastic** component, equal to ε_i. This is the part of the model that is **unpredictable**. It represents random variation around our predicted values. When this variation is operating, it will contribute to our prediction of y_i in a random fashion, unlike the linear predictor η_i, which contributes to the prediction of y_i in a very systematic way.

Now, here is how generalized linear models are similar to yet different from the simple linear regression model. In a generalized linear model, the response variable y_i is not a natural **linear** function. That is, its expected value $E(y_i)$ is usually equal to something that has nothing to do with linearity at all. A classic case again is that of the binary response variable. So the question becomes: **How can we adjust the left-hand side of the regression equation, y_i, so that it corresponds, in some sense, to the right-hand side?**

We accomplish this through what is known as a **link function**, which is nothing more than a **transformation** of the nonlinear response variable into a variable that is linear. Thus, in the generalized linear model, in addition to the above two components (systematic and random parts), we add a third component:

- A **link function** between the random component y_i and systematic component $\alpha + \beta x_i$. The linear predictor, $\alpha + \beta x_i$, is a function of the expectation of the parameter μ we are modeling (e.g., expectation of a binary variable) via a link function. This new link function we will denote as $g(x)$. What is g a function of? It is a function of μ_i, so the link function proper is given by $g(\mu_i)$ (Fox, 2016, p. 421).

We return to the binomial setting in clarifying the above. What is the expectation of a binomially distributed variable? Recall that a Bernoulli variable has values 0 and 1 and has an expectation equal to p, the probability of success for any given trial. For example, on the flip of a fair coin, p is equal to 0.5, and μ in this case is equal to p. But, of course, it is best to not model p using a **linear** model. Any variable that can assume only two values is almost at best "artificially linear." What we need to do is alter p so that it resembles something more realistically linear. We can do this by **transforming** p through a well-chosen link function. The link function of choice for a binomial setting is the **log of the odds**. The log of the odds is defined as:

$$\log_e \left(\frac{p}{1-p} \right) \tag{10.3}$$

What (10.3) means is that if we take what is a nonlinear expectation of p, compute the odds on p, then transform the odds into the natural log of the odds (i.e., to base e), we will have effectively linearized an otherwise binary variable. Now, as a result of this transformation, we can treat the model as a **linear model**. When we compute the **log of the odds**, we get a value that can range not from 0 to 1 as was the case for the Bernoulli variable, but rather from $-\infty$ to $+\infty$. This is what generalized linear models do, **they transform the expectation of a nonlinear-occurring variable into one that is linear**.

A quick review of what we have discussed thus far, as it pertains to the binary response:

- Our response variable is a binary variable with values 0 and 1. It is usually unreasonable to consider this variable linearly, since the only two options one could obtain on the variable are 0 or 1.
- The expectation of the random variable is p, which is the probability of a success (we are denoting a success by "1"). What we would like to do is transform this expectation into something that is linear in form.
- By using the link function $\log_e\left(\frac{p}{1-p}\right)$, this will effectively transform the original expectation of our binary variable into a variable that is a linear predictor. Notice that to do this we had to first take the odds, $\frac{p}{1-p}$, but the essential point is that through this link function, we have basically linearized an otherwise essentially nonlinear variable.

Having only thus far spoken about the generalized linear model with respect to a binary response, you might think at this point that the generalized linear model is **specific** to binary-occurring variables. This is not the case. The true contribution of Nelder and Wedderburn (1972) was in summarizing a framework that could handle not only binary variables, but a whole host of other response variables as well, all through their respective link functions. The binary case, which we will use in our development of the logistic model, is but one possibility. Had our response variable not represented a binary situation, but rather a distribution of **counts**, then a **Poisson distribution** with link function equal to the log of the counts, rather than the odds, may have been more appropriate.

10.3   CANONICAL LINKS

The canonical link is the link function that is **natural** to the family of distributions. For instance, drawing again on our discussion of the binary response variable, we said that the link function for a binomial variable is the log of the odds, which we will come to name the **logit**. We also said that for a Poisson variable, the appropriate link function is the log. In addition to the binomial and Poisson families, there are many other families of distributions, all with respective link functions.

It should be no surprise then that the general form relating the link function $g(\cdot)$ to the linear predictor, η_i, can be written as:

$$g(\mu_i) = \eta_i = \alpha + \beta_1 x_1 + \beta_2 x_2 + \cdots + \beta_k x_k \tag{10.4}$$

where,

- $g(\mu_i)$ is the new function $g(\cdot)$ of the original expectation μ_i of the response variable. To emphasize that μ_i is an expectation, we can write it instead as $E(y_i)$; that is, $g[E(y_i)]$.
- η_i is the symbol for the linear predictor.

- $\alpha + \beta_1 x_1 + \beta_2 x_2 + \cdots + \beta_k x_k$ is simply the systematic portion of (10.4), analogous to what we would have in an ordinary regression with k predictors.

Notice in (10.4) how $g(\cdot)$ serves as the "bridge" or "link" between the right-hand side and the left-hand side of the equation. This is how you can think of it as being the **link** function, in that it provides a kind of way for both sides of the equation to "communicate" with each other through a "translator," which is the link function appropriate to the modeling context.

10.3.1 Canonical Link for Gaussian Variable

If the generalized linear model framework is as "generalized" as it sounds, then we should be able to fit our previously studied regression models into this framework. What is the canonical link for the family of distributions having a normal (Gaussian) distribution? That is, what link did we impose on an ordinary regression model? The answer is, of course, that we did not directly impose any link at all. **The expectation for the response variable, $E(y_i)$, was already linear**. It did not require any transformation to make it linear (Agresti, 2002, p. 117). However, to say that it did not require **any** transformation would not be as general as we would like. For the sake of the generalized linear model framework, we still want to specify what being "untransformed" looks like. For the **Gaussian family**, the correct canonical link is the **identity function**. This is simply the function $f(x) = x$. What the function means is **what you put in, you get out**. No transformation takes place. That is, $g(\mu_i) = \mu$, which is known as the **identity link**.

10.4 DISTRIBUTIONS AND GENERALIZED LINEAR MODELS

We will spend the vast majority of this chapter discussing the logistic regression model at some length, but before we do so, it is useful to briefly survey a few of the more common distributions featured in generalized linear models. We also briefly survey the concept of a **dispersion parameter** along with that of **deviance**.

10.4.1 Logistic Models

The logistic regression model, useful for modeling binomial data, is given by

$$p = \frac{1}{1 + e^{-(\alpha + \beta x_i)}} = \frac{e^{(\alpha + \beta x_i)}}{1 + e^{(\alpha + \beta x_i)}} \tag{10.5}$$

where we can also write the numerator, $e^{(\alpha + \beta x_i)}$ in (10.5) as:

$$e^{(\alpha + \beta x_i)} = \frac{p}{1 - p}$$

To "deexponentiate" $\alpha + \beta x_i$, we take the log, which gives us

$$\alpha + \beta x_i = \log_e \left(\frac{p}{1 - p} \right) \tag{10.6}$$

where the right-hand side of (10.6) is the **log of the odds**. When we plot such a function, we obtain the curve in Figure 10.2 (`curve(pnorm(x), -5, 5)`).

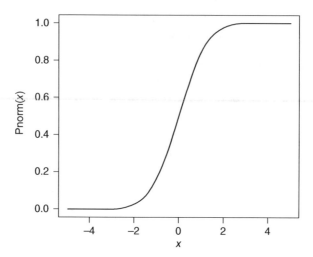

FIGURE 10.2   Logistic "sigmoid" function.

The shape of the resulting curve is the well-known **sigmoid** function. Since its shape is nonlinear, it makes fitting a least-squares regression line to such data typically inappropriate.

10.4.2   Poisson Models

The Poisson distribution (Poisson, 1837) is given by

$$p(y_i, \mu) = \frac{e^{-\mu}\mu^{y_i}}{y!}$$

where μ is the mean, often referred to in the Poisson distribution as the **rate parameter** and $y! = y(y - 1)(y - 2)...2 \cdot 1$. The Poisson distribution is useful for modeling count data that occur over a given period of time or space. Whereas the appropriate link for the logistic model is that of the logged odds, the appropriate link for the Poisson model is the log (Fox, 2016, p. 421).

One feature that is worth noting regarding the Poisson distribution is that the expectations of both its mean and variance are both equal to the mean, μ. That is, $E(y_i) = \mu$ and $E\left(\sigma_y^2\right) = \mu$. As μ gets larger, the Poisson distribution approximates that of a normal distribution (Fox, 2016). Just as the normal density was the limiting form of the binomial, so it is also the case that the normal is the limiting form of a Poisson variable. Further, when n is relatively large and p relatively small, the Poisson distribution is approximated by the binomial. Densities for Poisson distributions can be easily obtained using `dpois` in R. See Teetor (2011, p. 187) for details.

10.5   DISPERSION PARAMETERS AND DEVIANCE

A **dispersion parameter** is simply an index indicating a measure of spread in a distribution. For instance, in the normal distribution, $N(\mu, \sigma^2)$, the dispersion parameter is that of the variance, σ^2 (Fox, 2002, p. 232). Denoting the dispersion parameter more generally as ϕ, we can express the dispersion parameter of the normal distribution as $\phi = \sigma^2$. For the Poisson distribution, recall $E(y_i) = \mu$ and $E(\sigma_y^2) = \mu$. The dispersion parameter for Poisson distributions is fixed at a value of 1 (i.e., $\phi = 1$).

Overdispersion is exactly as the name suggests—the variance of a distribution is larger than some value we would expect or prefer under some idealized circumstance. To explain how dispersion parameters come about, we refer once more to the binomial distribution. Recall that the mean of a binomial distribution is given by $E(y_i) = np$ with variance equal to npq, which we can also write as $np(1-p)$ since $q = (1-p)$. Overdispersion, in this case, is said to exist if the computed variance on a sample of data exceeds $np(1-p)$. To measure overdispersion, we can write the variance of y_i as:

$$\sigma_y^2 = \phi np(1-p)$$

where ϕ is an unknown **dispersion**, or **scale parameter**. Dispersion parameters may be known, but most often are estimated. Dean and Lundy (2016) provide an excellent summary of overdispersion in generalized linear models.

If the dispersion assumption is satisfactory, **residual deviance** should be approximately equal to **residual degrees of freedom** for the model (Fox, 2016, p. 431), though as noted by Venables and Ripley (2002), this indicator is based on asymptotic theory, hence only being truly accurate for relatively large n. As they note:

> A common way to 'discover' over- or under-dispersion is to notice that the residual deviance is appreciably different from the residual degrees of freedom, since in the usual theory the expected value of the residual deviance should equal the degrees of freedom. This can be seriously misleading. The theory is asymptotic, and only applies for large $n_i p_i$ for a binomial and for large μ_i for a Poisson. (Venables and Ripley, 2002, p. 208)

Still, one general (and convenient) guideline for detecting overdispersion is a deviance that is at least twice as large as the number of degrees of freedom (Lindsey, 1999). We demonstrate later in an example using software how to compare the deviance statistic to degrees of freedom as a means for at least tentatively assessing overdispersion in a model.

The **deviance** of a model, first briefly mentioned in Chapter 5, is a measure used to estimate how different a given model is from the saturated model for some data (Everitt, 2002). It is defined as:

$$D = -2[\ln L_{\text{Model}} - \ln L_{\text{Saturated}}] \tag{10.7}$$

where L_{Model} is the likelihood of the hypothesized model and $L_{\text{Saturated}}$ is the likelihood under the saturated model. A **saturated** model is one in which degrees of freedom are equal to 0. Hence, the extent to which $L_{\text{Model}} - L_{\text{Saturated}}$ in (10.7) is **large** is indicative of a poor-fitting model. Conversely, the extent to which $L_{\text{Model}} - L_{\text{Saturated}}$ is **small** is suggestive of a well-fitting model. Thus, smaller values of D are preferable to larger ones. This is somewhat intuitive, in that if I told you a model had a high deviance, it would suggest it differs to a great extent from the "ideal" in this case, which is the saturated model, the model with the best fit.

10.6 LOGISTIC REGRESSION

10.6.1 A Generalized Linear Model for Binary Responses

Thus far, we have introduced the idea of nonlinear models and then the general class of models known as generalized linear models. We have said that these models can accommodate a variety of responses arising from a variety of families, including Gaussian, binomial, Poisson, etc. We now focus on and develop some of the theory behind a very popular model arising from the binomial family, that of the **logistic regression model**. We focus on this model for good reason. The model is quite popular in the social, biomedical, and natural sciences, because many times we wish to predict responses that intrinsically have a binary structure. Examples where a response is naturally binary include the following:

- Predicting survival versus death in a long-term medical trial for treating a disease.
- Predicting passing versus failure in an educational environment.
- Predicting marriage versus divorce in romantic relationships.

In logistic regression, just as we did for linear least-squares regression, we seek to find predictors, continuous, categorical, or both, that will successfully account for variance in a response variable. That is, we would like to model important predictors that help explain variance in survival versus death, passing versus failure, and marriage versus divorce. Does method of medical treatment explain variance in the binary variable of survival versus death? Does amount of study time predict whether a student will pass or fail a course? Does quality of communication predict whether couples remain married or divorced? The fundamental questions posed in a logistic regression parallel those asked of "ordinary" linear models. The only real distinction is in how these models are parameterized.

Logistic regression is a relatively popular technique for predicting group membership and is probably used more than its competitor, **discriminant analysis** (see Chapter 12). One reason for this is that discriminant analysis requires the assumption of normality. Logistic regression does not require this assumption, though it does make the assumption of **linearity in the logit,** which can be tested using the **Box–Tidwell test** (Hosmer and Lemeshow, 2000). One can also simply plot the sample logits against the predictor (e.g., in the one-predictor case) to obtain informal evidence that the assumption is satisfied (Agresti, 2002). Under some circumstances, discriminant analysis has been found to be more effective than logistic regression (Efron, 1975). For a general comparison of logistic regression to discriminant analysis, see Press and Wilson (1978) or Hastie et al. (2009). For a comparison of the classification errors made in each procedure, Lei and Koehly (2003) is a helpful read.

10.6.2   Model for Single Predictor

The logistic regression model for the one-predictor case is that given earlier in (10.5):

$$p = \frac{1}{1 + e^{-(\alpha + \beta x_i)}} = \frac{e^{(\alpha + \beta x_i)}}{1 + e^{(\alpha + \beta x_i)}}$$

where p is a probability with a possible range of 0 to 1, $\alpha + \beta x_i$ is, in the language of the generalized linear model, the **linear predictor,** e is a constant equal to approximately 2.718 and $e^{(\alpha + \beta x_i)}$ is the exponentiated logit, known also as the **odds.** We spend much of this chapter discussing the components of (10.5), so by the time this chapter is complete, they will be quite familiar.

When we take the natural log of the odds, that is, the log to base e, or ln (pronounced "lawn"), we obtain the **logit**:

$$\ln\left(\frac{p}{1-p}\right) = \alpha + \beta x_i$$

To better understand the relationship between odds, logarithms, and logits, a review of exponential and logarithmic functions is in order.

10.7 EXPONENTIAL AND LOGARITHMIC FUNCTIONS

You undoubtedly noticed that in the equations for the logistic distribution, and others for that matter, "*e*" appears repeatedly. But what is *e*? It is the **exponential function**. For a reasonable understanding of how logistic regression works, one must be at least somewhat familiar with two common functions in mathematics, the exponential function *e* and its inverse, the **logarithmic function**. An understanding of these functions is necessary for an understanding of what **odds**, **odds ratios,** and **logits** are all about, and how probabilities are generated in the logistic distribution.

To begin, recall from elementary mathematics (see Appendix) what constitutes a simple linear function:

$$f(x) = a + bx$$

where *a* is the intercept and *b* is the slope parameter. We define an **exponential function** to base *b* as:

$$f(x) = b^x$$

where *b* appears in the base and *x* appears in the **exponent** of the function. Consider a graph of the exponential function $f(x) = 2^x$ in Figure 10.3.

Notice that the curve increases at a faster rate for increasing values of *x*. This is what the expression **exponential growth** means in the context of an investment that promises to grow your money exponentially, or likewise some bacteria that are known to grow exponentially in biology. The more bacteria present, the more new bacteria are generated, analogous to how rolling over earnings from an investment can likewise help grow one's money at a faster rate. The opposite of exponential growth is that of **exponential decay**. As a recent example, recall the COVID-19 outbreak. When it was said the rate of diagnoses was growing "exponentially," it implied it was growing at an increasingly greater rate.

A few of the more common properties of exponents include the following:

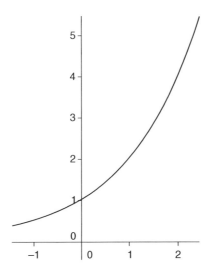

FIGURE 10.3   Example of a simple exponential function.

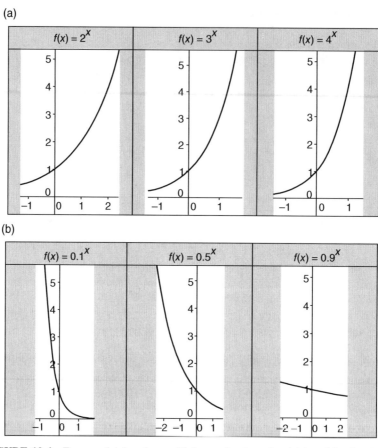

FIGURE 10.4   Exponential functions with increasing base $b > 1$ (a) and $0 < b < 1$ (b).

- $b^{-x} = \frac{1}{b^x}$ (the base raised to a negative exponent is equal to 1 divided by the base raised to that exponent).
- $x^n \cdot x^m = x^{n+m}$ (the product of two bases raised to different exponents is equal to the base raised to the sum of those exponents).
- When the base b is greater than 1, $b > 1$, the function will rise, called **exponential growth**. When the base is smaller than 1, but still greater than 0, $0 < b < 1$, the function will decrease, the so-called **exponential decay** (Figure 10.4) alluded to earlier. Notice that for $b > 1$, the function gets **steeper** as the base gets larger. For $0 < b < 1$, the function gets flatter as the base gets larger.

Aside from the linear function, the exponential and its inverse (the log) are two of the most popular functions for modeling change. For example, world population growth since 1800 (perhaps even earlier) follows an approximate exponential growth curve (Barnett, Ziegler, and Byleen, 2011). Investment growth at compound interest, radioactive decay, and animal learning trials in psychology are other examples. Exponential functions are also quite popular in physical applications as well (see Labarre, 1961, p. 425 for an example).

Some further characteristics regarding the exponential are worth noting. When the exponent x in $f(x) = b^x$ is negative, that is, $f(x) = b^{-x}$, then, as already mentioned, $f(x) = \frac{1}{b^x}$, and the curve reverses direction. For instance, compare the graphs of $f(x) = 2^x$ and $f(x) = 2^{-x}$ (Figure 10.5).

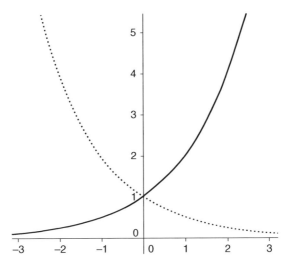

FIGURE 10.5   Graph of $f(x) = 2^x$ (solid line) and $f(x) = 2^{-x}$ (dotted line).

10.7.1   Logarithms

The logarithm of a number to a given base is the **exponent** to which the base must be raised in order to produce the number. For instance, consider a logarithm to base 10:

$$\log_{10} 100 = 2$$

The above reads that the base 10 raised to 2 equals 100, that is, $10^2 = 100$. We can choose a variety of bases for logarithms. As another example, consider the following log:

$$\log_2 8 = 3, \quad \text{that is,} \quad 2^3 = 8$$

The above reads that 2 to the exponent 3 equals 8. Graphically, logarithmic curves take on a quickly rising then **plateau** shape (Figure 10.6) of **diminishing returns**.

We see then that the **log** of a number is actually an **exponent**. We can generalize this to say that $y = \log_b x$ if and only if $x = b^y$ (Barnett et al., 2011, p. 108). We can say more generally that $\log_b x$ is the **index** to which b must be raised in order to get x. Logarithms to base 10 are called **common logarithms** and logarithms to base e are called **natural logarithms**, designated by the symbol "ln."

As shown in Figure 10.5, the curve for b^x increases rather dramatically. The opposite of the exponential function (the log) is one in which growth is substantial at the beginning, but then levels off at higher levels, as evidenced in Figure 10.6. For example, it is quite common for sedentary and overweight individuals to achieve great muscle gain and simultaneous weight loss during the first few months of an exercise program, only for progress to **level off** after many months. When this happens, the individual is nearing the height of the log function, since they are achieving much less weight loss for the same amount of unit-increase on x, where x may be quantified in terms of effort, amount of daily exercise, measured calorie deficit, and so on. That is, the **rate of return** is not what it was earlier in the diet or fitness program, and a new program must be implemented to achieve the same degree of progress. In psychology, the logarithm function is useful in describing such processes as **habituation to stimuli**. The new car you purchased today with all those fancy features is guaranteed to give you a happy feeling the first little while that you own the car, until, even in a relatively short period after

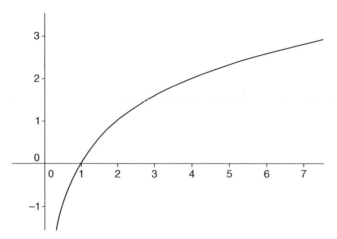

FIGURE 10.6   Graph of $y = \log_2(x)$. Note the steep rise then plateau.

the purchase, the car no longer arouses the same emotion (and a new competing emotion is aroused when regardless of your decreased excitement over the car, you are still having to make car payments). Stimuli often excite **initially**, but wear off over time, which is one of the most fundamental processes of learning in psychology. Marketing psychology exploits this principle in "hooking" us into purchases. **Habituation is a fundamental psychological law**. A logarithm function captures such processes very well. Indeed, with the COVID-19 outbreak in 2020, a logarithmic curve which would denote a "flattening" of new cases and transmissions was of course highly desired in the management of the virus.

10.7.2   The Natural Logarithm

Recall that the natural logarithm, denoted $\ln(x)$ and pronounced "lawn of x," is the logarithm for which the base b is equal to e, and equal to approximately 2.718281828459045. This is only an **approximate** number, since e is an irrational number (it is also transcendental), meaning that it cannot be written as the quotient of two integers a and b, a/b, where $b \neq 0$. Recall that if a number can be written as the quotient of two integers, then it is a **rational** number, and has a finite or recurring (i.e., periodic) decimal expansion.

The number e can be approximated in several ways. For example, it can be defined as the limit of the sequence $\left(1 + \frac{1}{n}\right)^n$. That is,

$$e = \lim_{n \to \infty} \left(1 + \frac{1}{n}\right)^n$$

which means that as n gets larger and larger and grows without bound, we get a better and better approximation for e. For example, consider the following approximations that get us closer and closer to the actual limiting value of e:

$$e = \lim_{n \to 1} \left(1 + \frac{1}{1}\right)^1 \quad e = \lim_{n \to 2} \left(1 + \frac{1}{2}\right)^2 \quad e = \lim_{n \to 100} \left(1 + \frac{1}{100}\right)^{100} \quad e = \lim_{n \to 1000} \left(1 + \frac{1}{1000}\right)^{1000}$$
$$= 2 \qquad\qquad\qquad = 2.25 \qquad\qquad\qquad = 2.70 \qquad\qquad\qquad\qquad = 2.7169$$

Notice that as n increases in cases from left to right, the limiting value of e gets closer and closer to 2.718 ….

10.8 ODDS AND THE LOGIT

Recall that in our discussion of the logistic function we had defined the ratio p to $1 - p$ as $e^{(\alpha + \beta x_i)}$. The ratio p to $1 - p$ is the odds in favor of an event with probability p. For example, suppose the probability of event A was equal to 0.70 and the probability of event B was equal to 0.30. The odds in favor of event A to event B are computed as

$$\frac{p}{1-p} = \frac{0.70}{1-0.70} = \frac{0.70}{0.30} = 2.33$$

That is, the odds in favor of A to B are 2.33 to 1. If the "1" is not mentioned explicitly in the statement about odds, then it is implied. So, if we say the odds of horse Charlie winning the race compared to horse Homestretch are 2.33 to 1, this implies that the probability that horse Charlie wins is 0.70 compared to a probability of only 0.30 that horse Homestretch will win. Suppose the odds of Charlie losing to Homestretch are 10 to 1. Then this implies there to be a $10/11 = 0.91$ probability that horse Charlie loses, and a $1 - 0.91 = 0.09$ probability that horse Charlie wins.

The odds are zero or undefined when either $p = 0$ or $p = 1$, since $\dfrac{p}{1-p} = \dfrac{0}{1-0} = 0$ and $\dfrac{p}{1-p} = \dfrac{1}{1-1} = \dfrac{1}{0}$, respectively. Recall that $\frac{1}{0}$ is not equal to 0. Rather, it is undefined (some might say the odds approach infinity in this case). The reason why $\dfrac{1}{0}$ is undefined is because there is no number x such that $0 \cdot x = 1$. Contrast this situation to, say, the fraction $\frac{2}{4}$. Since there exists a number 2 such that $2 \cdot 2 = 4$, $\frac{2}{4}$ is not undefined.

As mentioned previously, the natural logarithm of the odds, that of $\log_e\left(\dfrac{p}{1-p}\right)$, or $\ln\left(\dfrac{p}{1-p}\right)$, is called the **logit**. If we are given only the probability p of an event, it is a simple matter to obtain the logit. We can simply transform p into the odds $\dfrac{p}{1-p}$, then take the natural log. For example, suppose we find the probability of an event to be equal to $p = 0.98$. This means that the odds are equal to $\dfrac{0.98}{1-0.98} = \dfrac{0.98}{0.02} = 49$. The logit is thus, $\ln(49) = 3.89$. Conversely, given the logit, we can easily get the odds by simply exponentiating the logit:

$$\frac{p}{1-p} = e^{\ln\left(\frac{p}{1-p}\right)}$$

For our example in which $\ln(49) = 3.89$, the odds are therefore equal to $e^{\ln(49)}$. Of course, given the odds, we can easily also transform back to probabilities through generating a ratio of the odds (i.e., the exponentiated logit):

$$p = \frac{e^{\ln\left(\frac{p}{1-p}\right)}}{1 + e^{\ln\left(\frac{p}{1-p}\right)}}$$

10.9 PUTTING IT ALL TOGETHER: LOGISTIC REGRESSION

10.9.1 The Logistic Regression Model

Having reviewed concepts of exponential functions, logarithms, natural logs, odds, and probabilities, we are now ready to put all of these ingredients together to better understand the logistic regression model introduced earlier in (10.5):

$$p = \frac{1}{1 + e^{-(\alpha + \beta x_i)}} = \frac{e^{(\alpha + \beta x_i)}}{1 + e^{(\alpha + \beta x_i)}}$$

The formulation of the logistic model should now make good sense to you. Having reviewed how probabilities can be converted to odds, we see that $e^{(\alpha + \beta x_i)}$ in (10.5), the exponentiated logit, is nothing more than the odds, where $\alpha + \beta x_i$ is, in the language of the generalized linear model, the linear predictor. Let us now put all of these concepts together.

Where in a least-squares regression we obtained raw regression coefficients that we could interpret as **the expected change in the response variable for a one-unit change in the explanatory variable**, we will analogously be able to interpret the logit, only now, the linear predictor is not in any kind of "natural" units like it was in least-squares regression where no transformation was required. In the logistic model, the unit for the linear predictor is that of the log of the odds (i.e., logit). Hence, the interpretation for a given predictor variable will be the **expected change in the logit of the response variable for a one-unit change in the explanatory variable**. Notice that all that has changed, in this sense, going from OLS to the logistic model, is the **scale** of the units on which the response variable is interpreted. In least-squares, we did not have to transform the response variable, since the canonical link was that of the identity function, which amounted to no transformation at all. In logistic regression, we must perform a transformation in order to **linearize the odds**. However, once the transformation is complete, the interpretation of the coefficient associated with each predictor in the model will be essentially analogous in logistic regression as it was in a least-squares problem. The units on the response variable have simply changed.

10.9.2 Interpreting the Logit: A Survey of Logistic Regression Output

We survey some hypothetical small-scale data to demonstrate the interpretation of a logit in typical logistic regression output. Recall the Q–V data from Chapter 7 in which quantitative and verbal scores were obtained on nine subjects (Table 10.1). We adapt the data to include a **training group** variable

TABLE 10.1 Hypothetical Data on Quantitative and Verbal Ability for Those Receiving Training (Group = 1) Versus Those Not Receiving Training (Group = 0)

Subject	Quantitative	Verbal	Training Group
1	5	2	0
2	2	1	0
3	6	3	0
4	9	7	0
5	8	9	0
6	7	8	1
7	9	8	1
8	10	10	1
9	10	9	1
10	9	8	1

(coded 0, 1) corresponding to whether subjects received or did not receive a prior training program designed to improve their quantitative and verbal abilities (we also add an observation to make the data balanced, meaning equal numbers of subjects in each training group). For this analysis, we use only the quantitative scores in predicting group membership.

Our research question of interest is: **Can quantitative ability be used to predict group membership?** We run the analysis in R:

```
> Q <- c(5, 2, 6, 9, 8, 7, 9, 10, 10, 9)
> group <- c(0, 0, 0, 0, 0, 1, 1, 1, 1, 1)
> logistic <- glm(group ~ Q, family = binomial())
```

R will generate much more output than given below, and we will survey the full logistic output in our analysis of the **Challenger** data shortly. For now, we wish to only focus on the estimated coefficient for q to demonstrate its interpretation, and compare it to that of OLS regression:

```
> summary(logistic)
```

Coefficients:

	Estimate	Std. Error	z value	Pr(>|z|)
(Intercept)	-7.6466	5.2058	-1.469	0.142
q	0.9666	0.6220	1.554	0.120

Note that the coefficient for q is equal to 0.97 (we rounded up). It is not statistically significant, though that need not concern us here. We are interested now only in understanding the interpretation of the coefficient. If the output were from a linear least-squares regression, how would we interpret the coefficient for q? Our interpretation would be:

For a one unit increase in quantitative ability, we can expect, on average, Group to increase by 0.97 units.

Of course, the above interpretation is wrong and does not make sense in the current situation, since group is **binary**. The coefficient 0.97 is not in the "natural" units of the response variable. Rather, it is a transformed variable, the transformation being that of the natural log of the odds, or **logit**. Hence, the correct interpretation for the coefficient is the following:

For a one unit increase in quantitative ability, we can expect, on average, the logit of Group to increase by 0.97 units.

But what are the "units?" The coefficient 0.97 is in units of the logit and not the natural units of the variable as would be the case in OLS regression. Notice that both interpretations, that from OLS regression and from logistic regression, are quite similar. The difference is simply in the units of the actual estimated coefficient. To convince yourself of the necessity of the logistic model, in this case, consider what the least-squares interpretation would imply about our response variable. Expecting group to increase from an amount of 0.97 from group 0 to 1 makes no sense in this case since the response variable is not **linear**. We have to transform it to **linearity** for things to make sense. Of course, interpreting something called the "logit" is quite awkward. However, since we know that logits can be transformed back into probabilities and odds, we therefore have a solution to making the problem more interpretable. We first convert 0.97 to odds:

$$e^{\ln\left(\frac{p}{1-p}\right)} = 2.71^{0.97} = 2.63$$

Our interpretation is that for a one-unit increase in quantitative ability, the odds of being in group 1 versus 0 are, expectantly, 2.63 to 1. In this context, the number 2.63 is called the **odds ratio**. We could have also obtained this in R via

```
> exp(coef(logistic))
  (Intercept)                    q
0.0004776506 2.6289294420
```

Note that the value for q matches up to our computed value (we rounded up). Since often more intuitive than odds (unless you spend a lot of time at the horse track), we know from previous work that we can convert the odds into a probability. First, let us get the predicted logit for the one-unit increase in q. From the output we see that the intercept term is equal to -7.6466. Therefore, the regression equation is:

$$y_i' = -7.6466 + 0.9666(q_i)$$

Recall that we can use the regression equation just as we would in OLS regression, only that now, y_i will be in units of the logit. For example, for a subject who scores 5 on quantitative ability, that subject's predicted score (i.e., logit) is

$$y_i' = -7.6466 + 0.9666(q_i)$$
$$= -7.6466 + 0.9666(5)$$
$$= -2.8136$$

The predicted logit for such a subject is equal to -2.81. But these are logits, which are unintuitive and strange to interpret. We would much prefer interpret probabilities. What is the probability then of that subject being in group 1? To get the probability, we can demonstrate the full logistic function:

$$p = \frac{e^{(\alpha + \beta x_i)}}{1 + e^{(\alpha + \beta x_i)}}$$
$$= \frac{e^{(-7.6466 + 0.9666(q_i))}}{1 + e^{(-7.6466 + 0.9666(q_i))}}$$
$$= \frac{e^{(-7.6466 + 0.9666(5))}}{1 + e^{(-7.6466 + 0.9666(5))}}$$
$$= 0.057$$

Hence, we can see that for $q_i = 5$, the predicted probability of being in group 1 is equal to 0.057. That is, a subject scoring 5 on quantitative ability is probably not one coming from a population that received a training program. Predicted probabilities can be easily obtained in R:

```
> predict(logistic, type = "response")
         1          2          3          4          5          6          7
0.05658579 0.00329031 0.13620541 0.74126605 0.52148241 0.29305467 0.74126605
         8          9         10
0.88279164 0.88279164 0.74126605
```

Note that the first probability given, of 0.057, matches up with our computed probability. This is not coincidental, since the quantitative score for subject 1 was equal to 5.

10.10   LOGISTIC REGRESSION IN R

10.10.1   Challenger O-ring Data

On January 28, 1986, space shuttle **Challenger** lifted off from Cape Canaveral, Florida, and exploded in mid-air 73 seconds into its flight. The cause of the accident (aside from possibly a poor management decision to launch the shuttle in the first place, recall that **determining real and underlying causation is exceedingly difficult**) was the failure of a seal on one of the shuttle's O-rings that serves to keep fuel inside the booster instead of leaking out. The post-incident investigation revealed that the O-ring likely failed because the temperature at which Challenger was launched, 31° F, was a temperature much colder than in any previous launch. It is believed that the cold temperature caused the O-ring to expand and become dysfunctional, thereby leading to fuel leaking out of the booster and onto the main fuel tank of the shuttle, consequently causing the explosion.

The following are data on the occurrence of failures in O-rings on space shuttle data collected from launches prior to that of Challenger, where "1" is a failure, and "0" is a success. For each O-ring event is an associated temperature. It has been argued by many since the catastrophe that had NASA paid more attention to the relationship between temperature and O-ring failure, the disaster might have been averted. As Friendly (2000) remarked:

> The story behind the *Challenger* disaster is, perhaps, the most poignant missed opportunity in the history of statistical graphics. It may be heartbreaking to find out that some important information was there, but the graph maker missed it. (p. 208)

Hence, we ask the following question about this data—**Is temperature predictive of O-ring failure?** The challenger data follow, where "1" represents O-ring failure:

```
> oring <- c(1, 1, 1, 1, 0, 0, 0, 0, 0, 0, 0, 0, 1, 1, 0, 0, 0, 1, 0,
0, 0, 0, 0)
> temp <- c(53, 57, 58, 63, 66, 67, 67, 67, 68, 69, 70, 70, 70, 70,
72, 73, 75, 75, 76, 76, 78, 79, 81)
> challenger <- data.frame(oring, temp)
> some(challenger)
      oring temp
 [1,]     1   53
 [2,]     1   63
 [3,]     0   66
 [4,]     0   67
 [5,]     0   67
 [6,]     0   67
 [7,]     0   70
 [8,]     0   75
 [9,]     0   76
[10,]     0   79
```

The logistic regression of `temp` predicting `oring` is specified as follows:

```
> challenger.fit <- glm(oring ~ temp, data = challenger, family = binomial())
> summary(challenger.fit)

Coefficients:
            Estimate Std. Error z value Pr(>|z|)
(Intercept) 15.0429     7.3786   2.039   0.0415 *
temp        -0.2322     0.1082  -2.145   0.0320 *
---
Signif. codes:  0 '***' 0.001 '**' 0.01 '*' 0.05 '.' 0.1 ' ' 1

(Dispersion parameter for binomial family taken to be 1)

    Null deviance: 28.267  on 22  degrees of freedom
Residual deviance: 20.315  on 21  degrees of freedom
AIC: 24.315

Number of Fisher Scoring iterations: 5
```

About the output:

- The effect for `temp` is statistically significant ($p = 0.03$), suggesting that `temp` is predictive of O-ring failure in the population from which these data were drawn.
- The effect for temp of -0.23 is interpreted as **for a one-unit increase in temperature, on average, the expected change in logit is a decrease of 0.23**. When we exponentiate the logit, we find $e^{-0.23} = 0.795$ to be the odds ratio. That is, for a one-unit increase in temperature, on average, the expected odds of failure is 0.795. Since odds are "centered" at 1.0, the value of 0.795 indicates a **drop** in the odds of failure.
- The **null deviance** of 28.267 is computed with only the intercept in the model. The **residual deviance** of 20.315 includes the `temp` effect over and above the intercept. We could have also obtained the deviance through `deviance(challenger.fit)`. The drop in deviance from 28.267 to 20.315 is suggestive that `temp` may be useful, analogous to how residual sums of squares would drop if we included such a predictor in an OLS regression. The intercept itself of 15.04 is usually of little interest. As the model is currently parameterized, 15.04 is the expected logit for a temperature of **zero** (which of course, extrapolates significantly from the current database).
- The residual deviance of 20.315 is quite close to the residual degrees of freedom of 21, suggesting that we have an adequate model, and that overdispersion is likely not a problem. If overdispersion was a problem, the residual deviance would likely be quite larger than degrees of freedom. Recall that this is a convenient guide for assessing dispersion but is not fool proof (see Venables and Ripley, 2002).
- The AIC statistic is also provided and recall is useful for situations in which we wish to compare nested or nonnested models. As always, lower AIC values are preferable to larger ones.

In Figure 10.7 is what is referred to as an **effect plot** generated from the package `effects` in R (Fox, 2003). For our data, the effect plot relates the probability of a failure on the ordinate to temperature on the abscissa:

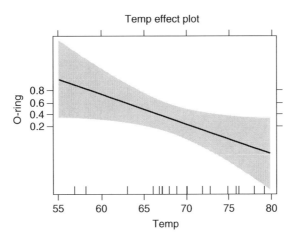

FIGURE 10.7   Effect plot for predicting O-ring failure as a function of temperature.

```
> library(effects)
> allEffects(challenger.fit)

model: oring ~ temp

 temp effect
temp
      53        60        67        74        81
0.939248 0.752713 0.374724 0.105539 0.022703

> plot(allEffects(challenger.fit))
```

Hence, we can see that as temperature increases, the probability of an O-ring failure gradually decreases. Notice the values generated by `allEffects(challenger.fit)` match up with those in the plot. See Fox and Weisberg (2011) for more details on constructing effect displays (including adjusting the default ordinate axis should one choose to do so).

We can also obtain confidence intervals for the model:

```
> confint(challenger.fit)
                  2.5 %        97.5 %
(Intercept)   3.3305848   34.34215133
temp         -0.5154718   -0.06082076
```

We interpret the above to mean that we are 95% confident that the true `temp` parameter likely lies between the lower limit of −0.52 and the upper limit of −0.06. Since the interval does not include 0, as also evidenced by the p-value for the predictor (recall, $p = 0.03$), we can reject the null hypothesis that the true population parameter is equal to 0.

We can obtain influence statistics for the fitted logistic model (we print the first five values only):

```
> influence.measures(challenger.fit)
```

```
Influence measures of
         glm(formula = oring ~ temp, family = binomial(), data = challenger) :
     dfb.1_  dfb.temp   dffit cov.r    cook.d     hat inf
1    0.1678   -0.1640  0.1733 1.305  0.007815  0.1675    *
2    0.2969   -0.2877  0.3174 1.340  0.027099  0.2078    *
3    0.3277   -0.3164  0.3555 1.329  0.034485  0.2090    *
4    0.3401   -0.3166  0.4539 1.141  0.064145  0.1429
5   -0.1502    0.1261 -0.3495 1.065  0.039120  0.0864
```

As recommended by Fox (1997), **hat values** that are more than two or three times the mean of the hat values may be worth looking at further in terms of their "leverage." Observations 2 and 3 above have hat values of 0.2078 and 0.2090, respectively, which are relatively large. We would not delete these observations, though it may be worth re-running the regression to see whether our findings would change without them included. Q–Q plots and plotting logit residuals against values of predictors can also be used as diagnostic aids following a logistic regression, though with some interpretative caveats as a result of the binary nature of the response variable. For details, see Fox (1997, p. 457).

10.11 CHALLENGER ANALYSIS IN SPSS

We now perform the analysis of the Challenger data in SPSS. We generate only output that is most essential in illustrating the analysis. We also obtain classification results and a plot to depict findings:

```
LOGISTIC REGRESSION VARIABLES oring * specifies oring as the response
variable
   /METHOD=ENTER temp * requests all variables be entered
simultaneously (the only variable is temp)
   /SAVE=PRED PGROUP RESID * requests to save predicted probabilities,
predicted group membership, and residuals
   /CLASSPLOT * requests a classification plot of results
   /PRINT=ITER(1) * requests an iteration history
   /ITERATE(20) CUT(0.5). * requests a maximum of 20 iterations, with
a classification cut-off point of 0.5
```

SPSS provides us with the classification rates in an intercept-only model, that is, a model excluding the predictor. A total of 69.6% of cases are correctly classified.

			Classification Table[a,b]		
				Predicted	
				O-ring	
	Observed		0.00	1.00	Percentage correct
Step0	O-ring	0.00	16	0	100.0
		1.00	7	0	0.0
	Overall percentage				69.6

[a]Constant is included in the model.

[b]The cut value is 0.500.

A summary of the model is given next, where in addition to the log-likelihood, so-called **pseudo-R^2** measures of model fit are also provided, which include the **Cox & Snell** R^2 along with the **Nagelkerke** R^2. These are known as pseudo-R^2 measures for the reason that though they attempt to "mimic" R^2 reported in OLS regression, they do not have "variance accounted for" interpretations. The Cox & Snell, for instance, does not have a maximum value of 1.0 such as is of course typically the case for R^2 in OLS regression (James et al., 2013). The Nagelkerke R^2 attempts to improve on the Cox & Snell through scaling it relative to the maximum value it can attain for a given problem, and hence, may be preferable to Cox & Snell. A drawback of both of these indices, however, is that they may not coincide with odds and odds ratios for a given problem in communicating the strength of evidence for a model (Cohen et al., 2003). Hence, we recommend, if these measures are to be interpreted at all, that they be used only as a "ballpark" indicator of overall effect rather than as a precise measure analogous to R^2 in OLS regression.

		Model Summary	
Step	−2 log likelihood	Cox & Snell R square	Nagelkerke R square
1	20.315[a]	0.292	0.413

[a]Estimation terminated at iteration number 5 because parameter estimates changed by less than 0.001.

The updated classification table is given next. As we can see, with the aid of temp, the model now correctly classifies 87% of cases.

			Classification Table[a]		
				Predicted	
				O-ring	
	Observed		0.00	1.00	Percentage correct
Step 1	O-ring	0.00	16	0	100.0
		1.00	3	4	57.1
Overall percentage					87.0

[a]The cut value is 0.500.

The significance tests for the predictor reveals the same as that which we noted in R. Temp is statistically significant at $p = 0.032$. The odds ratio is equal to 0.793 (i.e., Exp(B)), which, within rounding error, agrees with that reported in R.

			Variables in the Equation				
		B	S.E.	Wald	df	Sig.	Exp(B)
Step 1[a]	temp	−0.232	0.108	4.601	1	0.032	0.793
	Constant	15.043	7.379	4.156	1	0.041	3412315.418

[a]Variable(s) entered on step 1: temp.

SPSS next provides us with a **classification plot**, which corresponds to the numbers in the final classification table just discussed. On the abscissa is indicated the cut value of 0.5, where below and above this value corresponds to group classification. Within the plot, one does not count the actual

number of 0's and 1's to obtain the classification results. Rather, one counts the height of the bars to get a single frequency. For instance, at the far left of the plot, we can see one observation having a very low predicted probability. This one observation is represented by a bar height of four zeros. As another example, consider the number of 1's on the half greater than the cut value of 0.5. This number of 1's is equal to 4 (and not 16, by the **actual** number of 1's in the plot).

```
            Step number: 1

            Observed groups and predicted probabilities

       4 +                    1                                                      +
         I                    1                                                      I
         I                    1                                                      I
F        I                    1                                                      I
r      3 +                    1            0                                         +
e        I                    1            0                                         I
q        I                    1            0                                         I
u        I                    1            0                                         I
e      2 +        0 1         0            0                                         +
n        I        0 1         0            0                                         I
c        I        0 1         0            0                                         I
y        I        0 1         0            0                                         I
       1 +  000 0 0    0  0   0   0   0    0    0         1            1 1      1     +
         I  000 0 0    0  0   0   0   0    0    0         1            1 1      1     I
         I  000 0 0    0  0   0   0   0    0    0         1            1 1      1     I
         I  000 0 0    0  0   0   0   0    0    0         1            1 1      1     I
Predicted ---------+---------+---------+---------+---------+---------+---------+---------+---------+---------+----------
   prob:  0       .1        .2        .3        .4        .5        .6        .7        .8        .9        1
   Group: 000000000000000000000000000000000000000000000000011111111111111111111111111111111111111111111111111111111
           Predicted probability is of membership for 1.00
           The cut value is 0.50
           Symbols: 0 - 0.00
                    1 - 1.00
           Each symbol represents 0.25 cases
```

10.11.1   Predictions of New Cases

Recall that there are two general purposes for building a model from sample data. The first is to estimate whether one can obtain a well-fitting model that accounts for the empirical observations. If the model fits relatively well, a second purpose then might be to use the model to predict outcomes on **new** cases.

As an example of how this can be done in SPSS, suppose engineers at NASA obtained new temperature data and wanted to obtain predicted values on O-ring failures. We simulate such an example. We set up our data file as below, where the first five observations are the first from the original data, and the second five observations are the first from our new data, where we do not yet know predicted group membership (we are going to use temp to predict these values):

o-ring	temp	dataset
1.00	53.00	0.00
1.00	57.00	0.00
1.00	58.00	0.00
1.00	63.00	0.00
0.00	66.00	0.00
	55.00	1.00
	58.00	1.00
	51.00	1.00
	64.00	1.00
	65.00	1.00

Note once more that we only have O-ring data for the first data set, data set = 0. Our goal is to use the `oring ~ temp` model to predict new cases on the last five rows. Notice that our new data on `temp` consists of 55, 58, 51, 64, 65. The column **data set** is the designation of whether the data are the original modeled data (**data set = 0**) or the new data (**data set = 1**). The columns **prob** and **class** (in the SPSS output below) represent the probabilities and respective group classification associated with each case. The probabilities we obtain for the second half of the above file constitute probabilities that were computed on the first half.

Here is how we set up the syntax:

```
LOGISTIC REGRESSION VARIABLES oring
  /SELECT DATASET EQ 0
* tells SPSS to generate the model on the original data; it will nonetheless also
generate predictions on the second data set (i.e., dataset = 1).
/METHOD = ENTER temp
/SAVE PRED(prob) PGROUP (class)
/CRITERIA = PIN(0.05) POUT(.10) ITERATE(20) CUT(0.5).
```

The model output will be the same as that given earlier, so we do not reproduce it here. What we are interested in is the estimation of the new probabilities for the new data (based on the full data not only these 5 cases):

oring	temp	dataset	prob	class
1.00	53.00	0.00	0.94	1.00
1.00	57.00	0.00	0.86	1.00
1.00	58.00	0.00	0.83	1.00
1.00	63.00	0.00	0.60	1.00
0.00	66.00	0.00	0.43	0.00
	55.00	1.00	0.91	1.00
	58.00	1.00	0.83	1.00
	51.00	1.00	0.96	1.00
	64.00	1.00	0.55	1.00
	65.00	1.00	0.49	0.00

We see that SPSS has produced the two new columns, **prob** and **class**. We can now interpret the predicted values. For instance, on this new data, for a temperature of 55, the probability of failure (i.e., "1") is 0.91. This is the value under prob for temp of 55 (i.e., located in the sixth row). Because the probability is greater than 0.5, it is classified into class = 1. This prediction was generated by using the original coefficient weight of −0.232 from the regression:

$$\text{logit} = 15.043 - 0.232(\text{temp})$$
$$= 15.043 - 0.232(55)$$
$$= 15.043 - 12.76$$
$$= 2.283$$

Recall that the logit is transformed into the corresponding probability by:

$$p = \frac{e^{\text{logit}}}{1 + e^{\text{logit}}} = \frac{e^{2.283}}{1 + e^{2.283}} = 0.91$$

We note that the probability of 0.91 matches up with that generated by SPSS for the first unknown "test case" (i.e., row 6). We compute the remaining predicted cases in analogous fashion. Not surprisingly, failure is predicted for the lower temperatures, whereas for `temp = 65`, no failure for the O-ring is expected (indicated by prob = 0.49 and class = 0).

10.12 SAMPLE SIZE, EFFECT SIZE, AND POWER

One can estimate statistical power for logistic regression using `powerLogisticCon` within the `powerMediation` package (Qiu, 2013). To use it, we require the following inputs:

```
> install.packages("powerMediation")
> library(powerMediation)
> powerLogisticCon(n, p1, OR, alpha = )
```

where, n is estimated or desired sample size, p1 is the proportion of "successes" at the mean of the predictor (called the "event rate") on the binary dependent variable, OR is the expected or minimally desired odds ratio, and alpha is the desired type I error rate.

For example, suppose we wished to estimate power for a simple logistic regression with a continuous predictor with sample size equal to 100, event rate equal to 0.5, expected odds ratio (i.e., serving as a reasonable estimate of effect size (Chen, Cohen, and Chen, 2010a)) equal to 1.5, and we were willing to tolerate a 0.05 type I error rate. In R, we would compute:

```
> powerLogisticCon(100, 0.50, 1.5, alpha = 0.05)
[1] 0.5268531
```

Notice that estimated power is not very high. Suppose instead, we decided to sample 200 subjects:

```
> powerLogisticCon(200, 0.50, 1.5, alpha = 0.05)
[1] 0.817825
```

Note that estimated power has increased dramatically. To demonstrate the influence of the odds ratio on power, suppose that instead of hypothesizing one of 1.5, we hypothesized an odds ratio of 1.2:

```
> powerLogisticCon(200, 0.50, 1.2, alpha = 0.05)
[1] 0.251188
```

Notice how dramatically power has decreased. This simple demonstration once more illustrates the influence that effect size, which in this case is served by the odds ratio, has on statistical power. Our example featured a **continuous** predictor. If the predictor were binary instead, one could use `powerLogisticBin`, also in the `powerMediation` package. See Qiu (2013) for details.

10.13 FURTHER DIRECTIONS

As noted by Crawley (2013), if the distribution of the error term is uncertain enough that we are not comfortable with specifying a particular structure (e.g., Binomial, Poisson, etc.), we may use a robust alternative to estimation called **quasi-likelihood**. This can be requested in R by specifying the `family` as `quasi` (if the error term was distributed Poisson, one would specify `family = poisson`). Another alternative is to use a **generalized additive model** (or, "GAM"), which essentially makes fewer assumptions about the error term. For details, see Crawley (2013, pp. 565–566). Venables and Ripley (2002) also provide a discussion.

In this chapter, we surveyed the case of a single predictor in logistic regression. However, there is nothing preventing a researcher from entering multiple predictors, analogous to modeling multiple predictors in OLS regression so long as one notes the interpretive distinctions and transformations on the estimated coefficients. Interactions are therefore also quite easily modeled. Stepwise approaches can likewise be conducted.

A researcher could also use logistic regression to predict responses over time on the same individuals in a repeated-measures or longitudinal context. These models are often analyzed as generalized linear mixed models. See Venables and Ripley (2002, p. 292) for details. Nonlinear mixed models can also be fit relatively easily using R (see Venables and Ripley, 2002, p. 286). Indeed, most of the models suitable in ordinary analysis of variance and regression can be generalized to logistic regression with the assumption that the response variable is **binary** (or in the case of multinomial regression, **polytomous**). Modeling product terms in logistic regression is quite straightforward given some experience with modeling interaction terms in multiple regression. For relatively large models that contain numerous predictors and interaction terms, one could use R's `addterm` and `dropterm` functions in the `MASS` package, along with that of `stepAIC`, to test which terms in the model can be added or dropped to better the parsimony and fit. For details of how to use these functions, again see Venables and Ripley (2002, pp. 201–202).

Finally, recall that the chi-square test of independence for a 2×2 table, as discussed in Chapter 2, was easy to interpret because we were only modeling the association between **two** factors. When we classify counts by a **third** (and perhaps a **fourth** and **fifth**) variable, **log-linear models** are called upon to handle the multiway classification of counts. For details, see Agresti (2002). As well, instead of modeling such count data using Poisson regression requiring a link function, one can also model the data using a **multinomial** model approach. For details, see Venables and Ripley (2002, p. 203). **Support vector machines** are an alternative to logistic regression that are widely used in **machine learning** circles and **computer science**. The topic is relatively advanced and cannot be adequately understood without vectors and matrices. See Izenman (2008) for a thorough treatment.

10.14 CHAPTER SUMMARY AND HIGHLIGHTS

- A vast majority of statistical models make the assumption of **linearity in the parameters**. These include the well-known linear models. There are many cases, however, where a **nonlinear** relationship may be hypothesized, which necessitates a nonlinear model.

- **Linearity in the parameters** means that parameters, not necessarily variables, are raised to the exponent 1 in the model.

- Reasons for hypothesizing nonlinear models instead of linear ones include the fact that the actual **empirical relationship** between the response and predictor is thought to be nonlinear in form (e.g., Yerkes–Dodson inverted U curve) or the **operationalization of the response variable** necessarily makes the relationship nonlinear, such as in the case of a binary response.

- The class of models known as **generalized linear models**, first proposed by Nelder and Wedderburn (1972), can incorporate the classic linear model as a special case, but also features nonlinear cases either due to implicit or explicit nonlinearity.

- The classical linear model, $y_i = \alpha + \beta x_i + \varepsilon_i$, can be said to be generally composed of two components, one that is **systematic**, $\alpha + \beta x_i$, and one that is **unsystematic** or random (or stochastic), ε_i.

- The generalized linear model utilizes a **link function** to transform a nonlinear response variable (e.g., binary response) into one that is linear.

- **Canonical links** are those link functions that are **natural** to the family of distributions in a given context. For instance, the link function for a **binomial** variable is the **log of the odds**, known as the

logit. For a **Poisson** variable, the canonical link function is the **log**. For a **Gaussian** variable, the canonical link is the **identity** function.

- A **dispersion parameter** is an index measuring the degree of spread in a distribution. **Overdispersion** is said to exist if the dispersion parameter, either estimated or known, exceeds the dispersion expected for the given distribution.

- The **logistic regression** model is useful for modeling binary or polytomous response variables. Mathematically, it is somewhat different from the discriminant analysis model, but pragmatically, it can be interpreted as a competing alternative to discriminant analysis.

- In logistic regression, the **logit** can be **exponentiated** to provide a measure of the **odds**. Both the logit and the odds can be converted to **probabilities**. These transformations can be helpful if for no other reason than in helping one digest logistic regression results.

- **Power** and **sample size** can be easily estimated for logistic regression using `powerLogistic-Con` in R.

REVIEW EXERCISES

10.1.   Briefly discuss why the model $y_i = \alpha + \beta_1 x_1 + \beta_2 x_2^2 + \varepsilon_i$ can be regarded as a **linear model**, even with the presence of x_2^2 in the equation.

10.2.   Briefly discuss why the model $y_i = \alpha + \beta_1 x_1 + \beta_2^2 x_2 + \varepsilon_i$ can be regarded as **nonlinear**. Compare and contrast this model with the model in Exercise 10.1.

10.3.   Distinguish between **intrinsic** versus **nonintrinsic** linearity.

10.4.   How does **discontinuity** in a response variable introduce **nonlinearity** into a model? Explain.

10.5.   Give a research example in your field where you might expect the phenomena under investigation to follow an **exponential function**. How about a **logarithmic** one?

10.6.   Discuss what is meant by a **generalized linear model**, and how and why linear models are considered to be contained within, or be a special case of the wider and more inclusive generalized linear model framework.

10.7.   Describe the role of a **link function** in the generalized linear model.

10.8.   Discuss the nature of a **canonical link** in a generalized linear model. What is the canonical link for a Gaussian variable? Why is this so?

10.9.   What are the components to a **generalized linear model**? Explain each.

10.10. What is the appropriate **link function** for a binomial setting?

10.11. Discuss what is meant by the statement—**The limiting form of the Poisson distribution is that of the normal distribution**. More generally, what does it mean to say one distribution is the limiting form of another?

10.12. Run a **logistic regression** using Fisher's iris data (see Chapter 8) predicting group membership on species setosa and versicolor based on petal length and petal width. Are either petal length or petal width statistically significant predictors of species? If so, which predictor is more relevant? Why?

10.13. Use `powerLogisticCon` in R to estimate required **sample size** for a **logistic regression** with a continuous predictor in which desired power is set at 0.95, with an event rate equal to 0.3 and odds ratio equal to 1.1. Use a significance level of 0.05.

11

MULTIVARIATE ANALYSIS OF VARIANCE

A generalization of quite a different order is needed to test the simultaneous deviations of several quantities · · · Joint comparisons of correlated variates, and variates of unknown correlations and standard deviations, are required not only for biologic purposes, but in a great variety of subjects.

(Hotelling, 1931, pp. 361–362, The Generalization of Student's Ratio)

Multivariate analysis of variance (MANOVA) can be conceptualized as either an extension of univariate analysis of variance (ANOVA) or as a more general linear model with ANOVA constituting a "special case" of that more general model. The primary conceptual difference is that ANOVA features a **single** continuous dependent variable, while MANOVA features **multiple** dependent continuous variables, all considered **simultaneously** in the analysis. The dependent variable in a MANOVA is a **linear combination** (or linear **composite**) of response variables. As in univariate ANOVA where the user tests a null hypothesis of equality of population means, in MANOVA the researcher usually also tests null hypotheses about equality, although such a test is now on a **vector** of continuous dependent variables that make up the composite variable.

MANOVA is not suitable for every research context. Indeed, prior to constructing a linear composite of response variables, one should justify whether it is even substantively meaningful to hypothesize such a thing with regard to the variables under consideration. This decision must ultimately be made on grounds external to the statistical method. Statistical and software technology will allow us to analyze virtually anything and in any way. Whether what we analyze accords in any way with our scientific pursuits or theoretical interests, which is presumably **the reason why we are running statistical analyses in the first place**, is another matter entirely. Though this is true for virtually every statistical model, as we will see, the decision regarding whether or not to use a MANOVA provides a good context for having this discussion.

Applied Univariate, Bivariate, and Multivariate Statistics: Understanding Statistics for Social and Natural Scientists, With Applications in SPSS and R, Second Edition. Daniel J. Denis.
© 2021 John Wiley & Sons, Inc. Published 2021 by John Wiley & Sons, Inc.
Companion Website: www.wiley.com/go/denis/appliedstatistics2e

Johnson and Wichern (2007) is a good source for multivariate methods in general and includes a thorough chapter on MANOVA (Chapter 6). Rencher and Christensen (2012) as well as Rencher (1998) cover MANOVA in addition to multivariate multiple regression. For a treatment featuring more hands-on applications using software, Tabachnick and Fidell (2007) has become a classic resource. Tatsuoka (1971), Harris (2001), and Timm (2002) are also good sources, the last of these on the technical side. Hair et al. (2006) provide applications in business and marketing contexts. Anderson (2003) provides a deeper theoretical treatment and historically has been a classic resource among mathematical statisticians. Bilodeau and Brenner (1999) is a wholly theoretical, abstract, and very rigorous mathematical treatment.

11.1   A MOTIVATING EXAMPLE: QUANTITATIVE AND VERBAL ABILITY AS A VARIATE

To illustrate the context in which one may wish to conduct a MANOVA, recall the hypothetical Q–V data used to illustrate both regression analysis (Chapter 7) and logistic regression (Chapter 10). Recall that for these data, measures were taken of individuals' quantitative and verbal abilities using standardized tests, where scores on such tests could range from 0 (little ability) to 10 (maximum ability). In this chapter, we will treat both variables as **dependent variables**. Recall also that in our previous chapter on logistic regression, we extended the data to include a training group variable (coded 0, 1) corresponding to whether subjects received or did not receive a prior training program designed to improve their quantitative and verbal abilities. Five subjects received the training program (group = 1), while five did not (group = 0). The data are reproduced in Table 11.1.

Consider now how we might analyze these data. We could run one analysis with quantitative as our dependent variable and group as our independent variable, then another analysis with verbal as our dependent variable and once again group as our independent variable. The function statements and null hypotheses for each analysis against two-sided alternatives would therefore be:

$$\text{Quantitative as a function of } Group - H_0 : \mu_1 = \mu_2 \text{ against } H_1 : \mu_1 \neq \mu_2$$

$$\text{Verbal as a function of } Group - H_0 : \mu_1 = \mu_2 \text{ against } H_1 : \mu_1 \neq \mu_2$$

The appropriate analysis for each null hypothesis would be an independent-samples t-test or a between-subjects ANOVA on two groups. A rejection of the null hypothesis in the first analysis would suggest there to be mean population differences on quantitative. A rejection of the null hypothesis in the second analysis would suggest there to be mean population differences on verbal.

TABLE 11.1   **Hypothetical Data on Quantitative and Verbal Ability for Those Receiving Training (Group = 1) versus Those Not Receiving Training (Group = 0)**

Subject	Quantitative	Verbal	Training Group
1	5	2	0
2	2	1	0
3	6	3	0
4	9	7	0
5	8	9	0
6	7	8	1
7	9	8	1
8	10	10	1
9	10	9	1
10	9	8	1

Suppose now that instead of learning whether there are mean population differences on quantitative and verbal considered **separately**, we instead wanted to learn of any mean population differences on quantitative and verbal considered **jointly**. That is, suppose we were interested in generating a **composite score** on quantitative and verbal such that our new function statement becomes:

$$(\text{quantitative} + \text{verbal}) \text{ as a function of group}$$

Why might we want to consider these two variables simultaneously in such a manner? Why would we want to analyze them **together** instead of independently using separate ANOVAs? Perhaps it is because we believe quantitative and verbal ability "go together" in some sense, that they represent some underlying theme or **construct**. Indeed, if a student excels well in quantitative and verbal learning, we might be tempted to designate that student as **intelligent**. This is exactly the kind of rationale for why we might want to consider these variables jointly, because **taken together**, we believe, or hypothesize, that they represent some overall theme or construct. Had our variables been quantitative ability and height, for instance, we would probably not be interested in considering them as a linear sum as we would quantitative and verbal. **The consideration of variables jointly in such a manner usually only makes sense if their combination makes sense to us on a scientific, substantive level**. Exploratory searches are always good ideas, but combining quantitative ability and height would probably not be a meaningful linear composite, mostly because such a composite would likely be impossible to describe **substantively**.[1]

11.2 CONSTRUCTING THE COMPOSITE

Consider now how we might go about generating this linear sum. Naively, we could give each dependent variable equal weight, assigning scalars of "1" to each. Under this condition, our function statement would become:

$$(1)\text{quantitative} + (1)\text{verbal} \text{ as a function of group}$$

Or, perhaps in our theory of what we consider intelligence to actually be, we believe that verbal should get more **weight** in the linear composite, perhaps double the weight of quantitative. If this were the case, then we might weight our composite as:

$$(1)\text{quantitative} + (2)\text{verbal} \text{ as a function of group}$$

We could continue this process of pondering "ideal" scalars. But instead of simply guessing, let us instead set some criteria by which these weights should be chosen. What should be our goal in the choosing of such weights? It seems reasonable to select weights such that the ensuing linear combination provides **maximum separation** between groups on the independent variable. That is, we would like to weight our composite such that the linear sum $a_1(\textbf{quantitative}) + a_2(\textbf{verbal})$ provides more separation

---

[1] This is not to discourage exploratory research. It is only to point out that blindly including two variables such as quantitative ability and height into the same analysis would likely not be substantively or scientifically meaningful, unless of course it turns out that there **is** a correlation between these two variables and that we could attribute meaning to it. Otherwise, an appropriate guideline for generating a linear composite in MANOVA is to include variables that the investigator a priori theorizes "go together" in some theoretical sense. If, on the other hand, the investigator concedes his or her work to be 100% in the exploratory phase and does not wish to confirm any hypothesis whatsoever, then testing a variety of linear combinations may be more permissible, so long as the researcher indicates the nature of this exploratory search in any report or presentation of findings and encourages cross-validation of whatever findings may be "discovered."

between groups on the independent variable than any other weights a_1 and a_2 we could have chosen. The solution to this problem is essentially the technical basis on which MANOVA and discriminant analysis rest, that of **choosing a suitable linear combination of variables to satisfy a set of mathematical constraints**, usually already implied by the model parameterization. Discriminant analysis is featured in the following chapter, and as we will see, is essentially the "reverse" of MANOVA.

11.3   THEORY OF MANOVA

To properly develop the theory behind fixed effects multivariate analysis of variance, we first briefly review the concept of a **linear combination**. A linear combination y_i of variables $x_1, x_2, ..., x_n$ can be defined as

$$y_i = a_1 x_1 + a_2 x_2 + ... + a_n x_n$$

where y_i is the outcome of the linear combination (it is the set of values generated by the combination of $a_1 x_1 + a_2 x_2 + ... + a_n x_n$), $a_1, ..., a_n$ are scalars, typically all real numbers, and $x_1, ..., x_n$ are variables of which when weighted by respective scalars $a_1, ..., a_n$ make up the linear sum y_i.

Regardless of the context for which the given linear combination is being generated or estimated, the magnitude and sign of the scalars $a_1, ..., a_n$ plays a significant role in the determination of y_i. These scalars are of utmost importance, since, in combination with $x_1, ..., x_n$, determine the value for y_i. Note carefully that when we say MANOVA **analyzes a dependent variable that is a linear combination**, we mean that theoretically it analyzes a **single** variable y_i, only that y_i is now a **weighted composite** of other variables. The idea of a linear combination is definitely not new. This whole book, in one way or another, is about linear combinations. This commonality may at first glance not be apparent to the reader. Consider where linear combinations have already figured prominently:

- In simple linear regression, the response variable y_i is a linear combination of observed responses on a single predictor variable. That is, $y_i = \alpha + \beta x_i + \varepsilon_i$ is a linear combination with weights (i.e., parameters) α and β chosen (in OLS regression) such that they minimize $\sum \varepsilon_i^2$.
- In analysis of variance, the dependent variable y_i is a linear combination of the sort $y_{ij} = \mu + \alpha_j + \varepsilon_{ij}$ where weights (again, these are parameters) μ and α_j are chosen such that they again minimize $\sum \varepsilon_i^2$ (in ANOVA just as in least-squares regression, effects in both are usually estimated by ordinary least-squares).
- In considering contrasts and post-hocs in the ANOVA setting, we generated such linear combinations $C_i = c_1 \mu_1 + c_2 \mu_2 + ... + c_J \mu_J$, which were nothing more than linear sums of weighted population means. The job here too was to choose suitable scalars $c_1, ..., c_J$ that would weight the linear combination appropriately, generating a linear contrast that was substantively meaningful.
- Independent and paired samples t-tests were other examples of linear combinations (and contrasts) in which the computation of observed t was in part a weighted sum of sample means, where we implicitly weighted the numerator appropriately (1 and -1) to generate the contrast of interest:

$$t = \frac{(1)\bar{y}_1 + (-1)\bar{y}_2}{\sqrt{\dfrac{s_1^2}{n_1} + \dfrac{s_2^2}{n_2}}}$$

That is, though we typically cast the numerator of the independent-samples t-test as $\bar{y}_1 - \bar{y}_2$, it was always implied that such a difference was actually a contrast of the form $(1)\bar{y}_1 + (-1)\bar{y}_2$, where it was also clear that the sum of coefficients for this contrast was equal to 0 (i.e., $1 + (-1) = 0$). That is, recall from our previous discussion of contrasts, that the sum of weights equaled zero defined the linear combination as a legitimate contrast.

All of this is to say that when it comes to working with linear combinations, we have already dealt with them **aplenty** in one way or another in this book. In MANOVA, we again seek to generate a linear combination, but this time our linear combination will consist of a string of dependent variables and test null hypotheses about equality of **mean vectors** on this linear composite or "**variate**," as it is sometimes called.

11.4   IS THE LINEAR COMBINATION MEANINGFUL?

We mentioned earlier that if the linear combination of dependent variables is not hypothesized to be representative of some construct, or otherwise responses "go together" in some sense, then performing a MANOVA may not be your best analytical choice, unless of course you are simply on an unguided purely exploratory mission. Again, our goal is not to diminish the purpose or utility of unguided exploratory searches. But this is different than simply performing MANOVA because as a "recipe," you believe it appropriate since you have more than a single dependent variable in your data file. **The fact that you have multiple dependent variables at your disposal is not alone justification for running a MANOVA.** Too often, students (and sometimes researchers) "justify" their use of MANOVA based on the availability of several dependent variables. Such is not an ideal justification. As mentioned, a much more suitable rationale is a belief or theory held such that the variables you do have at your disposal are suitable to being **combined** into a composite variable. A compelling historical case for constructing such a linear composite is that of Charles Spearman's (1904a) theory of intelligence, the so-called "g-factor," variations of which have been studied ad nauseum since. But what is the **g-factor**? **It is a linear combination of observed abilities such when summed, is thought to reflect some construct of interest, that of IQ.**

Given this starting point and substantive rationale for MANOVA, there also turn out to be a few **statistical** benefits to the method when compared to running several independent univariate ANOVAs. These include **control over type I error rate**, **covariance among dependent variables**, and the fact that a multivariate effect can be observed even in the absence of univariate effects, known as **Rao's paradox**. We discuss each of these now.

11.4.1   Control Over Type I Error Rate

One reason to like MANOVA **statistically** is that it helps keep overall α (i.e., family-wise type I error rate) at a nominal level based on an omnibus test. Recall that with each running of a statistical test (for instance, an F-ratio in ANOVA), there is associated with it a type I error rate, equal to the significance level at which you set your decision criteria for rejection of the null hypothesis. For instance, if you run three separate ANOVAs, each at $\alpha = 0.05$, the overall error rate will not be 0.05, but rather will compound. This compounding is approximately additive across the three tests. It is not a simple sum of α_{PC} but is nearly so. As Hays (1994) notes, the probability of making no type I errors on three significance tests, each set at $\alpha = 0.05$, can be considered a binomial random variable with distribution:

$$p(r) = \binom{n}{r} p^r (1-p)^{n-r}$$

where recall from Chapter 2 that r is the number of successes, n is the number of trials, and p is the probability of a success on any given trial. For the case of $n = 3$ significance tests, if we set $r = 0$ (for zero type I errors) and the probability p of a success (i.e., "success" being a type I error) equal to 0.05, and we assume these three tests are independent of one another, then the probability of making zero type I errors is equal to:

$$p(0, 3, 0.05) = \binom{3}{0}(0.05)^0 (0.95)^3 = 0.86$$

Thus, the probability of making **at least one type I error** is equal to $1 - 0.86 = 0.14$. We see then that the overall probability of a type I error is roughly additive for chosen constant α across the three tests. A general rule for estimating the probability of one or more type I errors in a series of independent significance tests is given by:

$$p(\text{type I error} \geq 1) = 1 - (1 - \alpha_{PC})^k$$

where α_{PC} is the significance level of each test, and k is the number of successive tests being made. For example, with five dependent variables, should we choose to analyze each individually with its own ANOVA and not implement a correction on α_{PC} (e.g., Bonferroni adjustment), the estimated probability of making one or more type I errors across the five tests is

$$p\,(\text{type I error} \geq 1) = 1 - (1 - \alpha_{PC})^k = 1 - (1 - 0.05)^5 = 0.23$$

A probability of 0.23 of making at least one type I error across the five tests is quite high. If we performed a MANOVA on these five response variables simultaneously instead, we could **constrain the error rate** to be that of our nominal level (0.05 in this example). Hence, control over familywise type I error is a primary statistical reason why MANOVA is sometimes preferred over independent univariate ANOVAs. Again, to reiterate, this is a **statistical reason** for preferring MANOVA. If it does not make substantive sense to consider your dependent variables jointly in a MANOVA context, then the fact that MANOVA controls type I error rates is not in itself justification for forging on with the procedure. Otherwise, substantive considerations are taking a back seat to statistical benefits, instead of statistical benefits aiding in substantive discovery. Remember that in the realm of scientific application, statistics is used as a **tool** to help address scientific questions of interest. Just as one does not adapt the construction of a house to appease a hammer, **one should not adapt one's scientific mission to appease a statistical test or model**.

11.4.2 Covariance Among Dependent Variables

A second statistical reason for running a MANOVA over independent ANOVAs is that the MANOVA incorporates covariances that may exist among dependent variables that would otherwise go unaccounted for and unanalyzed in separate univariate analyses. For example, height and weight are examples of such measures likely to be correlated. Modeling this covariance into our analysis can possibly generate a more powerful test against the multivariate null hypothesis compared to if we were to conduct separate univariate tests on each dependent variable. Though, as summarized by Field (2009, p. 586), the power of MANOVA is both a function of the pattern of correlation among dependent variables and the size of multivariate effect, and hence such patterns can become quite complex. For further details, see Cole et al. (1994).

Regardless of what factors generate a most powerful test of a multivariate hypothesis, it must be emphasized again that the consideration of the covariance among dependent variables only makes sense if it first makes good research or scientific sense to lump these variables into the same design. As we will discuss with reference to Rao's paradox, one should not perform a MANOVA simply because it may be a more powerful test over univariate analyses. Such would make little **scientific** sense. On a substantive level, one should conduct a MANOVA because **one wishes to analyze a linear combination of dependent variables, regardless of whether it may be more or less powerful than univariate tests**. The fact that MANOVA accounts for correlations among dependent variables is, indeed, an analytical and statistical charm (who wouldn't want to boost power?), but this does not necessarily equate to scientific meaning.[2] As William James so adeptly noted, **we should not confuse data with the abstractions used to analyze such data**.

---

[2] Because of spurious correlations that are due to third variables, even a relatively substantial correlation among variables does not in the least imply they are suitable for MANOVA. For example, murder rate and ice cream cone sales are likely correlated, though combining such variables in a MANOVA would make little substantive sense. Even if seasonal temperature accounts for the correlation, it still does not help us know what murder rate + ice cream sales would mean as a "construct."

11.4.3 Rao's Paradox

A third reason for sometimes preferring MANOVA over successive ANOVAs can be summed up in a problem first brought to the forefront by C.R. Rao in 1966, known as **Rao's paradox**. The essence of the problem is that it is possible to reject a null hypothesis in a multivariate setting, but simultaneously not reject subset univariate hypotheses. Conversely, it is possible to reject univariately but not multivariately. As noted by Healy (1969), Rao's paradox is essentially equivalent to saying that for a given multivariate hypothesis, **there will be a univariate test of significance within the multivariate hypothesis that may be more powerful than this latter hypothesis**. The opposite scenario is also possible in that the multivariate test is more powerful than component univariate tests.

But how can this occur? Though the problem can be understood by drawing ellipses with corresponding rejection regions (see Healy, 1969, p. 412), relying on basics of probability theory can help us, in an approximate way at least, understand the paradox such that we may come to believe it not to be a "paradox" at all, but rather something quite logical. The essential principle is that a joint event (or more precisely, a **joint probability** on two events) can be more or less probable than each event, considered exclusively, that make up the joint event.

For example, the probability of selecting a person at random and that person being married is a probability equal to some number. The probability of selecting a person at random and that person being aged 19 years or less is also a probability. Both of these respective probabilities, that of being married and that of being aged 19 or less are each likely relatively high. After all, many people are married, and many people are 19 years of age or less. Each considered **univariately** would result in relatively substantial probabilities.

Now, consider the event **married and age 19 or less**. The probability of being married and aged 19 or less recall, is a **joint probability**, the probability of which undoubtedly would differ from each of the probabilities that make up the event. The probability of marginal events does not necessarily coincide with the probability of the corresponding joint event. Translated into the language of MANOVA, though there may be mean population differences on each dependent variable considered univariately, this does not guarantee mean differences on such dependent variables considered **jointly**. Likewise, mean differences on dependent variables considered jointly do not necessarily translate into mean differences on such dependent variables when considered univariately (Rencher and Christensen, 2012).

An understanding of Rao's paradox has very important implications for researchers. We summarize what the principle means in this regard:

> When you perform a MANOVA, you are testing a different null hypothesis than when you perform separate univariate ANOVAs, and as such, should not assume that the rejection of one hypothesis (e.g., multivariate hypothesis) automatically informs you of the status of other hypotheses (e.g., univariate hypotheses). Likewise, you should not assume that individual univariate findings on separate response variables will necessarily generate a multivariate effect if "combined."

Is this idea really unique to MANOVA? Not at all. We have emphasized it repeatedly in our discussions of multiple regression when considering partial and semipartial correlations, as well as emphasizing that effects in a multiple regression model are virtually always contingent on what other variables are included in the model. Recall this important principle, because it applies equally well to MANOVA as it did to multiple regression:

> Whenever you test a model, it is the MODEL that you are testing, not unique individual effects contained within the model.

Any effects evaluated in a model, no matter how simple or complex the model may be, must always be evaluated in the **context** of the model. This is as true for *t*-tests as for the most advanced and sophisticated of statistical models. Effects in models are always **context-dependent**.

11.5 MULTIVARIATE HYPOTHESES

Having surveyed some of the substantive issues germane to MANOVA, we now consider some of the specifics concerning testing a multivariate hypothesis. Recall that in a univariate setting, the typical null hypothesis under test in a one-way analysis of variance is

$$H_0 : \mu_1 = \mu_2 = \mu_3 \ldots = \mu_j$$

against the alternative H_1 that a pairwise or other contrast among means $\mu_1, \mu_2, \mu_3 \ldots \mu_j$ is **unequal** somewhere among the population means (e.g., $H_1 : \mu_1 \neq \mu_2$ as one possibility). In a multivariate setting, since we are now interested in testing a null hypothesis about equivalency on **mean vectors**, our null hypothesis becomes:

$$H_0 : \begin{pmatrix} \mu_{11} \\ \mu_{21} \\ \mu_{31} \\ . \\ . \\ . \\ \mu_{j1} \end{pmatrix} = \begin{pmatrix} \mu_{12} \\ \mu_{22} \\ \mu_{32} \\ . \\ . \\ . \\ \mu_{j2} \end{pmatrix} = \begin{pmatrix} \mu_{13} \\ \mu_{23} \\ \mu_{33} \\ . \\ . \\ . \\ \mu_{j3} \end{pmatrix} \ldots = \begin{pmatrix} \mu_{1p} \\ \mu_{2p} \\ \mu_{3p} \\ . \\ . \\ . \\ \mu_{jp} \end{pmatrix} \tag{11.1}$$

where $\mu_{11}, \mu_{21}, \mu_{31} \ldots \mu_{j1}$ are means for dependent variables 1 through j for group 1 of an independent variable. That is, the first column vector in (11.1) represents level 1 of the independent variable for dependent variables 1 through j. The second column vector represents level 2 of the independent variable for dependent variables 1 through j, etc.

The model for a one-way fixed-effects MANOVA can be given by:

$$\mathbf{Y}_{ij} = \boldsymbol{\mu} + \boldsymbol{\alpha} + \boldsymbol{\varepsilon}_{ij} \tag{11.2}$$

where $\mathbf{Y}_{ij}$ is a vector of response variables, $\boldsymbol{\mu}$ is a vector of grand means, $\boldsymbol{\alpha}$ is a vector of sample or treatment effects, and $\boldsymbol{\varepsilon}_{ij}$ is a vector of errors. In the absence of treatment effects, (11.2) reduces to simply $\mathbf{Y}_{ij} = \boldsymbol{\mu} + \boldsymbol{\varepsilon}_{ij}$. Notice that the one-way fixed-effects model of Chapter 3, $y_{ij} = \mu + \alpha_j + \varepsilon_{ij}$, can be regarded as a "special case" of model (11.2) in which we have only a single dependent variable. Both models generate $\boldsymbol{\varepsilon}_{ij}$. In the univariate case, this is a single column vector. In the multivariate case, if we conceive the variables contained in $\mathbf{Y}_{ij}$ to be distinct columns, then we will likewise have as many distinct columns in $\boldsymbol{\varepsilon}_{ij}$. However, in the spirit of MANOVA, if we conceive $\mathbf{Y}_{ij}$ to be a construct or "latent variable," then we can likewise conceive $\boldsymbol{\varepsilon}_{ij}$ to be a single column of errors analogous to the univariate case. For instance, if $\mathbf{Y}_{ij}$ is a linear combination representing IQ, then $\boldsymbol{\varepsilon}_{ij}$ is theoretically the error associated with generating the IQ variate. We develop the even "wider" matrix formulation of the general multivariate model later in this chapter, one that encompasses (11.2) along with that of the linear regression model studied earlier in the book. We also briefly demonstrate how to formally expand such models to include random effects in addition to fixed effects.

11.6 ASSUMPTIONS OF MANOVA

The assumptions of MANOVA are generally parallel to those made in the analysis of variance, such as **normality**, **independence of observations**, and **homogeneity of variance**, except that they now apply to linear combinations of the response vector rather than to single variables. Because we are now working in higher dimensions, the assumption of normality for each dependent variable is not sufficient. We must also make the assumption of **multivariate normality** for all linear combinations on the response vector. Such an assumption is important, for instance, in the T^2 distribution used as the multivariate generalization of univariate t. The assumption of multivariate normality in this case assures the independence of $\bar{y}$ and $\mathbf{S}$. Verifying multivariate normality, as discussed briefly in Chapter 2, is inherently difficult if not impossible due to the number of dimensions involved and what could be occurring in any **subset of those dimensions**. Verifying such normality is usually accomplished through plotting residuals and inspecting Q–Q plots. Fortunately, most multivariate tests are rather robust to modest violations of multivariate normality, especially if cell sizes are not too disparate.

While in ANOVA, we were required to make the assumption of equality (homogeneity) of variances, in MANOVA, we will need to assume **equality of covariance matrices**. We discuss this assumption at some length later in the chapter (Section 11.12).

11.7 HOTELLING'S T^2: THE CASE OF GENERALIZING FROM UNIVARIATE TO MULTIVARIATE

When discussing the nature of analysis of variance in Chapter 3, recall we said it could be considered an extension of the independent samples t-test. We follow a similar approach in developing the multivariate counterpart to the independent samples t-test, known as **Hotelling's T^2**, named after Harold Hotelling who derived its distribution in 1931.

Recall that in an independent samples t-test, we evaluate the tenability of the null hypothesis H_0 : $\mu_1 = \mu_2$ against the statistical alternative $H_1 : \mu_1 \neq \mu_2$. The multivariate counterpart, Hotelling's T^2, evaluates the tenability of whether two **population mean vectors are equal**:

$$H_0 : \begin{pmatrix} \mu_{11} \\ \mu_{21} \end{pmatrix} = \begin{pmatrix} \mu_{12} \\ \mu_{22} \end{pmatrix}$$

The statistical alternative is that there is an inequality between population mean vectors:

$$H_1 : \begin{pmatrix} \mu_{11} \\ \mu_{21} \end{pmatrix} \neq \begin{pmatrix} \mu_{12} \\ \mu_{22} \end{pmatrix}$$

How should we test a difference between mean vectors? In what follows, we develop Hotelling's T^2 by drawing on our knowledge of univariate t. The following derivation is of high importance, since through extending univariate t to its multivariate counterpart, we gain a general understanding as to how multivariate tests, in general, are distinct from univariate ones. The inclusion of a covariance matrix in what follows instead of simply variances as one would have in univariate ANOVA is a "staple" of the multivariate landscape.

Recall the univariate independent samples t-test:

$$t = \frac{E(\bar{y}_1) - E(\bar{y}_2) - \delta_0}{\sqrt{\dfrac{(n_1 - 1)s_1^2 + (n_2 - 1)s_2^2}{n_1 - n_2 - 2} \left(\dfrac{1}{n_1} + \dfrac{1}{n_2} \right)}} = \frac{\mu_1 - \mu_2 - \delta_0}{\sqrt{\dfrac{(n_1 - 1)s_1^2 + (n_2 - 1)s_2^2}{n_1 - n_2 - 2} \left(\dfrac{1}{n_1} + \dfrac{1}{n_2} \right)}} \tag{11.3}$$

where recall $\bar{y}_1$ and $\bar{y}_2$ are the sample means for each group, and their expectations $E(\bar{y}_1)$ and $E(\bar{y}_2)$ are equal to μ_1 and μ_2, respectively; δ_0 is a constant subtracted from the mean difference $\bar{y}_1 - \bar{y}_2$, n_1 and n_2 are the sample sizes of each group, s_1^2 and s_2^2 are the sample variances of each group used as estimators of the corresponding population variances σ_1^2 and σ_2^2. In most cases, the constant δ_0 will be equal to 0, as we implicitly assumed in our review of the t-test in Chapter 2, so that the null hypothesis tested is that of $H_0 : \mu_1 = \mu_2$, or, equivalently, $H_0 : \mu_1 - \mu_2 = 0$. In some research contexts, however, null hypotheses other than these may be useful. For example, in a medical setting, if we wished to demonstrate that our drug decreased cholesterol by more than five units, we might hypothesize a null such that $H_0 : \mu_1 - \mu_2 = 5$, in which case, the numerator of the t-test would be:

$$\bar{y}_1 - \bar{y}_2 - \delta_0 = \bar{y}_1 - \bar{y}_2 - 5$$

Cases in which null hypotheses other than the typical $H_0 : \mu_1 - \mu_2 = 0$, however, are nonetheless quite rare in practice, and statistical software will typically simply assume differences of zero under the null. However, it is good habit to never make assumptions about what the null hypotheses actually are that appear in a research report. If in doubt, confirm with the authors which null exactly was evaluated, since without knowledge of the null hypothesis, statements such as $p < 0.05$, along with effect sizes, carry with them little meaning. Statistical significance represents a surprising result (i.e., an unlikely event) under certain conditions as specified by the null. Without knowing these conditions or assumptions, it becomes impossible to evaluate the probability of the data.

As also reviewed in Chapter 2, the quantity

$$\frac{(n_1 - 1)s_1^2 + (n_2 - 1)s_2^2}{n_1 - n_2 - 2}$$

under the square root sign in the denominator denotes an estimate of the pooled variance, s_p^2 (or $\hat{\sigma}^2_{pooled}$ to emphasize it as an estimate of σ^2_{pooled}). Hence, we can simplify (11.3) to be:

$$t = \frac{\bar{y}_1 - \bar{y}_2 - \delta_0}{\sqrt{s_p^2 \left(\frac{1}{n_1} + \frac{1}{n_2} \right)}} \tag{11.4}$$

Recall that when $n_1 = n_2$, the denominator in (11.4) can be reduced to $\frac{s_1^2}{n_1} + \frac{s_2^2}{n_2}$ so that the standard error of the t statistic is now $\sqrt{\frac{s_1^2}{n_1} + \frac{s_2^2}{n_2}}$.

The multivariate counterpart T^2 is remarkably similar to univariate t, only that now it must encompass **vectors** rather than simply scalars. Following Stevens (2009, p. 147), we proceed first to square both the numerator and the denominator of t in (11.4), while also simultaneously allowing δ_0 to again drop out of the equation since its value will customarily be equal to 0:

$$t^2 = \frac{(\bar{y}_1 - \bar{y}_2)^2}{\frac{(n_1 - 1)s_1^2 + (n_2 - 1)s_2^2}{n_1 - n_2 - 2} \left(\frac{1}{n_1} + \frac{1}{n_2} \right)}$$

Next, we rewrite the equation by expressing the denominator as an inverse:

$$t^2 = (\bar{y}_1 - \bar{y}_2) \left[\frac{(n_1 - 1)s_1^2 + (n_2 - 1)s_2^2}{n_1 - n_2 - 2} \left(\frac{1}{n_1} + \frac{1}{n_2} \right) \right]^{-1} (\bar{y}_1 - \bar{y}_2) \tag{11.5}$$

We denote the pooled variance as an inverse to facilitate our generalization to the multivariate domain. We do not yet have matrices in our formulation, but will very soon, and recall from matrix theory that expressing a term as an inverse, for our purposes, is a way of denoting "division" using matrices. Hence, by expressing the denominator as an inverse now, we are "prepping the ground" so to speak, on our way to defining our new multivariate statistic.

The term $\left(\dfrac{1}{n_1} + \dfrac{1}{n_2}\right)$ in (11.5), when simplified, is equal to $\dfrac{n_1 + n_2}{n_1 n_2}$, yielding

$$t^2 = (\bar{y}_1 - \bar{y}_2)\left[\frac{(n_1 - 1)s_1^2 + (n_2 - 1)s_2^2}{n_1 - n_2 - 2}\left(\frac{n_1 + n_2}{n_1 n_2}\right)\right]^{-1}(\bar{y}_1 - \bar{y}_2) \tag{11.6}$$

Now, take a close look at (11.6) and ask yourself what is the multivariate matrix equivalent to the pooled variance? It is $\mathbf{S}$, the sample **covariance** matrix. And what is the equivalent of $(\bar{y}_1 - \bar{y}_2)^2$ in terms of matrices? It is $(\bar{\mathbf{y}}_1 - \bar{\mathbf{y}}_2)^2$ where $\bar{\mathbf{y}}_1$ and $\bar{\mathbf{y}}_2$ are mean vectors for each level of the independent variable, which recall for T^2, there are two. Hence, when we translate the univariate formulation in (11.6) to a multivariate one, we get the following, equation (11.7):

$$t^2 = (\bar{y}_1 - \bar{y}_2)\left[\frac{(n_1 - 1)s_1^2 + (n_2 - 1)s_2^2}{n_1 - n_2 - 2}\left(\frac{n_1 + n_2}{n_1 n_2}\right)\right]^{-1}(\bar{y}_1 - \bar{y}_2)$$
$$T^2 = \frac{n_1 n_2}{n_1 + n_2}(\bar{\mathbf{y}}_1 - \bar{\mathbf{y}}_2)'(\mathbf{S}^{-1})(\bar{\mathbf{y}}_1 - \bar{\mathbf{y}}_2) \tag{11.7}$$

The one-sample multivariate generalization of t is analogously obtained, and equal to:

$$T^2 = n(\bar{\mathbf{y}} - \boldsymbol{\mu}_0)'(\mathbf{S}^{-1})(\bar{\mathbf{y}} - \boldsymbol{\mu}_0)$$

where $\boldsymbol{\mu}_0$ now represents the mean population vector under the null hypothesis.

Analogous to univariate t for either the one-sample or two-sample case, T^2 is useful in situations where Σ is not known and must be estimated by $\mathbf{S}$. The corresponding Z^2 test statistics are thus $Z^2 = n(\bar{\mathbf{y}} - \boldsymbol{\mu}_0)'(\Sigma^{-1})(\bar{\mathbf{y}} - \boldsymbol{\mu}_0)$ for the one-sample case, and

$$Z^2 = \frac{n_1 n_2}{n_1 + n_2}(\bar{\mathbf{y}}_1 - \bar{\mathbf{y}}_2)'(\Sigma^{-1})(\bar{\mathbf{y}}_1 - \bar{\mathbf{y}}_2)$$

for the two-sample case. As usual, however, since Σ is rarely if ever known, we usually focus our development on T^2.

To compute Hotelling's T^2, we can compute a MANOVA and obtain **Hotelling's Trace**, and from there multiply this number by (N–L), where N is the sample size across groups, and L is the number of groups on the independent variable, which since we are performing Hotelling's, will be equal to 2. As a simple demonstration of computing T^2 in SPSS, consider the following data on dependent variables DV1 and DV2, and independent variable IV having two levels:

	DV1	DV2	IV
1	5.00	6.00	0.00
2	9.00	4.00	0.00
3	4.00	3.00	0.00
4	3.00	8.00	0.00
5	6.00	5.00	0.00
6	1.00	2.00	1.00
7	2.00	6.00	1.00
8	7.00	7.00	1.00
9	5.00	5.00	1.00
10	9.00	2.00	1.00

We conduct the MANOVA using the following code:

```
GLM DV1 DV2 BY IV
  /METHOD=SSTYPE(3)
  /INTERCEPT=INCLUDE
  /CRITERIA=ALPHA(.05)
  /DESIGN= IV.
```

This generates the output below:

Multivariate tests						
Effect		Value	F	Hypothesis df	Error df	Sig.
Intercept	Pillai's trace	0.925	43.124	2.000	7.000	0.000
	Wilks' lambda	0.075	43.124	2.000	7.000	0.000
	Hotelling's trace	12.321	43.124	2.000	7.000	0.000
	Roy's largest root	12.321	43.124	2.000	7.000	0.000
IV	Pillai's trace	0.064	0.238	2.000	7.000	0.794
	Wilks' lambda	0.936	0.238	2.000	7.000	0.794
	Hotelling's trace	0.068	0.238	2.000	7.000	0.794
	Roy's largest root	0.068	0.238	2.000	7.000	0.794

We note that Hotelling's trace for the IV (not the intercept) is equal to 0.068. To get Hotelling's T^2, we multiply this number by (N–L), equal to $(10 - 2) = 8$. Note that N is equal to 10 because there are 10 observations across groups, and L is equal to 2 because there are two levels on the independent variable. Hence, Hotelling's T^2 is equal to $0.068(8) = 0.544$. To get the statistical significance for T^2, we can simply refer to the above MANOVA output for Hotelling's Trace (IBM SPSS, Inc.). We see the result for our data is not statistically significant ($p = 0.794$). Though we computed T^2 for demonstration, in practice, one can simply perform the MANOVA since the statistical significance for the MANOVA will equal that for Hotelling's T^2.

To run a Hotelling's T^2 in R, we can use the package `ICSNP`. We code our data on each dependent variable, as well as the grouping variable on the IV:

```
> dv.1 <- c(5, 9, 4, 3, 6, 1, 2, 7, 5, 9)
> dv.2 <- c(6, 4, 3, 8, 5, 2, 6, 7, 5, 2)
> iv <- c(0, 0, 0, 0, 0, 1, 1, 1, 1, 1)
> y < cbind(dv.1, dv.2)
```

The statement `y < cbind(dv.1, dv.2)` defines the response variable as both dependent variables considered simultaneously. Now, we are all set to conduct Hotelling's T^2 (do not forget to install the package ICSNP first):

```
> library (ICSNP)
> y <- cbind(dv.1, dv.2)
> HotellingsT2(y ~ iv)

 Hotelling's two sample T2-test

data:  y by iv
T.2 = 0.23798, df1 = 2, df2 = 7, p-value = 0.7943
alternative hypothesis: true location difference is not equal to c
(0,0)
```

We can see that the p-value is the same as that produced in SPSS ($p = 0.7943$), though R by default is reporting the associated F statistic of 0.23798 rather than the Hotelling's T^2 we computed manually in SPSS. As was true for SPSS, as we will soon see, conducting the MANOVA instead of Hotelling's T^2 will yield the same decision on the null, and is usually the strategy employed by most researchers when comparing mean vectors on two groups (analogous to computing an ANOVA even when a two-sample t-test would suffice).

11.8 THE COVARIANCE MATRIX S

In seeking to understand any statistical equation, not unlike that of understanding the workings of an automobile or aircraft, it behooves one to literally **take it apart**, study its components, then **put it back together again**. We take a close look at what $\mathbf{S}^{-1}$ in Hotelling's T^2 represents. To better appreciate what **S** actually is, it is helpful to once more recall univariate t, specifically how the estimate of the pooled population variance was obtained. Recall once more, s_p^2 in (11.3):

$$\frac{(n_1 - 1)s_1^2 + (n_2 - 1)s_2^2}{n_1 - n_2 - 2}$$

The terms in the numerator, $(n_1 - 1)s_1^2$ and $(n_2 - 1)s_2^2$, are actually sums of squares terms for the respective groups on the independent variable. How are these sums of squares for each group? To understand how, recall how we formulated an unbiased estimate of the population variance (unbiased, since we are dividing the sum of squared deviations by $n - 1$):

$$s^2 = \frac{\sum\limits_{i=1}^{n}(y_i - \bar{y})^2}{n-1}$$

To eliminate the denominator "$n-1$," we can multiply both sides by $n-1$:

$$(n-1)s^2 = \frac{\sum\limits_{i=1}^{n}(y_i - \bar{y})^2}{n-1}(n-1)$$

Canceling out $(n-1)$, this yields simply $\sum\limits_{i=1}^{n}(y_i - \bar{y})^2$ on the right-hand side. Since we have two levels of the independent variable, we will have two sums of squares values, one for $(n_1 - 1)s_1^2$ and one for $(n_2 - 1)s_2^2$. Hence, we can write the pooled sample variance (s_p^2) as:

$$s_p^2 = \frac{ss_1 + ss_2}{n_1 + n_2 - 2}$$

where ss_1 and ss_2 are the sums of squares for each level of the independent variable in an independent samples t-test.

In the multivariate setting, we now have more than a single dependent measure. Hence, instead of having ss_1 and ss_2 alone to express within-group variability, we need to impose the matrix equivalent counterpart. We will have one matrix $\mathbf{E}_1$ to express the cross-products in level 1 of the independent variable, and another matrix $\mathbf{E}_2$ to express the cross-products in level 2 of the independent variable. The dimension (i.e., the number of rows and columns) of each matrix will be determined by how many dependent variables we include in the given model. For instance, in the case of three dependent measures, we define $\mathbf{E}_1$ for level 1 of the independent variable as

$$\mathbf{E}_1 = \begin{bmatrix} ss_{11} & ss_{12} & ss_{13} \\ ss_{21} & ss_{22} & ss_{23} \\ ss_{31} & ss_{32} & ss_{33} \end{bmatrix}$$

where ss_{11}, ss_{22} and ss_{33} represent the sums of squares for the first, second, and third dependent variables, respectively. Likewise, for the second level of the independent variable, we have the corresponding matrix $\mathbf{E}_2$. Then, just as is done in the univariate case, we pool $\mathbf{E}_1$ with $\mathbf{E}_2$ to get the matrix of sums of squares and cross-products:

$$\mathbf{E}_1 + \mathbf{E}_2 = \begin{bmatrix} ss_{11} & ss_{12} & ss_{13} \\ ss_{21} & ss_{22} & ss_{23} \\ ss_{31} & ss_{32} & ss_{33} \end{bmatrix} + \begin{bmatrix} ss_{11} & ss_{12} & ss_{13} \\ ss_{21} & ss_{22} & ss_{23} \\ ss_{31} & ss_{32} & ss_{33} \end{bmatrix}$$

Note again that for Hotelling's T^2, there will always only be **two** sums of squares and cross-product matrices to add, since Hotelling's T^2 is defined as having only two levels of the independent variable. In the general multivariate case, however, the number of matrices will depend on how many levels exist on the independent variable. For example, for the case of three levels on the independent variable, we would have:

TABLE 11.2   Cross-Product Matrices in 2 × 3 Multivariate Factorial Analysis of Variance

	Factor 2		
Factor 1	Level 1	Level 2	Level 3
Level 1	E_{11}	E_{12}	E_{13}
Level 2	E_{21}	E_{22}	E_{23}

$$\mathbf{E}_1 + \mathbf{E}_2 + \mathbf{E}_3 = \begin{bmatrix} ss_{11} & ss_{12} & ss_{13} \\ ss_{21} & ss_{22} & ss_{23} \\ ss_{31} & ss_{32} & ss_{33} \end{bmatrix} + \begin{bmatrix} ss_{11} & ss_{12} & ss_{13} \\ ss_{21} & ss_{22} & ss_{23} \\ ss_{31} & ss_{32} & ss_{33} \end{bmatrix} + \begin{bmatrix} ss_{11} & ss_{12} & ss_{13} \\ ss_{21} & ss_{22} & ss_{23} \\ ss_{31} & ss_{32} & ss_{33} \end{bmatrix}$$

In the case of a factorial MANOVA model (not considered in this book in any detail), it stands that each **cell** of the design will contain a matrix $\mathbf{E}_{jk}$ in row j, column k. For example, consider the case of the 2 × 3 factorial univariate model featured in Chapter 4. Within each cell will be a matrix $\mathbf{E}_{jk}$ (see Table 11.2).

11.9   FROM SUMS OF SQUARES AND CROSS-PRODUCTS TO VARIANCES AND COVARIANCES

In our matrices $\mathbf{E}_1$ and $\mathbf{E}_2$, elements of each matrix consist of sums of squares along the main diagonal, and cross-products on the off-diagonal. To get respective variances and covariances for $\mathbf{E}_1$, $\mathbf{E}_2$, $\mathbf{E}_p$, we divide each element of $\mathbf{E}_i$ by degrees of freedom, which are $n-1$ for each $\mathbf{E}_i$ in a balanced design. When we do so, we get the variance-covariance matrix $\mathbf{S}_i$ for level i of the independent variable. For example, for $\mathbf{E}_1$, we would have:

$$\mathbf{S}_1 = \begin{bmatrix} ss_{11} & ss_{12} & ss_{13} \\ ss_{21} & ss_{22} & ss_{23} \\ ss_{31} & ss_{32} & ss_{33} \end{bmatrix} \cdot \frac{1}{(n-1)} = \begin{bmatrix} \dfrac{ss_{11}}{(n-1)} & \dfrac{ss_{12}}{(n-1)} & \dfrac{ss_{13}}{(n-1)} \\ \dfrac{ss_{21}}{(n-1)} & \dfrac{ss_{22}}{(n-1)} & \dfrac{ss_{23}}{(n-1)} \\ \dfrac{ss_{31}}{(n-1)} & \dfrac{ss_{32}}{(n-1)} & \dfrac{ss_{33}}{(n-1)} \end{bmatrix}$$

Quantities along the main diagonal, $\dfrac{ss_{11}}{(n-1)}$, $\dfrac{ss_{22}}{(n-1)}$, $\dfrac{ss_{33}}{(n-1)}$ are now sample **variances**, and quantities on the off-diagonals, $\dfrac{ss_{12}}{(n-1)}$, $\dfrac{ss_{13}}{(n-1)}$, $\dfrac{ss_{23}}{(n-1)}$ are now **covariances**. Recall that since $\mathbf{S}_1$ is symmetric, the lower triangular will mirror that of the upper triangular. Likewise, for $\mathbf{E}_2$, $\mathbf{E}_3$,..., $\mathbf{E}_p$, we will have respective matrices $\mathbf{S}_2$, $\mathbf{S}_3$,..., $\mathbf{S}_p$ (where p here denotes the number of populations). Each of these sample variance-covariance matrices are estimators of their corresponding population variance-covariance matrices Σ_1, Σ_2,..., Σ_p (where the notation $\hat{\Sigma}_1$, $\hat{\Sigma}_2$,..., $\hat{\Sigma}_p$ is sometimes used to denote estimation). The pooled variance–covariance estimator is thus, for p populations:

$$\mathbf{S}_{pl} = \frac{(n_1 - 1)\mathbf{S}_1 + (n_2 - 1)\mathbf{S}_2 + \cdots + (n_p - 1)\mathbf{S}_p}{n_1 + n_2 \ldots + n_p - p}$$

where $\mathbf{S}_{pl}$ serves as an estimator of Σ. When the subscript pl in $\mathbf{S}_{pl}$ is not given, such that only $\mathbf{S}$ is shown, one might assume we are typically working with the pooled variance-covariance estimator.

11.10 HYPOTHESIS AND ERROR MATRICES OF MANOVA

Having considered the development of T^2 earlier, we now consider the case where there are more than two levels on the independent variable. Recall that Hotelling's T^2 only applies to the case of two groups. We need the requisite matrices that will allow for two or more levels on the independent variable.

Analogous to the univariate case where we have **between** and **within** sums of squares, in the general multivariate case we will have matrix counterparts also corresponding to between and within sources of variation. Only that now, in the language of the MANOVA model, these matrices will be customarily referred to as $\mathbf{H}$ for "hypothesis" and $\mathbf{E}$ for "error." These respective matrices are given by:

$$\mathbf{H} = n \sum_{i=1}^{p} (\overline{\mathbf{y}}_{i.} - \overline{\mathbf{y}}_{..})(\overline{\mathbf{y}}_{i.} - \overline{\mathbf{y}}_{..})'$$

and

$$\mathbf{E} = \sum_{i=1}^{p} \sum_{j=1}^{n} (\mathbf{y}_{ij} - \overline{\mathbf{y}}_{i.})(\overline{\mathbf{y}}_{ij} - \overline{\mathbf{y}}_{i.})'$$

A look at these matrices for $\mathbf{H}$ and $\mathbf{E}$ reveal that their computations are somewhat analogous to the computation of sums of squares in ANOVA, only that now, more than a single dependent variable is taken into account. The $\mathbf{H}$ matrix is one of potential treatment effects, that is, deviations of means $\overline{\mathbf{y}}_{i.}$ from a grand mean $\overline{\mathbf{y}}_{..}$, while the $\mathbf{E}$ matrix is a matrix corresponding to "within" variability computed by taking observations **within** cells, $\mathbf{y}_{ij}$, and subtracting corresponding means, $\overline{\mathbf{y}}_{i.}$. Analogous to univariate ANOVA, the total variation $\mathbf{T}$ in MANOVA can be partitioned into two parts, that of $\mathbf{H}$ and $\mathbf{E}$. Hence, $\mathbf{T} = \mathbf{H} + \mathbf{E}$. We use this identity next in developing test statistics for MANOVA. For a good arithmetical example of these computations for a two-group problem, see Stevens (2009, pp. 148–152).

11.11 MULTIVARIATE TEST STATISTICS

Recall that in one-way univariate ANOVA, we tested a null hypothesis of equality among population means by constructing an F-ratio, the ratio of two variances **MS between** to **MS within**, where MS between was a measure of between-group variance and MS within was a measure of within-group variance. This was the only "omnibus" test statistic in ANOVA. In MANOVA, because of the potentially complex configurations as a result of working in higher dimensions on mean vectors, **no single statistical test is uniformly most powerful under all circumstances** such as is true for the F-test in the ANOVA model. Hence, in MANOVA and most other multivariate techniques, there exist several test statistics that we may draw upon when evaluating statistical significance for a multivariate effect. We now briefly survey these test statistics.

Our first test, and undoubtedly most popular and of most historical significance, is that of **Wilk's lambda**, Λ, which bears a similar resemblance (but in reverse) to the univariate F-ratio. Wilk's Λ is given by:

$$\Lambda = \frac{|\mathbf{E}|}{|\mathbf{H} + \mathbf{E}|} = \frac{|\mathbf{E}|}{|\mathbf{T}|} \tag{11.8}$$

where $|\mathbf{E}|$ and $|\mathbf{H} + \mathbf{E}|$ indicate determinants of $\mathbf{E}$ and $\mathbf{H} + \mathbf{E}$, respectively. It is easy to see what Wilk's test accomplishes. Since $\mathbf{T} = \mathbf{H} + \mathbf{E}$, the extent to which there are treatment effects is the extent to which more of the variation is being accounted for by $\mathbf{H}$ relative to $\mathbf{E}$. If total variation is not being accounted for by treatment effects, then the size of $\mathbf{E}$ will dominate relative to $\mathbf{H}$. Note that Λ is an **inverse criterion**, meaning that smaller values of Λ are preferable to larger ones. To see this, consider the situation where the total variation $\mathbf{T}$ were completely accounted for by $\mathbf{E}$:

$$\Lambda = \frac{|\mathbf{E}|}{|0 + \mathbf{E}|} = \frac{|\mathbf{E}|}{|\mathbf{E}|} = 1.0$$

On the other hand, if all variation is accounted for by treatment effects, or between-group differences, then Λ would be

$$\Lambda = \frac{|\mathbf{E}|}{|\mathbf{H} + 0|} = \frac{|0|}{|\mathbf{H}|} = 0$$

Hence, the range on Λ is 0 for a perfectly fitting model to 1.0 under H_0 in which there is no multivariate effect. Unlike the F-ratio in univariate ANOVA, **smaller** values of Λ lead to a rejection of H_0 and an inference of the statistical alternative H_1. Indeed, when the number of dependent variables is reduced to one, Λ is equal to the ratio of **SS within** to **SS total** for univariate F, which also corresponds to $1 - \eta^2$ in the one-way model. Wilk's Λ can also be written as a function of eigenvalues $\lambda_1, \lambda_2, \ldots, \lambda_p$ of $\mathbf{E}^{-1}\mathbf{H}$:

$$\Lambda = \prod_{i=1}^{s} \frac{1}{1 + \lambda_i} \tag{11.9}$$

where λ_i denote respective extracted eigenvalues for the given MANOVA for $i = 1$ to s eigenvalues. This definition for Λ is more applicable when interpreting **discriminant function analysis**, which, as we will see in Chapter 12, is intimately related to MANOVA in that it defines the **eigenvector(s)** for which group separation on the independent variable is maximized. For the case in which there are only two levels on the independent variable, there will only be one eigenvalue extracted, and hence the product operator, $\prod_{i=1}^{s}$ in (11.9) becomes unnecessary, and Λ reduces simply to $\frac{1}{1+\lambda}$.

The statistical significance of Λ can be evaluated by

$$\chi^2 = -[(N-1) - 0.5(p + k)]\ln \Lambda$$

which is distributed approximately as a chi-square variable on $p(k - 1)$ degrees of freedom, where p in this case is the number of dependent variables and k is the number of populations (i.e., levels on the independent variable). This approximation is good for moderate to relatively large sample sizes (Stevens, 2009). For details, see Bartlett (1947).

11.11.1 Pillai's Trace

Pillai's trace (Pillai, 1955) is a second multivariate test statistic used for evaluating the statistical significance of a multivariate effect and is defined by:

$$V^{(s)} = \text{tr}\left[(\mathbf{E} + \mathbf{H})^{-1}\mathbf{H}\right] = \sum_{i=1}^{s} \frac{\lambda_i}{1 + \lambda_i}$$

where tr is the trace, and λ_i is the ith eigenvalue. For the case of two levels on the independent variable, since only a single eigenvalue is extracted, $V^{(s)}$ reduces to simply

$$V^{(s)} = \frac{\lambda}{1 + \lambda} \tag{11.10}$$

For the case of several λ_i, if we consider only the **largest** of these in Pillai's, then we have what is known as **Roy's largest root**:

$$\theta = \frac{\lambda_1}{1 + \lambda_1} \tag{11.11}$$

where λ_1 is the **maximum eigenvalue** extracted. Roy's test uses only the largest eigenvalue of $\mathbf{E}^{-1}\mathbf{H}$, and so it is more powerful than other multivariate test statistics under the condition that the mean vectors are **collinear**. What does it mean to say vectors are collinear? Recall that vectors "happen" in Euclidean spaces, whether in two, three, or higher dimensions. For example, consider two possibilities for three mean vectors (Figure 11.1).

Each of $\mathbf{\mu}_1, \mathbf{\mu}_2, \mathbf{\mu}_3$ represents mean vectors. In the first case (Figure 11.1a), mean vectors lay more or less in a straight line, and hence are **collinear**. Roy's largest root considers only the largest of extracted eigenvalues. The reason why the situation of even near collinearity of vectors is ideal for interpreting Roy's is that typically in such a case, the size of one eigenvalue will **dominate** the size of the others, because, as will be elaborated on in Chapter 12, it is suggestive that a **single** discriminant function suffices in accounting for group separation on the independent variable. Hence, it makes sense to consider only the largest of these eigenvalues. For situations in which mean vectors are not collinear, other tests such as Wilk's, Pillai's, or the Lawley–Hotelling trace (to be discussed), are generally recommended for use in place of Roy's. When mean vectors are spread out (i.e., not collinear), these other

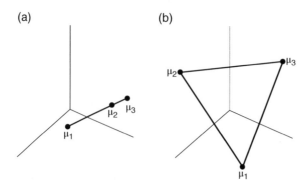

FIGURE 11.1 (a) Collinear vectors. (b) Noncollinear vectors. Source: Rencher and Christensen (2012). Reproduced with permission from John Wiley & Sons, Inc.

tests are generally more powerful. Note as well that for the case in which only a single eigenvalue is extracted, θ reduces to $V^{(s)}$ since there is only a single root that is extracted, and hence it must also be the largest root.

Pillai's is a function of the trace of $(\mathbf{E} + \mathbf{H})^{-1}\mathbf{H}$ and as such is a direct measure of how much greater $\mathbf{H}$ is relative to $\mathbf{T}$. The statistical significance of Pillai's trace can be evaluated as an approximate F-statistic. Pillai also suggested two alternative F-approximations (Pillai, 1956). Pillai's trace is quite robust and is therefore usually the statistic of choice when assumptions such as equality of covariance matrices are likely violated.

11.11.2 Lawley–Hotelling's Trace

The Lawley–Hotelling statistic (Lawley, 1938; Hotelling, 1951) is given by:

$$U^{(s)} = \text{tr}\left(\mathbf{E}^{-1}\mathbf{H}\right) = \sum_{i=1}^{s} \lambda_i \tag{11.12}$$

Consider the difference between $V^{(s)}$ and $U^{(s)}$. In $V^{(s)}$ we took the trace of $(\mathbf{E} + \mathbf{H})^{-1}\mathbf{H}$, or $(\mathbf{T})^{-1}\mathbf{H}$. In $U^{(s)}$ we are taking the trace of $\mathbf{E}^{-1}\mathbf{H}$. Hence, in the case of $V^{(s)}$, we are comparing the size of $\mathbf{H}$ relative to $\mathbf{T}$, whereas in $U^{(s)}$ we are comparing $\mathbf{H}$ relative to $\mathbf{E}$; thus, $U^{(s)}$ is a comparison of how much variation is accounted for by treatments in $\mathbf{H}$ relative to how much variation is "left over" in $\mathbf{E}$. Instead of comparing variation in $\mathbf{H}$ to the **total variation** in $\mathbf{T}$, as in $V^{(s)}$, $U^{(s)}$ compares this variation with that **unexplained**, that is, $\mathbf{E}$, analogous to comparing $SS_{between}$ to SS_{within} in the univariate case. Since $\mathbf{T} = \mathbf{H} + \mathbf{E}$, the value of $V^{(s)}$ will always be smaller than $U^{(s)}$.

11.12 EQUALITY OF COVARIANCE MATRICES

Recall that in univariate ANOVA, we required that variances across populations be equal, $\sigma^2_{j=1} = \sigma^2_{j=2} = \sigma^2_{j=J}$. In the MANOVA case, this assumption is also required, but in addition to it, we also require the assumption that the **covariances** across populations also be equal. Recall that we also required equality of covariance matrices in blocking and repeated-measures models, though occurring in a different context. For a problem in which there are p populations then, we require that

$$\Sigma_1 = \Sigma_2 = \Sigma_3 \ldots = \Sigma_p \tag{11.13}$$

For example, for a three-group MANOVA problem, what (11.13) implies is that the following matrix must be **constant** across populations as defined on the levels of the independent variable:

$$\Sigma = \begin{bmatrix} \sigma_v^2 & \text{cov}_c & \text{cov}_c \\ \text{cov}_c & \sigma_v^2 & \text{cov}_c \\ \text{cov}_c & \text{cov}_c & \sigma_v^2 \end{bmatrix}$$

where σ_v^2 is a common population variance and cov_c is a common population covariance. The null hypothesis is thus

$$H_0 : \Sigma_1 = \Sigma_2 = \Sigma_3 \tag{11.14}$$

tested against the alternative H_1 that at least two matrices of the set $\Sigma_1, \Sigma_2, \Sigma_3$ are **unequal**. A test of the null hypothesis of equal variances and covariances is available and known as the **Box M-test**. The test is described by Box (1949, 1950) in which he attributes it to Wilks (1946), who in turn references the likelihood ratio method of Neyman and Pearson (1928). The likelihood ratio test (see Johnson and Wichern, 2007 for details) required for testing (11.14), and which we will use for establishing the Box M-test, is given by

$$\Lambda = \prod_{g=1}^{n} \left(\frac{|\mathbf{S}_g|}{|\mathbf{S}_p|} \right)^{(n_g-1)/2} \tag{11.15}$$

where $\mathbf{S}_g$ is the sample covariance matrix for groups (or levels or populations) $g = 1, 2,\ldots,n$, and n_{g-1}. The pooled matrix across groups is given by $\mathbf{S}_p$, and as before is equal to the weighted sum of sample covariance matrices across groups (i.e., levels of the independent variable). What will make Λ small? If we assume the exponent $n_{g-1}/2$ is constant for a given problem, then the real "action" of Λ is in the ratio $|\mathbf{S}_g|$ to $|\mathbf{S}_p|$. The extent to which $(\mathbf{S}_{g=1}) = (\mathbf{S}_{g=2}) = (\mathbf{S}_{g=3}) = (\mathbf{S}_{g=g})$ holds is the extent to which the numerator $|\mathbf{S}_g|$ and the denominator $|\mathbf{S}_p|$ converge to the same value, under which case, $\frac{|\mathbf{S}_g|}{|\mathbf{S}_p|} \approx 1.0$. On the other hand, the extent to which $(\mathbf{S}_{g=1}) \cdot (\mathbf{S}_{g=2}) \cdot (\mathbf{S}_{g=3})\ldots \cdot (\mathbf{S}_{g=g})$ is different from the denominator is the extent to which an "imbalance" among $\mathbf{S}_g$ is occurring, and hence support for the alternative hypothesis that at least two $\mathbf{S}_g$ are different from one another.

Given Λ, we can now establish the **Box M-statistic**:

$$M = -2\ln\Lambda \tag{11.16}$$

Note that in the trivial case where Λ is equal to 0, M is undefined, since $\ln(0)$ is likewise undefined. For Λ equal to 1, M is equal to 0, since $\ln(1) = 0$. Under H_0, since we would expect $\frac{|\mathbf{S}_g|}{|\mathbf{S}_p|} \approx 1.0$, it stands that lower than not values for M are expected. Conversely, as $\frac{|\mathbf{S}_g|}{|\mathbf{S}_p|} \neq 1.0$, M gets larger, providing evidence against H_0 and in favor of H_1, suggesting at least one pairwise difference among population covariance matrices Σ.

It has been shown that the Box M-test is distributed as an approximate χ^2 distribution (Johnson and Wichern, 2007). An F approximation may also be used (see Box, 1950). The test is rather sensitive to nonnormality and kurtosis, and so a rejection of the null might occur even in situations for which the violation of equality of covariance matrices is minimal. For this reason, some authors (e.g., Johnson and Wichern, 2007) suggest interpreting MANOVA tests even in light of a statistically significant finding. We concur, but add also that should the Box-M test reject the null, a visual inspection of the covariance matrices may be in order to detect potentially problematic (or even interesting) patterns in the data. This highlights a general point emphasized throughout this book, and something to always keep in mind—as a researcher, **you are using these statistical tests to help you discern the presence or absence of a scientific finding, not for their statistical "lore" alone**. If the Box-M test issues a strong rejection, then not only may it be worth investigating further to satisfy the statistical test, but an unequal pattern of covariance matrices may also be of interest substantively. **Ask questions of your data**. Why are covariance matrices unequal in the first place? Is their inequality due to an interesting pattern in the data that may be of scientific value? Do not blindly keep trying to appease the test via statistical assumptions while missing out on what the test may be telling you about your data from a substantive point of view. If the null is rejected, seek value from its rejection from a **scientific** (in addition to **statistical**) perspective. In general, avoid "checklist data analysis" as much as possible where you simply tick off assumptions one by one. Explore the intricacies of your data, and what you find may lead to new scientific insights.

11.13 MULTIVARIATE CONTRASTS

Recall that whenever we perform a contrast, no matter the complexity of it, such always boils down to a **comparison** of two groups. Whether each group is a linear combination of means, or a single mean, we are nonetheless comparing groups. In the univariate case, we built our contrasts by hypothesizing comparisons among population means of the form:

$$C_i = c_1\mu_1 + c_2\mu_2 + \ldots + c_J\mu_J$$

When we move from the univariate case to the multivariate case, we are required to build our contrast not as a function of population means, but rather as a function of population **mean vectors**, $\boldsymbol{\mu}_1, \boldsymbol{\mu}_2, \ldots \boldsymbol{\mu}_p$. Hence, for the multivariate case, we can represent a contrast by:

$$C_i = c_1\boldsymbol{\mu}_1 + c_2\boldsymbol{\mu}_2 + \ldots + c_p\boldsymbol{\mu}_p$$

As was true for the univariate case, unless we are working specifically with population data, we will not ordinarily know the parameters of our contrast, and hence will be estimating them using functions of our sample data (i.e., estimators):

$$\hat{C}_i = c_1\bar{\mathbf{y}}_1 + c_2\bar{\mathbf{y}}_2 + \ldots + c_p\bar{\mathbf{y}}_p$$

A relevant null hypothesis is that population mean vectors are equal

$$H_0 : C_i = 0$$

against an alternative that population mean vectors somewhere in the set are unequal:

$$H_1 : C_i \neq 0$$

Now, just as was true for the univariate case, we will want to test our sample contrast for statistical significance. Hence, we need to know how to estimate the standard error of such contrasts. Recall how we estimated σ_C^2 in the case of a univariate contrast:

$$\sigma_{\hat{C}_i}^2 = \sigma_e^2 \sum_j \frac{c_j^2}{n_j}$$

for which an estimator for σ_e^2 was provided by MS error. We follow an analogous approach in the multivariate case. However, we cannot simply use MS error as our estimator. Why not? Because we are in the MANOVA context and there is no corresponding single MS error value as there was in the univariate context. To obtain our error estimate for the MANOVA case, we must also account, as usual, for the covariances among dependent variables, which means we will use $\mathbf{S}$ as our estimator of $\sum$. As was true for the univariate case, we will again compute a t-statistic, only now, because we are working with a linear combination of dependent variables instead of a single variable, we use T^2 as our test statistic. Hence, the contrast is given by

$$T^2 = \left(\sum_{i=1}^{p} \frac{c_i^2}{n_i} \right)^{-1} \hat{C}_i' \mathbf{S}^{-1} \hat{C}_i \tag{11.17}$$

Does (11.17) look familiar? It should, since it is somewhat analogous to Hotelling's T^2 featured earlier in our generalization of univariate t for the case of two levels on the independent variable. Only now, instead of the constant term for sample size equal to $\frac{n_1 n_2}{n_1 + n_2}$, the new term is now equal to $\sum_{i=1}^{p} \frac{c_i^2}{n_i}$ for the case of the contrast. However, note the similarity between Hotelling's and (11.17):

$$T^2 = \frac{n_1 n_2}{n_1 + n_2} (\bar{\mathbf{y}}_1 - \bar{\mathbf{y}}_2)'(\mathbf{S}^{-1})(\bar{\mathbf{y}}_1 - \bar{\mathbf{y}}_2) \quad \rightarrow \quad T^2 = \left(\sum_{i=1}^{p} \frac{c_i^2}{n_i} \right)^{-1} \hat{C}_i' \mathbf{S}^{-1} \hat{C}_i$$

In (11.17), we are not restricted to contrasts of the type $(\bar{\mathbf{y}}_1 - \bar{\mathbf{y}}_2)$. For equal n per group, as noted in Stevens (2009, p. 197), (11.17) reduces to $T^2 = \frac{n}{2} \hat{C}_i' \mathbf{S}^{-1} \hat{C}_i$ for a paired comparison.

11.14   MANOVA IN R AND SPSS

In now demonstrating a simple example of MANOVA in R, we amend our data slightly from Table 11.1 to now include three levels on the independent variable. Recall that there are two dependent variables, and since we can conceptualize quantitative and verbal ability as a **composite variable**, it makes sense to consider them simultaneously in a MANOVA. The independent variable for this example is whether or not subjects received prior training in courses that would foster the development of intellectual capacity and learning (1 = no specialized prior training, 2 = some training, 3 = extensive training). There are a total of $n = 3$ observations per group. The data are given in Table 11.3.

The null hypothesis we wish to test is

$$H_0 : \begin{pmatrix} \mu_{11} \\ \mu_{21} \end{pmatrix} = \begin{pmatrix} \mu_{12} \\ \mu_{22} \end{pmatrix} = \begin{pmatrix} \mu_{13} \\ \mu_{23} \end{pmatrix}$$

against a statistical alternative hypothesis of the kind that somewhere between mean vectors there is (or are) differences. For instance, one possibility is:

$$H_1 : \begin{pmatrix} \mu_{11} \\ \mu_{21} \end{pmatrix} \neq \begin{pmatrix} \mu_{12} \\ \mu_{22} \end{pmatrix} = \begin{pmatrix} \mu_{13} \\ \mu_{23} \end{pmatrix}$$

TABLE 11.3   Hypothetical Data on Quantitative and Verbal Ability as a Function of Training (1 = No training, 2 = Some training, 3 = Extensive training)

Subject	Quantitative	Verbal	Training
1	5	2	1
2	2	1	1
3	6	3	1
4	9	7	2
5	8	9	2
6	7	8	2
7	9	8	3
8	10	10	3
9	10	9	3

We generate the vectors Q, V, and T, for variables quantitative, verbal and training, respectively, then request R to generate the requisite data frame (iq.data):

```
> Q <- c(5, 2, 6, 9, 8, 7, 9, 10, 10)
> V <- c(2, 1, 3, 7, 9, 8, 8, 10, 9)
> T <- c(1, 1, 1, 2, 2, 2, 3, 3, 3)
> iq.data <- data.frame(Q, V, T)
```

Next, we bind the columns of Q and V to generate our dependent variable, which recall will be a linear combination of Q and V:

```
> Y <- cbind(Q, V)
```

We confirm that Y has been constructed correctly, printing only the first three observations:

```
> Y
       Q  V
  [1,]  5  2
  [2,]  2  1
  [3,]  6  3
```

We now generate factor levels for the independent variable training, naming our new factor T.f. We also identify T.f as having three levels, **none**, **some**, and **much** to reflect the extent of training received:

```
> T.f <- factor(T, levels = 1:3)
> levels(T.f) <- c("none", "some", "much")
```

For demonstration, we now proceed to run the **wrong** MANOVA, requesting Wilk's Λ as our multivariate test statistic, then requesting a summary of results:

```
> manova.fit <- manova(Y ~ T)
> summary(manova.fit, test = "Wilks")

          Df     Wilks    approx  F num Df   den Df   Pr(>F)
T          1   0.17871   13.787        2         6 0.005708 **
Residuals  7
---
Signif. codes:  0 '***' 0.001 '**' 0.01 '*' 0.05 '.' 0.1 ' ' 1
```

In the output, we note that our effect for T has only a single degree of freedom. However, since T is a grouping variable with three levels, the degrees of freedom should have been equal to 2 (i.e., $3 - 1 = 2$). What went wrong? What went wrong is that we used T instead of T.f, which recall we had designated as our factor. Using T is a good example of a mistake to avoid when fitting models. The failure to designate T as a factor caused R to treat it as a **continuous** variable, which of course it is not. It is a categorical variable with categories corresponding to the levels of the independent variable. That is, R fit a regression model above:

```
> reg <- lm(Y ~ T)
> anova(reg)
```

```
Analysis of Variance Table

            Df Pillai approx F num Df den Df   Pr(>F)
(Intercept)  1  0.974   113.1       2       6 0.000017 ***
T            1  0.821    13.8       2       6  0.0057 **
Residuals    7
---
Signif. codes:  0 '***' 0.001 '**' 0.01 '*' 0.05 '.' 0.1 ' ' 1
```

Notice that in the ANOVA table above, the p-value for T is equal to the same number that we obtained earlier (i.e., $p = 0.0057$). We now fit the correct model:

```
> manova.fit <- manova(Y ~ T.f)
> summary(manova.fit, test = "Wilks")

          Df    Wilks approx F num Df den Df   Pr(>F)
T.f        2 0.056095   8.0555      4     10 0.003589 **
Residuals  6
---
Signif. codes:  0 '***' 0.001 '**' 0.01 '*' 0.05 '.' 0.1 ' ' 1
```

Notice now that the degrees of freedom for T.f are equal to 2, which are the correct degrees of freedom for the factor. The p-value for Wilk's Λ is equal to 0.003589, leading to a rejection of the null hypothesis. We could have also instead requested Pillai's trace as our multivariate test:

```
> summary(manova.fit, test = "Pillai")

          Df Pillai approx F num Df den Df  Pr(>F)
T.f        2 1.0737   3.4775      4     12 0.04166 *
Residuals  6
---
Signif. codes:  0 '***' 0.001 '**' 0.01 '*' 0.05 '.' 0.1 ' ' 1
```

We note that while still calling for a rejection of H_0 at a significance level of 0.05, the observed p-value for Pillai is larger than that for Wilk's. We could have also obtained the Lawley–Hotelling test and Roy's test by specifying the option test = "Hotelling-Lawley" and test = "Roy" respectively, though we do not show the output of these tests here.

Analyzing these data in SPSS is straightforward and the output will mimic that generated in R. Consequently, we do not display its output. One can obtain the MANOVA in SPSS through manova Q V by T (1, 3). Of more interest, as it will relate to the material of the following chapter on discriminant analysis and canonical correlation, we can obtain the eigenvalues and canonical correlations from SPSS:

```
manova Q V by T(1,3)
/print = sig(eigen).
```

```
Eigenvalues and Canonical Correlations

  Root No.        Eigenvalue         Pct.      Cum. Pct.     Canon Cor.

         1          14.35158      98.88896      98.88896        .96688
         2            .16124       1.11104     100.00000        .37263
```

The eigenvalue for the first root can be computed by $(0.96688)^2/1 - (0.96688)^2 = 14.35$. This was calculated using the **canonical correlation** reported in the far-right column (`Canon Cor`) which, when squared, provides a measure of how much variance r^2 is accounted for by the given function. To get the squared canonical correlation r^2 from the eigenvalue, we compute $\dfrac{\lambda_i}{1 + \lambda_i}$ for each function. That is, for the first function, we compute:

$$r_1^2 = \frac{\lambda_1}{1 + \lambda_1} = \frac{14.35158}{1 + 14.35158} = \frac{14.35158}{15.35158} = 0.93$$

Notice that the above value of 0.93 matches up with that for the squared canonical correlation for the first function, $(0.96688)^2$. For the second function, we can likewise compute a squared canonical correlation:

$$r_2^2 = \frac{\lambda_2}{1 + \lambda_2} = \frac{0.16124}{1.16124} = 0.139$$

We discuss the canonical correlation more thoroughly in Chapter 12. The eigenvalue for the second root is equal to $(0.37263)^2/1 - (0.37263)^2 = 0.1612$. We can see then that the first root dominates the second in terms of size. What this suggests is that they both lay close to a single line (Rencher and Christensen, 2012, p. 189). Recall we had said that under such a case, Roy's test would be most powerful. When we compute Roy's on this data, we obtain:

```
> summary(manova.fit, test = "Roy")
            Df     Roy approx F num Df den Df     Pr(>F)
T.f          2 14.352    43.055      2      6 0.0002764 ***
Residuals    6
---
Signif. codes:  0 '***' 0.001 '**' 0.01 '*' 0.05 '.' 0.1 ' ' 1
```

As expected, the p-value for Roy's is smaller than that computed earlier for both Wilk's and Pillai. Roy's largest root in this case provides the most powerful test.

Two eigenvalues are extracted from this analysis because there are three levels on the independent variable. As we will learn in our discussion of discriminant analysis in Chapter 12, each eigenvalue extracted corresponds to a "proportion of variance" accounted for by each discriminant function. It is not "literally" a proportion of variance as it would be in a principal component analysis. However, calling it such does no harm so long as one realizes its limitation. Referring to it more as a measure of "importance" is more accurate of an interpretation (Rencher and Christensen, 2012). We will revisit this issue later in the book when we discuss principal components analysis.

The importance of the first extracted root (eigenvalue) is computed as $14.35/(14.35 + 0.16) = 14.35/14.51 = 0.989$, while the importance of the second extracted root (eigenvalue) is computed as $0.16/14.51 = 0.011$. **Clearly, the first discriminant function is responsible for most of the group separation between groups on the independent variable**.

Tests for **outliers** in MANOVA can be performed using the `mvoutlier` package (Filzmoser and Gschwandtner, 2014); for evaluating multivariate normality, **Mardia's test** in the MVN package (Korkmaz, Goksuluk, and Zararsiz, 2014). Multivariate normality can also be evaluated using the **Shapiro-Wilk test** in the `mvnormtest` package (Jarek, 2012) by calling the function `mshapiro.test`, which is a generalization of the Shapiro-Wilk test for univariate normality. Influential observations can also be detected by requesting **Cook's d** values (see Chapter 7 for details). Q–Q plots and histograms can also be generated.

For the covariance assumption, we can use the `boxM` test in the package `biotools` (da Silva, 2014) to test the assumption of equal covariance matrices. For a test of the homogeneity of variances only, one can use `bartlett.test`.

In SPSS, we can obtain Box's test for our Q–V data by:

```
/PRINT=HOMOGENEITY
```

Box's Test of Equality of Covariance Matrices[a]	
Box's M	9.949
F	0.849
df1	6
df2	897.231
Sig.	0.532

[a]Design: Intercept + T.

The test is not statistically significant, suggesting we do not have evidence to doubt the assumption of equal variance–covariance matrices between populations on the independent variable. Though recall as a result of the test being quite sensitive to even minimal assumption violations, it is generally recommended that we proceed with the MANOVA even in cases of a slightly statistically significant Box M-test. On the other hand, if the violation is rather severe, inspection of covariance matrices may be in order, both from a statistical point of view, but also more importantly from a scientific one.

11.14.1 Univariate Analyses

It is custom to follow up a statistically significant MANOVA with univariate ANOVAs. However, when one does so, one should be acutely aware of why one is doing such a thing. Oftentimes researchers will follow up with univariate analyses in an attempt to "break down" or otherwise "decompose" the multivariate effect. Recall from our earlier discussion of Rao's paradox, however, that **a multivariate effect may not always "decompose" into individual univariate effects, and vice versa, the presence of univariate effects does not necessarily indicate the presence of a multivariate effect**. Hence, the idea of performing univariate follow-up tests in an effort to "deconstruct" the multivariate effect is misguided. Furthermore, there was presumably a reason why you chose MANOVA over independent univariate analyses. If you find yourself more interested in the univariate effects than the multivariate findings obtained from your analyses, you might want to ask yourself why you performed the multivariate tests in the first place. Recall that the fact that you have numerous dependent variables at your disposal, by itself, should not be a rationale for why the MANOVA is performed.

Univariate tests can be obtained for our data via `summary.aov (manova.fit)`. Using `summary.lm` will provide us with contrasts on levels of `T.f` on each dependent variable. For instance, for quantitative, we can compute:

```
> summary.lm(aov(Q ~ T.f))

Coefficients:
            Estimate Std. Error t value Pr(>|t|)
(Intercept)   4.3333     0.7935   5.461  0.00157 **
T.fsome       3.6667     1.1222   3.267  0.01709 *
```

```
T.fmuch        5.3333      1.1222    4.753   0.00315 **
---
Signif. codes:  0 '***' 0.001 '**' 0.01 '*' 0.05 '.' 0.1 ' ' 1

Residual standard error: 1.374 on 6 degrees of freedom
Multiple R-squared:  0.7976,    Adjusted R-squared:  0.7302
F-statistic: 11.82 on 2 and 6 DF,  p-value: 0.008289
```

By default, and contrary to SPSS, R uses the first group (T.f = 1) as the reference group (to generate similar contrasts in SPSS, use /PRINT=PARAMETER). The intercept estimate is that of the mean of Q for $T = 1$ (4.33). The value of 3.67 for T.fsome is the mean difference between $T = 2$ and $T = 1$ (8.0 – 4.3). The final value of 5.3 is the mean difference between $T = 3$ and $T = 1$ (9.6 – 4.3).

As in univariate ANOVA, there is nothing stopping us from modeling several factors simultaneously in MANOVA and computing interaction terms. For example, had we a second factor in our Q–V data, we could have easily modeled the interaction with a second factor X by computing manova (Y ~ T.f + X + T.f:X). The interpretation of the interaction term parallels that in univariate ANOVA, except, of course, it is on the linear combination Y instead of a single dependent variable.

11.15   MANOVA OF FISHER'S IRIS DATA

We now demonstrate MANOVA on Fisher's classic iris data. Recall that the data consist of a total of 150 observations on three species of iris, 50 on *setosa*, 50 on *virginica*, and 50 on *versicolor*. On each sample, the length and width of both sepals and petals were recorded. In our analysis, we would like to learn whether there are "species" (*setosa, versicolor, and virginica*) differences on a linear combination of flower features.

We first generate a combination of Sepal.Length + Sepal.Width + Petal. Length + Petal.Width through cbind, then request the MANOVA:

```
> attach(iris)
> iris.manova <- lm(cbind(Sepal.Length, Sepal.Width, Petal.Length,
Petal.Width) ~ Species, data = iris)
> anova(iris.manova)

Analysis of Variance Table

              Df  Pillai approx F num Df den Df     Pr(>F)
(Intercept)    1 0.99313   5203.9      4    144 < 2.2e-16 ***
Species        2 1.19190     53.5      8    290 < 2.2e-16 ***
Residuals    147
---
Signif. codes:  0 '***' 0.001 '**' 0.01 '*' 0.05 '.' 0.1 ' ' 1
```

By default, R produces Pillai's test, yielding a value of 1.19 with associated *p*-value of 2.2e-16. Clearly, there is an effect of species on the linear combination of flower features. To get all four multivariate tests, as well as the sums of squares and products matrices, one could compute via the car package, summary (Anova(iris.manova), univariate = FALSE, digits = 4). Doing so would reveal that all four multivariate tests are statistically significant. Several options exist for plotting multivariate effects, among them include heplots (see Friendly, 2007, for details).

11.16   POWER ANALYSIS AND SAMPLE SIZE FOR MANOVA

Power analysis for MANOVA can be conducted using G*Power similar to how we conducted it in ANOVA, where one specifies in advance an estimated effect size (entered as f^2 in G*Power), desired significance level (α), desired power, and the number of groups on the independent variable. Because the analysis is multivariate, one needs to also specify the number of dependent variables for the analysis, since this number can no longer be assumed to equal 1 as in the univariate case. As an example, suppose a researcher estimates an effect size of $f^2 = 0.10$ at a significance level of 0.05, with power set to 0.95. Suppose the researcher has three groups on the independent variable and is interested in analyzing such group differences on a linear combination of two response variables. In G*Power, one specifies "MANOVA: Global effects," and then enters the parameters for effect size, error probability, power, number of groups, and number of response variables:

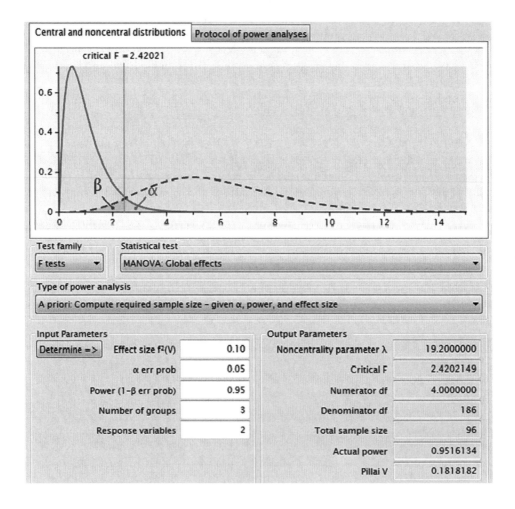

We see that for the inputted parameters, the total sample size required for this study is 96 (i.e., 32 per group, since there are three groups).

11.17   MULTIVARIATE ANALYSIS OF COVARIANCE AND MULTIVARIATE MODELS: A BIRD'S EYE VIEW OF LINEAR MODELS

As both analysis of variance and analysis of covariance can be considered special cases of the wider **general linear model**, so too can the multivariate analysis of variance and multivariate analysis of covariance be conceptualized as special cases of the wider **multivariate multiple regression model** or **multivariate general linear model**. That is, we expand the landscape from our earlier general linear model for regression in (7.7), $\mathbf{y} = \mathbf{X}\boldsymbol{\beta} + \boldsymbol{\varepsilon}$, to an even more inclusive general linear model, first introduced in Chapter 2:

$$\mathbf{Y} = \mathbf{XB} + \mathbf{E} \tag{11.18}$$

where $\mathbf{Y}$ is an $n \times m$ matrix of n observations on m response variables, $\mathbf{X}$ is the model matrix whose columns contain k regressors which includes the intercept term, $\mathbf{B}$ is a matrix of regression coefficients and $\mathbf{E}$ is a matrix of errors. Model (11.18) is adaptable to a variety of variable-types and can accommodate a wide number of variables. For example, in the multivariate regression model, because it is multivariate, $\mathbf{Y}$ in (11.18) could contain more than a single continuously distributed response, and $\mathbf{X}$ could contain a mix of continuous and polytomous predictor variables. If we expand (11.18) even further, we obtain model (5.11) discussed in the context of ANOVA models:

$$\mathbf{Y} = \mathbf{XB} + \mathbf{ZU} + \mathbf{E} \tag{11.19}$$

where $\mathbf{ZU}$ contains random effect terms (over and above the random effect $\mathbf{E}$). Recall model (11.19) is the matrix formulation of the mixed-model (Chapter 5), which contains a blend of fixed (in $\mathbf{XB}$) and random (in $\mathbf{ZU}$) effects. A researcher could also include one or more covariates in a MANOVA in an effort to reduce the error term (and hence, boost power) for testing treatment effects, yielding the MANCOVA model.

The point of this brief discussion is to emphasize the numerous possibilities for fitting a wide variety of models starting with a very **general** framework. As one becomes more familiar with more complicated models, one begins to see previously learned models as simply "special cases" of the wider landscape. Indeed, both models (11.18) and (11.19) could, in turn, be considered special cases themselves of the wider **generalized linear model**, as previously discussed in Chapter 10 in which our focus there was on the logistic regression model, but where several other "link functions" could have been used depending on the context.

This is the extent to which we discuss the comparison of parameterization options for the multivariate general linear model. A thorough discussion would require a chapter in its own right if not an entire book. The reader interested in learning about such parameterization options should consult Fox (1997) for an overview of such work with regard to the more global models briefly discussed here.

11.18   CHAPTER SUMMARY AND HIGHLIGHTS

- **Multivariate analysis of variance** (MANOVA) is a statistical method useful for situations in which one wishes to analyze group differences on a linear combination of dependent variables. The typical null hypothesis for MANOVA is that there are no population mean differences on a vector of dependent variables.
- MANOVA can be conceptualized as either an **extension of univariate ANOVA** or as a more **general linear model** of which ANOVA is a special case.

- For MANOVA to be a suitable model for a research problem, the hypothesized linear combination of dependent variables should usually be constructed based on theory, or at the very least, make some sense **a priori** to the investigator. If combining dependent variables is not logical for the given research context, then other than for strictly exploratory pursuits, MANOVA is not recommended.

- Statistical reasons for running a MANOVA include control over the **inflation of the type I error rate** and the capitalization of covariance between dependent variables, which in separate univariate ANOVAs would not be modeled.

- **Rao's paradox** describes the fact that one can fail to reject a multivariate hypothesis yet still reject individual univariate hypotheses, or fail to reject univariate hypotheses yet still discover a multivariate effect. The essence of the paradox is a reminder that whenever one tests a model, it is the **model** you are testing and never individual effects within the model. Even the simplest of models are still context-dependent.

- **Hotelling's T^2** tests for mean population differences between two groups on a linear combination of response variables. It can be understood as an extension of univariate t, or as a special case of the wider MANOVA model. Showing how Hotelling's T^2 is derived from univariate t is a powerful way to reveal how its structure is very much analogous to univariate t. Studying its makeup is also a powerful way of being introduced to the nature of multivariate methods.

- **Wilk's Λ** is historically the most popular multivariate test statistic. It is an inverse criterion, meaning that smaller values are more indicative of evidence against the null than are larger values. It is defined as the ratio of determinants of the error matrix relative to the total matrix, $\Lambda = \frac{|\mathbf{E}|}{|\mathbf{H} + \mathbf{E}|} = \frac{|\mathbf{E}|}{|\mathbf{T}|}$

- **Pillai's trace** is a multivariate test statistic defined as $V^{(s)} = \text{tr}\left[(\mathbf{E} + \mathbf{H})^{-1} \mathbf{H} \right] = \sum_{i=1}^{s} \frac{\lambda_i}{1 + \lambda_i}$, where λ_i are eigenvalues.

- **Roy's largest root** is another multivariate test statistic, given by $\theta - \frac{\lambda_1}{1 + \lambda_1}$, where λ_1 is the maximum eigenvalue extracted for the given problem. Roy's test is most powerful over competing multivariate tests when **mean vectors are collinear**. In problems where only a single eigenvalue is extracted, θ and $V^{(s)}$ will be the same.

- **Lawley–Hotelling's Trace** is another multivariate test statistic, given by $U^{(s)} = \text{tr}\left(\mathbf{E}^{-1} \mathbf{H} \right) = \sum_{i=1}^{s} \lambda_i$.

- MANOVA requires the assumption of **equality of covariance matrices**. This assumption can be tested using the **Box M-test**, though because the test is sensitive to distributional assumptions, even when Box reveals a statistically significant finding (thereby rejecting the assumption of equal variance-covariance matrices), it is recommended that in most cases one should nonetheless proceed with the MANOVA. Nonetheless, both statistically and substantively, the Box M-test is important because it provides insight into the structure of covariance matrices.

- **Contrasts** in MANOVA can be constructed in a similar manner as contrasts in univariate ANOVA, only that in MANOVA, mean vectors are being compared instead of simply univariate means.

- Following up a MANOVA with individual univariate ANOVAs on each dependent variable is acceptable if one has a theoretical reason for doing so. Because of **Rao's paradox,** however, one should not habitually conduct the follow-up univariate analyses in an effort to "decompose" the multivariate effect. **Multivariate models are distinct from univariate ones**.

REVIEW EXERCISES

11.1. Describe how **multivariate analysis of variance** differs from **univariate analysis of variance**. What are the distinguishing features? Discuss how MANOVA can be understood as either an extension of ANOVA or how ANOVA can be understood as a "special case" of MANOVA.

11.2. In the context of MANOVA, describe what is meant by a **linear combination** or **linear composite** of variables.

11.3. Discuss why it is that even if one has **several response variables** at his or her disposal, MANOVA may still not be a suitable statistical approach.

11.4. Discuss why using scalars of "1" in the linear combination (1)quantitative + (1)verbal = group would be considered a **naïve approach** to generating the given linear combination.

11.5. Discuss how the concept of **linear combinations** is not at all "new" to MANOVA and how it has been featured, sometimes indirectly, in virtually all statistical methodologies up to now (and further on) surveyed in this book.

11.6. Why is it that if the variate is not in some sense, even if minimally, "meaningful," then the fact that MANOVA helps to **regulate the type I error rate** is not in itself a "selling point" to using MANOVA.

11.7. Discuss what is meant by the **familywise error rate**.

11.8. Discuss the essential feature of **Rao's paradox**. As an applied researcher, how does this paradox influence the way you analyze your data? Why is the paradox especially relevant to your own research pursuits and your reading of the scientific literature?

11.9. Describe in words **Hotelling's** T^2 and draw as many parallels as you can between it and univariate t.

11.10. Describe the nature of the **H** and **E** matrices of MANOVA. How are they similar and different from scalar quantities used in ANOVA? Is the correspondence between **H** and **E** to such scalars in ANOVA a perfect one? Why or why not?

11.11. Why is Wilk's Λ referred to as an **inverse criterion**? Explain.

11.12. What does a Wilk's Λ of **1.0** mean? What does a Wilk's Λ of **0.0** imply?

11.13. When is it most appropriate to use **Roy's largest root** as a multivariate test? Give two scenarios where the value of Roy's is guaranteed to match that of Pillai's. Why is this so?

11.14. Explain how the **Box-M test** goes about testing the assumption of **equal covariance matrices**. Referring to its formula, explain the structure of the test.

11.15. Distinguish between a **univariate** and a **multivariate contrast**.

11.16. Discuss the components of the **multivariate general linear model** $\mathbf{Y} = \mathbf{XB} + \mathbf{E}$. Discuss how this model can be considered a special case of the model $\mathbf{Y} = \mathbf{XB} + \mathbf{ZU} + \mathbf{E}$ and how this latter model can be considered a special case of the generalized linear model.

11.17. Recall the data in which quantitative ability Q and verbal ability V were hypothesized as a function of training T. Make up data for a variable named **prior experience** P, with levels "none" and "at least some" accounting for how much prior educational experience individuals brought to the study. Generate the data such that there is a statistically significant training by prior experience interaction effect on the variate. Interpret its meaning, then conduct simple main effects and post-hocs to tease apart the interaction effect.

11.18 Consider data from Holzinger and Swineford (1939) (contained in package `lavaan` in R, (Rosseel, 2012). The complete data set contains 301 observations on the following 15 variables (note that tests x1–x9 constitute 9 of the 15): id (identifier), sex, ageyr (age in years), agemo (age in months), school (school attended by the child), grade, x1–x9 (9 tests of mental ability). We rename the data frame hs and print a few cases below:

```
> library(lavaan)
> hs <- data.frame(HolzingerSwineford1939)
> library(car)
> some(hs)

> some(hs)
    id sex ageyr agemo  school grade      x1   x2    x3       x4   x5
2    2   2     2    13       7 Pasteur  75.333333 5.25 2.125 1.666667 3.00
20  21   2     2    12       3 Pasteur  76.333333 8.75 3.000 3.666667 3.75
34  36   2     2    12       3 Pasteur  74.166667 6.00 2.375 3.333333 4.25

          x6       x7   x8       x9
2    1.285714 3.782609 6.25 7.916667
20   2.571429 3.478261 5.35 4.916667
34   1.857143 5.391304 4.35 5.638889
```

(a) Test the hypothesis that mental tests x1 through x9, considered as a composite, are a function of sex.

(b) Adapt the analysis in part (a) to include grade as a second independent variable in the model. Evaluate the potential effects for sex, grade, and comment on whether or not you have evidence to believe there is a sex by grade (i.e., two-way) interaction.

(c) Adapt the analysis in part (b) to include school as a third independent variable in the model. Evaluate the three-way interaction.

11.19. Anderson (2003) analyzed data on Egyptian skulls (p. 345) where it was hypothesized that change in skull size is a function of period of time (i.e., "epoch"). Skull size is operationally defined with four different variables: mb (maximum breadth of skull), bh (basibregmatic height of skull), bl (basialveolar length of skull), nh (nasal height of skull). Duplicate Andersen's analysis in R. The data is stored in the HSAUR package (Everitt and Hothorn, 2015). A few cases from the data frame appear below. Is there evidence that the linear combination mb + bh + bl + nh is a function of epoch?

```
> library(HSAUR)
> skulls

     epoch  mb  bh  bl nh
2   c4000BC 125 131  92 48
6   c4000BC 138 137  89 56
20  c4000BC 132 131 101 49
```

11.20. Reanalyze the data in 11.19, this time defining the linear combination of response variables as composed of only bh, bl, and nh, discarding mb. Is epoch statistically significant? Did the *p*-value for epoch change? Why might it have changed?

Further Discussion and Activities

11.21. We have learned that the multivariate analysis of variance is a relatively technically elegant statistical method. However, does application of the statistical method to empirical data "advance" science more than if we only had univariate methods (e.g., ANOVA) at our disposal? Do you believe it is reasonable, substantively, even if doable mathematically, to hypothesize linear combinations of dependent variables as representative of constructs? Are social scientists' claims made stronger or weaker by hypothesizing linear combinations rather than single variables as responses? What benefits or drawbacks can you think of to operationalizing variables in this way of linear combinations, that either advance, or delay the pursuits of scientific knowledge? Discuss.

12

DISCRIMINANT ANALYSIS

When two or more populations have been measured in several characters, $x_1, \ldots, x_s$, special interest attaches to certain linear functions of the measurements by which the populations are best discriminated ... In the present paper the application of the same principle will be illustrated on a taxonomic problem ... We shall first consider the question: What linear function of the four measurements

$$X = \lambda_1 x_1 + \lambda_2 x_2 + \lambda_3 x_3 + \lambda_4 x_4$$

will maximize the ratio of the difference between the specific means to the standard deviations within species?

(Fisher, 1936, p. 466)

Discriminant analysis is a statistical method first proposed by Fisher in 1936 for the purpose of classifying objects, subjects, or items into typically one of two or more mutually exclusive populations. Analogous to regression in which the task is to make predictions on a response variable based on a linear combination of predictors, the job of discriminant analysis is likewise to use a linear combination of explanatory variables (typically, continuous ones) to predict a response on a binary or polytomous dependent variable.

Recall that in MANOVA, we were interested in testing hypotheses about population differences on a mean vector. In linear discriminant analysis (LDA), we turn things around, and ask whether a linear combination of predictors might prove useful in predicting group membership. More formally, for a two-group problem, the **discriminant function** is the linear combination of predictors that maximizes the distance between the two group mean standardized vectors and ideally reduces errors of

Applied Univariate, Bivariate, and Multivariate Statistics: Understanding Statistics for Social and Natural Scientists, With Applications in SPSS and R, Second Edition. Daniel J. Denis.
© 2021 John Wiley & Sons, Inc. Published 2021 by John Wiley & Sons, Inc.
Companion Website: www.wiley.com/go/denis/appliedstatistics2e

classification. More technically, LDA seeks a projection (i.e., a mapping of vectors onto a vector sub-space) of observations such that the ratio of **between**-group variability to **within**-group variability is maximized. If a **cost function** is incorporated into the analysis, we can say as well that, in general, discriminant analysis seeks to minimize the cost of misclassification, though it is not necessarily the case that good discriminant functions are always one-to-one with good classification. As Timm (2002, pp. 419–420) notes, "While one may intuitively expect a good discriminant to also accurately predict group membership for an observation, this may not be the case. A classification rule usually requires more knowledge about the parametric structure of the groups."

Fisher first demonstrated the technique of discriminant analysis on the iris data first introduced in Chapter 8 and also analyzed as a MANOVA in the previous chapter. In the MANOVA, we were interested in learning whether mean species differences (setosa, virginica, and versicolor) could be inferred to exist on a linear combination of flower features sepal length, sepal width, petal length, and petal width. In the current chapter, we "reverse" the problem, and ask whether one can use a linear combination of these four features to maximize discrimination between species. Fisher's contribution in 1936 was to show how one could predict the species of iris based on characteristics of sepals and petals.

Many introductory to intermediate sources are available on discriminant analysis. Among them are Johnson and Wichern (2007), Rencher (1998), and Flury (1997).

12.1 WHAT IS DISCRIMINANT ANALYSIS? THE BIG PICTURE ON THE IRIS DATA

To motivate our development of discriminant analysis, we consider first the end result of Fisher's analysis, which we will also generate for ourselves later in the chapter. Consider Figure 12.1, which appeared on the final page of Fisher's seminal 1936 paper. The diagram is one of the final classification results based on using linear functions, called **discriminant functions**, to classify species of iris. Without even knowing what the discriminant functions are yet, through inspection of the diagram, we can informally assess how "good" the functions were that Fisher developed.

Consider first the classification of **setosa**. Notice how that distribution of classification scores is very well separated from both **versicolor** and **virginica**. Informally then, whatever function rules were derived from the LDA appear to have done a pretty good job at providing separation between setosa and the other two species. Are they discriminating well between versicolor and virginica? Not as well it seems, since between these there is substantial overlap in distributions. The goal of this chapter is to learn what these discriminating functions look like, how they are obtained, and how to assess their "goodness" in terms of how well they discriminate, analogous to how we evaluated regression equations in terms of how well they could make predictions.

As we will see, Fisher desired his discriminant functions to be ones that would "maximize the ratio of the difference **between** [emphasis added] the specific means to the standard deviations **within** [emphasis added] species" (Fisher, 1936, p. 466). That is, Fisher derived functions that **maximized between variation relative to within variation**. Does this sound familiar? This is an analogous idea to that encountered in our study of the analysis of variance in Chapter 3 through MS between and MS within, and even more so in the **multivariate** analysis of variance of Chapter 11 via matrices **H** and **E**. In MANOVA, the goal could be said to use a grouping variable to make predictions on a linear composite variable. We tested null hypotheses about population mean differences on a mean vector, but we did not identify what **function** was actually responsible for maximizing differences between groups. In discriminant analysis, we learn the nature of such functions, the so-called **discriminant functions**. Refer to Table 12.1 for an overview comparison of the two approaches in terms of the typical makeup of response and predictor variables in each case.

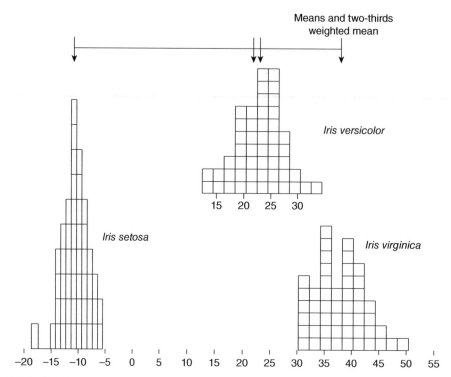

FIGURE 12.1   Fisher's discriminant function analysis of the *Iris* data. Source: Fisher (1936).

TABLE 12.1   Comparison of Typical Response and Predictor Variables for MANOVA versus Discriminant Analysis

	Response Variable	Predictor Variables
Multivariate analysis of variance	Continuous	Dichotomous (Hotelling's T^2) or polytomous (MANOVA)
Discriminant analysis	Dichotomous or polytomous	Continuous

12.2   THEORY OF DISCRIMINANT ANALYSIS

In surveying some of the theory behind discriminant analysis, we begin with the situation of predicting membership on two populations, then move on to considering LDA for several populations. Along the way, we demonstrate how features of the output obtained in LDA can be related to that obtained in MANOVA. This comparison is extremely useful so that the user recognize and appreciate the similarity of these models. Often in statistical analyses, researchers may be using very similar underlying models that simply go by different names and have different applied purposes. Appreciation of the **similarity between models** allows one to understand their underlying structures and the generality of statistical methods.

12.2.1 Discriminant Analysis for Two Populations

In a discriminant analysis for predicting membership on one of two mutually exclusive populations, we want to derive a linear combination of predictor variables that best discriminates group membership on a binary response variable. Recall what it means for categories to be **mutually exclusive**—practically, it means if you are in one category, you cannot simultaneously be in the other. In binary discriminant analysis, you are either in one group or the other, you cannot be in both simultaneously.

Recall from Chapter 2 the definition of a **linear combination** of the form:

$$\ell_i = a_1 y_1 + a_2 y_2 + \ldots + a_p y_p$$
$$= \mathbf{a'y} \tag{12.1}$$

where $\mathbf{a'} = (a_1, a_2, \ldots, a_p)$. These values are **scalars**, and serve to weight the respective values of y_1 through y_p, where p is the number of variables. The essence of discriminant analysis is to find a vector $\mathbf{a}$ that will **maximize** the standardized difference between groups on the response variable. That is, we want a vector that maximizes

$$d = \frac{(\bar{\ell}_1 - \bar{\ell}_2)}{s_\ell}$$

where s_ℓ is the pooled standard deviation of linear combinations. Because we know that $\bar{\ell}_1 - \bar{\ell}_2$ can be negative, we will use the **squared distance** instead:

$$d^2 = \frac{(\bar{\ell}_1 - \bar{\ell}_2)^2}{s_\ell^2} \tag{12.2}$$

But what vector will maximize the squared distance in (12.2)? It can be shown (see Rencher and Christensen, 2012, Chapter 8; Tatsuoka, 1971, Chapter 6) that the squared distance between mean vectors is a function of $\mathbf{a}$, given by:

$$d^2 = \frac{(\bar{\ell}_1 - \bar{\ell}_2)^2}{s_\ell^2} = \frac{[\mathbf{a'}(\bar{\mathbf{y}}_1 - \bar{\mathbf{y}}_2)]^2}{\mathbf{a'S}_p\mathbf{a}} \tag{12.3}$$

where $\mathbf{a'}$ is the transpose of estimated coefficients, $\bar{\mathbf{y}}_1 - \bar{\mathbf{y}}_2$ is the mean difference between vectors, and $\mathbf{a'S}_p\mathbf{a}$ is the variance of ℓ (i.e., $s_\ell^2 = \mathbf{a'S}_p\mathbf{a}$). It can be shown further that the maximum occurs when

$$\mathbf{a} = \mathbf{S}_p^{-1}(\bar{\mathbf{y}}_1 - \bar{\mathbf{y}}_2) \tag{12.4}$$

or when $\mathbf{a}$ is any multiple of (12.4). The maximizing vector $\mathbf{a}$ in this sense, is not **unique**. However, the **direction** of the vector is unique. What this means is that though we can multiply values of $\mathbf{a}$ by a scalar, the **ratios** of elements of $\mathbf{a}$ will remain the same (Rencher and Christensen, 2012). For example, if $a_1 = 10$ and $a_2 = 20$, the ratio of "2 to 1" (i.e., a_2 to a_1) remains even if we multiply by a scalar. Recall that if $\mathbf{a}$ is a vector, then when multiplied by a scalar of 3, for instance, $3\mathbf{a}$ simply **elongates** the vector without its changing direction. This is what we mean by saying the direction is unique. Consider a simple vector (1, 3) multiplied by a scalar of 2 (Figure 12.2):

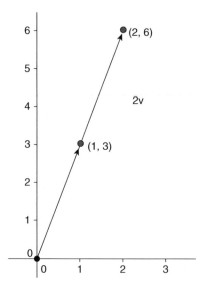

FIGURE 12.2   Multiplication of a vector by a scalar of 2.

The length of the vector increased by a factor of 2, yet the direction of the vector remained the same.

12.2.2   Substituting the Maximizing Vector into Squared Standardized Difference

We have said that the squared standardized difference between groups, that of (12.3), will be maximized by finding the vector (12.4). Substituting (12.4) into (12.3), we can state:

$$d^2 = \frac{\left(\bar{\ell}_1 - \bar{\ell}_2\right)^2}{s_\ell^2} = (\bar{\mathbf{y}}_1 - \bar{\mathbf{y}}_2)'\mathbf{S}_p^{-1}(\bar{\mathbf{y}}_1 - \bar{\mathbf{y}}_2) \tag{12.5}$$

We can see that structurally, we are "squaring" the mean difference in vectors, that of $(\bar{\mathbf{y}}_1 - \bar{\mathbf{y}}_2)$ (i.e., $(\bar{\mathbf{y}}_1 - \bar{\mathbf{y}}_2)'(\bar{\mathbf{y}}_1 - \bar{\mathbf{y}}_2)$) and dividing by the pooled covariance matrix, $\mathbf{S}_p$, where multiplying by its inverse, $\mathbf{S}_p^{-1}$, recall, acts here as division in matrix operations. To understand (12.5) better, imagine for a moment if we wrote it as:

$$\frac{\left(\bar{\ell}_1 - \bar{\ell}_2\right)^2}{s_\ell^2} = \frac{(\bar{\mathbf{y}}_1 - \bar{\mathbf{y}}_2)^2}{\mathbf{S}_p}$$

Though the above of course is not a technically correct way of displaying the equation, it does give a sense of what is going on. When we compute (12.5), all we are doing is generating a ratio of squared mean vector differences in the numerator relative to an overall measure of variance and covariance in ℓ. Interpreted geometrically, what (12.5) is "accomplishing" is making an adjustment on the original axes in the Cartesian plane that best accounts for mean vector separation by consideration of the contents of $\mathbf{S}_p$, which of course contain variances and covariances. By "standardizing" the squared distance

$(\bar{y}_1 - \bar{y}_2)^2$, this effectuates a new dimension (axis) along which group separation is maximized. For a simple demonstration of the geometrical interpretation, see Lattin, Carroll, and Green (2003, pp. 429–434).

We demonstrate the essential concepts of discriminant analysis through a simple and straightforward example. Consider the hypothetical generic data in Table 12.2 having response variable y_i and predictors x_1 and x_2.

Our goal is to find a **linear combination** of x_1 and x_2 such that the discrimination between groups 0 and 1 on y_i is **maximal**. That is, the task is to generate a linear combination of x_1 and x_2 that will aid in predicting group membership on y_i. But what does **maximal discrimination** mean? It is possible that our discriminant function, because of the nature of our data, will do less than a perfect job at discriminating between groups. What then does it mean to say the function will **maximally discriminate**?

To understand what this means, we draw on our knowledge of least-squares regression. What did it mean to say we were fitting the least-squares line? It meant that of all the possible lines we could theoretically fit to our sample data, the least-squares line is the one that minimized the sum of squared errors around the line (or plane in the case of multiple regression) better than any other line that could be fit to the given data. Did this fact guarantee that for the given set of data the sum of squared errors would be necessarily **small**? Not at all. For sloppy, high-variability data, the least-squares line will likewise provide a sloppy fit. But then, OLS cannot guarantee anything about **data**. It can only guarantee to minimize the sum of squared error **on that data**.

A parallel to OLS is somewhat evident in discriminant analysis. By the theory of LDA, we know we are maximally discriminating between groups, but whether this maximum discrimination is "good" or not will depend on the data we are computing the function on. Analogous to building a house with poor versus quality materials, LDA, as was true for OLS regression, will produce as good of a function (house) as the data (materials) allow, but no better. Regardless of the statistical model, it must always be remembered that no matter how sophisticated the model may be, **it can never work miracles on effects that are just not there to begin with**.

12.3   LDA IN R AND SPSS

We carry on now with a discriminant analysis on the data in Table 12.2. We enter our data in R by generating the relevant vectors and construct the data frame, which we name `discrim`:

TABLE 12.2   Hypothetical Data on Binary Response and Continuous Predictors

Subject	y	x_1	x_2
1	0	4	2
2	0	3	1
3	0	3	2
4	0	2	2
5	0	2	5
6	1	8	3
7	1	7	4
8	1	5	5
9	1	3	4
10	1	3	2

```
> y <- c(0, 0, 0, 0, 0, 1, 1, 1, 1, 1)
> x1 <- c(4, 3, 3, 2, 2, 8, 7, 5, 3, 3)
> x2 <- c(2, 1, 2, 2, 5, 3, 4, 5, 4, 2)
> discrim <- data.frame(y, x1, x2)
> discrim

     y x1 x2
1    0  4  2
2    0  3  1
3    0  3  2
4    0  2  2
5    0  2  5
6    1  8  3
7    1  7  4
8    1  5  5
9    1  3  4
10   1  3  2
```

We perform the discriminant analysis by calling the function lda ("linear discriminant analysis") denoting y as a function of x1 + x2:

```
> library(MASS)
> lda.fit <- lda(y ~ x1 + x2, data = discrim)
> lda.fit

Call:
lda(y ~ x1 + x2, data = discrim)

Prior probabilities of groups:
  0   1
0.5 0.5
```

By default, the prior probability of group membership is set at 0.5 for each group. For some problems, we may want to adjust this prior probability to differ from the default. This can be done for a similar reason why we may wish to incorporate baseline or base rate information in a wide variety of problems, such as the probability of surviving versus not surviving an operation, or the probability of passing or failing a university course. As discussed in the Appendix, these are **prior probabilities**, so they should to some extent reflect the "current status" with regard to the probability of a success or failure. Priors in this regard could also be selected as a function of sample size per group to represent proportional "baselines" in the population.

R provides us with the cell means across each group, then follows this up with the discriminant function coefficients:

```
Group means:
   x1  x2
0 2.8 2.4
1 5.2 3.6
```

```
Coefficients of linear discriminants:
          LD1
x1 0.4973955
x2 0.4310761
```

The **coefficients of linear discriminants** are the **raw** coefficients of the discriminant function esti-mated on our data. We can obtain the linear discriminant function scores quite easily (only the first three scores are given below):

```
> predict(lda.fit)

$class
 [1] 0 0 0 0 0 1 1 1 0 0
Levels: 0 1

$posterior
            0          1
1  0.67646834 0.32353166
2  0.91102713 0.08897287
3  0.83042643 0.16957357


          LD1
1  -0.4310761
2  -1.3595477
3  -0.9284716
```

The above are the posterior probabilities of group membership, along with the discriminant function scores for LD1. This information is summarized in Table 12.3.

The first two columns in Table 12.3, **Posterior 0** and **Posterior 1**, are the probabilities of being classified into group 0 or 1 based on the estimated discriminant function. Note that because they are probabilities, they are **continuously scaled** measures, not binary ones as required to predict

TABLE 12.3 Posterior Probabilities and Predicted Group Membership (Posterior G), along with Discriminant Scores

Posterior 0	Posterior 1	Posterior G	D Scores	Y	X_1	X_2
0.68	0.32	0	−0.43	0	4	2
0.91	0.09	0	−1.36	0	3	1
0.83	0.17	0	−0.93	0	3	2
0.92	0.08	0	−1.43	0	2	2
0.56	0.44	0	−0.13	0	2	5
0.03	0.97	1	1.99	1	8	3
0.04	0.96	1	1.92	1	7	4
0.09	0.91	1	1.36	1	5	5
0.53	0.47	0	−0.07	1	3	4
0.83	0.17	0	−0.93	1	3	2

membership on the response variable. The third column, **Posterior G**, is the predicted group membership based on the discriminant function. This group classification is based on the derived discriminant scores of the fourth column, **D Scores**. Notice that the sign of the discriminant scores, + or −, corresponds with whether an observation was classified into group 0 or 1, since zero is the balance point used for classification here where $n_1 = n_2$ (James et al., 2013, p. 141). We note from inspection of the column **Posterior G** and that of **D scores** that the function correctly classified all of the cases for $y_i = 0$, and 3 out of 5 for $y_i = 1$. That is, our function misclassified two cases. We can request R to produce the following classification table to summarize this fact, sometimes called a **confusion matrix**:

```
> table(discrim$y, predict(lda.fit)$class)

    0 1
  0 5 0
  1 2 3
```

Along the main diagonal of the matrix are the correct classifications. The misclassifications appear in the off-diagonal. The confusion matrix confirms the results of our analysis regarding the misclassified cases, that though two cases actually belong to $y_i = 1$, they were instead predicted into $y_i = 0$ (the two misclassified cases are in row 2, column 1).

We now conduct the analysis in SPSS. As we will see, the output essentially mirrors that of R, though SPSS will automatically also provide us with the corresponding **eigenvalue** for the discriminant function. The following syntax generates the discriminant analysis:

```
DISCRIMINANT * requests the discriminant analysis procedure
   /GROUPS=y(0 1) * specifies the binary grouping variable having levels "0" and "1"
   /VARIABLES=x1 x2 * the predictors
   /ANALYSIS ALL * includes all variables in the analysis
   /SAVE=CLASS SCORES * requests to save classification results and discriminant scores
   /PRIORS EQUAL * sets the prior probabilities as equal (in this case, 0.5)
   /STATISTICS=RAW * requests raw coefficients
   /PLOT=CASES * requests a plot of the discriminant function results
```

The eigenvalue for the discriminant function is 0.915 and accounts for 100% of the extraction (Table 12.4). Contrary to how "% of variance" may appear in SPSS, **the 100% figure is not an estimate of effect size or variance explained**. It would be incorrect to conclude that the discriminant function accounts for 100% of the variance in the response variable. The "100%" figure in this case denotes the fact that this is the only discriminant function extracted for the analysis, which, of course, makes sense, since there are only two groups on the response variable. Had we a problem where more than a single discriminant function were extracted, this figure would change, because not all of the "variance" (we will have more to say about the use of the word variance here later) would likely be accounted for by only the first discriminant dimension.

The **canonical correlation**, equal to 0.691, and when squared, $(0.691)^2 = 0.48$, provides us with a measure of **association** or **effect size** for the discriminant function. The squared canonical correlation is equal to the ratio $\dfrac{\lambda_i}{1 + \lambda_i} = \dfrac{0.915}{1 + 0.915} = 0.48$. Wilk's lambda of 0.522 yields a p-value of 0.103, and hence the function is not statistically significant at $\alpha = 0.05$; however, depending on our priority for

TABLE 12.4 Eigenvalue and Significance Test for Discriminant Function

	Eigenvalues			
Function	Eigenvalue	% of Variance	Cumulative %	Canonical Correlation
1	0.915[a]	100.0	100.0	0.691
		Wilks' lambda		
Test of function(s)	Wilks' lambda	Chi-square	df	Sig.
1	0.522	4.548	2	0.103

[a]First 1 canonical discriminant functions were used in the analysis.

TABLE 12.5 Unstandardized Coefficients (Left), Standardized Coefficients (Middle), and Structure Coefficients (Right) for Discriminant Analysis

Canonical Discriminant Function Coefficients		Standardized Canonical Discriminant Function Coefficients		Structure Matrix	
	Function		Function		Function
	1		1		1
x1	0.497	x1	0.854	x1	0.817
x2	0.431	x2	0.578	x2	0.523
(Constant)	−3.283				
Unstandardized coefficients					

minimizing type II errors, the function may still be of use, even if statistically significant at a more liberal level (such as 0.15).

The **standardized canonical discriminant function coefficients** (Table 12.5, middle) reveal which variables have the greatest "impact" on the discriminant function. For our data, variable x1 has the largest absolute weight (0.854) and hence can be said to be more "relevant" to the discriminant function than x2, which has a weight of 0.578. Note that these are the **standardized versions of the raw coefficients** generated in R earlier (Table 12.5, left). Also included is the intercept (constant) term of −3.283, which, just as in regression, is required for obtaining predicted values. The model equation for raw scores is thus:

$$y_i' = -3.283 + 0.497x_1 + 0.431x_2$$

The structure matrix (Table 12.5, right) provides the raw bivariate correlations between the given observed variables and the discriminant function. The relative magnitude of these coefficients does not always correspond with those of the standardized coefficients, since both coefficients measure something different. However, for these data, both the standardized and the structure coefficients generally tell the same story, in that x1, with a standardized coefficient of 0.854 and structure coefficient of 0.817, is more relevant to the discriminant function than x2 with a standardized coefficient of 0.578 and structure coefficient of 0.523.

SPSS next provides the classification results that are more or less parallel to those given earlier in R:

Casewise Statistics

Case Number	Actual Group	Predicted Group	Highest Group P(D>d | G=g) p	df	P(G=g | D=d)	Squared Mahalanobis Distance to Centroid	Second Highest Group Group	P(G=g | D=d)	Squared Mahalanobis Distance to Centroid	Discriminant Scores Function 1
Original 1	0	0	0.671	1	0.676	0.180	1	0.324	1.655	-0.431
2	0	0	0.614	1	0.911	0.254	1	0.089	4.907	-1.360
3	0	0	0.942	1	0.830	0.005	1	0.170	3.183	-0.928
4	0	0	0.568	1	0.920	0.325	1	0.080	5.205	-1.426
5	0	0	0.470	1	0.556	0.523	1	0.444	0.976	-0.133
6	1	1	0.257	1	0.968	1.286	0	0.032	8.095	1.990
7	1	1	0.286	1	0.964	1.140	0	0.036	7.722	1.923
8	1	1	0.614	1	0.911	0.254	0	0.089	4.907	1.360
9	1	0[a]	0.430	1	0.528	0.623	1	0.472	0.850	-0.066
10	1	0[a]	0.942	1	0.830	0.005	1	0.170	3.183	-0.928

[a]Misclassified case

Matching up the results with those of R is straightforward. The second column, **Actual Group**, and the third column, **Predicted Group**, provide the same information we generated earlier in Table 12.3. As we noted then, and as reported by SPSS, cases 9 and 10 were misclassified, but all other cases were classified correctly. By default, SPSS also provides us with Mahalanobis (squared) distances. These distances take the general form (in terms of mean vectors),

$$D^2 = (\bar{\mathbf{y}}_1 - \bar{\mathbf{y}}_2)' \mathbf{S}^{-1} (\bar{\mathbf{y}}_1 - \bar{\mathbf{y}}_2)$$

where $\bar{\mathbf{y}}_1$ and $\bar{\mathbf{y}}_2$ are mean sample vectors and $\mathbf{S}^{-1}$ is the inverse of the pooled sample covariance matrix of $\mathbf{S}_1$ and $\mathbf{S}_2$, respectively. Notice that values (in the SPSS output) of D^2 (now in terms of distances of points from class centers), coincide with the respective columns reporting predicted probabilities of group membership to the immediate left of D^2. This is not a coincidence, since $P(G = g \mid D = d)$ denotes the probability that a case belongs in a given group g given its respective Mahalanobis distance. Cases with relatively high distances relative to a given centroid (i.e., **mean of discriminant scores**) have a lower probability of being classified into that group. Conversely, cases with relatively small distances have an increased probability. The right-most column of the SPSS output provides the computed discriminant scores, analogous to those reported by R and given in Table 12.3. Entire chapters exist on so-called **classification analysis**, and a further discussion of these procedures is beyond the scope of the current chapter. However, an understanding of how to classify based on discriminant scores or D^2, as evidenced from R and SPSS output for our example, provides the essential ideas about what classification analysis is generally all about.

12.4 DISCRIMINANT ANALYSIS FOR SEVERAL POPULATIONS

Up to now, we have discussed discriminant analysis for the case of using a linear combination of predictors to predict group membership where there are only **two** groups on the response variable. If there are more than two groups defined on the response, we need to expand the theory underlying the method similar to how we did so in the case of MANOVA when extending it from univariate ANOVA.

For a problem in which there are more than two groups, we require more than a single discriminant function to account for group separation. The number of functions necessary to maximally discriminate between groups is referred to as the **dimensionality** of the separation (Timm, 2002). Referring once more to Fisher's iris data, while one discriminant function may distinguish species setosa from versicolor and virginica for instance, hypothetically, a second discriminant function may prove useful in differentiating between versicolor and virginica. Each discriminant function would then serve a purpose when it comes to the overall discrimination between species.

We survey some of the theory behind discriminant analysis for several populations by expanding the theory already discussed for the two-population discriminant function. We then illustrate how to conduct an analysis for the multi-population case.

12.4.1 Theory for Several Populations

Recall for the two-group problem, we sought a vector **a** that maximally separated $(\bar{\ell}_1 - \bar{\ell}_2)^2$. The separation criterion for the two-group case was given by (12.3). At its core, d^2 was expressing a very simple idea, that of a **squared difference between means relative to overall variance**. If the squared difference between means, $(\bar{\ell}_1 - \bar{\ell}_2)^2$ is large relative to overall variance s_ℓ^2, then separation is "better" than if $(\bar{\ell}_1 - \bar{\ell}_2)^2$ is relatively small or equal to overall variance, s_ℓ^2. Of course, the overtones to what is accomplished in ANOVA are glaring. What we found for the two-group problem was that the vector

$\mathbf{a} = \mathbf{S}_p^{-1}(\bar{\mathbf{y}}_1 - \bar{\mathbf{y}}_2)$ is what provides the **maximum** separation between groups on the binary response. When we made the relevant substitution, we got (12.5).

We now extend these principles to the p-group case for p populations where we have a polytomous response variable. For this, we will naturally invoke ideas and matrices from MANOVA, since recall that LDA and MANOVA are essentially inverses of one another. The LDA for several groups is quite easy to grasp if the principles of MANOVA were understood. We follow Rencher and Christensen (2012, p. 289) quite closely in our development.

Recall that MANOVA featured two primary matrices: the "hypothesis" matrix, or $\mathbf{H}$, and that of the "error" matrix, or $\mathbf{E}$. In going from the two-group case to the multigroup situation, we will use $\mathbf{H}$ in place of $(\bar{\mathbf{y}}_1 - \bar{\mathbf{y}}_2)(\bar{\mathbf{y}}_1 - \bar{\mathbf{y}}_2)'$ and $\mathbf{E}$ in place of $\mathbf{S}_p$. This substitution yields λ, a ratio of $\mathbf{a}'\mathbf{Ha}$ to $\mathbf{a}'\mathbf{Ea}$:

$$\lambda = \frac{\mathbf{a}'\mathbf{Ha}}{\mathbf{a}'\mathbf{Ea}} \tag{12.6}$$

Some algebra on (12.6) reveals that:

$$\lambda = \frac{\mathbf{a}'\mathbf{Ha}}{\mathbf{a}'\mathbf{Ea}}$$
$$\mathbf{a}'\mathbf{Ha} = \lambda\mathbf{a}'\mathbf{Ea}$$
$$\mathbf{a}'(\mathbf{Ha} - \lambda\mathbf{Ea}) = \mathbf{0}$$

We now ask the question, "What values of $\mathbf{a}$ (other than $\mathbf{a} = \mathbf{0}$, which is the trivial case) results in a maximum for λ?" We find solutions by:

$$(\mathbf{Ha} - \lambda\mathbf{Ea}) = \mathbf{0}$$
$$\left(\mathbf{E}^{-1}\mathbf{H} - \lambda\mathbf{I}\right)\mathbf{a} = \mathbf{0} \tag{12.7}$$

The solutions of (12.7) are the **eigenvalues** and corresponding **eigenvectors** of $\mathbf{E}^{-1}\mathbf{H}$ (Rencher and Christensen, 2012).

The number of nonzero eigenvalues is the rank of $\mathbf{H}$ and is the smaller of the number of predictors k or one less the number of populations p. The largest eigenvalue λ_1 is the maximum value of $\lambda = \frac{\mathbf{a}'\mathbf{Ha}}{\mathbf{a}'\mathbf{Ea}}$, with $\mathbf{a}_1$ being the coefficient vector that generates the maximum. In obtaining the eigenvectors $\mathbf{a}_1, \mathbf{a}_2,$ ..., $\mathbf{a}_s$ of $\mathbf{E}^{-1}\mathbf{H}$ corresponding to λ_1 to λ_s eigenvalues, we generate s discriminant functions of the like, $\ell_1 = \mathbf{a}_1'\mathbf{y}, \ell_2 = \mathbf{a}_2'\mathbf{y}, ..., \ell_s = \mathbf{a}_s'\mathbf{y}$, which reveals the dimensions or directions of differences among $\bar{\mathbf{y}}_1, \bar{\mathbf{y}}_2, ..., \bar{\mathbf{y}}_p$. As noted by Rencher and Christensen (2012), such discriminant functions are **uncorrelated** but are not **orthogonal** because $\mathbf{E}^{-1}\mathbf{H}$ is not a symmetric matrix. When we survey principal component analysis in Chapter 13, we will extract linear combinations that are both uncorrelated **and** orthogonal. This distinction is a crucial one when deciphering between discriminant functions and principal components. **Components have the stronger property of orthogonality**.

As an example of discriminant analysis for several groups, recall the Q–V data of Chapter 11 (Table 11.3), in which we performed a MANOVA testing the null hypothesis of no mean vector differences on training, having levels 1 = no training, 2 = some training, 3 = extensive training. Suppose instead we wished to learn whether the linear combination of Q and V can differentiate between training groups. We compute the discriminant analysis as follows:

```
> library(MASS)
> lda.fit <- lda(T.f ~ Q + V, data = iq.data)
> lda.fit
```

```
Call:
lda(T.f ~ Q + V, data = iq.data)

Prior probabilities of groups:
     none       some       much
0.3333333  0.3333333  0.3333333

Group means:
             Q V
none 4.333333 2
some 8.000000 8
much 9.666667 9
```

Note that the prior probabilities, by default, are set to 0.33 in each group. This is analogous to them being set to 0.5 by default in the two-group problem, only that now, because we have three groups, we divide the total probability of 1.0 by 3, yielding 0.33 per group. We also note the group means reported on Q and V for each level of the training factor. They are 4.33, 8.00, 9.67, and 2, 8, 9 for groups 1, 2, and 3, respectively.

Next are given the raw discriminant function coefficients. Recall that because we have three groups on the independent variable, this calls for two discriminant functions to be extracted:

```
Coefficients of linear discriminants:
          LD1            LD2
Q 0.02983363   0.8315153
V 0.97946790  -0.5901991

Proportion of trace:
   LD1     LD2
0.9889  0.0111
```

Plots of discriminant scores across LD1 (i.e., the first discriminant function) and LD2 are given below (to get the plots with the labels instead of numbers, levels(T.f) <- c("none", "some", "much")).

```
> plot(lda.fit)
```

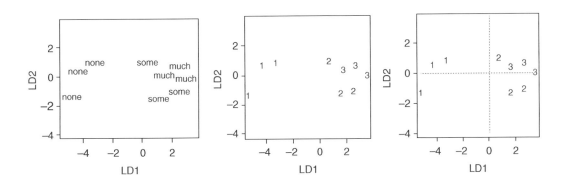

A helpful way to read these plots is to draw a vertical line at approximately LD1 = 0 and a horizontal line at approximately LD2 = 0. Looking at LD1, for instance, it is clear that the function is providing good separation between group 1 versus 2 and 3, while for LD2, it is not providing nearly as much separation. Of course, these are very small data so we might not put too much stock in the results, but the essence on how to read these plots is the same for larger data sets as well.

How "important" is LD1 compared with LD2? Recall that we can obtain a measure of the overall relevance of a discriminant function by contrasting its eigenvalue to the sum of eigenvalues extracted:

$$\frac{\lambda_i}{\sum\limits_{j=1}^{s} \lambda_j} \tag{12.8}$$

where λ_i is the ith eigenvalue for the ith discriminant function and $\sum\limits_{j=1}^{s} \lambda_j$ is the sum of all s eigenvalues from $j = 1$ (i.e., the first) to the last (i.e., s). Recall that the eigenvalues obtained through MANOVA for this problem were equal to 14.35 and 0.16 for functions 1 and 2 respectively. Hence, for the first function, (12.8) is equal to $14.35/(14.35 + 0.16) = 14.35/14.51 = 0.98897$, which is as noted in R's output under Proportion of trace. For the second function, $\lambda_i / \sum\limits_{j=1}^{s} \lambda_j$ is equal to $0.16/14.51 = 0.011$, which is reported by R as well for LD2. We already suspected by a look at the LD plots that the first function was "doing all the work," and indeed, our computation of (12.8) for each function confirms it. **LD1 is accounting for most of the group separation**.

12.5 DISCRIMINATING SPECIES OF IRIS: DISCRIMINANT ANALYSES FOR THREE POPULATIONS

We now demonstrate a discriminant analysis for three populations on the iris data. Recall that these data were analyzed as a MANOVA in the previous chapter. Since LDA is essentially the "reverse" of MANOVA, we remark once more on parallels between the two analyses. Our goal is to learn whether we can predict species membership (setosa, versicolor, virginica) based on knowledge of explanatory variables sepal length, sepal width, petal length, and petal width.

```
> lda.iris <- lda(Species ~ ., iris)
```

Note that since we are modeling all predictors, we can specify the model statement as " ~ ." to indicate this. The output now follows:

```
> lda.iris

Call:
lda(Species ~ ., data = iris)

Prior probabilities of groups:
    setosa versicolor  virginica
 0.3333333  0.3333333  0.3333333
```

```
Group means:
          Sepal.Length Sepal.Width Petal.Length Petal.Width
setosa           5.006       3.428        1.462       0.246
versicolor       5.936       2.770        4.260       1.326
virginica        6.588       2.974        5.552       2.026


Coefficients of linear discriminants:
                   LD1          LD2
Sepal.Length  0.8293776   0.02410215
Sepal.Width   1.5344731   2.16452123
Petal.Length -2.2012117  -0.93192121
Petal.Width  -2.8104603   2.83918785


Proportion of trace:
   LD1    LD2
0.9912 0.0088
```

As a result of having a total of three groups, the prior probabilities, by default, are again set at 0.33 for each. LDA has extracted two linear discriminant functions, LD1 and LD2. Recall that it extracted two as a result of the response variable having three levels. The `proportion of trace` figures reveal that the first eigenvalue and second eigenvalue extracted account for approximately 99.12% and 0.88% respectively of the total sum of eigenvalue. Clearly, the first discriminant function is much more relevant than the second. To gain an appreciation of their relative importance, we can plot the discriminant functions. Here, we plot the first function (`dimen = 1`):

```
plot(lda.iris, dimen = 1)
```

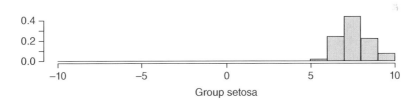

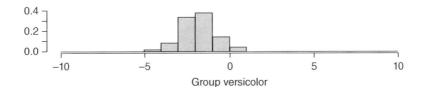

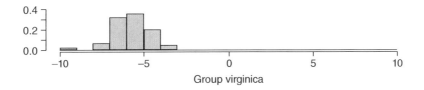

We can see that the first function appears to do a good job of separating setosa from versicolor and virginica. To visualize results for both functions, we request dimen = 2, and also request a density plot displaying the performance of the first function in distinguishing between species:

```
> plot(lda.iris, dimen = 2)
> plot(lda.iris, type = "density", dimen = 1)
```

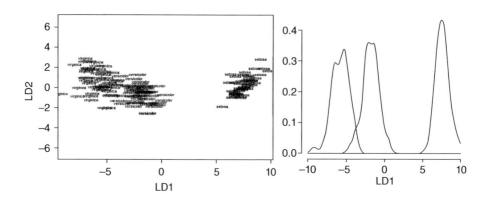

As confirmed by both plots, the first linear discriminant function, LD1, is doing a great job at differentiating versicolor and virginica from setosa. The second linear discriminant function, LD2, is not differentiating species very well. Of course, this aligns with the fact that the second eigenvalue extracted accounts for only 0.88% of the sum total of eigenvalues.

We request classification results:

```
> fit < lda(iris$Species ~., iris)
> table(iris$Species, predict(fit)$class)
```

```
             setosa versicolor virginica
setosa          50          0         0
versicolor       0         48         2
virginica        0          1        49
```

We can see that the discriminant analysis does a perfect job in the sample at classifying setosa (50 for 50), but makes two errors in classifying versicolor (48 for 50) and one error in classifying virginica (49 for 50).

12.6   A NOTE ON CLASSIFICATION AND ERROR RATES

While on the topic of classification, we should also emphasize that discriminant function analysis is usually distinguished by many authors as distinct from what is generally known as **classification analysis**. Their difference is best summarized by Timm (2002):

A classification rule usually requires more knowledge about the parametric structure of the groups. The goal of classification analysis is to create rules for assigning observations to groups that minimize the total probability of misclassification or the average cost of misclassification. Because linear discriminant functions are often used to develop classification rules, the goals of the two processes tend to overlap and some authors use the term classification analysis instead of discriminant analysis. (p. 420)

TABLE 12.6 Decision Table for Discriminant Analysis with Binary Dependent Variable

		Decision	
		D_1	D_2
Population	P_1	0	$C(2|1)$
	P_2	$C(1|2)$	0

In the current chapter, our focus has been on **linear discriminant analysis**. We have not covered classification analysis in any depth other than displaying select results of said analyses in R and SPSS output in the form of predicted values and confusion matrices. Linear classification functions have been developed in the pursuit of generating optimal rules that attempt to **minimize errors in classification**. **Quadratic classification functions** have also been developed for situations in which it is not reasonable to assume equality of covariance matrices across groups (James et al., 2013).

Though we do not discuss classification in any depth in this chapter, it behooves us nonetheless to consider how we can conceptualize the importance of costs associated with **misclassification**. For example, let us consider the simplest of cases, that of classification for two populations. In Table 12.6 shows a 2×2 table where **Decision** denotes the choice we are making with regard to classifying into populations P_1 and P_2. **Population** reflects the actual **true** population status. The costs of misclassification, denoted by $C(2|1)$ and $C(1|2)$ are given as follows:

- $C(2|1)$ is the cost associated with deciding on D_2 when in actuality, P_1 is the correct population.
- $C(1|2)$ is the cost associated with deciding on D_1 when in actuality, P_2 is the correct population.

The elements of $C(2|1)$ and $C(1|2)$ in Table 12.6 are the costs associated with making the **wrong** decision. The reader may have noticed that if we regard P_1 as standing for the null hypothesis and P_2 as the alternative hypothesis in a hypothesis-testing setup, then $C(2|1)$ represents the cost of making a **type I error**. Likewise, $C(1|2)$ represents the cost of making a **type II error**.

You might ask how and why estimating costs is relevant in the discriminant problem. Recall from our discussion earlier in the book, that in any decision, it behooves us to consider the costs of making the wrong decision, and, even if informally, quantifying these costs in terms of a **cost function**. True, for Fisher's iris data, the costs associated with misclassifying an observation into the wrong species class are likely in actuality not so great. That is, if a case of versicolor is misclassified as virginica, though to the plant biologist this may indeed be symbolic of a catastrophe, in the end, nobody gets hurt (though apparently iris flowers are somewhat toxic, so be careful!). However, it is quite another matter when considering the costs of misclassification in the treatment of mental illness, for instance. If a classification rule designates a client's suicidal probability as "low risk" when in fact the client is at high risk of committing suicide, the cost associated with making the **wrong decision** could indeed be catastrophic. Likewise, if a discriminant function predicts pre-flight wing de-icing on an airplane to not be required when in fact it is required, the costs of making a wrong decision in this context could likewise be life-threatening or catastrophic.

Decisions dominate science as they do everyday life. **Each decision has associated with it an error rate, whether known or unknown**. A point emphasized throughout this book is that decision criteria need to be set in an **intelligent** and **thoughtful** manner, relatively specific to the cost functions for a given problem. The problem of course inherent in any decision task is in estimating or otherwise quantifying the costs associated with making wrong decisions. Many times these costs are implicit in that we do not always acknowledge their existence, or are even **aware** of their existence.

12.6.1   Statistical Lives

As an example, consider the situation where on a stretch of highway, one or two lives have been lost in each of the preceding 10 years. Suppose that each of these deaths is quite easily attributed to not having a median separating the lanes of oncoming traffic. Each year, however, nothing changes. Deaths occur, but still, no median is constructed. Could these deaths be avoided? Of course. If they could be avoided by constructing the median, then why is the median not constructed? Quite simply, the energy, in terms of money, political motivation, and so on, is simply not "worth it" in terms of saving 1–2 lives per year. That is, statistically, losing 1–2 lives per year is, even if implicitly, regarded as **acceptable risk** and **associated cost** for this problem. Each death in this regard is generally what is referred to as a **statistical life**. But what if in the following 2 years, each year 10 persons die on the highway instead of only 1–2? What happens is that a median is constructed the following year. Why? Statistically speaking, because the risk and costs associated with the "status quo" decision of doing nothing have become too high. Under the new data of 10 deaths, the decision is changed to one of putting up a median, and consequently, the required finances and political motivation suddenly surface.

Our point then in our brief discussion of costs of misclassification is merely to emphasize an **awareness** of them in virtually every decision-context one may consider. The assessment of them, and hence their quantification, can be quite challenging. The point is, however, that they are usually still **there**, and any decision, made in an intelligent and rational manner, requires their consideration whether the problem is one of highway construction, career choices, family decisions, medical surgery choices, or spinning the wheel in roulette. Indeed, much of statistical and probabilistic analysis can be said to serve the end goal of **rational decision-making**. And as emphasized, it is impossible to make good decisions without at least some insight into the costs and benefits of making the right or wrong choices.

We have only skimmed the surface with regard to the field known as **decision analysis**, which is largely (but not exclusively) a branch of probability and statistics. The reader interested in learning more about how probabilities and other considerations (e.g., value appraisals) can be used in pragmatic decision-making is encouraged to consult Goodwin and Wright (2004). A useful source on decision-making in the medical sciences is Hunink et al. (2001). For the behavioral and psychological side of decision-making, see Kahneman and Tversky (2000).

12.7   DISCRIMINANT ANALYSIS AND BEYOND

Just as was true for logistic regression in which we said higher-order factorial and within-subjects designs could be used, such extensions can also be applied to discriminant analysis. For instance, suppose we wished to classify recovery (yes versus no) based on repeated health status measurements taken at several times over the course of a year. One could estimate classification rates based on this series of repeated measurements. One could also test the hypothesis that the repeated measure interacts with another variable (e.g., gender) in predicting recovery. The interested reader is referred to Lix and Sajobi (2010) for discriminant analysis for repeated measures.

All of the model-building strategies reviewed in Chapter 8 on multiple regression such as forward selection, backward elimination, and stepwise, are also available in discriminant analysis. Just as is true with the application of many other statistical models, cross-validation of a discriminant analysis is always a good idea if one has such availability in terms of sample size. The idea of statistical "learning" (Hastie, Tibshirani, and Friedman, 2009) often involves separating data first into a **training set** as well as a **test set**. Various sample-splitting approaches can be used such as the **holdout** (also known as **leave-one-out**) method in which all but one observation is used in generating a classification rule. The rule is then evaluated on how well it classifies the omitted observation. As discussed by Rencher and Christensen (2012), such a method can prove useful in estimating error rates. See James et al. (2013) for a good discussion of **training** versus **test** error rates in linear classification. Hastie, Tibshirani, and Friedman (2009) and Izenman (2008) are also excellent (if not very advanced) sources. The essential idea is that error rates based on training data are usually not expected to be the same as those on test data, and hence cross-validating results, in one form or another, is usually advised. Models are said to "learn" on training data, and then be evaluated on test data. Cross-validation, however, is not always employed, and as mentioned, its use will depend greatly on the availability of data and whether splitting a sample into two or more sets is even feasible. At minimum, researchers would do well to encourage cross-validation of their model at the conclusion of research papers if they are unable (or lack desire) to cross-validate their own studies. For a useful (yet technical) discussion of cross-validation in a medical context, see Houwelingen and Cessie (1990).

12.8   CANONICAL CORRELATION

We close this chapter with a brief survey of canonical correlation analysis (CCA). Recall that in results obtained both in MANOVA and LDA, software provided us with canonical correlation coefficients for respective discriminant functions. We demonstrated how these coefficients could be related to eigenvalues extracted in accounting for the respective discriminant functions. But what are these canonical correlations, exactly? In concluding this chapter, we survey the nature of this coefficient, as well as provide an example of one of the first canonical correlation analyses, Hotelling's analysis of 1936:

> Concepts of correlation and regression may be applied not only to ordinary one-dimensional variates but also to variates of two or more dimensions ... For example the scores on a number of mental tests may be compared with physical measurements on the same persons. The questions then arise of determining the number and nature of the independent relations of mind and body shown by these data to exist, and of extracting from the multiplicity of correlations in the system suitable characterizations of these independent relations. (Hotelling, 1936, p. 321)

CAA is a method for assessing the linear relationship between two sets of linear combinations. For example, suppose a researcher would like to evaluate the hypothesis that intelligence is related to achievement in school. The researcher could, in practice, collect data on a student's IQ using a single test, then use this to predict these same students' GPA in school. One could also test for a possible linear relationship using Pearson's correlation coefficient, evaluate it for statistical significance, and proceed to make an inference on the population.

However, on a substantive level, we must ask the question of whether that single IQ test truly "captures" the construct of intelligence. If we believe there is more to measuring IQ than the administration of a **single** test, then we might wish to use an additional measure in assessing one's IQ—perhaps another test that purports to measure another facet of intelligence. Likewise, we must ask whether GPA truly captures all there is to know about school achievement. In addition to GPA, we might use the general test from the GRE,[1] for instance, to assess one's mastery of verbal, quantitative, and analytical skills. It is in contexts such as these where canonical correlation may prove useful. We wish to generate a linear combination of one set of variables (e.g., IQ test 1 and IQ test 2) and use that linear combination to relate to another linear combination (e.g., GPA and GRE).

As another motivating example, suppose a researcher would like to assess the degree of relationship between reading and arithmetic. One might conceptualize the construct of "reading ability" by (1) reading speed, and (2) reading power, and define the construct of "arithmetic ability" by (1) arithmetic speed and (2) arithmetic power. By generating linear combinations of reading speed + reading power on the one hand, and arithmetic speed + arithmetic power on the other, the researcher could then correlate these two linear combinations. This was exactly the approach used by Hotelling (1936). We will reproduce his analysis toward the end of this chapter.

For canonical correlation to be substantively interpretable by the research scientist, each linear combination is usually hypothesized to represent some kind of construct or "variate." As was the case in our discussion of MANOVA, if the constructed variates do not carry with them some kind of theoretical meaning, then canonical correlation is usually not advised, except other than for exploratory purposes. Usually, the researcher employing canonical correlation should have at least **some** reason for wanting to combine variables into linear combinations other than the fact that he or she simply has many variables at his or her disposal. As always, theory should guide whatever statistical analyses you perform, not simply the availability of data.[2]

12.9   MOTIVATING EXAMPLE FOR CANONICAL CORRELATION: HOTELLING'S 1936 DATA

Harold Hotelling obtained data from Truman L. Kelley in which measurements on the aforementioned variables were recorded on 140 seventh-grade school children: reading speed, reading power, arithmetic speed, and arithmetic power. What Hotelling wanted to know from these data is whether reading speed and reading power, **considered together**, or as a **set**, were linearly related to arithmetic speed and arithmetic power, also considered together, or again, as a **set**. That is, Hotelling wanted to assess the linear relationship between the construct of reading and the construct of arithmetic, but knowing all too well that these constructs are multifaceted, incorporated a statistical method that would consider both speed and power **simultaneously** on each side of the equation. We can express the function statement for this problem as:

reading speed + reading power = arithmetic speed + arithmetic power

As we originally proposed when introducing MANOVA, one naïve way of computing the bivariate r between these constructs of reading and arithmetic would be to simply add reading speed to reading power, and arithmetic speed to arithmetic power. That is, naively, we could compute the bivariate correlation between the sums of

reading speed + reading power

---

[1] The GRE, or "Graduate Record Examination" is a standardized test published by **ETS**, the **Educational Testing Service**. It is taken by thousands of graduate school applicants each year as one of the many criteria used on which academic committees base entrance requirements into graduate school. The "general test" on the GRE tests skills such as verbal reasoning, quantitative abilities and aptitude, as well as analytical capacities.

[2] Once more, as discussed in previous chapters, this is not to discourage exploratory work. However, even the most rudimentary exploration is somewhat theory-guided. As a researcher, you should have some "reason" for wanting to correlate linear combinations over and above the fact that it can be done statistically.

and

arithmetic speed + arithmetic power

However, if we simply added them in this fashion, we would have no guarantee that these linear combinations would be **maximally** correlated and yield the largest R possible. That is, as we did in our initial "attempt" at MANOVA, our addition of reading speed to reading power implicitly weighted these variables as **(1) reading speed + (1) reading power**. Likewise, we implicitly weighted the construct of arithmetic with values of "1": **(1) arithmetic speed + (1) arithmetic power**. Does using weights of "1" result in the **maximum** correlation possible between the two linear composites? Probably not. What is needed is a method of estimating these coefficients that will weight the variables of each construct in such a way that the **maximum correlation** between linear combinations is achieved. This is the **canonical correlation** we seek.

Below is the correlation matrix analyzed by Hotelling, which we have reproduced in R:

	reading speed	reading power	arithmetic speed	arithmetic power
reading speed	1.0000	0.6328	0.2412	0.0586
reading power	0.6328	1.0000	-0.0553	0.0655
arithmetic speed	0.2412	-0.0553	1.0000	0.4248
arithmetic power	0.0586	0.0655	0.4248	1.0000

In what follows, we learn how to take such a correlation matrix and decompose it into canonical correlations. The canonical variates extracted will represent the dimensions along which the constructs reading and arithmetic are maximally linearly related. For a proof of why the canonical correlation is the maximum correlation, see Anderson (2003, pp. 495–496).

12.10 CANONICAL CORRELATION AS A GENERAL LINEAR MODEL

Canonical correlation can be interpreted as a technique that encompasses other techniques as special cases. Indeed, in learning statistical methods, it is advantageous to the learner to be able to see some analyses as "subcategories" of other analyses. In terms of function statements, canonical correlation can be expressed as

$$y_1, y_2 = x_1, x_2$$

where y_1, y_2 is one linear combination, and x_1, x_2 another. Notice that from this "wider" analysis can be identified many smaller analyses in the following function statements:

- If we use x_1, x_2 to **predict** y_1, y_2, both continuous, then the model can be conceptualized as a **multivariate multiple regression**. It is a multiple regression because we have more than a single explanatory variable. It is **multivariate** multiple regression because we have more than a single response variable.
- If we drop one of the dependent continuous variables, y_2, such that our model is $y_1 = x_1, x_2$, and we are interested in having continuous variables x_1, x_2 predict y_1 simultaneously, the analysis becomes a multiple regression. It is a multiple regression because we have more than one explanatory variable predicting a single response variable.
- If we keep y_1, y_2 as continuous but change x_1, x_2 to categorical predictors with "levels," then the model becomes a two-way factorial multivariate analysis of variance.

- If we again drop one of our dependent variables and make y_1 binary or polytomous (i.e., having several groupings), then our analysis could either be a discriminant analysis or logistic regression (or even a support vector machine). Recall that discriminant analysis and logistic regression, though differing in assumptions and interpretation, both use explanatory variables to predict group membership on a response variable.

We summarize the idea of model "generality" toward the end of this chapter. We also revisit it in Chapter 15 when we discuss path analysis and structural equation modeling (SEM). As we will see, canonical correlation itself can be conceived as a special case of the wider SEM framework.

12.11   THEORY OF CANONICAL CORRELATION

We begin the development of canonical correlation by first considering two sets of random variables $\mathbf{y} = (y_1, y_2, \ldots, y_p)$ and $\mathbf{x} = (x_1, x_2, \ldots, x_p)$. As mentioned earlier, the first set of p variables $y_1, y_2, \ldots, y_p$ might consist of a set of measures of intelligence, while the second set of p variables $x_1, x_2, \ldots, x_p$ might consist of scholastic achievement data. Or, in the case of Hotelling's data, they might consist of reading speed and reading power for $y_1, y_2, \ldots, y_p$ and arithmetic speed and arithmetic power for $x_1, x_2, \ldots, x_p$. It is important to recognize that both sets of measurements are on the **same** individuals. The goal of CCA is to measure the extent to which these two sets of variables are **linearly** related.

Notice that in the situation in which we were to reduce each set to a single variable, y_1 and x_1, the canonical correlation would reduce to the simple Pearson correlation coefficient r in which the linear relationship between two variables is assessed. Canonical correlation is simply the **maximum bivariate correlation**, but on **sets**, or **linear combinations**, of variables rather than on individual variables.

Technically, canonical correlation accomplishes something somewhat analogous to **principal components analysis** (see Chapter 13) in that it seeks to transform the first p_1 coordinate axes along with a transformation of the second p_2 coordinate axes to a new system $p_1 + p_2$ that depicts the correlations between vectors.

Recall the sample covariance matrix, $\mathbf{S}$ of Chapter 2:

$$\mathbf{S} = (s_{jk}) = \begin{bmatrix} s_{11} & s_{12} & \cdots & s_{1p} \\ s_{21} & s_{22} & \cdots & s_{2p} \\ \cdot & \cdot & & \cdot \\ \cdot & \cdot & \cdots & \cdot \\ \cdot & \cdot & \cdots & \cdot \\ s_{p1} & s_{p2} & \cdots & s_{pp} \end{bmatrix} \tag{12.9}$$

where s_{jk} are the covariances for variables j by k. Equation (12.9) can be partitioned as:

$$\mathbf{S} = \begin{bmatrix} \mathbf{S}_{yy} & \mathbf{S}_{yx} \\ \mathbf{S}_{xy} & \mathbf{S}_{xx} \end{bmatrix}$$

where $\mathbf{S}_{yy}$ and $\mathbf{S}_{xx}$ are the covariance matrices for y and x respectively, and $\mathbf{S}_{yx}$ and $\mathbf{S}_{xy}$ are covariance matrices between y and x. Now, suppose we have two linear combinations, $\ell_1 = \mathbf{a}'\mathbf{y}$ and $\ell_2 = \mathbf{b}'\mathbf{x}$. The sample correlation coefficient between two linear combinations is defined as

$$r_{\ell_1,\ell_2} = \frac{\text{cov}_{\ell_1,\ell_2}}{\sqrt{s_{\ell_1}^2 s_{\ell_2}^2}} = \frac{\mathbf{a}'\mathbf{S}_{yx}\mathbf{b}}{\sqrt{(\mathbf{a}'\mathbf{S}_{yy}\mathbf{a})(\mathbf{b}'\mathbf{S}_{xx}\mathbf{b})}}$$

Note the parallel between the correlation between linear combinations above and the "ordinary" Pearson correlation between two variables (rather than **variates**). In the numerator, we have $\mathbf{a}'\mathbf{S}_{yx}\mathbf{b}$, which is essentially (at least in concept) a **cross-product** of sorts, somewhat analogous (again, at least in concept) to the cross-product in an ordinary Pearson correlation. In the denominator, we have $\sqrt{(\mathbf{a}'\mathbf{S}_{yy}\mathbf{a})(\mathbf{b}'\mathbf{S}_{xx}\mathbf{b})}$ which is analogous to the product of standard deviations in ordinary Pearson correlation. The only conceptual (not technical) difference between canonical correlation and ordinary Pearson correlation is that the former is conducted on a linear combination of variables, while the latter is performed on much "simpler" linear combinations (i.e., consisting of only single variables). **If you forever think of simple correlation as a correlation of linear combinations consisting of only single variables, then canonical correlation will rightfully appear as an extension and expansion into more complex linear combinations made up of several variables.** That is, even single variables can be regarded as linear combinations; they are simply linear combinations of only one variable. Of course, linear combinations typically refer to a much longer string of variables, however, by reducing the concept to only a single variable, the generality of canonical correlation comes to light.

The goal of canonical correlation analysis is to find coefficient vectors $\mathbf{a}$ and $\mathbf{b}$ such that the correlation between linear combinations, r_{ℓ_1,ℓ_2}, is as large as possible. How can these coefficients be found? They are obtainable in several ways, one of which is appeal to a multiple $\mathbf{R}$-like statistic, of which $\mathbf{R}^2$ may be defined as:

$$\mathbf{R}^2 = \frac{|\mathbf{S}_{yx}\mathbf{S}_{xx}^{-1}\mathbf{S}_{xy}|}{|\mathbf{S}_{yy}|}$$

Notice that we are dividing by the determinant of $\mathbf{S}_{yy}$, $|\mathbf{S}_{yy}|$. We can also rewrite the above as:

$$\mathbf{R}^2 = |\mathbf{S}_{yy}^{-1}\mathbf{S}_{yx}\mathbf{S}_{xx}^{-1}\mathbf{S}_{xy}| \tag{12.10}$$

Why does this form of $\mathbf{R}^2$ make sense? One way to understand why, in an informal sense, is to consider what is contained in the product $\mathbf{S}_{yy}^{-1}\mathbf{S}_{yx}\mathbf{S}_{xx}^{-1}\mathbf{S}_{xy}$. Again, notice that what we are computing is somewhat analogous to what we compute when calculating Pearson r. That is, we are computing the product $\mathbf{S}_{yx}$ by $\mathbf{S}_{xy}$ and then "dividing" by the product $\mathbf{S}_{yy}$ by $\mathbf{S}_{xx}$, only we have to write $\mathbf{S}_{yy}^{-1}$ and $\mathbf{S}_{xx}^{-1}$ (i.e., using inverses) to denote the "division" because we are using matrices. The computation of $\mathbf{R}^2$ is somewhat conceptually analogous to Pearson r because the product is divided by the product of standard deviations. Recall Pearson r:

$$r = \frac{\dfrac{\sum_{i=1}^{n}(x_i-\bar{x})(y_i-\bar{y})}{n-1}}{\sqrt{s_x^2 \cdot s_y^2}} = \frac{\text{cov}}{\sqrt{s_x^2 \cdot s_y^2}}$$

Of course, it is not the same as Pearson r, since in (12.10) we are in a **multivariable** setting and there is a lot more going on in (12.10) than with r. But if you are able to spot parallels, or even generic inexact similarities between simpler statistical concepts and computations and more advanced ones, you will be well on your way to realizing that understanding advanced statistical procedures usually depends on your grasp of the simplest, most core essentials. Advanced statistical methods are usually expansions and extensions of such core, fundamental concepts, and many times these can be used as stepping stones to more sophisticated methodologies, or at minimum, informal ways to try to make sense of formulae.

The number of canonical correlations extracted will be the smaller of the number of y variables or x variables. Just as for ordinary Pearson r, canonical correlations are invariant to linear transformations on scales of the variables making up the correlation. That is, even if we linearly transposed the scale of x or y, the canonical correlation between variates would remain the same.

12.12 CANONICAL CORRELATION OF HOTELLING'S DATA

We perform a simple canonical correlation on Hotelling's data discussed at the outset of this section. We generate Hotelling's matrix in R:

```
> cancor <- c(1.0000, .6328, .2412, .0586,
+ .6328, 1.0000, -.0553, .0655,
+ .2412, -.0553, 1.0000, .4248,
+ .0586, .0655, .4248, 1.0000)
> cancor.matrix <- matrix(cancor, 4, 4, byrow = TRUE)

> cancor.matrix
         [,1]    [,2]     [,3]    [,4]
[1,] 1.0000  0.6328   0.2412 0.0586
[2,] 0.6328  1.0000  -0.0553 0.0655
[3,] 0.2412 -0.0553   1.0000 0.4248
[4,] 0.0586  0.0655   0.4248 1.0000
```

Because it is a correlation matrix, it is **symmetric**, meaning that the lower triangular is a mirror image of the upper triangular. The correlation between reading speed and reading power is the highest correlation (row 1, column 2, $r = 0.6328$), with the correlation between arithmetic speed and arithmetic power being the second highest (row 3, column 4, $r = 0.4248$). The correlation between arithmetic speed and reading power is quite small ($r = -0.0553$) as is the correlation between reading power and arithmetic power ($r = 0.0655$).

The relevant canonical correlations as found by Hotelling (1936, p. 342, (6.2)),

```
$cor
[1] 0.39450592 0.06884787
```

R next provides us with the raw coefficients for the extracted canonical correlations, both for the x variables and the y variables:

```
$xcoef
       Can.1      Can.2
1  1.256845  0.2970177
2 -1.025317  0.7852413

$ycoef
       Can.1        Can.2
3  1.1044722  -0.01818009
4 -0.4527216   1.00758746
```

Two canonical correlations are represented (Can.1 and Can.2) on four variables. The weights associated with the first canonical correlation are 1.2568 and −1.0253 for reading speed and reading power, and 1.1045 and −0.4527 for arithmetic speed and arithmetic power, respectively. The weights associated with the second canonical correlation are 0.2970 and 0.7852 for reading speed and reading power, respectively, and −0.0182 and 1.0076 for arithmetic speed and arithmetic power, respectively. Hence, the first canonical correlation is given by:

```
1.2568(reading speed) - 1.0253(reading power)    WITH    1.1045
(arithmetic speed) -0.4527(arithmetic power)
```

where "WITH" is replaced by the canonical correlation of 0.3945. That is, the above weighting of reading speed with reading power correlates to a degree of 0.3945 with the above weighting of arithmetic speed and arithmetic power. If we square the coefficient, we can say that approximately 16% [i.e., $(0.3945)^2 = 0.1556$] of the variance is accounted for by this first canonical dimension. The second canonical correlation is given by

```
0.2970(reading speed) + 0.7852(reading power)    WITH    -0.0182
(arithmetic speed) + 1.0076(arithmetic power)
```

where, this time, "WITH" is replaced by the canonical correlation of 0.0688. Again, we conclude that the above weighting of reading speed with reading power is correlated to a degree of 0.0688 with the above weighting of arithmetic speed and arithmetic power **given the extraction of the first canonical dimension**. That is, the canonical correlation of 0.0688 is the maximum correlation possible between these linear composites given the extraction of the first canonical correlation (i.e., uncorrelated to it, though not orthogonal, see Rencher and Christensen (2012, p. 408); Mukhopadhyay (2008, p. 369)). If we square the coefficient, we can say that approximately 0.005 [i.e., $(0.0688)^2 = 0.005$] of the variance is accounted for by this second canonical dimension.

12.13 CANONICAL CORRELATION ON THE IRIS DATA: EXTRACTING CANONICAL CORRELATION FROM REGRESSION, MANOVA, LDA

We close this chapter by computing a canonical correlation on the iris data analyzed through MANOVA of the previous chapter and through LDA of the current chapter. The correlation between variates that we are about to calculate is a strong way to conceptualize the underlying similarity among these multivariate techniques. As discussed, **MANOVA, LDA, and regression analysis can all be conceptualized as special cases of the wider canonical correlational model**.

When we run the MANOVA in SPSS, we obtain (where `species` has levels 0, 1, 2):

```
manova sepal_length sepal_width petal_length petal_width by
species(0, 2)
/print = sig(eigen).
```

Test Name	Value	Approx. F	Hypoth. DF	Error DF	Sig. of F
Pillais	1.19190	53.46649	8.00	290.00	.000
Hotellings	32.47732	580.53210	8.00	286.00	.000
Wilks	.02344	199.14534	8.00	288.00	.000
Roys	.96987				

Note.. F statistic for WILKS' Lambda is exact.

- -

Eigenvalues and Canonical Correlations

Root No.	Eigenvalue	Pct.	Cum. Pct.	Canon Cor.
1	32.19193	99.12126	99.12126	.98482
2	.28539	.87874	100.00000	.47120

The first canonical correlation between variates (species) and (iris features) is reported to be 0.98482. As a demonstration, recall from Chapter 11 that Pillai's is defined as

$$V^{(s)} = \text{tr}\left[(\mathbf{E} + \mathbf{H})^{-1}\mathbf{H}\right] = \sum_{i=1}^{s} \frac{\lambda_i}{1 + \lambda_i}$$

The respective eigenvalues for this problem, as noted above, are 32.19193 and 0.28539, with which $V^{(s)} = 1.19190$ can be easily verified. We can also compute $V^{(s)}$ as $\sum_{i=1}^{s} r_i^2$, where r_i^2 are respective squared canonical correlations, which for this problem can also be easily confirmed. And since LDA is essentially the "reverse" of MANOVA, the above canonical correlations provide a general "link" between these procedures. But what about regression? The corresponding analysis would be a **multivariate regression** since there are several response variables and a single predictor. However, the only analytical difference between such a model and that of the MANOVA model would be in coding the independent variable appropriately to accommodate a regression framework. Otherwise, the two analyses are essentially the same. Whether the model be ANOVA, MANOVA, LDA, or regression, **canonical correlation subsumes them all**.

12.14   CHAPTER SUMMARY AND HIGHLIGHTS

- **Discriminant analysis**, originally proposed by R.A. Fisher in 1936, is a procedure useful for classifying objects, subjects, or items into one of two or more mutually exclusive populations. The response variable is either dichotomous or polytomous, making ordinary least-squares regression typically inappropriate.
- **Discriminant analysis** is essentially the reverse of **multivariate analysis of variance**. In MANOVA, the linear composite is the response variable; in discriminant analysis, the linear composite is the predictor.

- Through the computation of **eigenvalues** and **eigenvectors**, discriminant analysis finds a vector that maximizes the ratio of the difference between population means to the standard deviations within. This vector is called the **discriminant function**.
- The **number of discriminant functions** extracted will be the smaller of the number of predictors or one less the number of populations on the response variable.
- Discriminant analysis in R can be performed using the `lda` function with an analogous model statement to that used for the `lm` function.
- **Prior probabilities** of population membership can be set to represent **a priori** base rate knowledge before conducting the discriminant analysis, analogous to how the probability of "success" versus "failure" can be set prior to modeling a binary variable with the binomial distribution. These prior probabilities can have an influence on the post-discriminant analysis classification results.
- For any discriminant analysis extracting more than a single function, the respective **eigenvalue** can be compared relative to the **sum of eigenvalues** to ascertain the relative "importance" that the function carries with discriminating populations.
- As was true in the case of MANOVA, **Wilks' lambda** can be obtained for each discriminant function along with an associated significance test. A consideration of error rates and their respective costs, even if informally, should be considered in any decision rule, not only those used in LDA.
- **Canonical correlation analysis** (CCA) is a method for assessing the linear relationship between two sets of linear combinations.
- Each **linear combination** is usually hypothesized to represent some kind of construct or "**variate**." If the variates are not meaningful, then other than for blind data reduction, canonical correlation is usually not advised.
- Scalars for the linear combinations are chosen such that they result in linear combinations that are maximally correlated. "**Maximally correlated**" does not equate to obtained canonical correlations being necessarily "large."
- Harold Hotelling's early use of **canonical correlation** was to correlate the linear combination of reading speed + reading power to arithmetic speed + arithmetic power.
- **Canonical correlation** can be seen as a technique that encompasses other techniques as "special cases." For instance, if $y_1, y_2 = x_1, x_2$ is the function statement for canonical correlation, then $y_1 = x_1, x_2$ is the function statement for a multiple regression.
- The **number of canonical correlations** extracted is equal to the lesser of the number of variables on the left-hand or right-hand side of the function statement.
- Canonical correlation can be derived in many ways, one of which is through a multiple R-like statistic. As is true of Pearson r, **canonical correlations are scale invariant**.
- **Canonical correlation subsumes ANOVA, MANOVA, LDA, and regression** and is pedagogically useful in linking and understanding such methods. In many statistical models, one is indirectly obtaining canonical correlations.

REVIEW EXERCISES

12.1. Briefly summarize the similarities and differences between the **multivariate analysis of variance** (MANOVA) and **linear discriminant analysis** (LDA). When is one analysis more suitable than the other?

12.2. Discuss the conceptual similarities between **regression analysis**, **discriminant analysis**, and **logistic regression**. On a conceptual, practical level, technicalities aside, what should be the motivating decision regarding which analysis a researcher should choose?

12.3. Compare the **discriminant analysis** on two populations to that on several populations. What are the primary technical distinctions?

12.4. Discuss the relevance of the following for the two-group discriminant problem:

$$\frac{\left(\bar{\ell}_1 - \bar{\ell}_2\right)^2}{s_\ell^2} = (\bar{y}_1 - \bar{y}_2)' S_p^{-1} (\bar{y}_1 - \bar{y}_2)$$

12.5. Interpret and discuss the statement "The maximizing vector **a** is not **unique**, however, the direction is." What does this mean, exactly?

12.6. Compare a **residual** in least-squares regression to that of one in discriminant analysis for two populations. How could they be considered conceptually similar? Different?

12.7. Why is it important to be aware of **costs of misclassification** in a discriminant analysis or any other procedure in which decisions are made regarding a case?

12.8. Distinguish between **raw** versus **standardized discriminant functions**. Which, in general, should be interpreted? Why?

12.9. Conduct an LDA on Fisher's 1947 data in which a linear combination of bodyweight and heartweight is used to differentiate between populations of sex. Summarize your overall findings.

12.10. Compare and contrast the **MANOVA** of Fisher's iris data to the **discriminant analysis** of the iris data. How does output from each procedure compare? What are the similarities and differences?

12.11. Recall the achiev data of Chapter 3. Perform a **discriminant analysis** using ac to predict group membership on teach. Summarize the overall findings of your analysis, and compare them to the fixed effects **ANOVA** analysis of the same data conducted in Chapter 3. Note as many parallels and differences between the two analyses as you can.

12.12. Give an example of a substantive application of **canonical correlation** from your research area of interest. That is, when might a researcher be interested in performing canonical correlation in your field?

12.13. For Hotelling's data, why is simply **correlating** (1)reading speed + (1)reading power to (1) arithmetic speed + (1)arithmetic power not going to give us the canonical correlation?

12.14. In what way can canonical correlation be considered a **general linear model** and encompass other techniques as "special cases?"

Further Discussion and Activities

12.15. It has been shown that a two-group discriminant analysis generates weights that are proportional to those estimated in the analogous regression analysis (e.g., see Flury and Riedwyl, 1985; James et al., 2013). Perform a regression analysis on the iris data where variables sepal length, sepal width, petal length, and petal width are used to predict categories on species **setosa** and **versicolor**. Then, perform the analogous discriminant analysis. Compare the results of regression to discriminant analysis in each analysis and comment on any similarities and differences.

13

PRINCIPAL COMPONENTS ANALYSIS

Of course the term "best fit" is really arbitrary; but a good fit will clearly be obtained if we make the sum of the squares of the perpendiculars from the system of points upon the line or plane a minimum …

(Pearson, 1901, p. 560)

Suppose a researcher has collected data on 100 variables and is interested in knowing whether the information in this collection of variables can be expressed in fewer than 100 dimensions. Perhaps the majority of the variability in this set of variables can be summarized in 4–5 dimensions without losing too much of the original information. These 4 or 5 dimensions could then potentially be used as predictors in a future analysis. The researcher may even wish to try to identify these new dimensions and give them names. An appropriate statistical tool for this purpose is that of principal components analysis (PCA).

Principal components analysis is a technique concerned with extracting information from a covariance or correlation matrix such that a group of p random variables can be represented by fewer than p component dimensions. PCA attempts to reduce the **dimensionality** of a group of correlated variables into a set of **mutually orthogonal linear combinations** of the variables of lower dimension (i.e., of lower **rank**) yet simultaneously attempting to explain most of the variance in the original variables. Substantively, PCA can be considered a **data reduction** technique.

Technically, principal components analysis involves the rotation of the original coordinate system to a new coordinate system with inherently desirable statistical properties. More precisely, we seek to define an **orthogonal transformation to a diagonal covariance matrix**. Recall that a diagonal matrix means that everywhere else other than the main diagonal are zeros, which implies a covariance among variables (or components, in this case) equal to zero. Principal components analysis is essentially, and

Applied Univariate, Bivariate, and Multivariate Statistics: Understanding Statistics for Social and Natural Scientists, With Applications in SPSS and R, Second Edition. Daniel J. Denis.
© 2021 John Wiley & Sons, Inc. Published 2021 by John Wiley & Sons, Inc.
Companion Website: www.wiley.com/go/denis/appliedstatistics2e

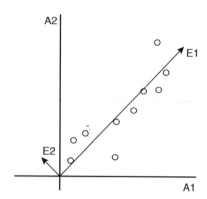

FIGURE 13.1   Basic principal components analysis where E1 and E2 are extracted components. Source: Modified from Pearson (1901).

quite simply, a **re-expression of the variance** encapsulated in a data matrix such that the reduction in dimensionality may provide more "insight" into patterns not immediately obvious by inspection of a covariance matrix alone. Computationally, PCA reduces to solving for the eigenvalues and eigenvectors of an at least semi-positive definite matrix by a process generally referred to as **eigenvalue analysis** or **spectral decomposition**.

The general idea of PCA is depicted in Figure 13.1, featuring a bivariate plot on variables A1 and A2. The vectors E1 and E2 are the principal components. Essentially, all that PCA does is establish new axes on the original data so that variance is **maximized**. Note that as is the case of the original A1 and A2 axes (i.e., representing the abscissa and ordinate, respectively), the angle between E1 and E2 is one of 90°. That is, the components extracted, E1 and E2, are **orthogonal** to one another.

A second related goal of PCA is to attempt to account for the **substantive structure** of these derived component variables, and if possible, to name these newly obtained variates. Identifying whether these linear combinations are substantively meaningful is sometimes a priority for the user of PCA, while other times, a primary goal is to estimate scores based on the newly obtained components and use these scores as inputs to other analyses. When the priority is to name underlying latent qualities of the input variables, **exploratory factor analysis** is a popular alternative to PCA, to be discussed in the following chapter.

13.1   HISTORY OF PRINCIPAL COMPONENTS ANALYSIS

The history of principal components analysis can be traced to Karl Pearson's work in 1901 in a paper published in **Philosophical Magazine** titled **On Lines and Planes of Closest Fit to Systems of Points in Space**. In the paper, Pearson outlined the essential method of PCA using a least-squares approach. His technique for obtaining components has generally come to be known as the **planes of closest fit** approach (Anderson, 2003, p. 466). Pearson introduced the problem as follows:

> In nearly all the cases dealt with in the text-books [sic] of least squares, the variables on the right of our equations are treated as the independent, those on the left as the dependent variables. The result of this treatment is that we get one straight line or plane if we treat some one variable as independent, and a quite different one if we treat another variable as the independent variable. There is no paradox about this; it is, in fact, an easily understood and most important feature of the theory of a system of correlated variables. (p. 559)

The quote from Pearson quite simply notes that the regression of y on x is not the same as the regression of x on y, and that each has its own regression line. The motivation for principal components analysis comes from Pearson's following words, where he considers the situation in which both independent and dependent variables comprise an entire set or **system**:

> In many cases of physics and biology, however, the "independent" variable is subject to just as much deviation or error as the "dependent" variable ... In the case we are about to deal with, we suppose the observed variables – all subject to error – to be plotted in plane, three-dimensioned or higher space, and we endeavor to take a line (or plane) which will be the "best fit" to such a system of points. (pp. 559–560)

Pearson then goes on to give an example of principal components and methods for finding roots, then specifies many algebraic and geometrical implications of the fitting of the new best-fit line, beginning first with telling us exactly what he considers to be a "best-fitting" line:

> Of course the term "best fit" is really arbitrary; but a good fit will clearly be obtained if we make the sum of the squares of the **perpendiculars** [emphasis added] from the system of points upon the line or plane a minimum ... We shall make $U = S(p^2)$ a minimum. If y were the dependent variable, we should have made $S(y'-y)^2$ a minimum.
>
> (Pearson, 1901, p. 560)

With these words, Pearson contrasted his method of principal components with that of the then fairly recent, but still relatively established, method of least-squares in which the sum of squared deviations about the regression line is minimized (i.e., $S(y'-y)^2$ in Pearson's quote). Instead of minimizing this sum, Pearson wanted to minimize the sum of squared **perpendiculars** (i.e., $U = S(p^2)$) and gave Figure 13.2 to illustrate what he was up to. A principal component is a line that minimizes the sum of these squared perpendicular distances.

Pearson then went on to derive the principal components and provided a geometrical representation of his derivation (Figure 13.3).

In Figure 13.3, Pearson drew 3 lines, EE', FF', and AA'. The lines EE' and FF' are the least-squares regression lines of y on x and x on y, respectively. The line AA' is the principal components line. As summarized by Pearson (p. 566):

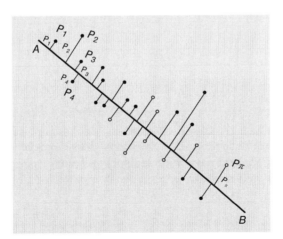

FIGURE 13.2 Pearson's 1901 depiction of minimizing perpendiculars. Each P_1, P_2 is the perpendicular distance from the component line (best-fit line in Pearson's use of the word) to the given data point. Source: Pearson (1901).© 1901 Taylor and Francis.

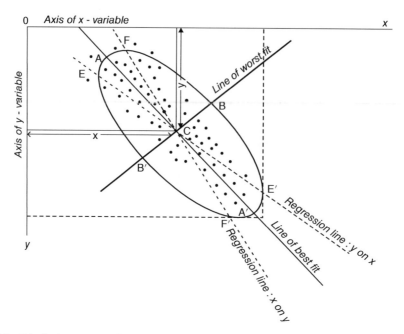

FIGURE 13.3 Principal component line of best fit versus least-squares regression lines. Source: Pearson (1901). © 1901 Taylor and Francis.

EE' is found by making $S(y' - y)^2$ a minimum,

FF' is found by making $S(x' - x)^2$ a minimum,

AA' is found by making $S(p^2)$ a minimum.

The line AA' is the **first principal component** accounting for maximum variance in the variables. The line BB', "line of worst fit" is the second principal component accounting for maximum variance unexplained by AA', but **orthogonal** to it since it is positioned at a 90° angle relative to the first component. Pearson's original representation of the principal components line of best fit will surely make more sense once we have surveyed a modern treatment of PCA in the remainder of this chapter.

13.2   HOTELLING 1933

Harold Hotelling is also historically recognized as an "inventor" of principal components analysis. In 1933, Harold Hotelling published a paper titled **Analysis of a Complex of Statistical Variables Into Principal Components**, in **The Journal of Educational Psychology**, emphasizing the technical similarities between it and factor analysis. Referring to a set of observed x variables, Hotelling began his paper:

The x's will ordinarily be correlated. It is natural to ask whether some more fundamental set of independent variables exists, perhaps fewer in number than the x's, which determine the values the x's will take. If $\gamma_1, \gamma_2,$ … are such variables, we shall then have a set of relations of the form $x_i = f_i(\gamma_1, \gamma_2, \ldots)$ $(i = 1, 2, \ldots, n)$.

Quantities such as the γ's have been called mental factors in recent psychological literature. However in view of the prospect of application of these ideas outside of psychology, and the conflicting usage attaching to the word "factor" in mathematics, it will be better simply to call the γ's **components** of the complex depicted by the tests. (p. 417)

Hotelling's reference to "mental factors" was as a result of much of the driving force behind components analysis and factor analysis occurring in the field of psychology in the early 1900s. Despite Hotelling's apparent alignment of PCA with the **factors** of **factor analysis**, PCA should not be interpreted nor viewed as a "special case" of factor analysis. Many writers have warned against equating PCA with FA (e.g., see Jolliffe, 2002). Hence, we emphasize from the outset—**principal components analysis is not equivalent to factor analysis**. PCA generates successive orthogonal linear combinations of the **variables**, whereas factor analysis, as we will see in the chapter to follow, generates linear combinations of the hypothetical **factors**. And whereas PCA models focus on variance, factor analysis models focus more on **commonality** among variables. How this distinction materializes in practice is that in PCA, the to-be-analyzed matrix in the case of a correlation matrix typically contains unit variances (1's) along the main diagonal. That is, the trace is equal to the sum of variables inputted into the analysis. Or equivalently, the trace is equal to the sum of the variances of the variables (for which in the case of a correlation matrix, that sum of variances is equal to the number of variables subjected to the analysis). In this way, each observed variable in the procedure contributes a single unit of variance before the correlation matrix is subjected to analysis. In factor analysis, however, the main diagonal consists of what are called **communalities**, or, more crudely, the amount of shared variance that the given variable has in common with other variables also subjected to the factor analysis.

As well, though PCA and FA often result in similar findings (assuming of course that components or factors are actually empirically present and not merely a wishful hope of the researcher), PCA is generally considered a relatively **atheoretical** technique when contrasted to FA. That is, a principal components analysis simply seeks to reduce the dimensionality of observed data. Factor analysis also seeks to reduce the dimensionality of data, but usually under the assumption that **unobserved** or **latent variables** subsume the observed correlation among the set of variables. The user of principal components often does not, at least to such an extent, assume an underlying "latent" scientific structure. Rather, she simply seeks to make the observed data more parsimonious through the extraction of components. In general, principal components analysis is a much simpler procedure than that of the factor-analytic methods, both in derivation and extraction, and also generally in substantive use and application. And though PCA is generally widely accepted across virtually all scientific fields, factor analysis carries with it a storied past of severe criticism, misuse, and rejection. One of the technical issues that plagues exploratory factor analysis (EFA), as we will discuss in the next chapter, is that **loadings for factors are not unique**, and hence all EFA solutions are subject to an orthogonal rotation and are contingent upon how many other factors are extracted. Principal components, on the other hand, derives loadings that do not change regardless of how many other components are "kept." Though eigenvectors which make up the given component are not unique in the strict mathematical sense of the word, the component is "stable" regardless of how many other components are interpreted or kept along with it. The "uniqueness" of the component is that it is **well-defined**. In factor analysis, extracted factors are typically not. This is a major problem when attempting to apply (and interpret) the factor-analytic technique to real research variables. Mathematically, it is not really an issue, but scientifically, it very much is.

We delay further discussion of factor analysis to the following chapter where we will also discuss some of the similarities and differences between EFA and PCA. For now, we focus our attention on components analysis.

13.3 THEORY OF PRINCIPAL COMPONENTS ANALYSIS

We now consider a summary of the formal development of principal components. Our treatment is very brief. For much deeper and thorough technical introductions to PCA the reader is encouraged to consult Johnson and Wichern (2007) and Jolliffe (2002) where the latter provides a book-length and thorough treatment of the topic along with applications. Izenman (2008) also gives a very good and deep overview of the technique.

We start by considering $\mathbf{x}$ to be a random vector of p observed variables with covariance matrix Σ. For ordinary PCA, the p random variables will usually be measurable on a **continuous** scale. PCA can also be performed on the standardized covariance matrix, that is, correlation matrix $\mathbf{R}$. The decision as to whether to analyze the covariance or correlation matrix for PCA is an important one, one which will be discussed later in this chapter.

We know from results in matrix theory (see Appendix) that associated with covariance matrix Σ are p eigenvalues and p eigenvectors. For example, supposing $p = 10$ observed variables in Σ, then it stands that we can extract 10 eigenvalues and 10 eigenvectors. We will see that each of these extracted eigenvalues and associated eigenvectors are associated with a principal component, and hence for $p = 10$ observed variables, there will be extracted a total of 10 components. **Generally, there are always as many principal components extracted as there are variables that serve as inputs to the principal components analysis**. The goal of PCA is to learn whether fewer than p components can be used to summarize the **variance** in the original p variables.

13.3.1 The Theorem of Principal Components Analysis

Principal components analysis is founded on a theorem that says if the expectation of the random vector $\mathbf{x}$ is equal to 0 (i.e., $E(\mathbf{x}) = 0$) and for a covariance matrix Σ, then one can conduct an orthogonal linear transformation to generate components that have maximum variance and are unrelated to successive components. See Anderson (2003, p. 464) for a proof of this theorem, though from an applied point of view, the proof of the theorem will not provide a great deal of insight.

To obtain components, we can perform **eigendecomposition**. For every square matrix $\mathbf{A}$, we can obtain a scalar λ and a vector $\mathbf{x}$ (other than zero) so that the following equality holds:

$$\mathbf{Ax} = \lambda\mathbf{x} \tag{13.1}$$

The scalar λ is called an **eigenvalue** of the matrix $\mathbf{A}$ and the vector $\mathbf{x}$ is called an **eigenvector** associated with λ. To solve for λ and $\mathbf{x}$, we can re-write (13.1) as

$$\mathbf{Ax} - \lambda\mathbf{x} = \mathbf{0}$$
$$(\mathbf{A} - \lambda\mathbf{I})\mathbf{x} = \mathbf{0}$$

It stands that if $|\mathbf{A} - \lambda\mathbf{I}| \neq 0$, then this implies that $(\mathbf{A} - \lambda\mathbf{I})$ has an inverse, which means that $\mathbf{x} = \mathbf{0}$ is the only solution. This is referred to as the **trivial solution**. To obtain nontrivial solutions, we deliberately set $|\mathbf{A} - \lambda\mathbf{I}| = 0$ and find values of λ that can be substituted into $(\mathbf{A} - \lambda\mathbf{I})\mathbf{x} = \mathbf{0}$ to then provide a solution for $\mathbf{x}$.

The equation $|\mathbf{A} - \lambda\mathbf{I}| = 0$ is called the **characteristic equation**. For a matrix $\mathbf{A}$ that is $n \cdot n$ (i.e., square, with n rows and n columns), the characteristic equation will have n roots, that is, n eigenvalues $\lambda_1, \lambda_2, \ldots, \lambda_n$, not all necessarily different from one another and not all nonzero. **Eigenvectors are unique only up to multiplication by a scalar**. That is, we can multiply the elements of a given eigenvector without "changing" the eigenvector in any fundamental way. As noted in Rencher and

Christensen (2012), this idea of "uniqueness up to multiplication by a scalar" can be more formally expressed by:

$$(\mathbf{A} - \lambda\mathbf{I})k\mathbf{x} = k\mathbf{0} = \mathbf{0}$$

where k is some scalar. What the above says is that if $\mathbf{x}$ is a vector, then so is $k\mathbf{x}$, where k is the factor by which we are multiplying (or "scaling") elements of the eigenvector. What this means fundamentally in matrix terms is that we can adjust the **length** of $\mathbf{x}$, but that the **direction** of the vector from the origin is unique (i.e., "unique" meaning that the direction remains the same even after the length adjustment). See Rencher and Christensen (2012, p. 33) for details.

If eigenvectors are unique up to multiplication by a scalar, the question then becomes one of having some way to set the values of the eigenvector in some consistent, normative way so that the variance of derived components cannot grow infinitely large depending on the size of the weights chosen. The way that is typically adopted in PCA is to **scale the eigenvector** such that $\mathbf{x}'\mathbf{x} = 1$. That is, we scale the eigenvector such that its length (i.e., $\sqrt{\mathbf{x}'\mathbf{x}}$) is equal to 1. An eigenvector of length 1 is said to be **normalized**.

13.4 EIGENVALUES AS VARIANCE

We have discussed the fact that each extracted eigenvector is associated with a respective eigenvalue. **Each eigenvalue represents the variance for the given component**. A given component of the p extracted components accounts for a certain amount of variance in the observed variables. This variance is encapsulated in the associated eigenvalue for this component so that, similar to what was done in LDA (but not exactly the same, since discriminant functions are typically not orthogonal (Rencher and Christensen, 2012, p. 408), if we would like to know the **proportion of variance** accounted for, we take the ratio of the given eigenvalue to the total of the eigenvalues extracted. For a covariance matrix, this total variance will be whatever the total variance is summing across the original variables subjected to the PCA. For a correlation matrix, since each variable is standardized to have a variance of 1, the total variance across variables is equal to simply the sum of variables. That is, in the case of a correlation matrix, p represents the total variance sought to be "explained" in the observed data. For example, suppose the PCA extraction revealed eigenvalues 1.5, 1.0, and 0.5 for a three-variable problem. The proportion of variance accounted for by the first extracted component would be $1.5/(1.5 + 1.0 + 0.5) = 1.5/3 = 0.50$, or, 50%. Ideally, in the spirit of data reduction, one hopes that most of the original variance in the data can be accounted for by as few components as possible.

13.5 PRINCIPAL COMPONENTS AS LINEAR COMBINATIONS

We have discussed that an extracted component is, in actuality, made up of an eigenvector associated with an eigenvalue. In this way, the elements of the extracted eigenvector simply represent the "weights" by which we attribute a measure of "importance" to the given observed variables. For example, for a three-variable problem, there will be three components extracted. The three linear combinations can be expressed as:

$$\ell_1 = \mathbf{a}_1'\mathbf{x} = \mathbf{a}_{11}x_1 + \mathbf{a}_{12}x_2 + \mathbf{a}_{13}x_3$$
$$\ell_2 = \mathbf{a}_2'\mathbf{x} = \mathbf{a}_{21}x_1 + \mathbf{a}_{22}x_2 + \mathbf{a}_{23}x_3$$
$$\ell_3 = \mathbf{a}_3'\mathbf{x} = \mathbf{a}_{31}x_1 + \mathbf{a}_{32}x_2 + \mathbf{a}_{33}x_3$$

where $\mathbf{a}_1'$ through $\mathbf{a}_3'$ are vectors of coefficients or "loadings" corresponding to each extracted principal component, and $\mathbf{x}$ is a vector of random variables, which in this example consists of three variables. It is easy to see that a **principal component is nothing more than a weighted sum, a linear combination of the observed variables**, each weighted by respective elements of the extracted eigenvector. What features are special to this extraction? What characteristics do these linear combinations possess? What is so unique about these linear combinations, these components? We discuss these issues next, starting with the extraction of the first component.

13.6   EXTRACTING THE FIRST COMPONENT

The goal of PCA is to extract the first component $\mathbf{a}_1'\mathbf{x}$ (i.e., ℓ_1) such that its variance is **maximized**. But what does this mean, exactly? This means that the component will account for as much of the variance in the original observed variables as possible. That is, out of all the linear combinations that could theoretically be computed on the observed variables, the "**principal**" **component** is the linear combination accounting for the **most** variance. However, since we could feasibly make the variance of $\mathbf{a}_1'\mathbf{x}$ (i.e., the linear combination) large by simply multiplying it by a constant, we must place a **constraint** on its maximization. What this means is that we cannot arbitrarily inflate the variance of a component without bound so that we account for increasingly larger amounts of variance. What we need is a guidepost, a benchmark of sorts from which to do our maximization. This benchmark is what we can refer to more generally as an **imposed constraint** on our analysis. These constraints are present in many statistical procedures where a maximization or minimization technique is applied. In the case of PCA, as we have already alluded, it is ordinarily the case to impose the constraint that the **sum of squared loadings for the component sum to 1.0**. That is, when we extract our first component, we are maximizing the variance of $\mathbf{a}_1'\mathbf{x}$ subject to the constraint $\mathbf{a}_1'\mathbf{a}_1 = 1$. This particular constraint is referred to as a **normalizing constraint**. We are seeking to maximize the variance of the linear combination relative to the length of $\mathbf{a}$ (i.e., the squared length of $\mathbf{a}$ is $\mathbf{a}_1'\mathbf{a}_1 = 1$).

   How does the maximization take place? The actual maximization procedure is usually accomplished by using **Lagrange multipliers**, which we will not detail here, but suffice to say is a widespread technique in linear algebra and the field of numerical analysis that is often used to find maximum or minimum values of a function when that function is first subjected to certain constraints (such as the normalizing constraint of $\mathbf{a}_1'\mathbf{a}_1 = 1$). For details, see Jolliffe (2002). For a lucid overview of Lagrange multipliers as used in structural equation models, see Mulaik (2009).

13.6.1   Sample Variance of a Linear Combination

We have said that the principal component is the linear combination of random variables extracted that has maximal sample variance out of all possible linear combinations that could have been extracted. But to know what this means, we need to know just what quantity it is actually maximizing. That is, **we need to know what the sample variance of a linear combination actually is**. In helping us arrive at the answer, recall first the "ordinary" sample variance for a variable:

$$s^2 = \frac{\sum_{i=1}^{n}(y_i - \bar{y})^2}{n-1}$$

What we need now is the equivalent variance computation for a linear combination. Recall that a linear combination ℓ, in its most general form, is equal to

$$\ell_i = a_1 y_1 + a_2 y_2 + \cdots + a_p y_p = \mathbf{a}' \mathbf{y}$$

and is simply a weighted sum (we use y_1 through to y_p here instead of x_1 through x_p as we did for the earlier components). That is, the composite variable ℓ_i is merely a weighted sum of the random variables $y_1, y_2, \ldots, y_p$. When we compute a linear combination, we are in actuality generating an **entirely new variable**. And just like any other variable, we want to be able to compute its mean and variance. Recall that the mean of ℓ_i is easily computed. We simply sum up the respective values of our new variable ℓ_i and divide by the number of pieces of information that went into the sum. The mean for the linear combination ℓ_i is thus:

$$\bar{\ell}_i = \frac{1}{n} \sum_{i=1}^{n} \ell_i$$

What is the variance of the linear combination? We can compute it the same way we computed the variance of the variable y_i above, but this time, with respect to ℓ_i:

$$s_{\ell_i}^2 = \frac{\sum_{i=1}^{n} \left(\ell_i - \bar{\ell} \right)^2}{n-1} \tag{13.2}$$

In addition to computing $s_{\ell_i}^2$ as in (13.2), the variance of ℓ_i can also be expressed through the following using matrix notation:

$$s_{\ell_i}^2 = \mathbf{a}' \mathbf{S} \mathbf{a} \tag{13.3}$$

That is, the variance of ℓ_i is a function of the weights $\mathbf{a}$ used in deriving the linear combination as well as the sample covariance matrix $\mathbf{S}$. So when we speak about the variance of a principal component in this chapter, we will be talking about (13.3). It is simply the variance of an optimally derived linear combination of variables having special properties, which we call the principal component.

Getting back to our discussion of extracting the first principal component, we can now put our understanding on a more solid footing. That is, the first principal component extracted is such that $s_{\ell_i}^2 = \mathbf{a}' \mathbf{S} \mathbf{a}$ is **maximized**. That is, the first principal component is that linear combination that accounts for maximum variance in the original variables subjected to the analysis.

13.7 EXTRACTING THE SECOND COMPONENT

Now that we have extracted the first component to account for maximal variance subject to the constraint that $\mathbf{a}_1' \mathbf{a}_1 = 1$, we now wish to extract the second and ensuing components. Similar to the first component, the second component, that of $\mathbf{a}_2' \mathbf{x}$, is extracted subject to the constraint that its variance again be maximized and that $\mathbf{a}_2' \mathbf{a}_2 = 1$. However, in addition to the constraint of $\mathbf{a}_2' \mathbf{a}_2 = 1$ imposed, the second component is extracted subject to a second constraint. That second constraint is that the **covariance of the second component with that of the first component be equal to 0**. That is, we extract and maximize the variance of $\mathbf{a}_2' \mathbf{x}$ subject to the constraints $\mathbf{a}_2' \mathbf{a}_2 = 1$ and $\mathrm{cov}\left(\mathbf{a}_1' \mathbf{x}, \mathbf{a}_2' \mathbf{x}\right) = 0$. We can also refer to this second condition more simply as $\mathbf{a}_2' \mathbf{a}_1 = 0$. The conditions $\mathbf{a}_2' \mathbf{a}_1 = 0$ and $\mathrm{cov}\left(\mathbf{a}_1' \mathbf{x}, \mathbf{a}_2' \mathbf{x}\right) = 0$ both can be used to represent the idea of **zero correlation between components** (Jolliffe, 2002, pp. 5–6). That is, the two vectors are geometrically **perpendicular**, and their dot product is equal to zero. More formally, the idea of orthogonality is that of a covariance matrix of derived components ℓ_i through ℓ_p that is a diagonal matrix with component variances along the main diagonal and zeros everywhere else. If $\mathbf{S}_\ell$ is the variance-covariance matrix of components, then we want $\mathbf{S}_\ell$ to be:

$$\mathbf{S}_\ell = \mathbf{ASA}' = \begin{pmatrix} s_{\ell_1}^2 & 0 & \dots & 0 \\ 0 & s_{\ell_2}^2 & \dots & 0 \\ 0 & 0 & s_{\ell_3}^2 & 0 \\ 0 & 0 & \dots & s_{\ell_p}^2 \end{pmatrix} \tag{13.4}$$

In other words, we are **diagonalizing the matrix** to one with only variances $s_{\ell_1}^2, s_{\ell_2}^2, \dots, s_{\ell_p}^2$ along the main diagonal. These, as we will see, are the respective eigenvalues, $\lambda_1, \lambda_2, \dots, \lambda_p$ of $\mathbf{S}_\ell$.

13.8   EXTRACTING THIRD AND REMAINING COMPONENTS

As a recap, the first component is extracted subject to the normalizing constraint. The second component is extracted subject to the normalizing constraint **and** the orthogonality constraint, that of $\mathbf{a}_2'\mathbf{a}_1 = 0$. The third principal component extracted, $\mathbf{a}_3'\mathbf{x}$, will be so subject to the normalizing constraint but will also be orthogonal to components one and two. That is, $\mathbf{a}_3'\mathbf{a}_1 = 0$ and $\mathbf{a}_3'\mathbf{a}_2 = 0$. This third component will exhibit maximal variance subject to these two constraints. Note as well that if the third component is the last component to be extracted, then we can also say that this component exhibits **minimal variance** out of the three components. That is, it is the least "relevant" (in the sense of variance) component in accounting for variance in the observed data.

Remaining components are extracted in an analogous fashion. That is, each remaining linear combination is extracted that accounts for maximal variance **given the already included extracted components before it**, which really means, in PCA, given that it is orthogonal to the previously extracted components.

13.9   THE EIGENVALUE AS THE VARIANCE OF A LINEAR COMBINATION RELATIVE TO ITS LENGTH

We have discussed the idea that when extracting linear combinations (i.e., components), we are doing so such that we extract the component that has maximal variance, but subject to the constraint that it does so relative to the squared length of the eigenvector (equal to $\mathbf{a}'\mathbf{a}$). We can express this idea of "relative to" through a ratio, essentially comparing the variance of the linear combination to its squared length (where $\mathbf{a}'\mathbf{a}$ is usually set at 1):

$$\frac{\mathbf{a}'\mathbf{Sa}}{\mathbf{a}'\mathbf{a}} \tag{13.5}$$

This is the **eigenvalue** of the linear combination. As usual, we denote the eigenvalue by λ ("lambda"), and write:

$$\lambda = \frac{\mathbf{a}'\mathbf{Sa}}{\mathbf{a}'\mathbf{a}}$$

The eigenvalue is also the **maximum** value of the ratio $\mathbf{a'Sa/a'a}$. For a proof of why the variances $\mathbf{a_1'x}, \mathbf{a_2'x}$ and $\mathbf{a_3'x}$ are given by the eigenvalues λ_1, λ_2 and λ_3, see Johnson and Wichern (2007, p. 432).

Theoretically, though seldom if ever in practice, a PCA could generate **eigenvalues that are approximately equal**. What this means substantively (i.e., from a scientific point of view) is that each extracted component is accounting for approximately the same amount of variance. The solution to this problem is, pragmatically speaking, to prioritize that component of the two which makes the most substantive sense (if either of them do). As we will see, the fact that a component explains a certain amount of variance is not always justification alone for keeping it or attempting to interpret it.

Furthermore, it sometimes happens that one or more eigenvalues are equal to zero. This is suggestive of a redundancy (or "dependency") among observed variables, which may imply that one variable is an exact linear combination of one or more other variables. As Jolliffe (2002) notes:

> Any PC with zero variance defines an exactly constant linear relationship between the elements of $\mathbf{x}$. If such relationships exist, then they imply that one variable is redundant for each relationship, as its value can be determined exactly from the values of the other variables appearing in the relationship ... Ideally, exact linear relationships should be spotted before doing the PCA, and the number of variables reduced accordingly. (p. 27)

Hence, as recommended, a potential solution is to examine the raw observed variables (not components) and delete variables as necessary to ease the dependency, then redo the components analysis.

13.10 DEMONSTRATING PRINCIPAL COMPONENTS ANALYSIS: PEARSON'S 1901 ILLUSTRATION

To demonstrate a very simple principal components analysis, we consider data featured in Pearson's 1901 paper (Pearson, 1901, p. 569). Pearson gave data on two variables, x and y, which we reproduce below:

```
> x <- c(0.0, 0.9, 1.8, 2.6, 3.3, 4.4, 5.2, 6.1, 6.5, 7.4)
> y <- c(5.9, 5.4, 4.4, 4.6, 3.5, 3.7, 2.8, 2.8, 2.4, 1.5)
> pc.data <- data.frame(x, y)
> pc.data
      x   y
1   0.0 5.9
2   0.9 5.4
3   1.8 4.4
4   2.6 4.6
5   3.3 3.5
6   4.4 3.7
7   5.2 2.8
8   6.1 2.8
9   6.5 2.4
10  7.4 1.5

> plot(x, y)
```

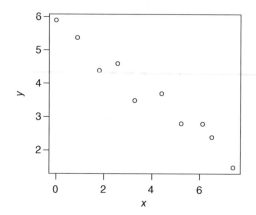

Obviously, the data are currently in two dimensions (x and y). However, do we really need these two dimensions to account for the variance in the data? The job of principal components analysis will be to learn whether the variance in the data can be accounted for primarily by the first component. Because there are two variables, PCA will extract two components. We are hoping, however, that the first component accounts for most of the variance and that we could conveniently discard the 2nd component as not worthwhile.

We will perform the PCA directly on the covariance matrix. First, we build the covariance matrix:

```
> A <- cov(pc.data)
> A
          x          y
x   6.266222  -3.381111
y  -3.381111   1.913333
```

We can verify that R has constructed the matrix correctly by computing variances and pairwise covariances to match the entries above:

```
> cov(x, y)
[1] -3.381111
> var(x)
[1] 6.266222
> var(y)
[1] 1.913333
```

We now run the PCA on the covariance matrix **A** using R's `princomp`, using `covmat = A` to identify the covariance matrix we want analyzed:

```
> pca <- princomp(covmat = A)
> summary(pca)
```

```
Importance of components:
                        Comp.1        Comp.2
Standard deviation      2.8479511 0.262164656
Proportion of Variance  0.9915973 0.008402695
Cumulative Proportion   0.9915973 1.000000000
```

R has extracted two components (Comp.1 and Comp.2), which should make sense, since there are two variables in Pearson's original data. The standard deviation of the first component is 2.848. When we square this number, we get the variance of the component, equal to $(2.8479511)^2 = 8.11$, which is the **eigenvalue** for the first component. The standard deviation of the second component is 0.262. When we square this number, we get the variance of the component, equal to $(0.262164656)^2 =$ 0.07, which is the eigenvalue (rounded up) for the second component.

The **proportion of variance** accounted for by the first component is equal to approximately 0.99, computed as the variance for the given component divided by the total sum of variances across both components (i.e., $8.11/(8.11 + 0.07)) = 8.11/8.18 = 0.99$. The proportion of variance accounted for by the second component is equal to 0.008, computed as the variance for the given component divided by the total sum of variances across both components (i.e., $0.07/8.18 = 0.008$).

We obtain the loadings in R by:

```
> loadings(pca)

Loadings:
  Comp.1 Comp.2
x -0.878 -0.479
y  0.479 -0.878
```

Recall that the loadings for a principal component are actually elements of the eigenvector that make up the component. That is, in a linear combination (component) of the form

$$\ell_i = a_1 y_1 + a_2 y_2 + \cdots + a_p y_p = \mathbf{a}' \mathbf{y}$$

the "loadings" are the values $a_1, a_2, \ldots, a_p$.

We can also easily obtain the same eigenvectors as above, as well as the corresponding eigenvalues, within rounding error by solving for them directly using R's eigen function:

```
> eigen(A)
$values
[1] 8.11082525 0.06873031

$vectors
            [,1]        [,2]
[1,] -0.8778562 -0.4789243
[2,]  0.4789243 -0.8778562
```

Note that the loadings and eigenvectors (i.e., $vectors) are identical. We can easily demonstrate the orthogonality of eigenvectors by obtaining their dot product:

```
> eigen.1 <- c(-.8778562, 0.4789243)
> eigen.2 <- c(-0.4789243, -0.8778562)
> eigen.1%*%eigen.2
     [,1]
[1,]    0
```

The product of eigenvectors computed in R is equal to 0, confirming that both extracted components are orthogonal to one another, as they are required to be.

We can also easily demonstrate that each eigenvector extracted must have a length equal to 1.0. That is, recall that the sum of squared loadings must equal 1.0 ($\mathbf{x'x} = 1$). We verify that this is indeed the case:

```
> sum(eigen.1*eigen.1)
[1] 1
> sum(eigen.2*eigen.2)
[1] 1
```

Recall that components analysis does not generate "new" variables, but rather simply transforms existing ones into new linear combinations. Because of this, the actual **total variance** in the sample data remains the same. This idea is encapsulated by a general property of principal components analysis:

$$\sum_{i=1}^{p} \lambda_i = s_1^2 + s_2^2 + \cdots + s_p^2 \tag{13.6}$$

That is, **the sum of eigenvalues will equal the sum of variances of the original variables**, where p is the number of observed variables, λ_i is the ith eigenvalue, and $s_1^2 + s_2^2 + \cdots + s_p^2$ is the sum of the respective variances for each observed variable. PCA does nothing more than summarize the variance of the original variables in a different way. It "repackages" the variance of the original variables onto new dimensions. The transformation does not fundamentally change the variability inherent in the data. It only **reorganizes** it. Indeed, property (13.6) can be used as a quick check of one's work in computing components in that if the sum of eigenvalues for a covariance matrix does not total the sum of observed variable variance, it could be indicative of a miscalculation or other more serious problem. In a standardized covariance matrix, otherwise known as a **correlation matrix**, the sum of eigenvalues should be equal to the sum of variables subjected to the analysis, since for a correlation matrix each observed variable contributes a single unit of variance (i.e., value of 1) at the outset of the analysis.

We can easily confirm (13.6) for Pearson's data. Recall the eigenvalues for components 1 and 2 were equal to 8.11 and 0.07, respectively, for a sum of 8.18. The original variances of variables x and y were equal to 6.27 and 1.91 respectively, for likewise a sum of 8.18. We can see then that the total variance in the data has been **preserved**. All the PCA has done is to find new axes, mutually orthogonal to one another, for which the first few (in our case, first only) hopefully accounts for as much of the total variance as possible. PCA does not "change" the amount of variance in a set of data, it merely **reconstructs the dimensions** on which this variance exists and is represented.

13.11 SCREE PLOTS

The **scree plot** is a graphical device used for helping to decide the number of worthwhile components to retain from a principal components or factor analysis. It is generally attributed to Cattell (1966), though as noted in Jolliffe (2002), scree plots were well in use before Cattell. In a scree plot, eigenvalues are plotted in order of decreasing magnitude. Generally, and quite subjectively, where one sees a "bend" or "elbow" in the plot, one uses this as a cut-off point for the number of components to retain.

We obtain a scree plot in R by following up the `princomp` function with the `plot` function:

```
> plot(pca,type="lines")
```

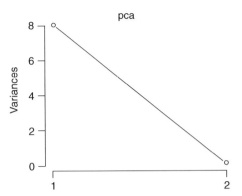

FIGURE 13.4   Scree plot for Pearson's data.

For Pearson's data, the plot clearly reveals that the first component is accounting for the majority of the variance, while the second component is accounting for little (Figure 13.4). Naturally, scree plots are more useful when the number of components is quite large. For instance, consider Figure 13.5, where a plot was made for an eight-component problem, one which we will feature toward the end of this chapter (where we analyze the generic matrix `cormatrix`).

Inspection of Figure 13.5 reveals that the "elbow" appears to occur at component number 2, which may suggest the retention of one to two components, the first explaining quite a bit more variance than that of the second.

Contrary to what some researchers profess about the scree plot along with Cattell's original enthusiasm for it, I personally do not find them very useful. There is nothing inherently significant about the elbow in the graph, and one can usually draw an identical conclusion about component retention with or without the plot. In the time before high-speed computers, the scree plot may have been a bit more helpful in assisting one to wade through numerical complexities and make sense of one's data (Tucker, 2009). As well, as we will discuss more so with regard to factor analysis, component or factor retention is somewhat of an art at best, and should be influenced more by researcher judgment than by a simple diagram such as the scree plot. And in applications of PCA to physical phenomena such as extracting components of digital images, choosing the number of components to retain will have less to do with a bend in a scree plot and much more to do with features of the image one wishes to account for. For an

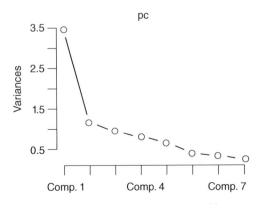

FIGURE 13.5   Scree plot for eight-component problem on cormatrix.

example of such an application to "eigenfaces" and how PCA can be used in facial recognition, see Izenman (2008, Chapter 7).

Hence, though the use of the scree plot is not to be discouraged, if you find yourself retaining a certain number of components based solely on the results of such a plot, seeking additional advice when interpreting your findings is probably in order. Scree plots should be used at most as **aids** to decision-making in this regard, and not relied on exclusively for the determination of the number of components or factors.

How many components should you then retain? The purpose of the analysis should, in part, influence your decision. Presumably, components analysis was performed for one of two reasons or both:

- You wished to reduce a large body of data into a few components that account for most of the variance in the original data, but you are not overly concerned with substantive interpretation of the components since you simply wanted to use the component scores in a future analysis (e.g., regression analysis) as a predictor of a response variable.

- You wished to reduce the large body of data into fewer components but were also very interested in the interpretation of these extracted components.

The question of component retention is rarely easily answered, and its answer depends in large part on how many components can be properly interpreted or otherwise valued by the researcher, unless the extraction was done for purely statistical variance-maximization reasons alone or for a task such as pattern recognition with images. The following guidelines may nonetheless prove useful:

- Using a scree plot, as discussed, detect where the primary "elbow" occurs, and base the retention of components on this indicator. However, recall that this must be considered a very crude and atheoretical way of proceeding in making the decision regarding component retention. As noted, we generally advise against using the scree plot exclusively as a decision tool in this regard. It can be useful as an initial screening of your components solution, however.

- Retain only those components that have eigenvalues associated with them that are greater than or equal to the average of eigenvalues. The rationale behind this rule is that because eigenvalues are corresponding variances that represent each component, the average eigenvalue can be considered the average variance of the observed variables subjected to the PCA. Hence, in this way, the most "important" components will be those that are "above average" relative to the set of extracted eigenvalues. In a correlation matrix, the average of eigenvalues is equal to 1. This is generally known as the **Guttman–Kaiser criterion**, which originated with Guttman's work in 1954 and was adapted and modified by Kaiser (1960, 1961). Yeomans and Golder (1982) summarize the decision-rule:

  The technique is justified in the original Guttman article in terms of it providing a lower bound for the number of common factors underlying a correlation matrix of observed variates having unities in the main diagonal. More intuitively the argument has been advanced that no component "explaining" less than the variance of an original variate can be deemed to represent a significant source dimension. (pp. 222–223)

However, as noted by these same authors, using the criterion as a decision rule, especially an exclusive one, is usually ill-advised. Under most circumstances they found the criterion to be a poor predictor of the number of factors or components inherent in a set of data. They also found that only when the number of factors is substantially lesser than the number of variables, and communalities are relatively high, does it make any sense to use the criterion at all. Hence, these authors advise that if one is to use the criterion, one be sure to also include information about estimated communalities. Consult Preacher and MacCallum (2003) for a detailed discussion of why the adoption of arbitrary decisions when deciding on the number of factors to retain can prove problematic.

Overall then, our general recommendation regarding the Guttman–Kaiser criterion is similar to that of using the scree test: **it is a poor decision tool if used exclusively to render a decision**, but is potentially useful if used in conjunction with researcher expertise and judgment. The moral of the story is clear, in that in most cases, **component retention cannot be made based on statistical evidence alone**.

13.12 PRINCIPAL COMPONENTS VERSUS LEAST-SQUARES REGRESSION LINES

The reader initially examining the principal component plot first produced by Karl Pearson in 1901 (Figure 13.2) might very well ask a good question: **What is the difference between a least-squares line and a principal component "line?"** After all, they look very similar in that they both seem to account for variation in the plane. However, as mentioned, they are constructed in a different manner. Recall Pearson's explanation discussed earlier in reference to his original 1901 article. In least-squares regression, we regress y on x so that the so-called "least-squares criterion" is satisfied. Recall that in ordinary least-squares, the objective is to fit a line subject to the minimization criteria that the sum of squared errors be as small as possible. That is, the line is fit subject to keeping $\sum_{i=1}^{n} \varepsilon_i^2$ to a minimum. What this amounts to geometrically is minimizing the squared **vertical distances** between observed values and fitted values along the regression line, as depicted in Figure 13.6.

In PCA on the other hand, we do not wish to minimize the vertical distances. Geometrically, we want to minimize the sum of squared **orthogonal** or **perpendicular** distances from the line. This is why Pearson minimized $S(p^2)$, which represented the sum of squared perpendicular distances, where S stood for "sum" and p^2 stood for "squared perpendiculars." That is, in PCA, we minimize the **perpendicular** distances rather than the **vertical** ones. We can see from Pearson's plot (Figure 13.2) that this is indeed what he had in mind.

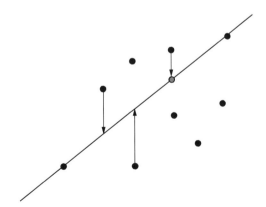

FIGURE 13.6   Vertical distances are minimized in least-squares regression.

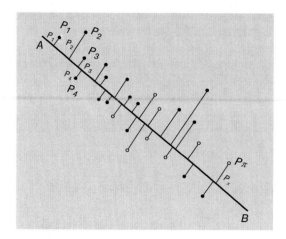

Notice the distances are **perpendicular to the line** instead of **vertical** as they were with the least-squares line. Another way to understand this idea is to consider the scatterplot given by Pearson in Figure 13.3 (reproduced below) where both regression lines and the first principal component are plotted. Note again that the three lines are different. This is because they satisfy different objectives.

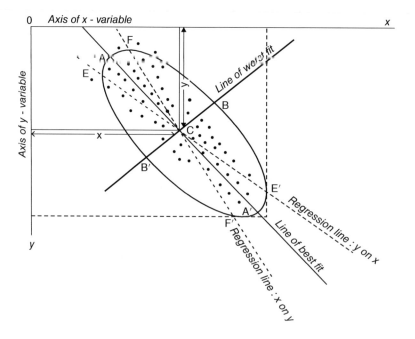

13.13   COVARIANCE VERSUS CORRELATION MATRICES: PRINCIPAL COMPONENTS AND SCALING

The most ideal situation for a PCA is that all variables subjected to the analysis are measured in the same units. That is, principal components is most suitable when all observed variables are measured on the same or at least similar **metric**. Eigenvalues and eigenvectors are not invariant to scale, which means that depending on whether one analyzes the covariance or correlation matrix, one will generally obtain different roots and vectors (Anderson, 2003). As summarized by Rencher and Christensen (2012):

> Generally, extracting components from **S** rather than **R** remains closer to the spirit and intent of principal components analysis, especially if the components are to be used in further computations. However, in some cases, the principal components will be more interpretable if **R** is used. For example, if the variances differ widely or if the measurement units are not commensurate, the components of **S** will be dominated by the variables with large variances. The other variables will contribute very little. For a more balanced representation in such cases, components of **R** may be used. (pp. 419–420)

Performing a PCA on R rather than S will not result in the same analysis. For one, the variance accounted for by each component is not guaranteed to remain constant across both matrices, nor will the coefficients of the eigenvectors remain the same. Also, though the components extracted from **S** are not scale invariant, the components extracted from **R** are so. This is simply because **R** itself is scale invariant, whereas **S**, of course, is not. Recall from Chapter 2 that two variables could have a very high covariance yet less linear **standardized** relationship simply due to the fact that one or both variables exhibit high variances, which is what may be in part making the covariance large. In the case of correlation, due to standardization (i.e., dividing the average cross-product by the product of standard deviations), correlations become scale invariant.

To summarize then, if variables have wildly different variances, then standardizing the covariance matrix to get the correlation matrix is a reasonable solution, so long as one is aware that differences exist when analyzing the one matrix versus the other and that different solutions for each may be obtained. Otherwise, in the case of analyzing **S**, variables with much higher variances will dominate the determination of components. For a demonstration and example of this effect, see Rencher and Christensen (2012, pp. 420–422). For further considerations on the differences between PCA on the covariance versus correlation matrix, consult Jolliffe (2002). If in doubt, there is nothing preventing a researcher from trying **both solutions** and then comparing the two. If the variances of one or more variables is having an influential impact, the comparison could help to shed light on the issue (e.g., see James et al., 2013, p. 381).

13.14   PRINCIPAL COMPONENTS ANALYSIS USING SPSS

We now demonstrate a PCA in SPSS on `cormatrix`, a hypothetical correlation matrix consisting of eight observed variables, tests 1–8 (T1–T8). Principal components analysis is considered an "option" in SPSS's factor analysis function (the so-called "components analysis" option of factor analysis). As already discussed, however, factor analysis should not be equated with that of components analysis proper.

In preparing the matrix, we specify in SPSS that the input data is of the form of a matrix, then list the observed variables of the matrix (first row of code below). We will base this components analysis assuming 1000 observations are available on each variable, hence the reason why the third line of the input reads "1000 1000 …" a total of eight times (once per variable). We also use the first column of the matrix to specify CORR for each row, which tells SPSS that correlations appear in each row:

```
MATRIX DATA VARIABLES=ROWTYPE_ T1 T2 T3 T4 T5 T6 T7 T8.
BEGIN DATA
N 1000 1000 1000 1000 1000 1000 1000 1000
CORR 1.00000
CORR .343      1.00000
CORR .505       .203   1.00000
CORR .308       .400    .398  1.00000
CORR .693       .187    .303   .205  1.00000
CORR .208       .108    .277   .487   .200  1.00000
CORR .400       .386    .286   .385   .311   .432  1.00000
CORR .455       .385    .167   .465   .485   .310   .365 1.00000
END DATA.
```

To run the components analysis, we request:

FACTOR MATRIX=IN(CORR=*) * specifies a correlation matrix is being inputted.
/PRINT= INITIAL EXTRACTION CORRELATION REPR * requests initial and extraction communalities for both original matrix and reproduced matrix.
/CRITERIA FACTORS(8) * requests that 8 factors (i.e., components) be extracted, which is the maximum number in this case.
/EXTRACTION=PC * specifies the extraction to be that of principal components.
/METHOD=CORRELATION. * requests the correlation matrix be analyzed.

The first part of the output is the correlation matrix we requested in our syntax. SPSS gives us the full correlation matrix, not only the lower triangular. We confirm that the correlation matrix matches that which we input into SPSS.

Correlation Matrix		T1	T2	T3	T4	T5	T6	T7	T8
Correlation	T1	1.000	0.343	0.505	0.308	0.693	0.208	0.400	0.455
	T2	0.343	1.000	0.203	0.400	0.187	0.108	0.386	0.385
	T3	0.505	0.203	1.000	0.398	0.303	0.277	0.286	0.167
	T4	0.308	0.400	0.398	1.000	0.205	0.487	0.385	0.465
	T5	0.693	0.187	0.303	0.205	1.000	0.200	0.311	0.485
	T6	0.208	0.108	0.277	0.487	0.200	1.000	0.432	0.310
	T7	0.400	0.386	0.286	0.385	0.311	0.432	1.000	0.365
	T8	0.455	0.385	0.167	0.465	0.485	0.310	0.365	1.000

Next are the communalities (given below), both the initial and the extracted.

Communalities

	Initial	Extraction
T1	1.000	1.000
T2	1.000	1.000
T3	1.000	1.000
T4	1.000	1.000
T5	1.000	1.000
T6	1.000	1.000
T7	1.000	1.000
T8	1.000	1.000

Extraction method: principal component analysis

Notice that all initial communalities are equal to 1.0. Recall that the reason why they are all equal to 1.0 is because we are requesting a **principal components** solution for this correlation matrix, and hence each variable is contributing unit variance to begin. In the typical exploratory factor analysis solution, as we will discuss next chapter, the initial communalities will no longer be equal to 1.0, and hence each variable will no longer contribute unit variance. For instance, in the case of **principal axis factoring**, initial communalities will reflect the degree to which the given observed variable shares variance with other variables in the model. Indeed, such measures will be a more accurate and representative depiction for what is meant by **communalities**. We will discuss this concept more thoroughly when we survey factor analysis in the following chapter.

The extraction communalities reflect the degree to which a given variable shares commonality across the extracted components. Because we are extracting the maximum number of components (eight) in this case, SPSS reports all extraction communalities equal to 1.0, the same as those for the initial communalities. Had we requested a smaller number of components to be extracted (i.e., seven or less), then all of the extracted communalities would not have been equal to the initial communalities.

Next, SPSS provides us with the breakdown of the eigenvalue extraction:

Total Variance Explained

	Initial Eigenvalues			Extraction Sums of Squared Loadings		
Component	Total	% of Variance	Cumulative %	Total	% of Variance	Cumulative %
1	3.447	43.088	43.088	3.447	43.088	43.088
2	1.157	14.465	57.554	1.157	14.465	57.554
3	0.944	11.796	69.349	0.944	11.796	69.349
4	0.819	10.237	79.587	0.819	10.237	79.587
5	0.658	8.226	87.813	0.658	8.226	87.813
6	0.390	4.873	92.686	0.390	4.873	92.686
7	0.336	4.201	96.887	0.336	4.201	96.887
8	0.249	3.113	100.000	0.249	3.113	100.000

Extraction method: principal component analysis.

The eigenvalue for the first extracted component is equal to 3.447 and is clearly the largest of all eigenvalues extracted. Since there are a total of eight possible components, the variance explained by component 1 is equal to 3.447/8 = 0.43088, or 43.088% as shown in the first row of the table. Component 2 has associated with it an eigenvalue of 1.157, which accounts for 1.157/8 = 0.1446, or 14.46% of the variance. SPSS also provides the cumulative percentage of variance explained, which for components 1 and 2, is equal to 43.088 + 14.465 = 57.55%.

Notice that the extraction sums of squared loadings, located in the right-hand side of the output, are identical to those on the left-hand side. The reason for this is because in a PCA, **whether we extract all possible components or a subset of all possible components, the extraction of eigenvalues for each component remains the same**. In factor analysis, however, as we will see in the following chapter, this is typically not the case, and the value of eigenvalues will change depending on the number of factors extracted. This is a key and important difference between principal components analysis and factor analysis.

Next are the component loadings in SPSS's component matrix. These are **scaled eigenvectors** corresponding to each **eigenvalue**. That is, the first eigenvector, for component 1, that of 0.766, 0.563, 0.591, 0.693, 0.663, 0.559, 0.680, and 0.707, is the scaled eigenvector corresponding to the first extracted eigenvalue of 3.447. **The sum of squared loadings for each eigenvector is equal to its respective eigenvalue**. Note that this is different than what we mentioned earlier about the sum of squared eigenvector weights equalling 1.0. Because SPSS treats PCA as a special case of factor analysis, the sum of squared loadings for each component is equal to respective eigenvalues.

	Component Matrix[a]							
	Component							
	1	2	1	4	5	6	7	8
T1	0.766	−0.492	0.096	0.080	0.054	0.084	−0.053	−0.377
T2	0.563	0.123	−0.619	0.427	0.072	0.293	0.076	0.072
T3	0.591	−0.074	0.531	0.526	−0.099	−0.120	0.214	0.132
T4	0.693	0.463	0.002	0.101	−0.382	−0.110	−0.371	−0.020
T5	0.663	−0.585	0.066	−0.284	0.004	0.137	−0.180	0.286
T6	0.559	0.531	0.370	−0.363	0.053	0.338	0.142	−0.029
T7	0.680	0.232	−0.059	−0.055	0.629	−0.277	−0.061	0.028
T8	0.707	−0.051	−0.353	−0.359	−0.310	−0.246	0.297	−0.012

Extraction method: principal component analysis.
[a]Eight components extracted.

Virtually all observed variables load relatively high on component 1, especially T1 (0.766), T4 (0.693) and T8 (0.707). Negative signs for loadings are interpreted to mean that the given variable is negatively associated with the component. For instance, the negative loading of −0.492 for T1 on component 2 indicates a moderate negative relationship between T1 and the given component (whatever we shall name it, if it indeed makes sense to give it a name for these data).

We can generate the respective scree plot in SPSS:

```
/PLOT EIGEN
```

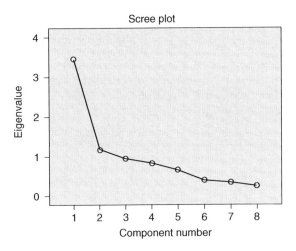

The scree plot confirms what the numerical output suggested, that the first component accounts for the majority of the variance in the variables.

13.15   CHAPTER SUMMARY AND HIGHLIGHTS

- Principal components analysis (PCA) is a statistical technique that extracts information from a covariance or correlation matrix such that the original information within the matrix may be represented in fewer dimensions hopefully without a loss of too much information.
- The number of principal components extracted from a data set will generally equal the number of variables in that data set and originally subjected to the PCA. The user retains the number of components that adequately account for as much of the original variance as possible. Should the user choose to keep all derived components, then no data reduction has occurred, and only a transformation of the original axes to new dimensions has taken place.
- PCA can be said to have originated with the work of Karl Pearson in 1901 and was extended with the work of Harold Hotelling in 1933.
- The primary technical distinction between **principal components analysis** and **factor analysis** is that the former generates successive orthogonal linear combinations of the variables, whereas factor analysis generates successive linear combinations of the factors. Principal components analysis and factor analysis often yield similar findings but should nonetheless be regarded as distinct techniques.

- The primary theorem of PCA states that for a vector of random variables with associated covariance matrix, an orthogonal linear transformation can take place that generates components having maximum variance and which are unrelated to successive components extracted. In the classic PCA, each component extracted is orthogonal to those previously extracted.

- PCA generally proceeds by the extraction of eigenvalues and eigenvectors from a covariance matrix. The extracted eigenvalues correspond to variances of the components. The extracted eigenvectors correspond to weights used to derive the components and are often and conveniently scaled such that the length of the component equals 1.0. This is typically referred to as the **normalizing constraint**.

- PCA is most ideally performed on variables measured on the same units. Caution should be exercised when conducting PCA on variables not of the same units. Analysis of the correlation matrix instead of that of the covariance matrix may be suitable in cases that feature such incommensurate variables.

- The **scree plot**, depicting eigenvalues in decreasing order, is a tool that may prove useful in helping the analyst decide on the number of components to retain. However, other than in a purely exploratory sense, it should never be used exclusively in deciding on the number of components to keep.

- A comparison of PCA to linear regression reveals that while ordinary least-squares regression seeks to minimize the sum of squared errors around the line of best fit, that is, the **vertical distances** from the line, PCA likewise seeks to minimize the sum of squared errors around the line of best fit, but this time, it is the **perpendicular distances** (not vertical) that are minimized. Pearson clearly distinguished between these two cases in his 1901 paper in which he originated PCA.

- The sum of eigenvalues for a PCA is equal to the sum of variances of the original variables, that is, the trace of the covariance matrix. This is because PCA does not "change" data, it merely projects it onto new axes as a way of "re-expressing" it. The original variance in the variables remains intact

REVIEW EXERCISES

13.1. Interpret Karl Pearson's quote to open this chapter that "**the term 'best fit' is really arbitrary**." What does this mean, exactly? And how did such thinking on his part reflect ingenuity in developing the principal components solution?

13.2. Provide two interpretations of the goal of principal components analysis. Which do you think is most relevant? Why?

13.3. Why is it said that PCA seeks to define an orthogonal transformation to a **diagonal** covariance matrix? What does this mean, exactly, and what does it mean for the covariance matrix to be diagonal?

13.4. We said that PCA reduces to solving for the eigenvalues and eigenvectors of an at least semipositive definite matrix. What does it mean for the matrix to be **semipositive definite**, and why does this matter in the context of PCA?

13.5. Recall Francis Galton's correlational ellipse (left). Compare and contrast Pearson's ellipse of 1901 (right). Can you identify similarities?

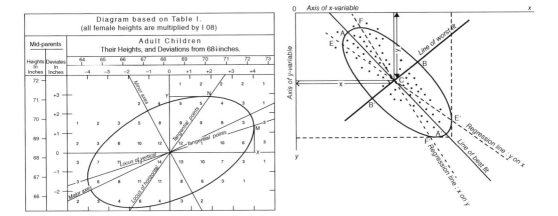

13.6. Explain why it is true that there are generally as many principal components extracted in PCA as there are variables input into the procedure. Why does this make sense?

13.7. Recall that the theorem of principal components states that **if the expectation of the random vector x is equal to 0 (i.e., $E(\mathbf{x}) = 0$) and for a covariance matrix Σ, then one can conduct an orthogonal linear transformation to generate components that have maximum variance and are unrelated to successive components**. Explain what it means for components to have "maximum variance."

13.8. What does it mean to say that eigenvectors are unique only up to multiplication by a scalar?

13.9. How is the eigenvector usually scaled in PCA, and what is this constraint typically called?

13.10. Discuss how a principal component is a **linear combination**. What does this mean, exactly? How are the linear combinations obtained in PCA similar or different to previous linear combinations encountered in this book (e.g., t-tests, regression, discriminant analysis)? That is, what distinguishes these different types of linear combinations for each setting?

13.11. What is the variance of a linear combination equal to, and how does PCA go about "normalizing" this variance?

13.12. Explain the process of extracting the first and second principal components. What condition must be satisfied when extracting the second component that did not (and could not) apply when extracting the first?

13.13. Compare the procedures of principal components analysis to that of least-squares regression. How are they similar? How are they different?

13.14. Under what conditions should the correlation matrix rather than the covariance matrix be analyzed in a PCA? Which matrix is more in the "spirit" of PCA?

13.15. Conduct a principal components analysis on variables x1 through x9 representing nine tests of mental ability of the Holzinger and Swineford (1939) data. Interpret results. How many components would you extract? Why? Without knowing more about the variables x1–x9, their meaning and nature, and the general substantive purpose for conducting the PCA, why might decisions regarding component retention be difficult if not impossible in a **substantive**, as opposed to, **statistical** context? That is, why might the statistical indicators of component retention not be enough for you to make a decision?

13.16. Anderson (2003, p. 471) performed a PCA on a subset of the iris data using only the species versicolor. Duplicate that analysis, and confirm that the first component extracted accounts for the majority of the variance in the original data.

Further Discussion and Activities

13.17. A useful strategy for better understanding and appreciating statistical methods is to compare methodologies to one another. In the case of principal components analysis, a comparison of the **geometry of PCA** to that of the analysis of variance (ANOVA) proves insightful. Read Smith (1980) and provide a brief account of how the two techniques compare geometrically.

13.18. The **biplot** is an innovative graphical technique for depicting multivariate data in spirit similar to how scatterplots are used to depict bivariate data. They are the multivariate analog to the scatterplot. The seminal publication for biplots is Gabriel (1971). Read this article and then for the PCA performed in Exercise 13.16, generate a biplot in R using `biplot()`. Interpret the plot.

14

FACTOR ANALYSIS

When the factors are partialled out from the observed variates there no longer remains any correlation between these.

(Jöreskog, 1978, pp. 453, 455)

Factor analysis is a statistical method used to uncover latent structures that may hypothetically underlay covariance or correlation among typically continuously observed variables. A more crude designation of the method is that it is a **data reduction technique**, similar in vein to that of principal components analysis in which linear combinations are extracted from empirical observations. As emphasized in the previous chapter, however, factor analysis is not equivalent to principal components analysis, and as argued by many, efforts to seemingly equate them are severely misguided (e.g., see Chatfield and Collins, 1981). More technically, we may say that the task of factor analysis is to approximate one matrix, whether it be the covariance or correlation matrix, by one of lower rank (Eckart and Young, 1936). More in the spirit of Jöreskog's quote, we can also interpret the factor-analytic model as addressing the question of whether there exist a number of factors, necessarily less than the number of observed variables, such that the partial correlations between every pair of observed variables equal zero. As noted by Gnanadesikan (1997, p. 16), this interpretation also gives rise to the factor-analytic model of the current chapter.

While principal components analysis seeks to generate linear combinations of observed random variables, in factor analysis, it is the observed variables that are hypothesized as linear combinations of hypothetical **underlying factors**. While the priority of PCA was to explain as much of the total variance of variables as possible, the priority of factor analysis is to explain the covariance or correlation, or more generally, the **commonality** among variables. And though factor analysis and principal components do share the goal of summarizing variation of a potentially large number of variables into a

Applied Univariate, Bivariate, and Multivariate Statistics: Understanding Statistics for Social and Natural Scientists, With Applications in SPSS and R, Second Edition. Daniel J. Denis.
© 2021 John Wiley & Sons, Inc. Published 2021 by John Wiley & Sons, Inc.
Companion Website: www.wiley.com/go/denis/appliedstatistics2e

smaller set of linear combinations, the assumptions that underlie each procedure are quite different, and their purposes and applications can equally be quite distinct.

14.1 HISTORY OF FACTOR ANALYSIS

Factor analysis is an enormous subject and has a turbulent history to say the least. Since its original inception with the work of Spearman (1904a), the method has been a favorite target of criticism. And though the mathematics and structural development of factor analysis has been, historically, generally agreed upon, it is its **usage** and ties to psychological theory, along with its philosophical difficulties that have been at the root of the majority of criticisms directed at the procedure (e.g., see Mulaik, 1987). As will be elucidated on later in this chapter, I believe the storied and extensive criticism of factor analysis to be somewhat misguided. The statistical procedure cannot be blamed for its misuse, just as a set of tools cannot be held responsible for the poor construction of a building. It is the **users** of factor analysis who are well deserving of attack for its extensive misuse, abuse, and too often arbitrary "discovery" of factors. Used correctly, and with an appreciation of its limitations, factor analysis is a helpful tool in the arsenal of the social or natural scientist on par with any other statistical method so long as one does not prescribe powers to it for which it does not possess. We thus find that factor analysis has much merit to it if used judiciously in the hands of a cautious and conscientious data analyst. Should the technique not be used with this sense of care, we strongly believe it can misguide more than it can serve, and under these circumstances should not be used at all.

One domain where factor analysis has especially (and successfully) flourished is in test construction and item analysis. For an excellent account of the history of factor analysis, along with its philosophical foundations, the reader is strongly encouraged to refer to Mulaik (1987).

14.2 FACTOR ANALYSIS AT A GLANCE

To motivate our technical development of factor analysis, we consider first a brief overview of the end-result of a factor analysis performed on the Holzinger and Swineford data of 1939, where recall a subset of the data consists of tests of mental ability. These tests were subjected to a factor analysis, in which the following output was obtained:

	Factor1	Factor2
visual perception	0.354	0.376
cubes	0.232	0.219
lozenges	0.364	0.293
paragraph comprehension	0.866	0.112
sentence completion	0.794	0.205
word meaning	0.815	0.114
addition	0.126	0.624
counting dots		0.864
straight-curved capitals	0.288	0.635

We can see, for instance, that the bivariate correlation between visual perception and factor 1 is 0.354, the correlation between visual perception and factor 2 is 0.376, and so on for the remainder of the variables. Note especially that there exist high correlations between paragraph comprehension, sentence completion, and word meaning on factor 1, and relatively high correlations between addition, counting dots, and straight-curved capitals on factor 2. These correlations between variables and factors are called **loadings**. They are used in part to help name the hypothetical factors. Note carefully that factor 1 and factor 2 are not "observed" variables. They are, from a technical standpoint, linear combinations, and from a substantive standpoint, **potentially** hypothetical constructs extracted from the factor-analytic procedure. Observed variables that helped generate the factors are then correlated to the factors for the purpose of attempting to name these **latent** (or "unobserved") constructs.

In this chapter we survey some of the theory and application of this controversial technique, as well as highlight some of the more salient issues and concerns that lay at the forefront of the data-analytic tool. It is important to understand that of all the statistical tools surveyed in this book, factor analysis, and its close relative **structural equation modeling**, their **subjects** at least, are perhaps the most philosophically difficult to disentangle. Factor analysis, its theory and application, is a subject on which many books have been written. Even rotational methods for FA alone can take up half a book. In this chapter, we genuinely only scratch the surface to provide a basic overview and introduction to some of its features. For a thorough, authoritative and book-length study of factor analysis, consult Mulaik (1972).

14.3 EXPLORATORY VERSUS CONFIRMATORY FACTOR ANALYSIS

In this chapter we consider only **exploratory factor analysis**, leaving the study of **confirmatory factor analysis** to the following chapter where we include our study of it as a special case of **structural equation modeling**. The distinction between exploratory versus confirmatory factor analysis is, by its nature, quite fuzzy. As noted by Jöreskog, pioneer in the field of factor analysis:

> Most studies are to some extent both exploratory and confirmatory since they involve some variables of known and other variables of unknown composition. The former should be chosen with great care in order that as much information as possible about the latter may be extracted. (1978, p. 444)

Though we refrain here from delving too deeply into philosophical discussions of their differences, it is enough to define the exploratory model as factor analysis performed with **fewer a priori theoretical predictions** about what one will discover or uncover from the outset of the analysis. Needless to say, this definition is limited, since whenever one undertakes a factor analysis of any kind, the researcher usually arrives at the scene with at least some idea of what he or she will find. However, in a confirmatory factor analysis one usually makes many more explicit predictions about virtually every facet of the hypothesized model, which often includes hypothesizing correlated error terms, constraining select parameters to certain values, or even testing improvement in model fit by slight modifications of the tested model.

One might summarize this distinction to say that what differentiates EFA from CFA is that in the former, the specification of the model is usually quite elementary and "automatic," whereas in the later, the specification of the model, essentially every part of it, is required and expected before the model-fitting process even begins. On the technical side, the difference between EFA and CFA is that through the identification of proper constraints, **CFA can generate uniquely estimated factor loadings** (Rencher and Christensen, 2012). As we will discuss, loadings in EFA are typically not uniquely determined. We survey why this is the case later in this chapter.

14.4 THEORY OF FACTOR ANALYSIS: THE EXPLORATORY FACTOR-ANALYTIC MODEL

As we have undoubtedly come to see thus far in our study of statistical modeling, understanding statistical analysis is very much facilitated by an awareness that for each and every statistical procedure, there is an underlying hypothesized statistical model that we, as researchers, must implicitly or explicitly propose at the outset. Other than for simple descriptive statistics computed on samples, statistical modeling is the process of arriving at the data with an **a priori** imposed hypothetical structure, however simple or complex. This is true whether we are working in a binomial situation in which we have to hypothesize a value for p (probability of success) or in more complex modeling situations in which we are confronted with many more parameters. Whatever the model in each case, we impose the given model onto empirical data, and the extent to which it fits or "accounts" for the data, generally speaking, it is regarded as "well-fitting." The extent to which our model does not fit is a sign that our theory needs revision.

Be sure to note that whenever you test a model, you want **it** to be under test, not the **data**, otherwise there is little point to model-fitting at all. Sometimes researchers, upon learning that their model does not fit, blame the **data** for not being as it somehow "should" be. **Before any model-fitting occurs, you need to ensure that your data was collected in a scientifically standardized fashion, and have a clear understanding of the population to which you seek to generalize**. If your model then does not fit the data, your theory should be to blame, not the data.[1] Just imagine where physics would be if all along we decided our theories were correct and the atoms were wrong.

14.5 THE COMMON FACTOR-ANALYTIC MODEL

The model traditionally assumed in most exploratory factor-analytic work is the following, usually referred to as the **common factor-analytic model**:

$$\mathbf{x} = \boldsymbol{\mu} + \boldsymbol{\Lambda}\mathbf{f} + \boldsymbol{\varepsilon} \tag{14.1}$$

where $\mathbf{x}$ is a vector of random variables that are assumed observable (or "manifest"), $\boldsymbol{\mu}$ is a vector of means for the random variables in $\mathbf{x}$, $\boldsymbol{\Lambda}$ is a matrix of factor loadings, $\mathbf{f}$ is a vector of unobservable common factor random variables,[2] and $\boldsymbol{\varepsilon}$ is a vector of specific factors. These specific factors comprise of variation unexplained by $\boldsymbol{\mu} + \boldsymbol{\Lambda}\mathbf{f}$ for each observed variable in vector $\mathbf{x}$.

The model in (14.1) is customarily written in terms of mean deviations, that is,

$$\mathbf{x} - \boldsymbol{\mu} = \boldsymbol{\Lambda}\mathbf{f} + \boldsymbol{\varepsilon}$$

When we "unpack" the model for p variables and m factors, we find:

$$\mathbf{x} - \boldsymbol{\mu} = \boldsymbol{\Lambda}\mathbf{f} + \boldsymbol{\varepsilon}$$
$$x_1 - \mu_1 = l_{11}f_1 + l_{12}f_2 + \cdots + l_{1m}f_m + \varepsilon_1$$
$$x_2 - \mu_2 = l_{21}f_1 + l_{22}f_2 + \cdots + l_{2m}f_m + \varepsilon_2$$
$$\vdots$$
$$x_p - \mu_p = l_{p1}f_1 + l_{p2}f_2 + \cdots + l_{pm}f_m + \varepsilon_p$$

---

[1] Of course, should you discover after the fact that you collected data from the wrong population, then yes, in this sense, your data may be to "blame" for the model not fitting. I am not meaning to discount or disregard this possibility. What I am saying is that if you have collected data from your intended population and your hypothesized model does not fit, the most likely explanation for the nonfit is your model, not your data.

[2] In some models, we may wish to consider the vector as fixed instead of random. However, such a model, as noted by Anderson (2003), would imply that the individuals specifically sampled are of interest, instead of regarding them as a random sample from a wider population. See Anderson (2003, p. 571) for details.

where $x_1, x_2, ..., x_p$ are observed variables, $\mu_1, \mu_2..., \mu_p$ are the means of the p variables, $l_{11}, l_{21}, ..., l_{p1}$ are loadings for the p variables across $f_1, f_2..., f_m$ factors, and $\varepsilon_1, \varepsilon_2..., \varepsilon_p$ are errors associated with each observed variable $x_1, x_2, ..., x_p$.

Notice that the model of (14.1) is strikingly similar to the regression models discussed earlier in the book. Consider a side-by-side comparison of these models:

$$\mathbf{x} - \boldsymbol{\mu} = \boldsymbol{\Lambda}\mathbf{f} + \boldsymbol{\varepsilon} \quad \text{versus} \quad \mathbf{y} = \mathbf{x}\boldsymbol{\beta} + \boldsymbol{\varepsilon}$$

Consider for now only the right-hand sides of each model $\boldsymbol{\Lambda}\mathbf{f} + \boldsymbol{\varepsilon}$ versus $\mathbf{x}\boldsymbol{\beta} + \boldsymbol{\varepsilon}$. We see that for both models, observed vectors $\mathbf{x}$ and $\mathbf{y}$ are linear functions of observed estimated coefficients, $\boldsymbol{\Lambda}$ in the factor model, and $\boldsymbol{\beta}$ in the regression model. These weights are both applied to $\mathbf{f}$ and $\mathbf{x}$, respectively. But here is where the factor-analytic model and the regression model differ. Whereas in the regression model $\mathbf{x}$ is a vector of **observed** manifest variables, in the factor model, $\mathbf{f}$ is a vector of **unobserved latent variables**. That is, as the theory goes in EFA, there are no real, true, empirical variables contained in the vector $\mathbf{f}$ as there are in the vector $\mathbf{x}$ in the regression model. This is the key distinction between these two models. In the factor analysis model, we are essentially hypothesizing that $\mathbf{x}$ is a function, in part, of **unobserved variables**. In the regression model, other than for the error term (which can be regarded as an "unobservable" variable) we make no such assumption, instead hypothesizing that $\mathbf{y}$ is a function of observed variables (i.e., the predictor variables chosen by the researcher).

Both models also have a vector of errors, $\boldsymbol{\varepsilon}$, which in each case can be thought of as variation unaccounted for by the systematic portion of each model. In observed data, these will assume the name of **errors** in regression, whereas in factor analysis, these generally assume the name of **unique variances**. Some factor analysts further delineate unique variance into two components, one of **specific variance** and the other of **error variance**. The distinction is that specific variance is thought to be variance that is generally uncorrelated with other variables. In this regard, it is considered to be "true" variance, which is a reliable part of a variable that is found in no other variable. It is "particular" to the given observed variable. Error variance, on the other hand, is variance that is not necessarily unique to the given variable. While specific and error variances may indeed exist at a theoretical level, they can be inexorably difficult to distinguish in a pragmatic manner, and we at times use these terms interchangeably. Our position is that the error term in factor analysis can be more or less thought of as analogous (or at minimum, similar) to the error term in the regression model, at least in the attempt to conceptually "bridge" the two methodologies. Both essentially represent variation unaccounted for by the given systematic portion of the model, that being $\mathbf{x} - \boldsymbol{\mu} = \boldsymbol{\Lambda}\mathbf{f}$ in factor analysis and $\mathbf{y} = \mathbf{x}\boldsymbol{\beta}$ in regression.

A model is defined by the assumptions it makes. This is especially true of the factor-analytic model. We consider shortly the many assumptions made by the exploratory factor analysis model. We restrict our discussion for now to the **orthogonal** model, in which "orthogonal" in this context implies a requirement that estimated factors be **uncorrelated**, and hence each factor accounts for a distinct amount of covariance among observed variables. There is no "overlap" in the orthogonal model. This assumption can be relaxed somewhat, and later in this chapter we will briefly discuss factor models in which orthogonality among factors is not a requirement. We refer to such models as **oblique** factor models.

Before considering the assumptions, it is worth asking whether the assumptions we are about to discuss are necessarily correct when applied to empirical data. The answer is that for some assumptions, given the way in which the covariance or correlation matrix is analyzed and decomposed, we can be assured that these assumptions will hold. For instance, the fact that we are beginning with the orthogonal model and specifying it as such guarantees that the assumption of no covariance among factors will be satisfied. We are assured it is satisfied because we are parameterizing our model as such. In other cases however, assumptions are not guaranteed in this way, and we make them oftentimes out of **convenience**, or as a means to **identify**[3] the given model. The situation is not unlike that of assumption-making in regression models. We regularly assume, for instance, that errors are normally distributed, but we are required to perform residual analyses (as an imperfect approximation to the errors) in order to verify this assumption to learn if it is even plausible in practice. In other cases, the "assumption" of OLS regression that the sum of squared errors will be kept at a minimum value is not a true assumption at all but is more of a **constraint** imposed by the method of estimation. This is a condition imposed by the method. The assumption that errors are not related, on the other hand, though convenient, may not hold in practice after a further investigation into residual plots and such.

14.6 ASSUMPTIONS OF THE FACTOR-ANALYTIC MODEL

Perhaps more than any other statistical method, an understanding of the assumptions of factor analysis is vital. We now survey the most salient of these. In our discussion of each, we try as much as possible to "unpack" the assumption so that the reader becomes aware of what these assumptions actually mean in practice.

1. It is typically assumed that the mean of the latent vector $\mathbf{f}$ is equal to $\mathbf{0}$. That is, $E(\mathbf{f}) = \mathbf{0}$. Why is this assumed? It is assumed as such mostly for convenience, and since the actual vector $\mathbf{f}$ is latent and thus unobserved, it is especially easy to make this assumption (how easy it is to make assumptions about variables that do not exist!). This idea of imposing a constraint of sorts on a model may seem quite arbitrary within an EFA framework, but when we consider structural equation models in the following chapter, we will see that the act of constraining parameters to specific values is common practice, and the assumption that $E(\mathbf{f}) = \mathbf{0}$ in the exploratory model will no longer seem so bold an assumption.

2. In the orthogonal factor-analytic model, it is assumed that factors do not covary. That is, in a covariance matrix of estimated factors, we would expect pairwise covariances between factors to equal zero. And because we are still dealing with an unobserved random vector $\mathbf{f}$, it is convenient to assume that the variances of these derived factors will equal 1.0, $E(\mathbf{ff'}) = \mathbf{I}$. That is, the covariance matrix of factors we would expect to be an **identity matrix**. For instance, for a two-factor model:

$$\text{cov}(\mathbf{f}) = \begin{bmatrix} 1.0 & 0 \\ 0 & 1.0 \end{bmatrix}$$

The fixing of these variances in a factor model provides another assumption or constraint and hence a parameter we do not need to estimate. In an oblique factor model, one in which we allow the extracted factors to correlate, this assumption will no longer hold, and we will expect the off-diagonal elements to be **unequal** to zero. As a general reference to what we are talking about, note that in a multiple regression we do not constrain the covariances between observed variables to equal 0. We deal with such issues with measures such as the variance inflation factor (VIF) discussed in Chapter 7. However, when we standardize variables, we do in a manner of speaking "constrain" the variances to equal 1.0 by the simple fact that standardizing a variable gives it a

---

[3] Identification will be discussed in more detail in the following chapter on structural equation modeling. For now, it is enough to know that a parameter is deemed "identified" if there is enough information in the data to estimate it. If all parameters in a model are identified, the model is said to be identified.

mean of 0 and a variance of 1.0. Again, it is pedagogically useful to appreciate the parallels and differences between the factor analysis model and the classic multiple regression model in terms of what assumptions are made (and not made) in each. **Statistical modeling, in general, is very much the imposing and freeing of constraints on parameters before we overlay it onto empirical data**.

3. It is assumed that the errors in the factor-analytic model have a mean equal to **0**. Since the matrix of errors, $\boldsymbol{\varepsilon}$, is a random vector, the assumption is that $E(\boldsymbol{\varepsilon}) = \mathbf{0}$. This assumption is not unlike that of the assumption for the error term in the classic multiple regression model.

4. We assume that the errors, the specific variances in $\boldsymbol{\varepsilon}$, do not covary. That is, the unexplained variation for each observed variable has nothing in common with the unexplained variation for another observed variable. More generally, in matrix terms, we assume the covariance matrix of errors to have zeros everywhere except the main diagonal (i.e., a diagonal matrix) where are located the variances of the errors. We can state this assumption more concisely to say that the expectation of the cross-product of errors is equal to a matrix $\boldsymbol{\psi}$ (e.g., $\boldsymbol{\psi} = \begin{bmatrix} \psi_1 & 0 \\ 0 & \psi_2 \end{bmatrix}$).

   Does this seem like an unreasonable assumption? Is it unreasonable to assume that specific variances (contained in the errors) are unrelated? It may be, which is one reason why **confirmatory factor analysis** is useful in contexts in which we do wish to hypothesize a covariance among errors. It would seem reasonable in many substantive scenarios that the unexplained variance in a given observed variable is related to the unexplained variance in a second observed variable. In the following chapter on structural equation models, we discuss how we can allow the matrix $\boldsymbol{\psi}$ to be something other than a diagonal matrix, thus freeing up errors to covary. Such decisions are usually heavily steeped in the theory of the researcher.

5. Finally, it is assumed that the estimated factor and its specific (or "unique") variance do not covary. That is, $E(\boldsymbol{\varepsilon}\mathbf{f}') = \mathbf{0}$. The analogous assumption in regression (though incorporating an "observed" variable rather than a latent one) is that $E(\boldsymbol{\varepsilon}\mathbf{x}') = \mathbf{0}$, that is, the covariance between the error term and the predictor variable is equal to 0. As we have seen throughout this book, this is a common assumption made in linear models, and hence it is a relatively convenient one to also make in the case of the factor-analytic model.

We also expect that observed variables are linearly related in the common factor model. More specifically, we assume **linearity in the common factors**. If there is little to no correlation among observed variables, then it should be obvious that performing a factor analysis will not make much sense, analogous to computing a variance on a set of data consisting of constants would likewise make little sense. And if nonlinear relationships are hypothesized among variables instead of linear ones, then options such as **nonlinear factor analysis** are also available (e.g., see Yalcin and Amemiya (2001)), though not discussed in this book.

14.7   WHY MODEL ASSUMPTIONS ARE IMPORTANT

In light of the aforementioned assumptions and constraints, recall again why explicitly stating model assumptions is important and relevant, not only in the factor-analytic model but generally in all statistical models. Recall that we use a model to provide a rational and coherent theoretical representation of the observed data. In identifying the model, we need to make certain basic assumptions about the structure we are imposing, otherwise, it is virtually impossible to fit **any** model to data since it is impossible to even begin the process. **We must start somewhere**, and that "somewhere" takes the form of a statistical model along with its implied and imposed assumptions and constraints. This idea is not unlike that of a psychologist coming to the careful analysis of human behavior with a priori "background" assumptions already in place, whether emanating from psychoanalytic, behaviorist, or humanistic traditions. The point is that when we seek to explain or model, or otherwise provide a **narrative** to empirical data, we usually have to come to the data with at least some theoretical stance, even if minimal. Even in the age of "Big Data" and the popular mantra "let the data speak for themselves," the choice of algorithm and other facets of the search are still typically selected in advance of seeking out patterns in data. **The eyes one uses to see often plays a significant role in what is seen**. Not only is this psychologically true, but it is paramount to statistical modeling. In both cases, we are fitting narratives to data.

Indeed, in the early stages of the COVID-19 outbreak in 2020, it was often said that predicting (i.e., "modeling") the path of the disease depended very much on the assumptions one incorporated into the modeling. Now, these assumptions can often be well-grounded (e.g., the behavior of similar viruses in the past, and their known trajectories), but the point is that to set up virtually any statistical model, one must "set the stage" for it by incorporating certain baseline assumptions or beliefs. These sometimes take the form of constraints imposed on the analysis, and other times come from expert judgment. Either way, whatever model you are using (whether factor analysis or other), you can be sure there are "starting points" to it in some form.

14.8 THE FACTOR MODEL AS AN IMPLICATION FOR THE COVARIANCE MATRIX Σ

The assumptions that we made for the factor-analytic model imply a structure of the covariance matrix among observed variables. We follow Johnson and Wichern (2007) in showing how the assumptions made in (14.1) imply the observed covariance matrix to be a function of factor loadings plus an error variance containing uniqueness. That is, (14.1) implies:

$$\sum = \Lambda\Lambda' + \psi$$

To see why this is true, consider once more the factor model thus stated:

$$\mathbf{x} - \mu = \Lambda\mathbf{f} + \varepsilon$$

When we "square" $\mathbf{x} - \mu$, or in matrix terms, multiply $\mathbf{x} - \mu$ by its transpose $(\mathbf{x} - \mu)'$ to get the covariance matrix, we perform the same operation on the right-hand side, and get

$$\mathbf{x} - \mu = \Lambda\mathbf{f} + \varepsilon$$
$$(\mathbf{x} - \mu)(\mathbf{x} - \mu)' = (\Lambda\mathbf{f} + \varepsilon)(\Lambda\mathbf{f} + \varepsilon)' \tag{14.2}$$

When we expand the right-hand side of (14.2), we obtain

$$(\Lambda\mathbf{f} + \varepsilon)(\Lambda\mathbf{f} + \varepsilon)' = \Lambda\mathbf{f}(\Lambda\mathbf{f}') + \Lambda\mathbf{f}\varepsilon' + \varepsilon(\Lambda\mathbf{f})' + \varepsilon\varepsilon'$$

Hence, what all the above means simply is that we can write $(\mathbf{x} - \mu)(\mathbf{x} - \mu)'$ as equivalent to $\Lambda\mathbf{f}(\Lambda\mathbf{f}') + \Lambda\mathbf{f}\varepsilon' + \varepsilon(\Lambda\mathbf{f})' + \varepsilon\varepsilon'$.

When we take the relevant expectations, we get:

$$E(\mathbf{x} - \mu)(\mathbf{x} - \mu)' = \Lambda E(\mathbf{f}\mathbf{f}')\Lambda' + E(\varepsilon\mathbf{f}')\Lambda' + \Lambda E(\mathbf{f}\varepsilon') + E(\varepsilon\varepsilon') \tag{14.3}$$

Now, recall the assumptions we began with in developing the factor model. Since $E(\mathbf{f}) = \mathbf{0}$, and $E(\mathbf{ff'}) = \mathbf{I}$, (14.3) reduces to:

$$
\begin{aligned}
E(\mathbf{x} - \boldsymbol{\mu})(\mathbf{x} - \boldsymbol{\mu})' &= \boldsymbol{\Lambda} E(\mathbf{ff'})\boldsymbol{\Lambda}' + E(\boldsymbol{\varepsilon}\mathbf{f}')\boldsymbol{\Lambda}' + \boldsymbol{\Lambda} E(\mathbf{f}\boldsymbol{\varepsilon}') + E(\boldsymbol{\varepsilon}\boldsymbol{\varepsilon}') \\
&= \boldsymbol{\Lambda}(\mathbf{I})\boldsymbol{\Lambda}' + (\mathbf{0})\boldsymbol{\Lambda}' + \boldsymbol{\Lambda}(\mathbf{0}) + E(\boldsymbol{\varepsilon}\boldsymbol{\varepsilon}') \\
&= \boldsymbol{\Lambda}\boldsymbol{\Lambda}' + 0 + 0 + \boldsymbol{\psi} \\
&= \boldsymbol{\Lambda}\boldsymbol{\Lambda}' + \boldsymbol{\psi}
\end{aligned}
\tag{14.4}
$$

Equation (14.4) is very important. In words, it says that the covariance matrix $\boldsymbol{\Sigma}$ is equal to "squared" factor loadings ($\boldsymbol{\Lambda}\boldsymbol{\Lambda}'$) plus error associated with each observed variable ($\boldsymbol{\psi}$). Recall that since $\boldsymbol{\psi}$ is a diagonal matrix containing only error variances along the main diagonal and covariances equal to 0 everywhere else, the factor model under consideration does not allow error terms to covary.

14.9 AGAIN, WHY IS $\boldsymbol{\Sigma} = \boldsymbol{\Lambda}\boldsymbol{\Lambda}' + \boldsymbol{\psi}$ SO IMPORTANT A RESULT?

Result (14.4) is of significance in defining the factor-analytic model because it reveals that one can essentially **reproduce** or **regenerate** the covariance matrix by knowledge of the factor loadings and specific variances. Hence, if we can estimate loadings for $\boldsymbol{\Lambda}$, and likewise consider the unique variances in $\boldsymbol{\psi}$, we can, in theory, account for the makeup of the covariance matrix $\boldsymbol{\Sigma}$. Just as the goal of building a model in regression is to **regenerate** existing data through specification of a suitable regression model, in factor analysis, the goal is to specify a model such that the covariance matrix $\boldsymbol{\Sigma}$ is reconstructed. **The trick in factor analysis is, of course, to specify a matrix $\boldsymbol{\Lambda}$ that is best suited in reproducing $\boldsymbol{\Sigma}$.** Should the matrix $\boldsymbol{\Lambda}$ contain a single factor? Should it contain two factors? Three? If a factor-analytic solution is to be deemed somewhat sensible and reasonable, determining the appropriate or correct number of factors is one of the most significant challenges faced by the analyst. Technically, however, the challenge reduces down to reproducing the covariance matrix $\boldsymbol{\Sigma}$. Theoretically and substantively, the job is that much more difficult, and the factor analyst must also justify the reproduction of the covariance matrix based on a **meaningful**, **substantive** solution, not only one that adequately reproduces $\boldsymbol{\Sigma}$.

14.10 THE MAJOR CRITIQUE AGAINST FACTOR ANALYSIS: INDETERMINACY AND THE NONUNIQUENESS OF SOLUTIONS

The primary criticism against factor analysis since its inception is that the derived factor loadings in the matrix $\boldsymbol{\Lambda}$ are not **unique**. Recall that for a solution to be unique implies there to be a **single** solution to the equation. The estimation of $\boldsymbol{\Lambda}$ actually does, in a manner of speaking, provide a unique solution, **but only up an orthogonal matrix**. The implication of this is that **regardless of the solution obtained in the estimation process, we are able to quite freely rotate the factor solution yet still provide the same reproduction of the covariance matrix as with the original solution**. As Mulaik (1972) summarizes:

> Thus factor analysis, at least in the traditional sense, is concerned with the problem of analyzing a variable into components … from a strictly mathematical point of view, there is an infinite number of potential sets of components which might be determined for a given set of variables. Because of this mathematical ambiguity, the history of factor analysis has been filled with controversy over what are the appropriate components into which to analyze a set of variables. These controversies are by no means over today. (pp. 95–96)

The two components Mulaik speaks of are the common factors and unique factors previously discussed. Mulaik's words were written in 1972 but are just as relevant today.

In what follows, we work through the technical argument to show that the factor solution is only unique only up to an orthogonal matrix. Recall what the orthogonality of a matrix implies. If matrix $\mathbf{T}$ is an orthogonal matrix, it implies that $\mathbf{TT}' = \mathbf{T}'\mathbf{T} = \mathbf{I}$, where $\mathbf{I}$ is the identity matrix with values of 1 along the main diagonal. Consider now introducing this matrix into the factor-analytic model we have been working with. We post-multiply the loading matrix $\mathbf{\Lambda}$ by $\mathbf{TT}'$, and get:

$$\mathbf{x} - \mathbf{\mu} = \mathbf{\Lambda f} + \mathbf{\varepsilon}$$
$$= \mathbf{\Lambda TT'f} + \mathbf{\varepsilon}$$

Notice that we haven't "changed" the model per se, since if $\mathbf{T}$ is orthogonal, then it must be true that $\mathbf{TT}'$ is equal to $\mathbf{I}$, and so we could have just as easily written the above model as:

$$\mathbf{x} - \mathbf{\mu} = \mathbf{\Lambda TT'f} + \mathbf{\varepsilon}$$
$$= \mathbf{\Lambda If} + \mathbf{\varepsilon}$$

So, what is the big deal? What is the problem then with introducing the orthogonal matrix $\mathbf{T}$? The problem is that $\mathbf{\Lambda TT}'\mathbf{f} + \mathbf{\varepsilon}$ in $\mathbf{x} - \mathbf{\mu} = \mathbf{\Lambda TT}'\mathbf{f} + \mathbf{\varepsilon}$ can be written as $(\mathbf{\Lambda T})(\mathbf{T}'\mathbf{f}) + \mathbf{\varepsilon}$. That is, we may consider our original matrix of loadings $\mathbf{\Lambda}$, subject to the transformation $\mathbf{T}$ to be equal to $\mathbf{\Lambda T}$, and the vector of factors $\mathbf{f}$ to be equal to $(\mathbf{T}'\mathbf{f})$. What this means is that through an orthogonal transformation matrix $\mathbf{T}$, we are able to define a **new** loading matrix $\mathbf{\Lambda T}$ and a new factor vector $(\mathbf{T}'\mathbf{f})$.

The question now becomes, does this new definition of $\mathbf{\Lambda}$ and $\mathbf{f}$ change things in terms of the model? We check this through taking the relevant expectations once more. We take the expectation of $(\mathbf{T}'\mathbf{f})$, and since $E(\mathbf{f}) = \mathbf{0}$, then it stands that $E(\mathbf{T}'\mathbf{f}) = \mathbf{0}$. That is, **the expectation of the transformed factor vector is identical to that of the untransformed vector.** In both cases, before transformation and after, the expectation is equal to $\mathbf{0}$. In plain English, this means that **both the original factor vector and the transformed factor vector have the same mean**.

The next question that needs to be addressed is whether the expectation of the covariance of $(\mathbf{T}'\mathbf{f})$ changes as a result of its transformation by $\mathbf{T}$. With the transformation, the covariance term becomes $\mathbf{T}'\,\text{Cov}(\mathbf{f})\mathbf{T}$. Since $\mathbf{T}$ is an orthogonal matrix, and by definition $\mathbf{TT}' = \mathbf{I}$, this means that the covariance of the transformed factor matrix $(\mathbf{T}'\mathbf{f})$ remains $\mathbf{I}$. Hence, the point is that the transformation of $\mathbf{f}$ by the orthogonal matrix $\mathbf{T}$ does not alter the expectation for the covariance matrix of $\mathbf{f}$. Again, in plain English, this means that both the original covariance matrix and the transformed covariance matrix have expectations equal to the identity matrix $\mathbf{I}$.

In summary then, the process of transforming the factor-analytic solution by an orthogonal matrix does not change its expectations, and as such, whether we use $\mathbf{\Lambda f} + \mathbf{\varepsilon}$ or $\mathbf{\Lambda TT}'\mathbf{f} + \mathbf{\varepsilon}$, we are able to reproduce the same covariance matrix. Be sure you understand this result at least in concept because it is extremely important and has wide implications for factor analysis and the kinds of conclusions one draws from it. To summarize, the "big deal" is this: **both the original loadings and the transformed loadings generate the same covariance matrix**, and hence, if one were to apply an orthogonal transformation to loadings estimated in a factor analysis, one could still just as well reproduce the original covariance matrix. The question then naturally arises:

Which loadings are the "correct" loadings, the ones originally derived, or the ones derived through an orthogonal rotation?

The answer to this question does not arise from mathematical analysis, derivation or deduction. And hence, we have come to the primary critique charged against factor analysis. The answer must come from the **subjective consideration regarding which loadings make the most sense to the researcher**. It is the researcher who must select the solution he or she **prefers**. This is often the reason cited for the disdain for factor analysis by some, since some perceive it as a statistical methodology one can "adjust" until one arrives at a solution that agrees with one's factor-analytic hypothesis. To these critics, factor analysis is little more than the wishful projection of the scientist.

14.11   HAS YOUR FACTOR ANALYSIS BEEN SUCCESSFUL?

If a factor analysis does not provide a meaningful structure to your data, it may very well be because **there is no structure to your data**. It does not necessarily mean something went "wrong" in your factor analysis. It is the stream of consciousness of the self-absorbed narcissistic researcher who subjects his data to factor analysis, not to find the common structure hypothesized to exist (even after rotation after rotation), and concludes that something must have gone "amiss" with the statistical technique or that "weird" data was obtained. What is much more likely is that there are no underlying factors that theoretically gave rise to observed correlations. In other words, **it did not work**. Conservatism in scientific discovery is unfortunately not publishable, but it makes one a better scientist.

Though significance tests for factor loadings do exist, one should usually not rely on such things to establish whether a factor analysis has been successful. Oftentimes researchers adopt a sequential testing strategy and test a number of hypothetical factors until a given solution "meets their expectations." It is reasonable to suspect that error probabilities will accumulate under this process. However, these error probabilities are generally difficult to get a handle on (Anderson, 2003), which further complicates the process. More recently, computationally intensive approaches have been proposed for factor selection. For example, Chen, Huang, and Tu (2010) proposed a new approach based on unbiased risk estimation, which has shown to recover factors better than some traditional approaches.

Even if an apparent structure does reveal itself from the analysis, the question of the size of loadings on estimated factors remains. How large is "large enough" to consider a variable loading on a given factor? Though there have been rough rules of thumb-type guidelines advanced on this issue, it is not a hard science, and some flexibility in decision rules must be granted. Generally, if a loading is greater than 0.3 or 0.4, it is probably worth looking at. However, one can also envision a situation where if theoretically meaningful, a loading of 0.2 should also make its way into the determination of an overall factor solution. The point is that there is no absolute cutoff, and efforts to establish such cutoffs from a **substantive** point of view are, at best, imperfect.

Our advice for deciding whether a factor solution is meaningful or not parallels that of Johnson and Wichern (2007), which they call the "WOW" criterion: "If, while scrutinizing the factor analysis, the investigator can shout '**Wow, I understand these factors,' the application is deemed successful**" (p. 526). If you sit with your factor solution for weeks splitting hairs, it may be time to concede the absence of a solution and move on. This can be said to be true of any statistical model that you run. Too often researchers advance models that account for such small proportions of variance under the presumption that because the model explains **any** variance, it must somehow be worthwhile. Random data can also account for variance, which is something researchers should always keep in mind. As is true of most statistical findings, factor-analytic results and solutions should be cross-validated if possible to help confirm (or disconfirm) the existence of a solution. **Jackknife** or **bootstrap validation** (see Lattin, Carroll, and Green (2003) for a brief discussion) may be used to help confirm results found in one sample to another.

14.12   ESTIMATION OF PARAMETERS IN EXPLORATORY FACTOR ANALYSIS

Recall that in any statistical model, one must estimate parameters. For instance, recall in the simple linear regression model,

$$y_i = \alpha + \beta x_i + \varepsilon$$

we estimated parameters α and β with estimators a and b, respectively. These parameters in regression were usually obtained using ordinary least-squares regression or maximum likelihood (ML). In factor analysis, we likewise need to estimate parameters, which consist of loadings and communalities.

There are numerous options available for estimating parameters in factor analysis. We survey only two, that of principal factor (or **principal axis factoring**) and **maximum likelihood,** since they are the most widely used. For details on their differences, see Winter and Dodou (2012). For literature on other methods of estimation, the interested reader is encouraged to consult Mulaik (1972). For a comparison of results based on different extraction methods, see Tabachnick and Fidell (2007, pp. 633–635).

After conducting a factor analysis, the researcher may wish to estimate **factor scores** based on the extracted solution. Factor scores are in principal somewhat analogous (at least in concept) to predicted values in regression analysis, only that factor scores must be estimated from unobservable variables rather than observable ones as in regression. There are numerous methods for estimating factor scores. The so-called **regression method** is among the more popular choices. For a discussion and numerical example of the procedure, see Johnson and Wichern (2007, pp. 516–518). Factor scores are easily estimated in R and SPSS.

14.13   PRINCIPAL FACTOR

A common method of estimating factors is that of **principal factor**, which as mentioned also goes by the name of **principal axis factoring**. This is a least-squares estimation technique that makes few distributional assumptions (Gnanadesikan, 1997), and accomplishes its job by minimizing the unweighted least squares (ULS) or ordinary least squares (OLS) of the residual matrix (Winter and Dodou, 2012). In this method, initial communalities are estimated and inputted into the diagonal of the covariance or correlation matrix. For example, consider the following correlation matrix on five observed variables. An initial visual inspection reveals that variables 1 and 3 are highly correlated (0.96) along with variables 2 and 5 (0.85) as well as 4 and 5 (0.79). Other bivariate correlations are quite small, for instance 1 and 2 (0.02) as well as 1 and 5 (0.01).

$$\begin{bmatrix} 1.00 & 0.02 & 0.96 & 0.42 & 0.01 \\ 0.02 & 1.00 & 0.13 & 0.71 & 0.85 \\ 0.96 & 0.13 & 1.00 & 0.50 & 0.11 \\ 0.42 & 0.71 & 0.50 & 1.00 & 0.79 \\ 0.01 & 0.85 & 0.11 & 0.79 & 1.00 \end{bmatrix}$$

Note that 1.00 appears in the diagonal of the correlation matrix, since these are essentially correlations of variables with themselves. Recall that in principal components analysis, values of 1.00 appeared along the diagonal of the correlation matrix to indicate that each variable contributed an even 1 unit of variance to the problem. Since factor analysis is typically interested in analyzing **commonality** rather than **total** variance, these numbers along the main diagonal will be different in EFA than they were in PCA. Principal axis factoring replaces these 1's with initial communality estimates before estimating relevant parameters. A popular estimator for these diagonal elements in the correlation matrix is

$$\hat{h}_i^2 = R_i^2 = 1 - \frac{1}{r^{ii}}$$

where $\hat{h}_i^2$ is the estimated communality for variable i, R_i^2 is the coefficient of determination for regressing variable i on all other observed variables, and r^{ii} is the ith diagonal element of the inverse of the correlation matrix $\mathbf{R}$ (Rencher and Christensen, 2012).

If you consider for a moment what the coefficient of determination is telling us, it makes good sense to name it "communality." The communality for a given variable is estimated using all other variables in the data set to predict the variable under consideration. In this way, we are interested in learning how much variance for the given variable is shared with other variables in the set. Be sure to note that in this way of estimating communalities, a variable's estimated communality will be a function of **what other variables are under consideration in the factor analysis**. Hence, as can be said for virtually all multivariate models, the results one obtains are contingent upon what other variables are simultaneously considered in the model. This is why, for instance, we emphasized a careful interpretation of partial regression coefficients in Chapter 8 as distinct from zero-order coefficients. **All models are context-dependent**.

If the covariance matrix $\mathbf{S}$ is analyzed instead of the correlation matrix, an appropriate estimator is

$$\hat{h}_i^2 = ss_{ii}R_i^2$$

where ss_{ii} is now the ith diagonal element of $\mathbf{S}$ (Rencher and Christensen, 2012).

Other ways of estimating communalities are also available. The key point to retain, however, is the concept that these numbers are being estimated as a "starting point" to conducting the factor analysis. The estimated communalities, in a sense, **initiate** the factor-analytic procedure.

14.14 MAXIMUM LIKELIHOOD

One of the most common methods of estimating parameters in the context of factor analysis is that of maximum-likelihood estimation. The log-likelihood function for a multivariate normal distribution is given by:

$$\ln L(\boldsymbol{\mu}, \boldsymbol{\Sigma}) = -np \ln \sqrt{2\pi} - \frac{1}{2} n \ln |\boldsymbol{\Sigma}| - \frac{1}{2}(n-1)\mathrm{tr}(\boldsymbol{\Sigma}^{-1}\mathbf{S}) - \frac{1}{2}n(\bar{\mathbf{y}} - \boldsymbol{\mu})'\boldsymbol{\Sigma}^{-1}(\bar{\mathbf{y}} - \boldsymbol{\mu})$$

Substituting the maximum likelihood estimator $\bar{\mathbf{y}}$ for $\boldsymbol{\mu}$, along with substituting the assumed structure of $\boldsymbol{\Lambda}\boldsymbol{\Lambda}' + \boldsymbol{\Psi}$ for the population covariance matrix $\boldsymbol{\Sigma}$, the log-likelihood function reduces to the following:

$$\ln L(\boldsymbol{\Lambda}, \boldsymbol{\Psi}) = -np \ln \sqrt{2\pi} - \frac{1}{2} n \ln |\boldsymbol{\Lambda}\boldsymbol{\Lambda}' + \boldsymbol{\Psi}| - \frac{1}{2}(n-1)\mathrm{tr}\left[(\boldsymbol{\Lambda}\boldsymbol{\Lambda}' + \boldsymbol{\Psi})^{-1}\mathbf{S}\right]$$

Maximizing the log-likelihood is equivalent to minimizing the following fit function:

$$F_{ML} = \ln \mid \sum(\boldsymbol{\theta}) \mid + \text{tr}\left(\mathbf{S}\sum{}^{-1}(\boldsymbol{\theta})\right) - \ln \mid \mathbf{S} \mid -p \qquad (14.5)$$

where p is the number of observed variables. As will be discussed in the following chapter, the fit function of (14.5) is very **general** in that it specifically does not define the exact nature of the covariance matrix of model parameters in $\sum(\boldsymbol{\theta})$. In the orthogonal factor model under consideration in this chapter, we can rewrite F_{ML} to be:

$$\begin{aligned} F_{ML} &= \ln \mid \sum(\boldsymbol{\theta}) \mid + \text{tr}\left(\mathbf{S}\sum{}^{-1}(\boldsymbol{\theta})\right) - \ln \mid \mathbf{S} \mid -p \\ &= \ln \mid \boldsymbol{\Lambda}\boldsymbol{\Lambda}' + \boldsymbol{\Psi} \mid + \text{tr}\left(\mathbf{S}\mid\boldsymbol{\Lambda}\boldsymbol{\Lambda}' + \boldsymbol{\Psi}\mid^{-1}\right) - \ln \mid \mathbf{S} \mid -p \end{aligned}$$

Notice that similar to how we did for the log-likelihood function, in place of $\sum(\boldsymbol{\theta})$, we now have inserted the nature of the covariance matrix for the estimation problem of the factor model under consideration, that of $\boldsymbol{\Lambda}\boldsymbol{\Lambda}' + \boldsymbol{\Psi}$. In Chapter 15 on structural equation and latent variable models, we most often use the general form $\sum(\boldsymbol{\theta})$ in our notation, even though for any particular problem the covariance matrix may be different. For instance, in some models, we will impose restrictions and constraints on the covariance matrix that in other models are not present.

As will also be discussed more fully in Chapter 15, the fitting function will equal zero when $\sum(\boldsymbol{\theta}) = \mathbf{S}$. To the extent that $\sum(\boldsymbol{\theta}) \neq \mathbf{S}$, the value of the fitting function will be greater than zero (i.e., $F_{ML} > 0$).

Both the minimizing of F_{ML} and the maximizing of the log-likelihood amount to the same thing under the assumption of normality (Everitt, 2007, p. 69) and will yield identical parameter estimates. The minimization or maximization is achieved through numerical iteration methods of which the study is a field onto itself and is a rather specialized area in mathematical statistics and numerical algebra. For a readable introduction to the field of numerical algebra, consult Trefethen and Bau (1997). The reader interested further in the computational iterative details on maximum-likelihood estimation as it relates to factor analysis is encouraged to consult Johnson and Wichern (2007, pp. 527–530) and Lawley and Maxwell (1971). Historically significant papers can be found in Lawley (1958) and Jöreskog (1967, 1969).

As mentioned, there are several other methods of estimation in factor analysis. These include **image analysis**, **alpha factoring**, the **centroid method**, among others. Regarding which method to use, it is probably best to heed the recommendation of Rencher (1998):

> **The various methods of estimating factor loadings will generally yield different solutions. However, for samples from populations in which the basic factor analysis model is valid, most methods yield similar loadings, at least after rotation. Thus if the researcher has data to which a factor analysis model can be successfully fit with large communalities, the choice of technique is not important. To a lesser extent, if the number of variables is large, the various methods will also yield similar results, regardless of the adequacy of fit.** (pp. 385–386)

14.15   THE CONCEPTS (AND CRITICISMS) OF FACTOR ROTATION

Oftentimes in factor analysis, interpretation of the extracted solution can prove difficult in terms of whether or not it defines **true** factors. **Factor rotation** is a procedure used for the purpose of facilitating interpretation of derived factors and in an effort to achieve what Thurstone called **simple structure**, best described by Mulaik (1972):

> Thus if in a factor analysis of n variables r common factors were obtained, Thurstone deemed the factor solution ideal when each variable required fewer than r factors to account for its common variance. By the same token, when it came to interpreting the common factors by noting the observed variables associated with each respective factor, parsimony of interpretation could be obtained when each factor was associated with only a few of the observed variables. A factor solution displaying these properties of parsimony was designated a **simple-structure factor solution**. (pp. 218–219)

Rotation of factors is one way in which a factor analyst attempts to ameliorate the solution so as to approximate simple structure as closely as possible. However, as with many decisions in the factor-analytic process, choosing the **correct** rotation can likewise come down to a subjective choice. As noted by Jöreskog (1967):

> Though Λ^* and Λ are equivalent from the mathematical point of view, they may not be so from the psychological point of view. The problem of choosing one particular psychologically meaningful Λ out of the infinite set $\{\Lambda T^{-1}\}$ has been called the **problem of rotation**, although the **problem of transformation** would be a better term, since it includes also the transformation to oblique factors, in which case the transformation matrix T is not orthogonal and hence does not represent only a rotation. (p. 166)

Rotations are generally divided into two broad categories, **orthogonal** rotations and **oblique** rotations. Orthogonal rotations transform factors to new axes but keep them at a 90° angle. That is, orthogonal rotations do not allow factors to correlate. Two popular methods of orthogonal rotation include **varimax** and **quartimax**, both of which will be discussed shortly. Oblique rotations allow factors to correlate, and hence overlap will exist between factors. Examples of oblique rotations include **promax**, **oblimin**, **quartimin**, and **covarimin**. We do not detail oblique methods in this chapter, though if one has a general understanding of what rotation means for the orthogonal case, it is easy enough to generalize this understanding, at least conceptually, to nonorthogonal rotations.

Before surveying the idea of factor rotation, we must once more recall what is meant by an **orthogonal transformation**. We briefly discussed the concept earlier in our discussion of the nonuniqueness of solutions. A square matrix $\mathbf{T}$ is orthogonal if the following condition holds:

$$\mathbf{TT}' = \mathbf{T}'\mathbf{T} = \mathbf{I}$$

Recall that the idea of orthogonality was one that allowed a variety of solutions (up to an orthogonal matrix) to be found for a given covariance matrix when considering the fundamental equation for factor analysis, $\sum = \mathbf{\Lambda\Lambda}' + \mathbf{\psi}$. When we introduced an orthogonal matrix into the original factor model, we found we were still able to produce the same covariance matrix as before. This is why factor rotation in factor analysis is allowable and mathematically "permissible," because despite the rotation, the generation of $\sum$ does not change. We now detail to some extent how rotation works in factor analysis. After estimating loadings in the factor-analytic routine, we wish to rotate the coordinate axes by an angle that will typically **maximize** or **minimize** some quantity.

Consider now the rotation of the loading matrix $\mathbf{\Lambda}$. In rotating the loading matrix $\mathbf{\Lambda}$, we multiply this matrix by $\mathbf{O}$, which is an orthogonal matrix. Upon multiplying by this matrix, we obtain $\mathbf{\Lambda}_R$, which is the rotated loadings matrix. As an example, consider a rotation for a simple (and arithmetically easy) two-dimensional structure:

$$\mathbf{\Lambda}_R = \mathbf{\Lambda O}$$

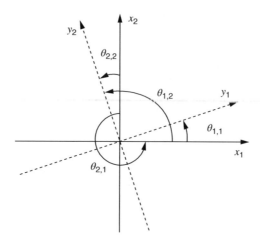

FIGURE 14.1 Orthogonal rotation in two dimensions where the new axes are indicated by dashed lines. Source: Reproduced with permission from Abdi (2003).

To rotate the matrix $\mathbf{\Lambda}$ clockwise, we multiply $\mathbf{\Lambda}$ by coordinates of rotation in $\mathbf{O}_1$:

$$\mathbf{O}_1 = \begin{bmatrix} \cos\phi & \sin\phi \\ -\sin\phi & \cos\phi \end{bmatrix}$$

To rotate the matrix $\mathbf{\Lambda}$ counterclockwise, we multiply $\mathbf{\Lambda}$ by $\mathbf{O}_2$:

$$\mathbf{O}_2 = \begin{bmatrix} \cos\phi & -\sin\phi \\ \sin\phi & \cos\phi \end{bmatrix}$$

It is of course very useful to depict a factor rotation geometrically, as done in Figure 14.1. As can be seen from the figure, the rotation, in this case, is simply transforming the original axes of x_1 and x_2 to y_1 and y_2.

14.16 VARIMAX AND QUARTIMAX ROTATION

There have been numerous rotations for factor analysis proposed. We here only briefly survey two of the most popular ones, **varimax** and **quartimax**. Both are **orthogonal** rotations. These are the ones that the reader will likely perform when conducting a factor analysis. An understanding of these makes it easier to then grasp other rotations one might encounter. We do not discuss nonorthogonal, so-called **oblique** rotations in this book. For a relatively detailed and thorough account of rotational methods, see Mulaik (1972, pp. 249–292).

The goal of the varimax rotation (Kaiser, 1958) is to maximize "within factor" variance of squared loadings for extracted factors. What will make the variance of squared loadings large? All else equal, they will be large when there is much dispersion among the loadings. In other words, loadings that are either high (near |1.0|) or small (near 0.0) in a correlation matrix are preferred over "mediocre" loadings since loadings approaching the upper and lower limits of |1.0| and 0.0 serve to maximize the variability of loadings. The varimax rotation essentially **drives large loadings to be even larger, and small loadings to be even smaller**. This makes sense, since the minimum variance of factor loadings occurs when all loadings are identical, in which case, the variability in loadings is equal to zero. Hence, maximizing high loadings and minimizing smaller ones "disrupts" this "minimal variance" pattern. The disruption of this homogeneity often helps in the interpretation of the factor-analytic solution.

The question then is: What kind of transformation matrix should the original solution be rotated against so that the variance of factor loadings is maximized? Kaiser (1958) found that the transformation matrix should be chosen such that the following quantity is maximized:

$$V = \sum_{i=1}^{p} \left(l_{ij}^2 - \overline{l_j^2} \right)^2$$

where l_{ij}^2 are the squared loadings i to p for a given factor and $\overline{l_j^2} = \sum_{i=1}^{p} \dfrac{l_{ij}^2}{p}$ is the average squared loading across the p variables. The varimax technique works to emphasize **within factor** distribution of loadings to high versus low on each observed variable. V is usually maximized using iterative techniques and hence its maximization is dependent upon a computing algorithm for computational ease and efficiency.

Whereas the varimax rotation attempts to maximize the variance of squared loadings by focusing on the **columns** of factor loadings (i.e., the columns representing each factor), the **quartimax** rotation focuses on the **rows** of the loading matrix, seeking to maximize the variance of loadings **across** factors instead of within factors. The quartimax criterion is given by:

$$Q = \sum_{j=1}^{m} \left(l_{ij}^2 - \overline{l_j^2} \right)^2$$

where now the sum is **across** factors $j = 1$ through m. Different, yet somewhat analogous to V in what it accomplishes, Q drives loadings across factors toward either 0 or |1| in a correlation matrix instead of within factors.

14.17 SHOULD FACTORS BE ROTATED? IS THAT NOT CHEATING?

The question of whether factors should be rotated has existed since the advent of factor analysis. Sometimes students, upon first learning of factor rotation, mistakenly believe that it "cheapens" the value of one's data, or at minimum, constitutes "dishonest" data analysis. However, this view is misguided. One must realize that the original axes upon which the factor solution was derived are not, in themselves, "special" in any way, at least not when it comes to the **substantive** interpretation of the factor solution. Since factors are hypothetical structures which do not have "true" axes, the fact that we are rotating axes simply represents a different way of interpreting these unknown structures.

Once we agree that rotation is acceptable for factor analysis, the next question that usually presents itself is whether orthogonal rotations are more valid compared to oblique rotations. Recall that in an oblique rotation, factors are allowed to covary, whereas in orthogonal rotations, they are not. There are differing views on this topic, but in the end, it largely comes down to subjective opinion regarding the nature of what the factors **should** be or how they are idealized by the analyst. Indeed, according to Thurstone (1947), "It seems just as unnecessary to require that mental traits shall be uncorrelated in the general population as to require that height and weight be uncorrelated in the general population." In our view then, allowing factors to correlate is acceptable so long as it coincides with one's theory. Indeed, in **confirmatory factor analysis**, the topic of the following chapter, researchers quite often estimate correlations between factors, and hence it would seem acceptable to allow them to likewise correlate in EFA models.

14.18   SAMPLE SIZE FOR FACTOR ANALYSIS

Factor analysis is inherently a large sample technique, meaning that one usually requires a reasonably strong sample size in order to have any "trust" in the findings of the analysis. How large is large enough? How small is too small? Historically, writers on factor analysis have offered many "rules of thumb" recommendations regarding adequate sample size. Recommendations are not consistent across the board, and according to MacCallum et al. (1999), they are not equally applicable from study to study:

> Our theoretical framework and results show clearly that common rules of thumb regarding sample size in factor analysis are not valid or useful. The minimum level of N, or the minimum $N{:}p$ ratio, needed to assure good recovery of population factors is not constant across studies but rather is dependent on some aspects of the variables and design in a given study. Most importantly, level of communality plays a critical role. (p. 96)

What MacCallum et al. (1999), generally found was that when communalities are high (e.g., greater than 0.6), and factors are well determined (and the computational algorithm converges), samples even smaller than 100 may be enough to conduct the analysis. However, as communalities get smaller, in general, a greater sample size is required. For instance, with communalities ranging in the neighborhood of 0.5, sample size in the general range of 100–200 is recommended. If communalities are quite low for estimating a small number of factors with very few indicators for each factor (e.g., 3–4), sample sizes in the neighborhood of 300 or more are preferred. For cases in which communalities are very low and factors are poorly determined, sample sizes of 500 are generally required.

Overall, since factor analysis is a large-sample technique (t-tests, in comparison, are a small sample technique), in general, the greater the sample size, the more confidence one can have in the stability of the factor solution. **When performing factor analysis then, more times than not researchers should aim for large sample sizes.**

14.19   PRINCIPAL COMPONENTS ANALYSIS VERSUS FACTOR ANALYSIS: TWO KEY DIFFERENCES

As mentioned at the outset of this chapter, factor analysis is a method distinct from that of principal components analysis. We summarize two primary distinctions we have already touched upon throughout our discussion up to now.

14.19.1   Hypothesized Model and Underlying Theoretical Assumptions

This first distinction is perhaps the most important one when comparing EFA with PCA. In EFA, a definitive **model** is subsumed, whereas in PCA, no such model is ever hypothesized. In PCA, the analyst is not so often seeking to usually "uncover" underlying latent structures to his or her data, even in challenging pursuits as exhibited in the psychological sciences. Rather, she is simply wanting to know if her data can be expressed in a simpler form while still accounting for most of its variance. In this way, principal components analysis can be considered simply an **empirical transformation** of observed data. We are not fitting a theoretical model to empirical observations, nor do we require model assumptions such as multivariate normality in the typical components analysis (Timm, 2002) especially if formal inference is not the goal. For a discussion of large sample inferences where normality assumptions are beneficial, consult Johnson and Wichern (2007, p. 456). As is true of most statistical methods, however, PCA has been shown to be sensitive to outliers.

Factor analysis, on the other hand, is by its very nature much more theory-driven, with the focus on uncovering hypothesized **lurking variables** that subsume observed correlations or covariances. What this technically translates into is an emphasis on different parts of the covariance or correlation matrix, as summarized by Jolliffe (2002):

> Both factor analysis and PCA can be thought of as trying to represent some aspect of the covariance matrix Σ (or correlation matrix) as well as possible, but PCA concentrates on the diagonal elements, whereas in factor analysis the interest is in the off-diagonal elements. (p. 158)

The above materializes, as Jolliffe notes, into the fact that an extracted principal component can result from a **single variable** being independent of remaining variables in the sample. That is, if one variable "does the job," then it is possible in a PCA that it be designated as a component to account for some of the variance in the sample. In a factor analysis (i.e., versions of which aren't simply glorified components analysis), since commonality between variables is the focus, the given factor must be "determined" by at least two or more observed variables. This is an example of how the emphasis in factor analysis is on **covariance** whereas the emphasis in principal components is on **variance**. This distinction is critical in understanding the difference between PCA and EFA and is one that clearly distinguishes the two methodologies.

While it is also true that analysts will sometimes seek to name derived components in PCA, it is not a typical assumption of the analysis. Factor analysis, or perhaps more appropriately the **tradition of factor analysis**, usually encourages the researcher to make at least some sense out of derived factors, to assign them names or meaning, otherwise, the procedure is usually considered "unsuccessful" in the sense that underlying latent variables were not discovered. Of course, the mechanics of EFA do not care whether you name or not name linear combinations of factors. So in this sense, this characteristic has nothing to do with the actual procedure, but rather more to do with the substantive use, application, and tradition of the factor-analytic method. When a researcher is performing a factor analysis, it is usually not simply for data reduction, otherwise, they would likely be performing a PCA instead.

14.19.2   Solutions Are Not Invariant in Factor Analysis

A second key and very important difference between EFA and PCA is the contingency issue of loadings on the number of derived components or factors. In a principal components analysis, whether the analyst decides to derive or keep two or three components, for instance, will not have an effect on the loadings for such components. In factor analysis, however, whether the analyst decides to extract two or three factors will typically have an effect on the loadings. For those in opposition to the factor-analytic method, this issue provides them with much ammunition. **What it means is that the very nature of a given factor usually depends on how many other factors were extracted along with it**. At a philosophical level, this is a real problem for factor analysis. "I found a two-factor solution" says the analyst, is never the "full story." As always with statistical models, context matters, and perhaps nowhere is this more true than in factor analysis.

The situation is similar, though by no means identical, to the effect one predictor may have on the estimated partial regression weight of a second predictor in a multiple regression model in that the model is specified by the inclusion of **all** predictors, with the interpretation of each predictor **contingent** upon other predictors in the model. Likewise in EFA, the solution of one factor is contingent upon the solution to others in the model. This leads some to look upon EFA with great suspicion. However, so long as researchers are aware of this issue, and **communicate it to the audiences to which they present their results**, it should not be regarded as an "obstacle" to factor analysis any more than the interpretation of predictors in a wider multiple regression model should be considered an obstacle. In both cases, and as emphasized throughout this book with regard to multivariate methods in general, estimated coefficients are contingent upon the model tested and should never be interpreted independent of this context.

As long as one is also aware of such limitations when interpreting solutions to factor analysis, the "subjectivity" of solution-selection should not be a barrier to using EFA. So long as one has some understanding of the tool they are using, the tool may prove quite useful. Of course, if one is making nonsensical conclusions about the existence of factors that make little substantive sense (e.g., all one-factor solutions could be named "jello" and the fitted models would not object!), then this **is** reason for proper critique, but that critique is more appropriately targeted toward the (mis)user than the technique. Factor analysis itself is quite innocent. It is the **users** who usually have blood on their hands.

14.20   PRINCIPAL FACTOR IN SPSS: PRINCIPAL AXIS FACTORING

We now demonstrate exploratory factor analysis in SPSS. We later feature an example in R. Recall that the method of principal axis factoring in SPSS, or "PAF," is a method of common factor analysis that uses the squared multiple correlation coefficient as its estimate of communality for each variable. For our example, we factor analyze `cormatrix`, first featured in Chapter 13 on which we conducted a PCA:

```
MATRIX DATA VARIABLES=ROWTYPE_ T1 T2 T3 T4 T5 T6 T7 T8.
BEGIN DATA
N 1000 1000 1000 1000 1000 1000 1000 1000
CORR 1.00000
CORR .343     1.00000
CORR .505      .203   1.00000
CORR .308      .400    .398  1.00000
CORR .693      .187    .303   .205  1.00000
CORR .208      .108    .277   .487   .200  1.00000
CORR .400      .306    .206   .305   .311   .432  1.00000
CORR .455      .385    .167   .465   .485   .310    .365 1.00000
END DATA.
```

Recall that the columns are designated by T1 through T8, and there are 1000 observations per variable.

Just as we did for the PCA example of the previous chapter, we can have SPSS reproduce the correlation matrix of observed variables by appending CORRELATION to the PRINT command:

```
/PRINT CORRELATION
```

		T1	T2	T3	T4	T5	T6	T7	T8
					Correlation Matrix				
Correlation	T1	1.000	0.343	0.505	0.308	0.693	0.208	0.400	0.455
	T2	0.343	1.000	0.203	0.400	0.187	0.108	0.386	0.385
	T3	0.505	0.203	1.000	0.398	0.303	0.277	0.286	0.167
	T4	0.308	0.400	0.398	1.000	0.205	0.487	0.385	0.465
	T5	0.693	0.187	0.303	0.205	1.000	0.200	0.311	0.485
	T6	0.208	0.108	0.277	0.487	0.200	1.000	0.432	0.310
	T7	0.400	0.386	0.286	0.385	0.311	0.432	1.000	0.365
	T8	0.455	0.385	0.167	0.465	0.485	0.310	0.365	1.000

The statistical significance of the above correlations can also be requested by `/PRINT SIG`. We resist printing the statistical significance of these correlations, however, mostly for the reason that they are not required for us to push on with the factor analysis. In EFA, we are not interested in making inferential statements about any particular pairwise correlations. Recall as well that in the previous chapter on PCA, we designated `EXTRACTION = PC` to request a components analysis on this data. By specifying `EXTRACTION = PAF` now, we are requesting a **principal axis factoring** solution in which we extract two factors (i.e., `/CRITERIA FACTORS(2)`):

```
FACTOR MATRIX=IN(CORR=*)
/PRINT= INITIAL EXTRACTION
/CRITERIA FACTORS(2)
/EXTRACTION=PAF
/METHOD=CORRELATION.
```

SPSS provides us with the following **initial** and **extraction** communalities:

	Communalities	
	Initial	Extraction
T1	0.619	0.916
T2	0.311	0.236
T3	0.361	0.256
T4	0.461	0.678
T5	0.535	0.551
T6	0.349	0.340
T7	0.355	0.382
T8	0.437	0.398

Extraction method: principal axis factoring.

How were the initial communalities obtained? The initial communality of 0.619 was computed by regressing T1 (i.e., T1 is the response variable in this case) on variables T2 through T8. Likewise, the initial communality of 0.311 for T2 was computed by regressing T2 on variables T1, T3, through to T8. Recall what the extracted communalities represent. They are a measure of how much the given observed variable has in **common** with the derived factors after the factor-analytic routine has done its job. For instance, note that the communality for T1 rose from 0.619 to 0.916, which in turn suggests that the given variable is, overall, relatively correlated with one or more of the derived factors. In other words, T1 appears to be a key contributor to the derived factors. The balance of the communality of 0.916 is equal to $1 - 0.916 = 0.084$, which is the **unique variance** associated with T1. That is, this is the proportion of variance in the observed variable **unaccounted** for by the derived factors.

SPSS next provides us with a breakdown of the eigenvalues extracted by the factor analysis. We note that these are identical to the eigenvalues extracted by the PCA of Chapter 13, and which also sum to a total of 8 corresponding to the number of variables in the EFA. They are named "initial" eigenvalues because they are the eigenvalues extracted under a principal components model (i.e., with 1.0's in the main diagonal) rather than a common factor analysis model in which communalities have been estimated. We see that the first factor (component) accounts for 43.09% of the variance. This number was obtained by the ratio 3.447/8. Additional initial eigenvalues are reported in decreasing value.

		Total Variance Explained				
		Initial Eigenvalues		Extraction Sums of Squared Loadings		
Factor	Total	% of Variance	Cumulative %	Total	% of Variance	Cumulative %
1	3.447	43.088	43.088	2.974	37.171	37.171
2	1.157	14.465	57.554	0.784	9.799	46.970
3	0.944	11.796	69.349			
4	0.819	10.237	79.587			
5	0.658	8.226	87.813			
6	0.390	4.873	92.686			
7	0.336	4.201	96.887			
8	0.249	3.113	100.000			

Extraction method: principal axis factoring.

The right-hand side of the above table contains the **extraction sums of squared loadings**, and reports figures based on **common variance** rather than **total variance**. SPSS also only reports the extraction sums of squared loadings for a two-factor solution, since this is what we requested. These then are the values of interest to us, since they are based on the estimated communalities inserted in the main diagonal of the correlation matrix. Formally speaking, these are not actual "true" eigenvalues, but it does not hurt to see them as such in comparison to the eigenvalues extracted in the PCA, so long as one realizes they are different things. One can informally and crudely think of them as the eigenvalues based on the EFA solution in contrast to those based on the PCA solution. We see that the first factor, yielding an "eigenvalue" of 2.974, accounts for 37.17% of the variance, while the second factor, with a corresponding "eigenvalue" of 0.784, accounts for 9.80% of the variance. As discussed in this chapter, the choice of how many factors to retain must wholly be a decision made by the researcher given requisite substantive interpretation of the factor solution. Both the PCA and EFA results suggest keeping one or two factors. The first two factors, considered together, account for a total percentage of the variance of 46.97% of the original variance in the set of variables.

Next in SPSS's output is the **factor matrix**, also known as the **loading matrix**. These weights correspond to how much an observed variable loads on, or correlates with, a given factor. These weights are thought to reveal the "structure" of the given factor. From this matrix, we can also compute the extraction communalities previously discussed. For "T1," the extraction communality of 0.916 is obtained as $(0.819)^2 + (-0.496)^2$, and as mentioned, represents the proportion of variance contributed by the observed variable across estimated factors or, equally, how well the factor structure explains the observed variable.

	Factor Matrix[a]	
	Factor	
	1	2
T1	0.819	-0.496
T2	0.472	0.115
T3	0.506	-0.013
T4	0.666	0.485
T5	0.633	-0.389
T6	0.480	0.331
T7	0.596	0.163
T8	0.629	0.040

Extraction method: principal axis factoring.
[a]Two factors extracted. Thirty iterations required.

As an aid in defining the substantive existence and nature of a factor, we generally look for observed variables having relatively high-magnitude correlations with the given factor. We can see that for our data, most variables load at least moderately well on the first factor. Based on the factor matrix in conjunction with the aforementioned eigenvalues, the EFA seems to be dominated by a one-factor solution. Indeed, a scree plot appears to confirm this, though depending on the substantive context, an argument can be made for a two-factor solution as well. It should be noted that the eigenvalues that follow are based on the PCA solution, not the EFA sums of squared loadings. Nonetheless, they usually help the investigator in determining or estimating the number of factors to extract, at least in a preliminary sense:

```
/PLOT EIGEN
```

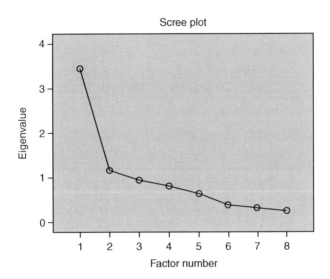

We now rotate the solution via both varimax and quartimax:

```
/PRINT = ROTATION
/ROTATION VARIMAX (/ROTATION QUARTIMAX)
```

SPSS first provides us with a brief summary and overview of how the variance was redistributed (shown here for varimax rotation only):

	Total Variance Explained		
	Rotation Sums of Squared Loadings		
Factor	Total	% of Variance	Cumulative %
1	1.891	23.638	23.638
2	1.867	23.332	46.970

Extraction method: principal axis factoring.

We note that the proportion of variance explained by the two factors is still the same as in the unrotated solution (i.e., 46.97%); however, the sums of squared loadings have been recomputed to reflect a

TABLE 14.1   Estimated Loadings for Varimax and Quartimax Rotations of Cormatrix

	Varimax Rotated Factor Matrix			Quartimax Rotated Factor Matrix	
	Factor			Factor	
	1	2		1	2
T1	0.931	0.223	T1	0.604	0.742
T2	0.255	0.413	T2	0.483	0.051
T3	0.369	0.346	T3	0.472	0.182
T4	0.132	0.813	T4	0.790	−0.233
T5	0.723	0.168	T5	0.465	0.579
T6	0.108	0.573	T6	0.564	−0.151
T7	0.309	0.535	T7	0.616	0.046
T8	0.419	0.471	T8	0.606	0.174

change of axes. Whereas in the unrotated solution they were 2.974 and 0.784 for factors 1 and 2, respectively, they are now 1.891 and 1.867. Notice that the sum of these eigenvalues has not changed, in that they still sum to 3.758 in each case. Only the distribution of variance has changed among the two factors.

In Table 14.1, factor matrices for varimax and quartimax rotations are given. Recall that there is no magic formula regarding which rotation should be interpreted. Rotations that make most **substantive sense** are those typically adopted by the researcher.

SPSS also provides us with the transformation matrices for each rotation:

	Factor Transformation Matrix	
Factor	1	2
1	0.711	0.704
2	−0.704	0.711

Extraction method: principal axis factoring.
Rotation method: varimax with Kaiser normalization.

	Factor Transformation Matrix	
Factor	1	2
1	0.942	0.335
2	0.335	−0.942

Extraction method: principal axis factoring.
Rotation method: quartimax with Kaiser normalization.

Using these transformation matrices, we can demonstrate, for instance, the computation of the varimax rotation in R. We first generate the column vectors of the original, unrotated factor matrix:

```
> f1 <- c(.819, .472, .506, .666, .633, .480, .596, .629)
> f2 <- c(-.496, .115, -.013, .485, -.389, .331, .163, .040)
> f.sol <- cbind(f1, f2)
> f.sol
```

```
          f1      f2
[1,]   0.817  -0.493
[2,]   0.472   0.114
[3,]   0.506  -0.013
[4,]   0.666   0.485
[5,]   0.633  -0.392
[6,]   0.480   0.331
[7,]   0.596   0.163
[8,]   0.630   0.039
```

The vectors f1 and f2 are the respective loadings for the two extracted unrotated factors. We named the object f.sol simply to bind f1 and f2 into columns. We now generate the varimax transformation matrix:

```
> tm.1 <- c(.711, -.704)
> tm.2 <- c(.704, .711)
> t.matrix <- cbind(tm.1, tm.2)
> t.matrix
          tm1    tm2
[1,]    0.711  0.704
[2,]   -0.704  0.711
```

Finally, we post-multiply the two-factor solution by the varimax transformation matrix:

```
> varimax <- f.sol%*%t.matrix
> varimax
            tm1        tm2
[1,]  0.931493   0.223920
[2,]  0.254632   0.414053
[3,]  0.368918   0.346981
[4,]  0.132086   0.813699
[5,]  0.723919   0.169053
[6,]  0.108256   0.573261
[7,]  0.309004   0.535477
[8,]  0.419059   0.471256
```

Note that the above "manually" computed transformed factor loadings are equal to those generated by SPSS for the varimax solution. The quartimax rotation can likewise be demonstrated. Demonstrating quantities numerically like this goes a ways to "de-mystifying" what they represent computationally and is a great way of practicing computations in R or other software.

Finally, we can request the reproduced matrix of correlations. These are the correlations **implied** by the specification of the factor-analytic model. Recall that from a technical vantage point, the goal of EFA is to regenerate the observed correlation matrix by way of the factor solution. SPSS also provides us with the residuals of the reproduced correlations, which are merely the differences between the empirical correlations observed in our data and the correlations implied by the model. A perfectly fitting model would have a residual matrix filled with zeros. Inspection of the matrix can help us identify on a substantive level correlations for which the model did well at regenerating. For instance, the two-factor model in this case did a nice job at reproducing the correlation of T1 and T4, since it yields a quite small residual of only 0.004.

```
/PRINT REPR
```

					Reproduced Correlations				
		T1	T2	T3	T4	T5	T6	T7	T8
Reproduced	T1	0.916[a]	0.330	0.420	0.304	0.711	0.229	0.407	0.495
correlation	T2	0.330	0.236[a]	0.237	0.370	0.254	0.265	0.300	0.302
	T3	0.420	0.237	0.256[a]	0.330	0.325	0.239	0.299	0.318
	T4	0.304	0.370	0.330	0.678[a]	0.232	0.480	0.476	0.438
	T5	0.711	0.254	0.325	0.232	0.551[a]	0.175	0.313	0.383
	T6	0.229	0.265	0.239	0.480	0.175	0.340[a]	0.340	0.316
	T7	0.407	0.300	0.299	0.476	0.313	0.340	0.382[a]	0.382
	T8	0.495	0.302	0.318	0.438	0.383	0.316	0.382	0.398[a]
Residual[b]	T1		0.013	0.085	0.004	−0.018	−0.021	−0.007	−0.040
	T2	0.013		−0.034	0.030	−0.067	−0.157	0.086	0.083
	T3	0.085	−0.034		0.068	−0.022	0.038	−0.013	−0.151
	T4	0.004	0.030	0.068		−0.027	0.007	−0.091	0.027
	T5	−0.018	−0.067	−0.022	−0.027		0.025	−0.002	0.102
	T6	−0.021	−0.157	0.038	0.007	0.025		0.092	−0.006
	T7	−0.007	0.086	−0.013	−0.091	−0.002	0.092		−0.017
	T8	−0.040	0.083	−0.151	0.027	0.102	−0.006	−0.017	

Extraction method: principal axis factoring.
[a]Reproduced communalities.
[b]Residuals are computed between observed and reproduced correlations. There are 10 (35.0%) nonredundant residuals with absolute values greater than 0.05.

14.21   BARTLETT TEST OF SPHERICITY AND KAISER–MEYER–OLKIN MEASURE OF SAMPLING ADEQUACY (MSA)

As you might imagine, one must have at least some initial correlation among observed variables for factor analysis to have any hope of providing a sensible solution. But how much correlation is enough? Ordinarily, a visual inspection of the correlation matrix provides us with enough detail on the sizes of correlations to make an informal decision as to whether or not it is worth proceeding. If correlations are quite small (e.g., if most of them are less than 0.10 to 0.20, for instance), then conducting the analysis may not be worthwhile (though there is certainly no reason to not try the procedure and see what happens, you definitely will not "break" the software). Typically, we would prefer relatively sizeable correlations to justify pushing forward with the analysis.

The **Bartlett test of sphericity** (Bartlett, 1950, 1954) evaluates the null hypothesis that the correlation matrix is an **identity matrix**, which recall implies there to be values of "1" along the main diagonal and zeros everywhere else. Bartlett's test is given by

$$\chi^2 = -\left[(n-1) - \frac{(2p+5)}{6}\right] \ln |\mathbf{R}|$$

where p is the number of variables, n is the number of observations, and $\ln |\mathbf{R}|$ is the natural logarithm (i.e., $\log_e$) of the determinant of the correlation matrix. The most relevant component of the test is $\ln |\mathbf{R}|$. For a constant value of $-\left[(n-1) - \frac{(2p+5)}{6}\right]$, we note that what will make the value for Bartlett increase or decrease is entirely a function of $|\mathbf{R}|$, which can be regarded as a measure of **generalized**

variance. A rejection of the null hypothesis suggests that overall and across the board, pairwise correlations are not equal to zero. The test, however, is very much a function of sample size, and therefore a rejection of the null hypothesis should not be taken too seriously under most circumstances. According to Tabachnick and Fidell (2007), the test should ordinarily only be interpreted if there are relatively few cases per variable (e.g., 5–10).

For the analysis of `cormatrix`, we obtain:

```
/PRINT= INITIAL EXTRACTION KMO
```

KMO and Bartlett's Test		
Kaiser–Meyer–Olkin Measure of Sampling Adequacy		0.741
Bartlett's test of sphericity	Approx. Chi-square	2702.770
	df	28
	Sig.	0.000

SPSS first reports the **Kaiser–Meyer–Olkin** test, which is a ratio of the sum of squared correlations to that of the sum of squared correlations and squared partial correlations (Tabachnick and Fidell, 2007). Values generally exceeding 0.7–0.8 are preferred, although so long as values are not too low (e.g., 0.6 or lower), it is typically not cause for concern. Bartlett's test of sphericity is statistically significant, suggesting that the correlation matrix is not an identity matrix, though as just discussed, this is hardly surprising in this case since sample size is that of 1000 for `cormatrix`. To demonstrate the influence of sample size on this test, we rerun the analysis with the exact same correlation matrix but with sample size equal to 10 for each variable. We obtain:

KMO and Bartlett's Test		
Kaiser–Meyer–Olkin Measure of Sampling Adequacy.		0.741
Bartlett's test of sphericity	Approx. Chi-square	14.932
	df	28
	Sig.	0.979

We note that with a sample size of $n = 10$, KMO remained the same, although Bartlett's test of sphericity is no longer statistically significant due to using such a small sample size. Hence, as with all significance tests, one must be cautious when drawing conclusions as a function of Bartlett's p-value.

We can also request the determinant of the correlation matrix in SPSS:

```
/PRINT= DET
```

Correlation Matrix[a]

[a]Determinant = 0.066

A determinant unequal to zero indicates the matrix is not **singular**. Had the correlation matrix been singular, SPSS would have provided an error message and halted the factor analysis. Singularity can occur when two vectors of a matrix (variables, in this case) are perfect linear combinations of one another. See the Appendix for further details.

14.22 FACTOR ANALYSIS IN R: HOLZINGER AND SWINEFORD (1939)

We demonstrate a factor analysis in R using the Holzinger and Swineford data. Recall that we featured the output to this analysis at the very outset of this chapter. We now come full circle and perform the factor analysis to obtain those results. Recall that this classic data set consists of mental ability tests of seventh- and eighth-grade children from two different schools. The data here is a subset of the original Holzinger and Swineford data featuring 15 variables:

- id is an identifier
- sex
- ageyr (age in years)
- agemo (age in months)
- school (school attended by the child)
- grade
- x1–x9 (nine tests of mental ability; x1 = visual perception, x2 = cubes, x3 = lozenges, x4 = paragraph comprehension, x5 = sentence completion, x6 = word meaning, x7 = addition, x8 = counting dots, x9 = straight-curved capitals.)

We use the `factanal` function in R to perform maximum-likelihood factor analysis on psychological tests x1–x9. These tests are stored in the form of a correlation matrix based on a total of 145 observations in the object `Holzinger.9`:

```
> library(psych)
> Holzinger.9
```

	vis_perc	cubes	lozenges	par comp	sen comp	wordmean	addition
vis_perc	1.00000	0.325800	0.448640	0.34163	0.30910	0.31713	0.104190
cubes	0.32580	1.000000	0.417010	0.22800	0.15948	0.19465	0.066362
lozenges	0.44864	0.417010	1.000000	0.32795	0.28685	0.34727	0.074638
par_comp	0.34163	0.228000	0.327950	1.00000	0.71861	0.71447	0.208850
sen_comp	0.30910	0.159480	0.286850	0.71861	1.00000	0.68528	0.253860
wordmean	0.31713	0.194650	0.347270	0.71447	0.68528	1.00000	0.178660
addition	0.10419	0.066362	0.074638	0.20885	0.25386	0.17866	1.000000
count_dot	0.30760	0.167960	0.238570	0.10381	0.19784	0.12114	0.587060
s_c_caps	0.48683	0.247860	0.372580	0.31444	0.35560	0.27177	0.418310

	count_dot	s_c_caps
vis_perc	0.30760	0.48683
cubes	0.16796	0.24786
lozenges	0.23857	0.37258
par_comp	0.10381	0.31444
sen_comp	0.19784	0.35560
wordmean	0.12114	0.27177
addition	0.58706	0.41831
count_dot	1.00000	0.52835
s_c_caps	0.52835	1.00000

We request a two-factor solution on these data:

```
> fa <- factanal(covmat = Holzinger.9, factors = 2, n.obs = 145, rotation =
"varimax")
```

```
Call:
factanal(factors = 2, covmat = Holzinger.9, n.obs = 145)

Uniquenesses:
 vis_perc    cubes  lozenges  par_comp  sen_comp  wordmean  addition count_dot
    0.733    0.899     0.781     0.237     0.327     0.323     0.595     0.253
  s_c_caps
     0.514

Loadings:
           Factor1 Factor2
vis_perc   0.354   0.376
cubes      0.232   0.219
lozenges   0.364   0.293
par_comp   0.866   0.112
sen_comp   0.794   0.205
wordmean   0.815   0.114
addition   0.126   0.624
count_dot          0.864
s_c_caps   0.288   0.635

               Factor1 Factor2
SS loadings      2.455   1.882
Proportion Var   0.273   0.209
Cumulative Var   0.273   0.482

Test of the hypothesis that 2 factors are sufficient.
The chi square statistic is 61.7 on 19 degrees of freedom.
The p-value is 2.08e-06
```

The "Uniquenesses" that begin the output are the specific variances corresponding to each variable across both factors. For instance, for visual perception (vis_perc), the sum of squared estimated factor loadings is equal to $(0.354)^2 + (0.376)^2 = 0.267$, which when we subtract from 1, we obtain the specific variance of 0.733. Other values for "Uniqueness" are computed in an analogous manner.

Regarding the actual factor solution, based on the loadings for each factor, it would appear that the first factor is composed of variables **paragraph comprehension**, **sentence completion,** and **word meaning**, which are all verbal tasks. Hence, we might name the first factor by **verbal ability** or some-such. The second factor appears to be "made up of" addition, counting dots, and straight-curved capitals, which are all quantitative-like or analytic-related tasks. Hence, we might name the second factor by **quantitative ability** or similar. The loading for counting dots under the first factor is suppressed because it is very small and thus negligible.

14.23 CLUSTER ANALYSIS

We conclude this chapter with a brief and rather cursory survey of the statistical method of **cluster analysis**. Cluster analysis is a method with the aim of grouping cases or individuals that are in some sense, **similar**. Using a measure of similarity, cluster analysis groups cases into mutually exclusive sets. Once this partitioning of cases into sets is complete, identifying or otherwise naming these clusters often is a priority for the cluster analyst.

Why conclude this chapter with a discussion of cluster analysis? While cluster analysis is a technique distinct from that of factor analysis, it does share with it some parallels. At their core, both methods are concerned with generating groups based on an index of similarity. In traditional EFA, as we have seen, it is customarily the case where variables are grouped into factors. The index of similarity in EFA is that of **covariance** or **correlation**. In cluster analysis, **cases** are the typical objects of classification on which similarity may be defined and conceptualized in many different ways. Having made this distinction, we could, in theory, also "cluster analyze" variables in addition to cases. For our purposes, however, we assume clustering procedures to operate on cases rather than variables. Indeed, most applications of cluster analysis are concerned with the clustering of cases. In all that follows, we assume to be working with continuous data. For a discussion of cluster analysis applied to dichotomous data, see Finch (2005).

Generating groupings is rather easy. Generating **good** groupings is much more difficult. It is **how** these groupings should be made that is the topic of research in the cluster analysis field. As a problem of combinatorics, it is well known (by those who specialize in this area) that the number of ways of partitioning n cases into g clusters is given by:

$$N(n, g) = \frac{1}{g!} \sum_{k=1}^{g} \binom{g}{k} (-1)^{g-k} k^n$$

where n is the number of cases and g is the number of clusters (Anderberg, 1973; Hastie, Tibshirani, and Friedman, 2009). The number can also be approximated by $\frac{g^n}{g!}$. This number can get unreasonably large very fast, and hence the challenge of cluster analysis algorithms is to be able to shrink the problem to a manageable size. As Johnson and Wichern (2007) note, "Even fast computers are easily overwhelmed by the typically large number of cases, so one must settle for **algorithms** that search for good, but not necessarily the best, groupings" (p. 672). Ward (1963) summarized this idea perfectly:

> Situations often arise in which it is desirable to cluster large numbers of objects, symbols, or persons into smaller numbers of mutually exclusive groups, each having members that are as much alike as possible. Grouping in this manner makes it easier to consider and understand relations in large collections; hence it often increases efficiency of management. Grouping, however, ordinarily results in some loss of information that may be quantified in a "value-reflecting" number. (Ward, 1963, p. 236)

14.24 WHAT IS CLUSTER ANALYSIS? THE BIG PICTURE

A simple way to conceptualize cluster analysis, even if somewhat crudely, is to consider the following swarm of data points:

Suppose these points represent that of 1000 human beings. Theoretically, on each observation is associated an infinite number of characteristics, some measured, some unmeasured. For instance, some of the possible characteristics on these individuals include height, age, gender, temperament, personality, motivation, brain chemistry, etc. We could indeed go on to list an infinite number of characteristics. The question cluster analysis asks about this undefined, messy swarm, is the following:

Can any of the measured characteristics in this swarm be useful in establishing a group (or "cluster") structure of any kind?

For instance, if we considered individuals' heights in the swarm, do any **patterns of cases** emerge? Perhaps we might find a pattern of shorter people versus taller people. Likewise, if we considered gender, perhaps we might discover a pattern emerge of males versus females. As an example, consider the hypothetical data of Table 14.2 of heights and weights across four persons.

What cluster analysis will generally try to do with the data in Table 14.2 is use variables height and weight and search for **similarity** among cases. For example, consider a plot of height and weight in Figure 14.2.

What do we notice? Even in such a simple plot of four individuals, already a group structure seems to be emerging, as evidenced in Figure 14.2b. Cluster analysis essentially looks at the data in Table 14.2 (and corresponding plot in Figure 14.2) and asks the question: **Which points are closest to one another, and can this closeness be used to potentially define different "groups" or "clusters" of observations?** Notice that even if cluster analysis is able to identify groups, it does not purport in any way to be able to identify **what** those groups are, apart from actually fitting nicely into the given cluster structure. We thus ask the question: **Why are Mark and Mary similar, yet both different from Bob and Julie, who are themselves similar to each other?**

TABLE 14.2   Fictional Data for Simple Cluster Analysis

Person	Height	Weight (lb)
Mary	4 ft 2 in.	120
Bob	5 ft 2 in.	190
Julie	5 ft 8 in.	180
Mark	4 ft 3 in.	130

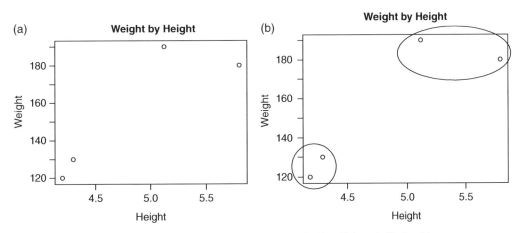

FIGURE 14.2   Plot of height and weight (a). Identifying similarity (b).

Of course, there are literally an infinite number of reasons why the data in Figure 14.2 might group the way it does. Individuals are complex and have many characteristics. However, can we theorize on at least one of those reasons? It may very well be that Mark and Mary are in the group called "children" and Bob and Julie belong to a cluster called "adults." Indeed, if we happened to have these individuals' ages at our disposal, it would not take long to identify this as a **covariate** to the group separation. At its core, this is how cluster analysis generally works. In the machine learning field, cluster analysis is often referred to as one of the "unsupervised" learning techniques, since we are not supplying the problem with a predefined grouping structure as we do in ANOVA or discriminant analysis, for instance. And though cluster analysis is often spoken of as **discovering natural groupings**, it must be emphasized that groupings are only as "natural" as we designate them to be. For instance, it is also quite possible that Mary and Mark eat more candy than do Bob and Julie, but since age is a more "natural" distinguisher, we would hardly be interested in a cluster solution separating those who eat a lot of candy versus those who eat less. In other situations, as with the iris data which we will cluster analyze later, cluster groupings are indeed more naturally defined (e.g., by species).

Just as the factor analyst must guard against the possibility of seeing structure where none is apparent, the cluster analyst must also be wary of finding clusters simply because mathematical optimization has been successful. As eloquently put by Everitt, Landau, and Leese (2001):

> The problem is, of course, that since in most cases the investigator does not know **a priori** the structure of the data (cluster analysis is, after all, intended to help to uncover any structure), there is a danger of interpreting **all** clustering solutions in terms of the existence of distinct (natural) clusters. The investigator may then conveniently 'ignore' the possibility that the classification produced by a cluster analysis is an artefact of the method and that actually she is **imposing** a structure on her data rather than discovering something about the actual structure. (pp. 7–8)

Hence, cluster analysis will, under most circumstances, generate corresponding clusters. As is true in factor analysis, however, whether or not the solution **means** anything is usually left up to the judgment of the researcher. As R.A. Fisher once said about ANOVA, cluster analysis, as is true of virtually all statistical methods, is simply an exercise in **arranging the arithmetic**. Cluster analysis, or any other method, cannot purport by itself to make scientific discoveries. It may be "unsupervised" learning, but a scientist is still required to interpret the solution.

14.25   MEASURING PROXIMITY

We have said that cluster analysis is based on identifying similarity among cases. But how should we define such proximity, or its opposite, that of **distance**? The choice of a good proximity measure is by no means obvious in all cases and is often quite discretionary and subjective. As was true for factor analysis, this subjective component can generate some skepticism. Indeed, **if what you see is dependent on the microscope you use to see it, then it becomes difficult to define what you are looking at as distinct from the tool you are using to do the viewing**. Which objects in a data set are more similar or distant than others? The answer depends on what definition of similarity or distance we use. We survey some of the more popular distance measures.

Euclidean distance is that which we think of when we want to compute the distance between two points on a straight line. It is by far the most easily recognizable and historically relevant distance measure. The Euclidean distance between two p-dimensional objects, $\mathbf{x} = [x_1, x_2, \ldots, x_p]$ and $\mathbf{y} = [y_1, y_2, \ldots, y_p]$ is defined as:

$$d(\mathbf{x}, \mathbf{y}) = \sqrt{(x_1 - y_1)^2 + (x_2 - y_2)^2 + \cdots + (x_p - y_p)^2}$$
$$= \sqrt{(\mathbf{x} - \mathbf{y})'(\mathbf{x} - \mathbf{y})}$$

where $d(\mathbf{x}, \mathbf{y})$ is the distance between vectors, with differences between observations denoted by $(x_1 - y_1)^2$, $(x_2 - y_2)^2$, ..., $(x_p - y_p)^2$. As remarked by some (e.g., Johnson and Wichern, 2007; Rencher and Christensen, 2012), one might be tempted to use the statistical distance:

$$d(\mathbf{x}, \mathbf{y}) = \sqrt{(\mathbf{x} - \mathbf{y})'\mathbf{S}^{-1}(\mathbf{x} - \mathbf{y})}$$

where $\mathbf{S}^{-1}$ is the inverse of the variance-covariance matrix. However, the argument against this is that if so-called "natural groupings" do end up emerging from the data, then computing $\mathbf{S}$ on the entire sample might be misleading and not provide accurate estimates of variances and covariances. This is because presumably, in computing $\mathbf{S}$, we would be pooling across the data without any attention to the potential existence of groups. For this reason, Euclidean distance is usually preferred over a measure of statistical distance.

To illustrate a simple example of Euclidian distance, consider the two vectors $\mathbf{x}$ and $\mathbf{y}$:

$$\mathbf{x} = [2, 4]$$
$$\mathbf{y} = [4, 7]$$

We compute the Euclidean distance between the two vectors to be 13:

$$d(\mathbf{x}, \mathbf{y}) = \sqrt{(x_1 - y_1)^2 + (x_2 - y_2)^2 + \ldots + (x_p - y_p)^2}$$
$$= \sqrt{(2 - 4)^2 + (4 - 7)^2}$$
$$= 3.61$$

We can visualize the vectors in R by computing

```
> plot(c(2,4), c(4,7))
> arrows(2, 4, 4, 7)
```

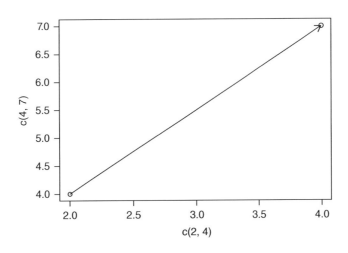

TABLE 14.3   Hypothetical Data for Cluster Analysis on Two Variables Using Euclidean Distance

Object	Variable 1	Variable 2
1	5	4
2	6	6
3	8	7
4	2	2
5	3	1

Note that the vector begins at the point (2,4) and ends at the point (4,7). We can easily compute Euclidean distances in R. For instance, consider data for two variables on five cases in Table 14.3. We first generate a matrix for these two variables:

```
> v1 <- c(5, 6, 8, 2, 3)
> v2 <- c(4, 6, 7, 2, 1)
> M <- cbind(v1, v2)
> M
     v1 v2
[1,]  5  4
[2,]  6  6
[3,]  8  7
[4,]  2  2
[5,]  3  1
```

We compute the distances using the dist function by requesting "euclidean":

```
> dist(M, method = "euclidean")
          1        2        3        4
2  2.236068
3  4.242641 2.236068
4  3.605551 5.656854 7.810250
5  3.605551 5.830952 7.810250 1.414214
```

Note that the matrix is a symmetric matrix, since the upper part, above the main diagonal, is a mirror image of the lower part. By inspecting the distance matrix, we can informally survey which objects are similar to other objects. For example, objects 4 and 5 appear to be rather similar, with a distance of only 1.41, while objects 3 and 4 and 3 and 5 appear to be quite dissimilar, with distances of 7.81. This process of inspecting a distance matrix and attempting to spot similarities parallels that of initially inspecting another distance matrix, that of the correlation matrix in factor analysis where we got a first glance at which variables might "go together" in the sense of potentially indicating underlying constructs.

The **Minkowski metric** is given by:

$$d(\mathbf{x}, \mathbf{y}) = \left[\sum_{i=1}^{p} |x_i - y_i|^m \right]^{1/m}$$

where when $m = 2$, the distance reduces to Euclidean distance:

$$d(\mathbf{x}, \mathbf{y}) = \left[\sum_{i=1}^{p} |x_i - y_i|^m \right]^{1/m}$$

$$= \left[\sum_{i=1}^{p} |x_i - y_i|^2 \right]^{1/2}$$

$$= \sqrt{(x_1 - y_1)^2 + (x_2 - y_2)^2 + \ldots + (x_p - y_p)^2}$$

That is, the only difference between the Minkowski distance and that of Euclidean is that m in Minkowski is not fixed at $m = 2$ as it is in Euclidean. A third and similar distance measure, **city-block** (or **Manhattan**) is given by:

$$d(\mathbf{x}, \mathbf{y}) = \sum_{i=1}^{p} |x_i - y_i|$$

where $|x_i - y_i|$ is now simply the absolute difference. The choice of which distance measure to use for which type of problem has been investigated (e.g., see Gower, 1988). It was found, in general, that the choice of measure can lead to somewhat different findings. According to Everitt, Landau, and Leese (2001), more research is required before a conclusion regarding which measure is most **optimal** (in some sense) is reached, and the conditions for such optimality. For instance, if multicollinearity is inherent between variables, using a distance measure that compensates for this, such as Mahalanobis, is usually advised. Mahalanobis distances (1936) are given by:

$$D^2 = (\mathbf{y}_1 - \mathbf{y}_2)' \mathbf{S}^{-1} (\mathbf{y}_1 - \mathbf{y}_2)$$

where $\mathbf{y}_1$ and $\mathbf{y}_2$ are sample vectors and $\mathbf{S}^{-1}$ is the pooled sample variance-covariance matrix of $\mathbf{S}_1$ and $\mathbf{S}_2$, respectively. Mahalanobis distances can be considered a **generalized Euclidean measure** since it adjusts for the covariance among variables through $\mathbf{S}^{-1}$. Relatively large values of D^2 can also be used in spotting outliers in multivariate analysis in general.

In other cases, and as a working rule, researchers might do well to run a data set under a variety of proximity options, and if substantively meaningful clusters emerge under each choice, this may then lend credibility to the given cluster solution.

It must be emphasized that regardless of the distance measure chosen, magnitudes in data can at times simply be reflective of the scales used in the analysis. A solution to this problem is to **standardize** the data to z-scores when working with continuous measures. Standardization is not a panacea for all problems in cluster analysis, but it does help guard against the possibility of one or two variables contributing most heavily toward the final solution relative to other variables. Hence, standardizing data prior to performing cluster analysis is sometimes advised.

14.26   HIERARCHICAL CLUSTERING APPROACHES

Generally, there are two ways in which one can cluster observations. On the one hand, we can begin with single observations considered uniquely then proceed to join "alike" observations on the road to building clusters. This is generally known as the **agglomerative** or **hierarchical** approach. Alternatively, one can start with a single cluster containing all cases then partition these cases into respective clusters in a stepwise fashion. This latter form of clustering is generally known as the **divisive** approach. Regardless of adopted approach, generally, once two cases are "fused," they cannot be unfused.

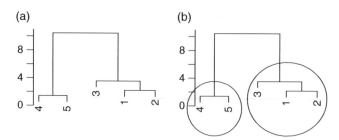

FIGURE 14.3   Dendrogram of clustering (a). Possible final solution (b).

The consequence of such a constraint to cluster formation was highlighted in Ward's depiction of cluster analysis featured earlier, in which recall he noted, "Grouping, however, ordinarily results in some loss of information" (Ward, 1963, p. 236). Akin to developing a circle of friends, not **everyone** can be included in the group, and the fact that you are friends with one person might exclude you being friends with another. At some point, some degree of simplification has to occur. Decisions, to some extent, "burn bridges," but to some degree, this is required if progress is to be made. Hence, no cluster algorithm will ever provide a "perfect" solution. Some will be simply more **preferable** than others for a particular context.

The linking of objects by their similarity or dissimilarity can be pictorially represented using what is known as a **dendrogram** (Figure 14.3). A dendrogram reveals the chain of linkage of objects deemed similar enough to be fused. The dendrogram can be conceived as a kind of historical record of the clustering process. **Icicle plots** can also be used to visualize results from hierarchical cluster analysis, though we do not feature their use here. See Kruskal and Landwehr (1983) for details.

In Figure 14.3a, (using a technique called Ward's method) we can see that objects 4 and 5 were fused together into one cluster, while objects 1 and 2 were fused into another cluster. Next, the cluster of 1 and 2 was fused with object 3. We can see then that by the top of the dendrogram, the hierarchical cluster routine for Figure 14.3b appears to have generated two groups. Of course, if one's substantive theory suggests that object 3 should be kept distinct from merging with objects 1 and 2, then one could conclude that 3, not 2 groups were generated by the cluster solution, that is, (4,5) versus (3) versus (1,2). Note carefully then that without knowing what objects 1 through 5 actually **are**, one cannot, other than possibly in an algorithmic sense, make a determination about the final cluster solution. Hence, so-called **stopping rules** in cluster analysis can be quite subjective. Just as in factor analysis, one requires an understanding of the substantive nature of the variables one is working with in order to draw meaningful conclusions from the procedure.

Within the hierarchical clustering paradigm, several different methods exist for fusing objects. These include **single linkage**, **complete linkage,** and **average linkage**. In single linkage, also known as **minimum distance** or **nearest neighbor**, cases having the smallest distance between them are merged at each successive step. After each merger, the smallest distance between clusters is once more computed, and those clusters having the smallest distance are then likewise merged together. As an example of single linkage, recall the Euclidean distance matrix featured earlier:

```
    1                   2           3           4
2   2.236068
3   4.242641   2.236068
4   3.605551   5.656854    7.810250
5   3.605551   5.830952    7.810250    1.414214
```

The first step in the cluster process via single linkage is to merge the two closest objects, which for this matrix are objects 4 and 5 (distance of 1.41). These two objects constitute the "nearest neighbors" in the given matrix of distances. The merger of objects 4 and 5 is what gave us our first "cluster" in the dendrogram of Figure 14.3 (though using Ward's method in this case). Next, the algorithm evaluates distances between our newly formed cluster (4,5) and remaining objects. It continues in this fashion until it arrives at the final cluster solution.

Complete linkage, also known as **maximum distance** or **farthest neighbor**, is opposite to that of single linkage. Unlike single linkage in which proximity was preferred, in complete linkage, we are interested in observations between clusters that are most **distant**, the smallest of which distances are merged. Finally, for **average linkage**, distance between clusters is defined as the average between all pairs of objects where one member of a pair belongs to each cluster (Johnson and Wichern, 2007).

We note then that as a result of different ways of defining distance, mergers between clusters will likewise be different depending on one's choice of linkage. The decision of which linkage to use is often made on subjective grounds and hence the extent to which one "finds" something in the data is at least somewhat dependent on one's choice of linkage. These are by no means the only methods for linking or joining clusters. Another quite common method is that of **centroid linkage**, where distance between cluster centroids (i.e., means) is employed as the metric. **Minimax linkage** (Bien and Tibshirani, 2011) has also been evaluated as an alternative, boasting some favorable properties. This method defines what is called the **minimax radius** between clusters, and bases its assessment of distance on this radius. We do not discuss this nor other additional methods of linkage in this chapter. The take-home point for the reader is simply to realize that there exists a whole literature on methods of hierarchical linkage, and hence those typically offered by mainstream software are by no means the only ones available. A further point is to understand that when you perform a cluster analysis, to never assume the cluster analysis procedure is an "objective" process. To the contrary, given so many ways of cluster-analyzing, the inputs and decisions regarding metrics and such are often just as important as whether the data actually reveal clusters.

14.27 NONHIERARCHICAL CLUSTERING APPROACHES

Thus far we have surveyed some of the more common hierarchical methods for clustering objects as a function of distance matrices. Nonhierarchical approaches, on the other hand, do not operate on distance matrices and hence are typically less computationally demanding than hierarchical approaches. As was the case for the hierarchical case, numerous nonhierarchical methods have been proposed. Of these, the **K-means** approach, or variants thereof, is typically the most popular. MacQueen (1967):

> The main purpose of this paper is to describe a process for partitioning an N-dimensional population into k sets on the basis of a sample. The process, which is called 'k-means,' appears to give partitions which are reasonably efficient in the sense of within-class variance … Stated informally, the k-means procedure consists of simply starting with k groups each of which consists of a single random point, and thereafter adding each new point to the group whose mean the new point is nearest. After a point is added to a group, the mean of that group is adjusted in order to take account of the new point. Thus at each stage the k-means are, in fact, the means of the groups they represent (hence the term k-means). (pp. 281, 283)

As summarized by MacQueen, the general algorithm for K-means begins by partitioning cases into k initial clusters. This can be accomplished through a random process or by specifying **seeds** to initiate the clustering algorithm. Once these seeds are chosen, remaining cases in the data are assigned to the cluster having the nearest seed, usually based on a measure of Euclidean distance. That is, once a cluster has more than a single case, the initial starting seed is replaced with that of the **centroid**, which is the mean of the given cluster. Each time a new case is added to the cluster, the centroid is recalculated. The process is repeated until no new assignments are made.

Arguably, the most challenging part of K-means, as was the case for hierarchical methods, is still in identifying or naming the clusters generated. Again, be sure to note that **K-means will typically generate clusters**. Whether such clusters have any inherent meaning is to a large extent a **substantive** decision, not a statistical one. Analogous to factor analysis, when one performs a cluster analysis, one should generally be prepared to experience an "aha!" moment upon looking at the output. If you need to spend hours and hours contemplating the "nature" of the solution, it becomes more and more difficult to argue that so-called **natural** groups were produced.[4]

14.28   *K*-MEANS CLUSTER ANALYSIS IN R

We demonstrate a *K*-means cluster analysis in R on the iris data. For pedagogical purposes, this data set is ideal for demonstrating cluster analysis since we already know in advance of a suitable cluster solution, that of **species**. We begin by first identifying the variables on which we wish to cluster the cases:

```
> attach(iris)
> iris.data <- cbind(Sepal.Length, Sepal.Width, Petal.Length, Petal.
Width)
> library(car)
> some(iris.data)
```

```
      Sepal.Length Sepal.Width Petal.Length Petal.Width
[1,]          4.7         3.2          1.3         0.2
[2,]          5.0         3.4          1.5         0.2
[3,]          5.0         3.0          1.6         0.2
```

We could have also used data.frame (Sepal.Length, Sepal.Width, Petal.Length, Petal.Width) to generate the data set. Obtaining a scatterplot matrix is helpful in being able to visualize initial degrees of separation:

```
> pairs(iris.data)
```

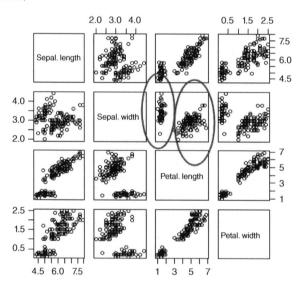

[4] Of course, perhaps the cluster analysis did generate natural groupings but that as of yet, you are unaware of what those groups could be. I am not meaning to discount the use of exploring possibilities about why cluster analysis solutions come out the way they do. Indeed, a huge part of scientific discovery is the pondering of data in hopes that eventually insight into the structure of such data is achieved. This practice is fine, so long as one is aware that numerous theories can be fit to the same data and that the theory of one's preference is not necessarily the right one.

Though scatterplots will not tell us what the **final** solution will look like, initial insights into potential clusters can nevertheless be gleamed by a cursory exploration of such plots. For instance, we can see that the scatterplot of sepal width and petal length seems to suggest the presence of two groups or clusters (i.e., the groups circled in ovals in row 2, column 3). By performing a cluster analysis, we wish to "discover" a variable on which such separation might be based. Note that if we were already aware of the species grouping structure, then the task might be to learn how well a linear combination of iris features **predicts** type of species. Such would call for the **linear discriminant function analysis** of Chapter 12 or the logistic regression of Chapter 10. Of course, a MANOVA or a dummy-coded multivariate multiple regression could also be performed to test a multivariate hypothesis of equality among mean vectors across species. In this sense, in the machine learning domain, these techniques would be considered "supervised" learning methods, because we already have knowledge of the grouping structure (i.e., the classes that we wish the statistical method to "learn"). In the case of cluster analysis, however, **we do not yet know the groups**. It is in this sense that cluster analysis is usually considered more exploratory than these other more "confirmatory" approaches.

We proceed to now fit the *K*-means cluster solution:

```
> set.seed(20)
> k.means.fit <- kmeans(iris.data, 3, nstart = 20)
> k.means.fit

K-means clustering with 3 clusters of sizes 50, 62, 38

Cluster means:
  Sepal.Length Sepal.Width Petal.Length Petal.Width
1     5.006000    3.428000     1.462000    0.246000
2     5.901613    2.748387     4.393548    1.433871
3     6.850000    3.073684     5.742105    2.071053

Clustering vector:
  [1] 1 1 1 1 1 1 1 1 1 1 1 1 1 1 1 1 1 1 1 1 1 1 1 1 1 1 1 1 1 1 1 1 1 1 1 1 1
 [38] 1 1 1 1 1 1 1 1 1 1 1 1 1 2 2 3 2 2 2 2 2 2 2 2 2 2 2 2 2 2 2 2 2 2 2 2 2
 [75] 2 2 2 3 2 2 2 2 2 2 2 2 2 2 2 2 2 2 2 2 2 2 2 2 2 3 2 3 3 3 3 2 3 3 3 3
[112] 3 3 2 2 3 3 3 3 2 3 2 3 2 3 3 2 2 3 3 3 3 3 2 3 3 3 3 2 3 3 3 2 3 3 3 2 3
[149] 3 2

Within cluster sum of squares by cluster:
[1] 15.15100 39.82097 23.87947
 (between_SS / total_SS =  88.4 %)
Available components:

[1] "cluster"     "centers"    "totss"     "withinss"   "tot.withinss"
[6] "betweenss"   "size"
```

Provided in the output are the cluster means for each of the input variables. The clustering vector tells us which observation each cluster has been assigned. For instance, the first observation is in cluster 1, the second observation is in cluster 1, the 75th observation is in cluster 2, etc. R also provides us with **within cluster sum of squares** for each cluster as an overall indicator of the degree of homogeneity within each cluster grouping, along with a ratio of **SS between** to **SS total** for an estimate of how much variance is accounted for by cluster membership.

As mentioned, since we are in the rather unique position of actually knowing **a priori** clusters for these data, we can request classification results similar to how we did so for the discriminant analysis of Chapter 12, only that now the numbers 1, 2, 3 across the first row represent **cluster membership** and not species:

```
> table(Species, k.means.fit$cluster)

Species        1   2   3
  setosa      50   0   0
  versicolor   0  48   2
  virginica    0  14  36
```

Cluster 1 (50 cases) is made up of the species *setosa*, with no cases from *versicolor* or *virginica*. Cluster 2 is made up of 48 cases of *versicolor* and 14 cases of *virginica*. Cluster 3 is made up of 2 cases from *versicolor* and 36 cases from *virginica*, with no cases from *setosa*. Ideally, for perfect classification, the two cases going into cluster 3 should have gone into cluster 2, and the 14 cases that went into cluster 2 should have gone into cluster 3.

We next obtain plots of petal width against petal length and sepal width against sepal length to reveal the three clusters:

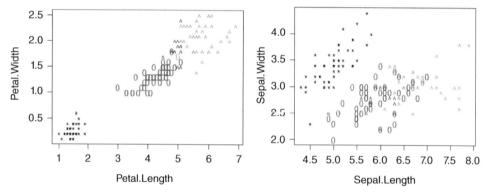

Cluster membership in both plots is evident. Clearly, and in agreement with the classification table, species *setosa* (i.e., "*") distinguishes itself from species *versicolor* and *virginica*.

To perform a hierarchical cluster analysis in R, we first define the distance matrix, and follow this up with a contrast of single versus complete linkage cluster solutions (Figure 14.4).

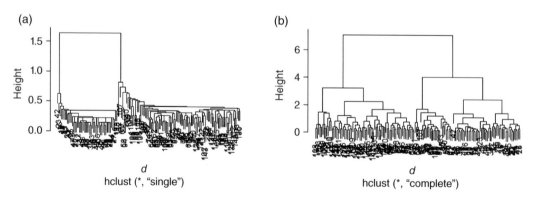

FIGURE 14.4   Dendrograms for single linkage (a) and complete linkage (b) cluster analysis of iris data.

```
> d <- dist(iris.data)
> clust.single <- hclust(d, method = "single")
> clust.complete <- hclust(d, method = "complete")
> plot(clust.single)
> plot(clust.complete)
```

Though the dendrograms generated from these respective solutions are of little practical "visual" use due to the number of objects being fused (they require a much larger screen for deciphering object numbers), for pedagogical purposes, one can nonetheless appreciate a comparison of single versus complete linkage dendrograms. One can see how the final cluster solution differs depending on which approach is adopted. Which, then, is the "true" cluster solution? That is up to the analyst to decide, presumably informed by scientific experience and insights.

14.29   GUIDELINES AND WARNINGS ABOUT CLUSTER ANALYSIS

Cluster analysis is generally considered a rather crude exploratory technique. As discussed, performing a cluster analysis is relatively straightforward using high-powered computers. The more difficult part is, of course, making sense of the clusters that do emerge. In concluding our brief discussion of cluster analysis, we issue some general guidelines and warnings about its use:

- Cluster algorithms will usually be quite sensitive to **outliers**. Before running a cluster analysis, be sure that you have properly recorded data and that no observations are extremely distant from all other observations. Using criteria such as **Mahalanobis distances** can be useful in identifying multivariate outliers.

- Always consider the final clustering solution to see if it is sensible. If it is not, one possible "verdict" of a given cluster analysis must be that **there does not appear to be any "natural" groups in this data**. That is, the possibility of there being **no substantive solution** must exist as a potential outcome to the analysis.

- For any given data, it is advisable to try several clustering methods and calculate distances in various ways for each method. If the outcomes from the several methods are roughly consistent with one another, this might help in "triangulating" an argument for "natural" groupings. As warned by Venables and Ripley (2002), "Do not assume that 'clustering' methods are the best way to discover interesting groupings in the data; in our experience, the visualization methods are often far more effective. There are many different clustering methods, often giving different answers, and so the danger of over-interpretation is high" (p. 316).

- Statistical significance testing in cluster analysis is generally inappropriate. Since the goal of cluster analysis is to maximize group differences, the probability of the data given the null, if low (e.g., $p < 0.05$) is hardly surprising, since you actually put to work an algorithm to accomplish just this! Hence, resist the temptation to run inferential tests on your cluster solution to "support" a claim of naturally-occurring clusters.

14.30   A BRIEF LOOK AT MULTIDIMENSIONAL SCALING

Having surveyed the methods of exploratory factor analysis and cluster analysis, we close this chapter with a brief look at the technique of **multidimensional scaling** (**MDS**), which, similar to factor analysis, can be regarded as a dimension-reduction technique. It is also considered a data visualization technique, where, somewhat analogous to cluster analysis, seeks to identify clusters of points (Izenman, 2008). Given an understanding of factor analysis and the concepts of distance and proximity surveyed in cluster analysis, comprehending the principles behind MSD is relatively easy. This is not meant to suggest the theory and details behind MDS are elementary, but only to suggest that one can get a cursory understanding of the technique by relating it to previously learned methods. In a strong sense, MDS cannot be that much different from these, as, in the end, **there is only so much one can do with a set of data**. Indeed, one of the goals of the current book has been to convey just as much **what statistical techniques cannot do, as much as what they can do**.

MDS is typically more flexible than EFA, however, as it is not restricted to analyzing covariance or correlation matrices as in factor analysis, but rather can handle a wider array of **dissimilarity matrices**. Whereas in factor analysis the dissimilarity matrix was that of the covariance or correlation matrix, multidimensional scaling allows for increased flexibility on the nature of the dissimilarity matrix. And, just as in factor analysis, MDS seeks to explain or account for corresponding distances (i.e., covariances or correlations in EFA) between objects. In this sense, MDS will try to "uncover" that which "gave rise" (in some sense) to the observations we are seeing in a data matrix. Hence, like factor analysis, it seeks to reduce or "recover" dimensionality. As summarized by Hastie, Tibshirani, and Friedman (2009), "The idea is to find a lower-dimensional representation of the data that preserves the pairwise distances as well as possible" (p. 570). This idea, then, is definitely not new to us.

To demonstrate MDS, we consider airline distance data between 10 major cities in the United States, where distances between cities is recorded in miles, adapted from Kruskal and Wish (1978):

Flying Mileage between 10 US Cities (Kruskal and Wish, 1978)

	Atl	Chi	Den	Hou	LA	Mia	NY	SF	Sea	DC
Atlanta	0	587	1212	701	1936	604	748	2139	2182	543
Chicago	587	0	920	940	1745	1188	713	1858	1737	597
Denver	1212	920	0	879	831	1726	1631	949	1021	1494
Houston	701	940	879	0	1374	968	1420	1645	1891	1220
Los Angeles	1936	1745	831	1374	0	2339	2451	347	959	2300
Miami	604	1188	1726	968	2339	0	1092	2594	2734	923
New York	748	713	1631	1420	2451	1092	0	2571	2408	205
San Franc.	2139	1858	949	1645	347	2594	2571	0	678	2442
Seattle	2182	1737	1021	1891	959	2734	2408	678	0	2329
Wash. DC	543	597	1494	1220	2300	923	205	2442	2329	0

As we can see from the matrix, the distance between Atlanta and Chicago, for instance, is 587 miles, while the distance between Atlanta and Denver is 1212 miles, and so on for other cities. Note as well that the distance matrix has a series of zeros along the main diagonal, indicating that the distance between a city and itself, as would be expected, is zero miles.

Using this data matrix as an input, multidimensional scaling, as was true for factor analysis, will essentially attempt to **extract underlying dimensions that best account for the pairwise distances between objects**, which in this case, are the corresponding city distances. The technique does this by, not surprisingly by now, **minimizing a function**. For classical MDS, the function that is minimized is referred to as the **stress function** (see Hastie, Tibshirani, and Friedman, 2009, for details).

But what does all this mean, exactly? We already know what it means because it is a somewhat analogous concept to what is done in principal components analysis and factor analysis (and discriminant analysis to some extent), and that is to account for observed data (in this case, distances, often Euclidean) by uncovering so-called "latent" dimensions. Not surprisingly, classic or metric multidimensional scaling (Izenman, 2008; Rencher and Christensen, 2012) works by extracting eigenvalues and eigenvectors and resulting in a plot of this dimensionality similar in spirit to what we obtained with these prior procedures. "Nonmetric" MDS, which features the use of ranks instead of distances, is another alternative strategy. See Hastie, Tibshirani, and Friedman (2009) for details.

When we apply MDS to the above airline data, we obtain the following two-dimensional plot:

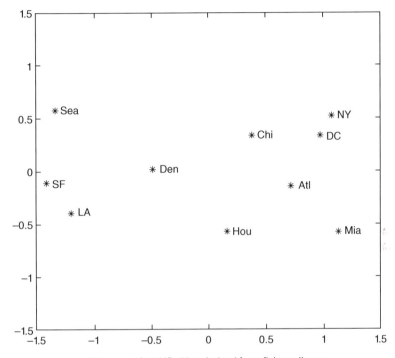

The map of 10 US cities derived from flying mileage.

What are we seeing in the plot? Something remarkably similar to what we saw in PCA and discriminant analysis, where each axis represents a new "dimension" on which the data are now plotted. The way to read the plot is likewise already familiar. Look at the x-axis, and observe the location of the city data points as you would a scatterplot. Do we notice any separation between cities? Absolutely, but the nature of the separation we will need to surmise from how the cities group themselves along that dimension. Notice to the left of the plot are cities Seattle, San Francisco, Los Angeles, Denver, whereas on the rightmost area of the plot are cities New York, Washington DC, Miami, etc. What dimension might the x-axis therefore represent? It would seem that our newly derived dimension along the x-axis corresponds to an "East versus West" underlying latent variable, whereas the newly created dimension along the y-axis may correspond to a "North versus South" latent variable. To see this, simply spin the plot around or draw a horizontal line at approximately $y = 0$ on the y-axis, and notice that northern versus southern cities appear to form groupings on this axis or dimension.

Hence, we can say that via multidimensional scaling, we have extracted two dimensions that underlie the distances between cities, just as we drew similar conclusions when conducting techniques such as PCA and discriminant analysis earlier. And, as we did in cluster analysis and EFA, we used a distance metric as the source of input to the analysis. Multidimensional scaling is a prime example of how understanding the fundamentals of other techniques makes MDS (and many other multivariate techniques) much easier to comprehend. The technical details underlying the procedures are not equivalent of course (hence, the reason statisticians spend their careers studying these things), however the conceptual base (i.e., the "ideas" that give them life) are quite similar. This is the overriding point to take away from our discussion here.

Though we have only scratched the surface, as mentioned, the technique is quite rich theoretically. Izenman (2008) provides a thorough and technical introduction, and Hastie, Tibshirani, and Friedman (2009) provide a good discussion of the method in the context of unsupervised learning techniques in general. These sources should be consulted for a more in-depth understanding, as well as to appreciate how MSD distinguishes itself from techniques such as factor analysis, PCA, and cluster analysis. Conceptual similarity is one thing, but technical similarity is quite another.

14.31   CHAPTER SUMMARY AND HIGHLIGHTS

- **Factor analysis** is a statistical method useful for uncovering latent structures that are thought to underlay covariance or correlation among observed variables. It may also be conceived as a **data reduction technique** similar to though still quite different from that of **principal components analysis**.

- While the priority in PCA can be said to account for as much **total variance** among variables as possible, the priority of factor analysis can be said to explain as much **commonality** among variables as possible.

- **Exploratory factor analysis**, or EFA, originating with Charles Spearman in 1904, has had a turbulent history. Due to its **nonuniqueness** of estimated loadings, along with its misuse by researchers, it has been a favorite target of criticism.

- Whereas principal components are **linear functions of observed variables**, outcomes in EFA are **linear combinations of factors**.

- Comparing and contrasting the common factor model of $\mathbf{x} - \boldsymbol{\mu} = \Lambda \mathbf{f} + \boldsymbol{\varepsilon}$ to the multiple regression model $\mathbf{y} = \mathbf{x}\boldsymbol{\beta} + \boldsymbol{\varepsilon}$ is pedagogically useful. In the regression model, $\mathbf{x}$ is a vector of observed manifest variables, whereas in the factor model, $\mathbf{f}$ is a vector of unobserved **latent variables**.

- The EFA model implies that the structure of the **observed covariance matrix** is a function of loadings and specific variances. That is, $\sum = \Lambda\Lambda' + \boldsymbol{\psi}$.

- In EFA, the factor solution is only **unique** up to an orthogonal matrix. What this means substantively is that deciding on the **correct** rotation usually reduces to a **substantive** concern, not a statistical one.

- Common methods of estimation in EFA include **principal axis factoring** (PAF) and **maximum likelihood** (ML).

- **Varimax** is an orthogonal rotation in which the variance of **within factor** loadings is maximized. **Quartimax**, also an orthogonal rotation, maximizes the variance of loadings **across** factors.

- Factor analysis is generally a **large-sample** technique, though research suggests that required sample size can be in part a function of the magnitude of communalities.

- **Bartlett's test of sphericity** and the **Kaiser–Meyer–Olkin** measure of sampling adequacy can be used to help determine whether sufficient correlation exists among a set of variables for it to be suitable for factor analysis.

- **Cluster analysis** is a statistical method based on the idea of grouping cases or individuals that are in some sense, similar.

- Cluster analysis, although distinct from factor analysis, can nonetheless be likened to it in that both methods seek to exploit similarities in data. Both can be said to use **distance measures** for this purpose.

- The number of ways in which n cases can be partitioned in g clusters is usually exceedingly large; hence, one goal of cluster analysis algorithms is to shrink this large number into a **partitioning** that is more manageable, while not losing too much information in the process.

- Cluster analysis does not "discover" groupings any more than factor analysis "discovers" underlying factors. In both cases, **similarity** is simply exploited to reveal potential structures in data. These structures are only as meaningful as they represent something of interest to the researcher.

- **Euclidean distance** is the most common approach to defining distance and is usually preferred over any type of statistical distance that incorporates the covariance matrix.

- Other ways of defining distance in cluster analysis include the **Minkowski metric** and **city-block** (or "Manhattan") distance, among others.

- When variables are measured on different units, distance measures may reflect magnitude simply because of the inflated variance of particular variables. **Standardization** of data is sometimes advised to solve this problem.

- Approaches to clustering include **agglomerative** or **hierarchical** clustering, which begin by considering each observation separately before building up clusters, and **divisive**, which begins with all objects in a single cluster then proceeds to partition these into separate clusters at each step of the procedure.

- A **dendrogram** is a convenient picture that displays the linkage history in a hierarchical cluster analysis.

- **Nonhierarchical clustering** methods, such as the *K*-means approach, typically do not require the computation of a matrix of distances or similarities. Consequently, these methods are usually less demanding computationally when compared to competing hierarchical methods.

- **Multidimensional scaling** is a technique similar in spirit to other dimension-reduction techniques such as principal components analysis and factor analysis, though more general, as it allows greater flexibility in the types of distance matrices that can be used.

REVIEW EXERCISES

14.1. Discuss two goals of factor analysis. Though they mechanically amount to the same thing (i.e., factor solutions), are these goals **substantively** equivalent? Why or why not?

14.2. Discuss one important way in which **factor analysis** is different from **principal components analysis**.

14.3. What does it mean to say that principal components analysis seeks to explain mostly **variance** but that factor analysis seeks to explain mostly **covariance**? How does this difference distinguish the two procedures, both mathematically and substantively?

14.4. Do you agree that factor analysis **uncovers** latent variables? Why or why not? What are some of the philosophical issues inherent in such a statement?

14.5. Distinguish between **exploratory** versus **confirmatory** factor analysis. How are they different? Is the distinction always evident in a practical setting? How so?

14.6. Describe the components of the **common factor-analytic model** $x = \mu + \Lambda f + \varepsilon$.

14.7. Compare the **factor analysis** model $x = \mu + \Lambda f + \varepsilon$ to the **regression model** of previous chapters $y = x\beta + \varepsilon$, noting their similarities and differences.

14.8. **"A model is defined by the assumptions it makes."** Discuss this statement, and explain what it means.

14.9. State and summarize the assumptions for the **orthogonal factor-analytic model**.

14.10. What does it mean to say that the factor model implies a **structure to the covariance matrix**? How might this idea help you understand statistical modeling in general?

14.11. State precisely how EFA is **parameterized** to imply a covariance or correlation matrix.

14.12. What is the major critique targeted against **factor analysis**? Do you believe it is justified? Why or why not?

14.13. What does it mean to say that factor analysis suffers from the problem of **indeterminacy** and **nonuniqueness** of solutions?

14.14. How can you tell whether your factor analysis has been successful? Do you agree with the **"WOW" criterion** recommended by Johnson and Wichern? Why or why not?

14.15. Briefly describe the **principal axis factoring** method of factor analysis, then briefly compare it to the **maximum-likelihood** method of estimating factors.

14.16. Interpret Jöreskog's quote:

> **Though $\Lambda*$ and Λ are equivalent from the mathematical point of view, they may not be so from the psychological point of view. The problem of choosing one particular psychologically meaningful Λ out of the infinite set $\{\Lambda T^{-1}\}$ has been called the problem of rotation, although the problem of transformation would be a better term, since it includes also the transformation to oblique factors, in which case the transformation matrix T is not orthogonal and hence does not represent only a rotation.** (p. 166)

In your interpretation, be sure to comment on the "mathematical point of view" versus "psychological point of view" distinction Jöreskog highlights. What do you think he means by this?

14.17. Distinguish between **varimax** and **quartimax** rotations.

14.18. Do you believe factors should be **rotated**? Or, do you believe that rotating factors is "fudging the data" so to speak? Why or why not?

14.19. In this chapter, we conducted a two-factor solution on the Holzinger data. Request a **three-factor solution** and compare your findings to that of the two-factor solution.

14.20. Consider the following **correlation matrix** depicting the correlations between disciplines on the GRE.

Intercorrelations Among The G.R.E. Tests Of General Education

	Math	P.S.	B.S.	Soc.	Lit.	Arts	Exp.	Voc.
Mathematics		.55	.44	.51	.36	.35	.52	.38
Physical Science	.55		.49	.43	.20	.40	.32	.29
Biological Science	.44	.49		.57	.42	.42	.46	.50
Social Studies	.51	.43	.57		.54	.40	.61	.59
Literature	.36	.20	.42	.54		.39	.53	.54
Arts	.35	.40	.42	.40	.39		.42	.52
Effecive Expression	.52	.32	.46	.61	.53	.42		.66
Vocabulary	.38	.29	.50	.59	.54	.52	.66	

Conduct an exploratory factor analysis on this data, requesting a **two-factor** and then a **three-factor solution**. Rotate the factors in each case, and summarize the main findings. Can you name the factors?

14.21. Describe the goal(s) of **cluster analysis**.

14.22. Interpret Joe H. Ward's statement made in 1963 that "**Grouping, however, ordinarily results in some loss of information that may be quantified in a 'value-reflecting' number.**" More specifically, how does clustering result in a loss of information?

14.23. How is cluster analysis similar to and different from factor analysis? How do they both utilize measures of **distance**?

14.24. In how many ways can 20 cases be partitioned into five clusters? Provide both the **exact number of ways**, as well as an **approximation** to this number.

14.25. Consider the statement "**In discriminant analysis, we know the grouping structure. In cluster analysis, we do not yet know it.**" Interpret the statement, emphasizing how cluster analysis can be seen as a more "primitive" technique when compared to discriminant analysis or ANOVA.

14.26. Comment on whether or not cluster analysis **discovers natural groupings**. What might this statement mean, and do you agree with it?

14.27. Provide a verbal interpretation or definition of **Euclidean distance**.

14.28. Distinguish between **hierarchical** versus **divisive** methods of clustering.

14.29. Distinguish between **single linkage** and **complete linkage** as methods of hierarchical clustering. How are these two different from **average linkage**?

14.30. Discuss how *K*-**means** clustering goes about generating clusters, and how this process generally differs from hierarchical methods.

14.31. Using SPSS, perform and interpret a *K*-**means cluster analysis** on the iris data originally analyzed using R in this chapter. Use the following syntax to generate the cluster analysis. You will have to first build the data set in SPSS (you can obtain the data from > iris in R).

```
QUICK CLUSTER sepal_length sepal_width petal_length petal_width
   /MISSING=LISTWISE
   /CRITERIA=CLUSTER(3) MXITER(25) CONVERGE(0)
   /METHOD=KMEANS(NOUPDATE)
   /SAVE CLUSTER DISTANCE
   /PRINT ID(species) INITIAL ANOVA.
```

14.32. Visualize your cluster solution in 14.31 using:

```
GRAPH
   /SCATTERPLOT(BIVAR)=petal_width WITH petal_length BY QCL_1
   /MISSING=LISTWISE.
```

14.33. Briefly describe how **multidimensional scaling** is similar (in concept at least) to principal components and factor analysis.

Further Discussion and Activities

14.34. It was mentioned in this chapter that the user of factor analysis, in addition to acquainting oneself with its technical limitations, should also be somewhat familiar with its **philosophical foundations**. Refer to Mulaik (1987) and summarize some of the more salient philosophical issues surrounding the interpretation of solutions in factor analysis.

15

PATH ANALYSIS AND STRUCTURAL EQUATION MODELING

The path coefficient, measuring the importance of a given path of influence from cause to effect, is defined as the ratio of the variability of the effect to be found when all causes are constant except the one in question, the variability of which is kept unchanged, to the total variability.

(Wright, 1920, p. 329)

Any correlation between variables in a network of sequential relations can be analyzed into contributions from all of the paths (direct or through common factors) by which the two variables are connected, such that the value of each contribution is the product of the coefficients pertaining to the elementary paths.

(Wright, 1934, p. 163)

This terminology is unfortunate, since most models do not establish causality, but only establish an empirical linear association among the latent and manifest variables under study.

(Timm, 2002, p. 557)

Path analysis is a statistical technique useful for modeling simple to complex networks of relationships among observed variables. **Observed** variables in path analysis are often referred to as **manifest** variables, because it is assumed they are, in general, readily **measurable**. The models considered in this chapter generally assume all variables are more or less continuous in nature.

In many respects, path analysis is similar to multiple regression, though unlike multiple regression, path analysis allows the user more freedom in specifying and hypothesizing models that may more closely mimic correlational reality than is possible with multiple regression. For instance, in regression, the model typically "ends" with the prediction of a response variable. In path analysis, one can use that very response as a predictor of further responses. Path analysis allows for the specification of **networks**

Applied Univariate, Bivariate, and Multivariate Statistics: Understanding Statistics for Social and Natural Scientists, With Applications in SPSS and R, Second Edition. Daniel J. Denis.
© 2021 John Wiley & Sons, Inc. Published 2021 by John Wiley & Sons, Inc.
Companion Website: www.wiley.com/go/denis/appliedstatistics2e

of observed variables and hence widens the multiple regression landscape. As some would argue, this "widening" better represents social reality.[1]

Structural equation modeling (or "SEM") is a rather sophisticated statistical methodology that incorporates elements of both factor analysis and regression or path analysis to test hypotheses about relationships among manifest **and** unobserved variables alike. These unobservable variables often go by the name of **latent** variables. Such variables are generally assumed to not be easily or readily measurable, and hence their existence is usually inferred by manifest variables.

In this chapter, we provide but a cursory overview and introduction to path and structural equation models. Authoritative sources on the subject include Bollen (1989) and Mulaik (2009) and should be consulted for more thorough introductions. For a very readable introduction to path models with applications to biology, consult Shipley (2002). Byrne (2009) provides applications of SEM models using AMOS, while a useful introduction to such models using LISREL is that by Schumacker and Lomax (2010). Structural equation modeling is a book-length topic even when considering special cases of such models. To say that their scope of application is vast is an understatement. With SEM, researchers gain virtual unlimited flexibility in generating models tailored to their research hypotheses. Among the possibilities include the modeling of longitudinal data (see Timm, 2002, pp. 600–604) and latent curve models (Bollen and Curran, 2006), as well as multilevel or mixed models (Bauer, 2003). They have also proved useful in the fitting of nonlinear polynomial structures (Wall and Amemiya, 2000). Because path and SEM models are sometimes referred to as "causal models," it is imperative that we survey some of the **historical roots** of these models as to gain an appreciation of how these models came to be labeled as such. As we will see, in most cases, inferring causal forces in an SEM model is usually unjustified based on the statistics alone. If one is working in a causal context, or can be assured in some sense that the entire network of variables has been considered in the given model, then inferring causality from a correlational network becomes slightly more plausible, if not still in most cases, philosophically unlikely.

15.1 PATH ANALYSIS: A MOTIVATING EXAMPLE—PREDICTING IQ ACROSS GENERATIONS

It is easiest to introduce path analysis through a simple substantive example that will help highlight some of its features and point out how it builds on, but ultimately differs from, multiple regression models.

A classic question in the late nineteenth and early twentieth century was that of determining the mechanism by which genetic characteristics were transmitted from one generation to subsequent generations. General cognitive ability was among the mental characteristics thought to be inherited by children based on their parentage and familial history. The path diagram, or **directed graph** (Mulaik, 2009), in Figure 15.1 shows a simple model in which parental IQ is hypothesized to predict offspring IQ, which in turn is hypothesized to predict the next generation's IQ (i.e., IQ 2).

The goal of path analysis for this example is to estimate respective coefficients along each arrow from parental IQ to offspring IQ to offspring IQ 2. As we will see, these weights are analogous to **regression coefficients** and as such are interpreted in similar fashion to those featured in our study of multiple regression (Chapter 8). The coefficients are often standardized, but need not be so (Bollen, 1989). Variables $d1$ and $d2$, both assumed to be latent and unobservable, are referred to as **disturbances**, and in addition to measurement error, comprise the sum of all other influences extraneous to the model but unaccounted for by exogenous or otherwise directed variables in the system of

---

[1] An equally plausible argument is that social reality is not complex at all, and that we should commit to representing it in as simple a manner as possible.

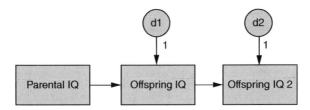

FIGURE 15.1   Path diagram modeling predictability of IQ across generations.

equations. For example, $d1$ pointing to offspring IQ would denote the sum of all variables other than parental IQ that play a role in predicting offspring IQ. Likewise, $d2$ pointing to offspring IQ 2 denotes the sum of all influences other than offspring IQ that can be assumed to predict or determine (in this case) offspring IQ 2. Such sources can usually be assumed to be infinite in number but are not currently observed in the given model. In Wright's original manuscript, as we will see, he defined these "d" terms in path models as representing other factors, "largely ontogenetic irregularity" (Wright, 1920, p. 328) since he was working in the area of genetic transmission of traits. Variables offspring IQ and offspring IQ 2 are named **endogenous** in path models, a term commonly used in econometric models, and indicating that they have at least one predictor pointing to them. That is, endogenous variables are determined by variables within the model (Bollen, 1989). **Exogenous** variables are those variables featured as predictors of endogenous variables and have no arrows pointing toward them, making them "external" (i.e.,"exogenous") to the system of variables. Its presumed "causes" lie outside of the system (Bollen, 1989).

We summarize a couple of the key differences between path analysis and multiple regression:

- **Path analysis** allows one to model a dependent (endogenous) variable as a predictor variable of one or more other dependent variables. Multiple regression models typically do not allow this.
- Path analysis allows one to specify models more precisely than one could ever do in a multiple regression framework. For example, one can estimate relationships among disturbance terms, or model **reciprocal prediction** among variables where two (or more) variables are predictive of each other. For example, it is theoretically possible to adjust the model in Figure 15.1 and test the following path model:

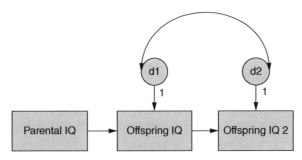

The two-headed arrow joining $d1$ and $d2$ specifies a covariance among disturbance terms. Substantively, what this modeling of the covariance would suggest is that all that predicts offspring IQ that is unaccounted for by parental IQ is related to all that predicts the next generation of offspring IQ that is unaccounted for by the previous generation's IQ. In such a context, the modeling of the disturbance terms in this way makes good sense. Drawing on the nature–nurture debate for instance, one may hypothesize that both *d*1 and *d*2 contain the influence of **nurturing** factors, and hence may help in accounting for both IQs. Path analysis allows one to easily model such relationships, and it is in this sense that path analysis allows a greater sense of model **specificity** than could ever be possible in the typical multiple regression.

15.2   PATH ANALYSIS AND "CAUSAL MODELING"

Perhaps more than any other statistical method discussed in this book, understanding the history of path analysis and structural equation modeling is crucial to gaining an appreciation of its strengths, and more importantly, its **limitations**. Path analysis owes its origins to the geneticist Sewall Wright (1889–1988), who developed the technique roughly between 1918 and 1921. The history of the development of path analysis is well documented elsewhere (e.g., see Denis and Legerski (2006)) and we do not survey its history in any depth here. For our purposes, it is enough to know that path analysis originated with Wright's studies of heredity in which he wished to learn of the genetic transmission of biological traits. One of Wright's first publications introducing the technique was **The Relative Importance of Heredity and Environment in Determining the Piebald Pattern of Guinea-Pigs** (Wright, 1920). In this publication, he included the historically significant path diagram in his discussion (Figure 15.2).

Without detailing every aspect of Wright's diagram, one can nonetheless achieve a basic understanding of how path analysis fit into Wright's goals in his study of these animals. Each one-headed arrow in the figure represents a directional **influence** of one characteristic onto another. For instance, consider a subset of his diagram, the lower right quadrant where lay one of the offspring guinea pigs:

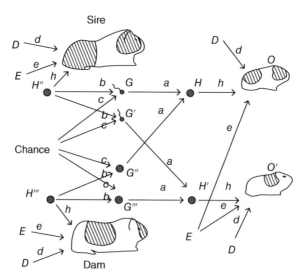

FIGURE 15.2   Sewall Wright's guinea pig path diagram of 1920. Source: Wright (1920). © 1920 Proceedings of the National Academy of Sciences.

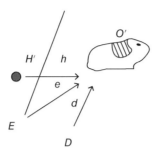

Letting $\mathbf{H}'$ represent influences of heredity, $\mathbf{E}$ environmental influences, and $\mathbf{D}$ a disturbance term, it becomes clear that Wright was hypothesizing the overall "makeup" of the baby guinea pig. Wright asked such questions as "**To what extent did heredity vs. environment contribute to the color of the offspring?**" What made path analysis so useful to him is that he was able to model a dependent variable both as a response and as a predictor to another variable. Notice that in Figure 15.2, $\mathbf{G}'''$ has an arrow pointing to $\mathbf{H}'$, which then has an arrow pointing to the guinea pig. Allowing dependent variables to serve simultaneously as predictors of other dependent variables was at the time an advancement over multiple regression. Wright named these coefficients **path coefficients**. And though these paths **a**, **h**, **e**, and **d** were fundamentally analogous to regression coefficients, in part since it also made sense that heredity and environment **caused** color in the offspring, Wright referred to these coefficients simultaneously by the name of **causal coefficients**. Such coefficients could be used to estimate **presumed** causal pathways.

Because this chapter is in no way intended as a historical analysis of Wright's contributions, we cut to the chase rather quickly and give you the bottom line: **Wright developed path analysis in a context in which "cause and effect" was an assumption that was quite reasonable, if not obvious**. That heredity and environment contributed to characteristics in guinea pig offspring was very much biologically apparent, and hence referring to cause and effect when defining a path coefficient made at least **reasonable** sense in the context in which Wright was working. The system of variables he was working with was more or less "complete." Beyond that, however, there was nothing at all "causal" about his coefficients, and Wright himself acknowledged this in a contentious debate about causation with Henry Niles of John Hopkins University beginning in 1922 (see Denis and Legerski (2006) for details). For Wright, it made good sense that he had a system in which causality could be deduced. **But this was in genetics**. When we take Wright's idea about a complete system and inferred causality on coefficients to other models where the system is much weaker, incomplete, or simply implausible, **assigning causation to these coefficients is simply unreasonable**. If the system is relatively narrow and strong, however, then causality may be "do-able" even if still philosophically quite challenging.

Regardless of Wright's attempt to clear the record, path analysis has since become linked to the misnomer "causal modeling," and has unfortunately been misused and misinterpreted in a wealth of substantive areas where causal "intuition" could never be considered obvious. This is unlike the study of the genetic transmission of color in guinea pigs, or the mechanism of heredity by which parents are hypothesized to transmit intellectual capacities to their children. **The statistical technique, whether that of path analysis or structural equation modeling, has no more claim to causality than any other statistical method.** Both methods are best considered simply as extensions to regression and factor analysis where causality enters the discussion only if it is warranted by factors extrinsic to the model (e.g., experimental design). For a discussion of causation in the context of structural equation modeling, see Mulaik (2009, pp. 63–110).

15.3   EARLY POST-WRIGHT PATH ANALYSIS: PREDICTING CHILD'S IQ (BURKS, 1928)

One of the earliest uses of path analysis following Wright was in modeling children's IQ as a function of both parental intelligence and environment (i.e., the classic nature–nurture debate mentioned previously). The work appeared in Burks (1928). The path diagram in Figure 15.3 was featured in Burks' work.

We note the following from Burks' path diagram:

- A one-headed arrow from parental intelligence to child's IQ is indicated, representing the hypothesis that parental intelligence is a partial predictor (or even "cause," in such a context) of child's IQ. The arrow is pointing to child's IQ to indicate the direction of the hypothesized relationship. The coefficient of $r = 0.6036$ is the Pearson correlation between parental intelligence and child's IQ. The coefficient "a" is called a **path coefficient**, and as such is equivalent to a partial regression coefficient (typically, standardized).
- A one-headed arrow from environment to child's IQ is indicated, representing the hypothesis that environment is a partial predictor (again, perhaps even **cause**) of child's IQ. The coefficient of $r = 0.4771$ is also the Pearson correlation coefficient, and the coefficient "c" another path coefficient, also equivalent to a partial regression coefficient (again, typically standardized).
- A two-headed arrow is indicated between parental intelligence and environment, representing the hypothesis that these two variables are linearly related. The coefficient of $r = 0.7653$ is the correlation between parental intelligence and environment, corresponding to the modeled path "b".

Burks' analysis is significant for a few reasons. First, it was one of the first applications of path analysis to a problem in social science since Wright's introduction of the method in genetics and biology. Second, Burks' use and application of path modeling evidenced a keen awareness of what the method **could do** versus what it **could not do** in terms of its ability to deduce causal claims. As Burks noted early in his paper:

> **The method [of path coefficients] is limited by the rarity with which we have actual knowledge of causal relations; but it provides a toll of the nicest precision in such situations as do offer an adequate basis for postulating causation. It cannot, itself, uncover what is cause and what is effect, though in the absence of definite knowledge regarding causal relationships between variables, the method 'can be used to find out the logical consequences of any particular hypothesis in regard to them.' Conservatively stated, in any situation in which we feel justified in drawing conclusions regarding the effects of certain phenomena upon others, the Wright method provides a numerical expression of such conclusions.** (Burks, 1928, p. 299)

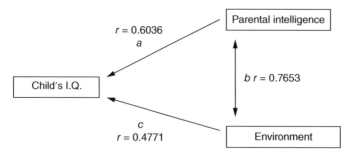

FIGURE 15.3   Burks' path diagram hypothesizing parental intelligence and environment as partial causes of child's IQ. Source: Burks (1928). © 1928 Stanford University.

Every student, researcher, and user of path analysis and structural equations would do well to memorize the above quote from Burks and repeat it to themselves each and every time they fit such a model. In Burks, we find the **correct** contextualization of path modeling. It is a statistical technology, which no more than regression, makes **any** claims about **uncovering** or otherwise **establishing** causality. As Burks correctly emphasizes, should one be working with variables for which causal relations among them may be safely **assumed**, then naming a path coefficient that of a **causal coefficient** perhaps makes more sense, assuming one can make sense of what such a **partial cause** actually means on a philosophical level (good luck with that one!). What permitted Burks to associate any element of causality with her model was not the fact that she employed path analysis. It was the fact that it made methodological sense, given the paradigm at the time, that intelligence was a hereditary trait, and thus "smart parents" often had "smart kids." Why? **Because of a genetic causal link**. Without evidence for a presumed causal link, speaking of causation makes little sense.

What has unfortunately happened since the advent of path analysis (and its overachieving offspring, **structural equation modeling**) is that the term **causality** has made its way into models that have absolutely no evidence of being causal extrinsic to the method. One can model the causal coefficients linking self-esteem to life satisfaction all one wants, but unless evidence exists to suggest the pathway is in fact causal, the so-called "causal coefficient" is more akin to an ordinary regression coefficient and should be interpreted as such. Our **theories** may be causal, but our coefficients are not. Causality is simply not that easy.

15.4 DECOMPOSING PATH COEFFICIENTS

When we speak of "decomposing" a path coefficient, what we mean is learning what the coefficient is a function of. That is, we want to know ways in which the coefficient can be generated by reference to other pathways. The decomposition of path coefficients is the essence of path analysis, so we begin with a simple example from Wright's original work.

Consider another of Wright's diagrams in Figure 15.4, also featured in his classic 1920 paper.

In the diagram, X and Y are designated as response variables, which recall in path models are typically known as **endogenous** variables. A, B, C, and D are the explanatory variables in Wright's model, which also recall are known as **exogenous** variables in path analysis. Path coefficients in the model are given by a, b, c, b', c', and d'. The two-headed arrow joining B and C represents the correlation between these two variables, denoted by r_{BC}.

Following Wright, we would like to know the influences on the endogenous variable X. Notice that X has three arrows pointing to it: one from A, one from B, and one from C. However, since B and C are correlated, this correlation must also be taken into account when determining the influences on X.

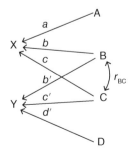

FIGURE 15.4 Wright's generic path diagram. Source: Wright (1920).

Squared path coefficients yield, as Wright put it, "the degree of determination by each cause." (Wright, 1920, p. 329). How can we then write the equation for the determination of X?

Wright noted the following:

$$a^2 + b^2 + c^2 + 2bcr_{BC} = 1 \tag{15.1}$$

That is, the determination of X is a function of the sum of all squared path coefficients pointing to it, that is, $a^2 + b^2 + c^2$. However, we must also account for the correlation between B and C (i.e., bcr_{BC} in (15.1)). Note that if the correlation between B and C were equal to zero, then the equation would reduce to

$$a^2 + b^2 + c^2 + 2bcr_{BC} = 1$$
$$a^2 + b^2 + c^2 + 2bc(0) = 1$$
$$a^2 + b^2 + c^2 = 1$$

We see then that under the condition of zero correlation between B and C, in Wright's model, the determination of X is simply a function of the sum of squared coefficients.

You might ask why it makes sense, conceptually, to add $2bcr_{BC}$ to the sum of squared coefficients should there be a correlation between B and C. Suppose we did not add this term, even under the condition that $r_{BC} \neq 0$. Without this correlation accounted for, however, can we really say we are **determining** the contributing factors to X? The reason we are adding $2bcr_{BC}$ is to account for the fact that there is **shared variation** between these variables. In doing, we are attempting to model the "system" of variables, however far-reaching that system may be. This was the very essence of Wright's path analysis, to try to account for a network of variables, and attempting to model as much as possible every aspect of that system.

Of note as well in Wright's analysis is the assumption that the correlation between A and B is equal to 0, or at minimum, simply not modeled. How do we know this? We do not actually know it is equal to 0, but we do know that Wright was not interested in modeling it, otherwise a two-headed arrow connecting A and B (and C and D) would have been included. As Wright wrote (Wright, 1920, p. 329), "… to illustrate a system (i.e., the system in Figure 15.4) in which the variations of two quantities X and Y are determined in part by independent causes, such as A and D, respectively, and in part by common causes such as B and C. These common causes may be correlated with each other as in the figure."

This idea of explicitly **not** modeling a path is an essential feature of path analysis and structural equation modeling. It is very important to understand that **not modeling** something still constitutes an **act of modeling**. Choosing **not** to correlate A and B should be a product of one's theory. It is in such ways that path and SEM models demand the investigator think carefully and clearly about the model he or she is subjecting to test. One can appreciate then how with path and SEM models, the otherwise constrained environment of regression and factor analysis has just been expanded to allow more flexibility in modeling possibilities.

15.5 PATH COEFFICIENTS AND WRIGHT'S CONTRIBUTION

If a path coefficient is essentially nothing more than a regression coefficient, what then was Wright's contribution? It certainly was not simply that of calling a regression coefficient by the name of a path coefficient. The contribution lay in demonstrating how coefficients along pathways could be **decomposed**, essentially revealing that correlations and the like could be written as a series of alternative pathways in a given model. This gave us the mathematics to compute, for instance, the **effect** of one variable on another **through an intervening variable**. By a series of rules, we could now trace paths in a system to determine the effects one variable has on others through intermediary pathways. As Wright noted:

> The path coefficients in a system of causes and effects can be calculated, if a sufficient number of simultaneous equations can be made, expressing the known correlations in terms of the unknown path coefficients ... and expressing **complete determination** [emphasis added] of the effects by their causes. (Wright, 1920, p. 330)

Ordinary regression models, even multivariate ones, do not allow for this, since the regression typically "ends" with the given endogenous variable(s). These variables are not given the opportunity to predict other variables in the system. Path analysis and structural equation modeling provide the user more flexibility in modeling a wider variety of hypotheses and more control over the fixing or freeing of parameters. Virtually all regression models can be considered as special cases of the wider path-analytic framework, just as many statistical models can be considered special cases of the wider structural equation modeling framework. If for no other reason, structural equation models are useful as a pedagogical tool for conceptualizing statistical models in general, a point we return to later in this chapter.

15.6 PATH ANALYSIS IN R—A QUICK OVERVIEW: MODELING GALTON'S DATA

Several software programs are available for fitting path and SEM models (e.g., AMOS, R, LISREL, EQS, SAS). We illustrate a very simple path analysis using R's lavaan package (Rosseel, 2012), using for now only a chi-square goodness of fit test to assess model fit (we discuss additional indicators of model fit later). We once again use data from the package HistData (Friendly, 2014), this time on the heights of mothers and fathers and their offspring. The data are located in `GaltonFamilies`:

```
> library(car)
> some(GaltonFamilies)
```

family	father	mother	midparentHeight	children	childNum	gender	childHeight
46	014	73.0	67.0	72.680	2	2	male	67.0
190	050	71.0	64.5	70.330	2	1	male	73.0
219	056	71.0	62.0	68.980	5	3	male	70.5

About the data:

- `family` is simply an index number identifying a given family in the data.
- `father` is the height of the father.
- `mother` is the height of the mother.
- `midparentHeight` is the mean height (computed as father + 1.08 * mother)/2).
- `children` is the number of children spawn by the family.
- `childNum` is the number of the child within the family (ordered by height for boys followed by girls).
- `gender` is the sex of the child.
- `childHeight` is the height of the offspring child.

We first try a model in which `childHeight` is a function of both `mother` and `father` heights:

```
> library(lavaan)
> gf.model <- 'childHeight ~ mother + father'
> sem.fit <- sem(gf.model, data = GaltonFamilies)
> summary(sem.fit)

lavaan (0.5-16) converged normally after    1 iterations

  Number of observations                                   934

  Estimator                                                 ML
  Minimum Function Test Statistic                        0.000
  Degrees of freedom                                         0
  P-value (Chi-square)                                   0.000

                Estimate  Std.err  Z-value  P(>|z|)
Regressions:
  childHeight ~
    mother        0.291    0.048    5.996    0.000
    father        0.368    0.045    8.218    0.000

Variances:
    childHeight  11.451    0.530
```

The model is based on a total of 934 observations and was estimated using maximum likelihood (ML). Note the model has **zero** degrees of freedom, which means it is **saturated**, implying that it will fit **perfectly** yielding a chi-square value of 0.000. A model with zero degrees of freedom regenerates the data fully, and hence **has no opportunity to be wrong**. Since the model is saturated, we do not interpret parameter estimates and move on to specifying a model that is not saturated. We accomplish this by imposing a **constraint**. We choose to constrain the path from `mother` to `childHeight` to be equal to 1.0 (1*mother). Doing such frees up a degree of freedom. When fitting this model, we obtain:

```
> gf.model <- 'childHeight ~ 1*mother + father'
> sem.fit <- sem(gf.model, data = GaltonFamilies)
> summary(sem.fit)

lavaan (0.5-16) converged normally after    11 iterations

  Number of observations                                   934

  Estimator                                                 ML
  Minimum Function Test Statistic                      193.065
  Degrees of freedom                                         1
  P-value (Chi-square)                                   0.000

                Estimate  Std.err  Z-value  P(>|z|)
```

```
Regressions:
  childHeight ~
    mother            1.000
    father            0.329     0.050     6.626     0.000

Variances:
    childHeight       14.080    0.652
```

We note the following regarding the output:

- Since the path from `mother` to `childHeight` is now fixed at 1.0, the model gains a single degree of freedom.
- The parameter estimate for `mother` is reported as 1.000, since we fixed it as such, and hence is not evaluated for statistical significance.
- Other parameter estimates have changed as a result of fixing the path to 1.0. For instance, note that the variance for `childHeight` has increased from 11.451 to 14.080.
- Likewise, the parameter estimate for `father` has changed from 0.368 to 0.329 as a result of fixing the `mother` to `childHeight` parameter at 1.0.

The following is the path diagram corresponding to the fitted model:

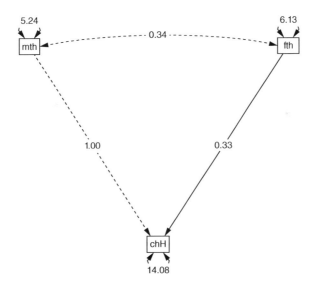

Our example featured here is simply a cursory overview of fitting a path model and an introduction to the concept of fixing parameters. Later, we will discuss a variety of fit indices for models such as that just featured, which will include the chi-square test as only one possibility. As we will see, the *p*-value obtained for Galton's model suggests we **not** deem the model well-fitting. Problems abound with the chi-square test as a measure of model fit, however, and other statistics will be discussed that seek to improve on it.

15.6.1   Path Model in AMOS

SPSS's AMOS program is a specialized program for fitting path and structural equation models that is relatively easy to use, and has a graphical interface that is helpful for drawing and visualizing path models. A full tutorial on how to use AMOS is well beyond the scope of this chapter. For an excellent tutorial and precise software guidance, see Byrne (2016). We wish here only to demonstrate a simple generic path model using the software and survey partial output in the context of fixing and freeing parameters as we did in the prior example using R. For this example, suppose we have data on variables Y, X, and Z:

	Y	X	Z
1	5.00	4.00	1.00
2	2.00	6.00	3.00
3	4.00	9.00	5.00
4	6.00	5.00	7.00
5	3.00	3.00	8.00
6	9.00	2.00	6.00
7	8.00	6.00	2.00
8	7.00	8.00	4.00
9	4.00	7.00	5.00
10	1.00	9.00	4.00

In AMOS, we draw the following path model that we wish to fit:

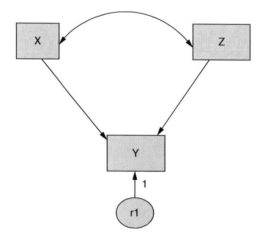

It is evident in the path diagram that variables X, Y, and Z are observed or "manifest" variables, since they are enclosed in squares/rectangles. Had they been latent variables, we would have enclosed them in ovals, and had observed variables "indicate" them. We can see that "$r1$" is enclosed in an oval, and represents the residual error term associated with observed variable Y. When we run this model using maximum likelihood as our estimator, we obtain the following output:

Notes for model (Default model)

Computation of degrees of freedom (Default model)

Number of distinct sample moments: 6

Number of distinct parameters to be estimated: 6

Degrees of freedom (6-6): 0

Result (Default model)

Minimum was achieved

Chi-square = .000

Degrees of freedom = 0

Probability level cannot be computed

Note that in the above output, there are a total of six sample moments available to be estimated, and we are estimating them all as indicated by "Number of distinct parameters to be estimated: 6". Hence, the degrees of freedom are equal to 0, indicating that the model is saturated. Recall that **saturated models fit the data perfectly** and hence have no way of being "wrong." Consequently, on a scientific level, they are useless to us. This is confirmed by the results of the chi-square statistic, equal to 0. AMOS reports that the probability level cannot be computed, since the model has zero degrees of freedom.

To make the model have positive degrees of freedom, we will again impose a constraint on one of the parameters. Instead of estimating the path from X to Y, we will fix it at a value of 1.0:

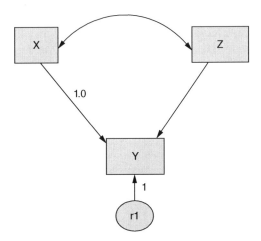

Notice that the path from X to Y now has on it a value of 1.0, and hence is not being freely estimated. When we once again run the model, we obtain the following:

Notes for model (default model)

Computation of degrees of freedom (default model)

Number of distinct sample moments:	6
Number of distinct parameters to be estimated:	5
Degrees of freedom (6 − 5):	1

Result (default model)

Minimum was achieved

Chi-square = 10.381

Degrees of freedom = 1

Probability level = 0.001

Maximum Likelihood Estimates

Regression weights: (group number 1—default model)

	Estimate	S.E.	C.R.	P	Label
Y <--- X	1.000				
Y <--- Z	0.259	0.641	0.404	0.686	

Covariances: (group number 1—default model)

	Estimate	S.E.	C.R	P	Label
X <--> Z	−1.250	1.635	−0.765	0.444	

Variances: (group number 1—default model)

	Estimate	S.E.	C.R.	P	Label
X	5.290	2.494	2.121	0.034	
Z	4.250	2.003	2.121	0.034	
r1	15.715	7.408	2.121	0.034	

As we can see, the number of parameters to be estimated is now equal to 5 instead of 6, yielding positive degrees of freedom (equal to 1). We see the chi-square statistic is now unequal to 0, yielding instead a value of 10.381, with associated p-value of 0.001. Recall from our brief discussion of the model fit in R that we actually desire a non-statistically significant p-value here, so if we were to conclude model fit based on the chi-square statistic alone (which as we will see is typically not a good evaluative strategy), this would be considered a negative research result and a rather poor-fitting model. On the underside of the output, AMOS reports the maximum likelihood estimates which are the estimated regression weights. We see the path from X to Y fixed at 1.000, while the path from Z to Y yielding an estimate of 0.259, and not statistically significant ($p = 0.686$). The covariance between X and Z was estimated equal to −1.250, and the variances of X, Z and the residual term $r1$ are reported underneath. All are statistically significant at $p = 0.034$, indicating evidence that they are all unequal to 0 in the population from which the data was drawn.

15.7 CONFIRMATORY FACTOR ANALYSIS: THE MEASUREMENT MODEL

Having briefly introduced the main ideas of path analysis, we now move on to a brief consideration of what will form a central component of a structural equation model, that of the **measurement model**. The measurement model in SEM usually takes the form of hypothesizing latent factors or hypothetical structures. The **structural model**, on the other hand, usually denotes relationships among such hypothetical constructs (Stevens, 2009), although structural models can, in theory at least, be as simple as the path models featured earlier where no latent variables were hypothesized. SEM has a terminology all its own, and specialists in the area may be adamant about terms used, but there is in reality nothing that special or unique about naming something a "structural" model. In this book, it simply denotes a modeled relationship between variables. In SEM models, those modeled relationships often involve latent variables. "Structural" for some writers implies a "causal" notion (e.g., see Bollen, 1989, p. 4), but as we have noted, there is nothing inherently "causal" about a structural equation model (Bollen would agree with us), and hence we prefer not to impose any meaning on the term "structural" other than a statistically modeled relationship.

Recall that in our presentation of the exploratory factor analysis model of Chapter 14, we made a cursory attempt to distinguish an **exploratory** model from that of a **confirmatory** model, and provided a tentative conclusion that at best, the distinction is, at least from a substantive point of view, quite **fuzzy**. From a technical point of view, confirmatory factor analysis (CFA) distinguishes itself from EFA in that the former overcomes the **rotational indeterminacy problem** and that given proper constraints, usually imposed by the investigator, will yield an identified model with unique parameter estimates. As Rencher and Christensen (2012, p. 482) note, "Every two-factor EFA model is underidentified because the factor loadings can be rotated without affecting the implied covariance matrix ... In the CFA setting, our intent is to impose phenomenologically based constraints on the model in order to ensure the model is identified. That is, we wish to guarantee that there exists a **unique solution** [emphasis added] for the parameter vector."

Even so, as Jöreskog noted, the distinction between CFA and EFA from a **substantive** vantage point is quite imprecise. For instance, when we choose to extract two instead of three factors in EFA, are we not engaged in confirmatory work? Of course we are. The fact that we specified *a priori* the extraction of two factors instead of three implied, however imprecise or ill-defined, an underlying **hypothesis**. In this sense then, the EFA was not really "exploratory" at all.

A convenient definition for our purposes then might be that when we start imposing additional constraints on a model, we might be said to be entering, at a **technical** level, the confirmatory stage of model-building, whether that be a factor analysis, multiple regression, or any other type of model. But surely, at a substantive or scientific level, this distinction is not really important. The difference then between exploratory and confirmatory models is one more of **flavor** and **degree** than it is one of absolute difference. One never finds oneself at a computer station, unsure of how he got there and totally naïve about the data before him, and proceeds to engage in exploratory modeling. In the end, all models are confirmatory, some less so than others. In models increasingly confirmatory in nature, the investigator is typically more **aware** of the fixing and freeing of parameters.

We turn again to Jöreskog in defining the nature of confirmatory factor analysis:

> **We shall describe a general procedure for performing factor analysis in the following way. Any values may be specified in advance for any number of factor loadings, factor correlations and unique variances. The remaining free parameters, if any, are estimated by the maximum likelihood method.** (Jöreskog, 1969, p. 183)

As previously mentioned in relation to path analysis, and as emphasized by Mulaik (2009), estimated parameters are no more "important" than parameters either not estimated or constrained to particular values. As we featured in Wright's analysis, that we not modeled a relation is nevertheless a choice to model it. Hence, in the full range of SEM models, there is nowhere for the investigator to "hide" or simply relegate decisions about parameter estimates to the "computer." One must know one's model inside and out.

The CFA model is given by

$$\mathbf{x} = \Lambda \mathbf{f} + \varepsilon$$

where as before, $\mathbf{x}$ is a vector of manifest observed variables, Λ is a matrix of factor loadings corresponding to the latent variables in $\mathbf{f}$, and ε is a matrix of "unique factors" or "specific variances" unique to each manifest variable in $\mathbf{x}$. As an example of a simple CFA model, consider the following model consisting of four observed variables and two hypothesized factors:

$$
\begin{array}{ccccc}
\mathbf{x} & = & \Lambda & \mathbf{f} & + \quad \varepsilon \\
\begin{bmatrix} x_1 \\ x_2 \\ x_3 \\ x_4 \end{bmatrix} & = & \begin{bmatrix} \lambda_{11} & 0 \\ \lambda_{21} & 0 \\ 0 & \lambda_{32} \\ 0 & \lambda_{42} \end{bmatrix} & \begin{bmatrix} f_1 \\ f_2 \end{bmatrix} & + \begin{bmatrix} \varepsilon_1 \\ \varepsilon_2 \\ \varepsilon_3 \\ \varepsilon_4 \end{bmatrix}
\end{array}
$$

Multiplying through matrices we obtain the following:

$$
\begin{aligned}
x_1 &= \lambda_{11} f_1 + 0 f_2 + \varepsilon_1 \\
x_2 &= \lambda_{21} f_1 + 0 f_2 + \varepsilon_2 \\
x_3 &= 0 f_1 + \lambda_{32} f_2 + \varepsilon_3 \\
x_4 &= 0 f_1 + \lambda_{42} f_2 + \varepsilon_4
\end{aligned}
$$

Understanding the meaning of each of the above equations is important. Consider, for instance, the first equation. We are hypothesizing that observed variable x_1 can be written as a function of latent variable f_1 (weighted by the loading λ_{11}), plus a disturbance term ε_1 (or "specific variance"). For the second equation, we hypothesize that observed variable x_2 can also be written as a function of latent variable f_1 plus its own unique disturbance term ε_2. Notice then that both of these first two manifest variables load onto latent variable f_1. The final two manifest variables load onto the latent factor f_2 only, since $0f_1 = 0$ in each case. That is, it is hypothesized that x_3 can be written as a function of f_2 plus its own disturbance term, ε_3. Finally, it is hypothesized that x_4 can be written as a function of f_2 plus its own disturbance term, ε_4. Note that a loading set to zero, such as with $0f_1 = 0$, still constitutes a **confirmatory move. Such a fixing should still be supported by theory.** This is what we are referring to when arguing that freely estimated parameters are no more "important" than those fixed. In most "nonconfirmatory" modeling contexts, a path may be constrained without the researcher having any awareness of the given parameterization of the model. In this way, the user is never given the opportunity to consider the theoretical implications of the implicit constraining of such paths, analogous to a cell phone with automated features that you cannot disable and may not even be aware of. Unmodeled paths are somewhat akin to a disbelief in something. **The disbelief is nonetheless a belief, analogous to how not modeling a path is still an act of modeling. In the confirmatory model, the researcher is implicitly accountable for virtually every parameter of the model, whether that parameter is freely estimated or a priori fixed.**

15.7.1 Confirmatory Factor Analysis as a Means of Evaluating Construct Validity and Assessing Psychometric Qualities

As we discussed when surveying exploratory factor analysis, latent variable modeling is not solely about statistics. Rather, it is equally about **measurement**. Confirmatory factor analysis modeling is popular in part because it allows researchers to hypothesize **unobservable variables** that may be exceedingly difficult to measure in practice and require several manifest indicators in order to get a "handle" on them. One of many areas where CFA has proved useful is in the area of pain measurement in chronically ill patients. The **Brief Pain Inventory (BPI)** is one such measure often used to assess self-reported pain of patients, and CFA has been used to assess the **construct validity** of the instrument (Lapane et al., 2014). The following is a CFA path diagram depicting how observed manifest variables load onto a hypothesized two-factor structure:

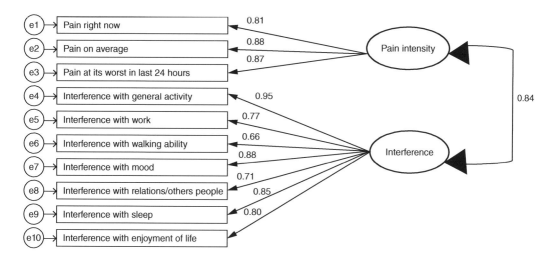

For this model, researchers hypothesized the two-factor structure having latent variables "Pain Intensity" and "Interference" along with corresponding manifest variables loading onto each respective construct. It is important to note that these two latent variables are considered "latent" for the reason that they are not easily measured, and must be inferred by measuring the series of manifest (or more easily measured) variables on the left, which are the items of the Brief Pain Inventory. Loadings appear along the paths indicating each construct, along with a two-sided arrow correlating both latent variables to an extent of 0.84. Error terms **e1** through **e10** are also present pointing to each observed variable.

As we will survey shortly with regard to SEM fit statistics, this model was deemed an adequately well-fitting model based on model fit statistics commonly used in the SEM field. What is perhaps most relevant to our discussion here in the spirit of factor analysis in general, however, is that the latent constructs of pain intensity and interference are **inferred constructs**, named as such only because research specialists in the field of pain measurement choose to call these constructs by these names. This is to say that the CFA model did not "discover" the latent variables "Pain Intensity" or "Interference." These names had to be inferred by the research community, analogous to what is done in the typical exploratory factor analysis. The difference between EFA and CFA, however, is that CFA requires us to identify the number of hypothesized latent variables in advance, along with the number of manifest variables thought to load on each construct, as is done in the above pain CFA model. What makes this particular model "confirmatory" is that the researchers evaluated an exact and precise a priori model thought to reveal factor structure, instead of following a more exploratory route as is customarily done in EFA. However, the naming and identification of the factors is not all that different than in exploratory models. The key distinction is that, presumably, the research field on which the CFA is being performed is "advanced" enough that researchers are evaluating specific factor structures named prior to testing the actual model, rather than still being in state of "exploration" attempting to figure out what factor structure is revealed. Many psychometric instruments such as this one are designed specifically with particular factor structures in mind, but statistically speaking, all the underlying statistics are "doing" is **evaluating the fit between model and data**, not unlike what is done in classical regression analysis with an R-squared statistic, for instance. If that fit is reasonable, then the model becomes a plausible account for the observed data, one of potentially many.

15.8 STRUCTURAL EQUATION MODELS

Having surveyed both the essentials of path and CFA models, we now introduce the full structural equation model, of which both path analysis and CFA, as well as many other statistical modeling techniques, can be considered **special cases** of this wider framework.

The classic structural equation model is given by

$$\boldsymbol{\eta} = \mathbf{B}\boldsymbol{\eta} + \boldsymbol{\Gamma}\boldsymbol{\xi} + \boldsymbol{\zeta}$$

where $\boldsymbol{\eta}$ is a vector of latent endogenous variables (appearing on both sides of the equation to allow endogenous variables to predict one another), $\boldsymbol{\xi}$ is a vector of latent exogenous variables, $\boldsymbol{\zeta}$ is a vector of latent errors or "disturbances," $\mathbf{B}$ is a coefficient matrix for latent endogenous variables, and $\boldsymbol{\Gamma}$ is a coefficient matrix for latent exogenous variables. The assumptions underlying a structural equation model are many and typically include (depending on the given parameterization), $E(\boldsymbol{\eta}) = 0$, that is, the mean of endogenous variables is equal to 0, $E(\boldsymbol{\xi}) = 0$, the mean of latent exogenous variables is equal to 0, $E(\boldsymbol{\zeta}) = 0$, the mean of latent errors or "disturbances" is equal to 0, and that $\boldsymbol{\zeta}$ are uncorrelated with $\boldsymbol{\xi}$, that is, latent errors are uncorrelated with latent exogenous variables. For additional assumptions, see Bollen (1989, p. 20).

The assumptions underlying a structural equation model in large part parallel those underlying the classic multivariate linear model, with the key exception being, of course, that the multivariate linear model does not **explicitly** feature such things as latent variables. Indeed, this distinction was also paramount as you may recall when comparing the EFA model to regression in the previous chapter. As we did then, it is pedagogically meaningful to compare the two models

$$\mathbf{Y} = \mathbf{XB} + \mathbf{E} \tag{15.2}$$

versus

$$\boldsymbol{\eta} = \mathbf{B}\boldsymbol{\eta} + \boldsymbol{\Gamma}\boldsymbol{\xi} + \boldsymbol{\zeta} \tag{15.3}$$

To highlight the similarities and differences, consider the simple SEM model in Figure 15.5.

Differences between the model in Figure 15.5 and that of the classic multivariate linear model (15.2) include the following:

- **Y**, rather than $\boldsymbol{\eta}$, is the response variable in the regression model. It is an **observed** vector of responses, whereas $\boldsymbol{\eta}$ is a vector of endogenous response variables. In contrast to the classic linear model, $\boldsymbol{\eta}$ is **unobserved**. Note that $\mathbf{B}\boldsymbol{\eta}$ in (15.3) also allows for the possibility of relating endogenous variables. In the model of (15.2), no such allowance is made for **Y**.

- In both models, **E** and $\boldsymbol{\zeta}$ are measures of unexplained variation, which includes measurement error and random "shocks," as well as all those influences acting on **Y** or $\boldsymbol{\eta}$ that are not accounted for in each model, or perhaps are even "unknowable."

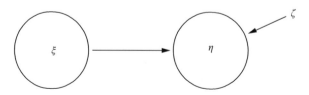

FIGURE 15.5 Simple structural equation model.

Just as was the case in EFA where we learned that the covariance matrix could be "decomposed" into $\sum = \Lambda\Lambda' + \psi$, the general structural equation model, of which recall the EFA model may be considered a special case, can also be decomposed into $\sum$. See Bollen (1989, pp. 323–326) for details.

15.9 DIRECT, INDIRECT, AND TOTAL EFFECTS

Three types of effects can be distinguished in a path or structural equation model. A **direct effect** is the prediction of one variable on another, unmediated by any other model variables. An **indirect effect** is the prediction of one variable on another but mediated by at least one other intervening variable. A **total effect** is the sum of direct and indirect effects. As noted by Bollen (1989, p. 36), "The decomposition of effects always is with respect to a specific model. If the system of equations is altered by including or excluding variables, the estimates of total, direct, and indirect effects may change." Of course, this isn't simply true of SEM models. As noted several times in this book, the **"context"** of a model is a powerful thing, in that a given variable may be predictive of another variable in one model, but the strength of that prediction might change in another when one or more new variables are added. Statistical models do not pretend to be perfect reflections of physical processes. The extent to which a model reflects "reality" usually depends more on design issues and such things as **ecological validity**, that is, the extent to which the experiment or study is actually reflective of what goes on in nature outside of the laboratory or setting in which the correlational study is being performed. Fitting a model is only the first step. **Relating that model to a scientific process is where things get philosophically very difficult**.

Returning to our defining of effects in SEM, we illustrate these effects by considering a now classic structural equation reproduced from Bollen (1989, p. 37) on industrialization and political democracy (Figure 15.6). For our purposes here, the actual substantive meaning of the variables is not relevant. What we wish to demonstrate is simply how direct, indirect, and total effects can be interpreted.

Some immediate features of Bollen's model are as follows:

- There are three latent variables, η_1, η_2, and ξ_1.
- η_1 is indicated by manifest variables y_1 through y_4, with errors ε_1 through ε_4.
- η_2 is indicated by manifest variables y_5 through y_8, with errors ε_5 through ε_8.
- ξ_1 is indicated by x_1 through x_3, with errors δ_1 to δ_3.
- Disturbances are associated with η_1 and η_2 (i.e., ζ_1 and ζ_2 respectively).
- Path parameters include λ_1 through λ_{11}, γ_{11}, γ_{21}, and β_{21}.

Recall that when we speak of a **direct effect**, as the name suggests, it is the effect of one variable onto another which does not go **through** any other variables. For example, the effect of ξ_1 on η_1 in Figure 15.6 is a direct effect, represented by parameter γ_{11}. Notice that ξ_1 on η_1 does not go through any other intermediary path. If we compare this to the effect of ξ_1 on η_2, the distinction between a direct effect and an indirect effect becomes immediately apparent. Note that like ξ_1 on η_1, ξ_1 has a direct effect on η_2 modeled by coefficient γ_{21}. However, it also has an indirect path. That path is $\xi_1 \rightarrow \eta_1 \rightarrow \eta_2$. We say that ξ_1 "acts on" η_2 **through** η_1. Again, using the words "acts on" or "goes through" is fine, so long as one knows what one means by such **physical-sounding** statements. **What would be incorrect to assume is that our semantics alone somehow give these coefficients "powers" they do not possess**. Stemming from our earlier discussions of Wright and those of mediation and moderation earlier in the book, coefficients along paths in any model are simply functions of a calculating machine. Any assignment of **substantive powers** must be a function of factors external to the modeling process. If you conclude that ξ_1 truly "acts on" η_2, for instance, this conclusion must be defended not with reference

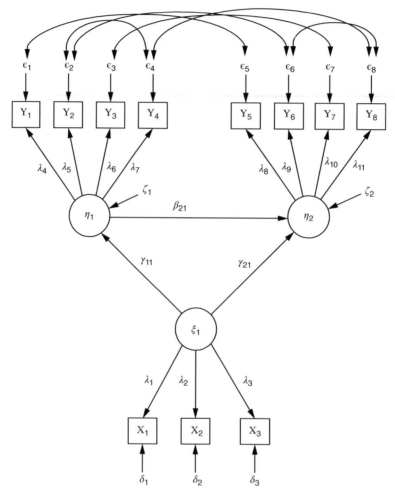

FIGURE 15.6 Bollen's classic structural equation model on industrialization and political democracy. Source: Bollen (1989). With permission from John Wiley & Sons, Inc.

to the model, but rather with reference to the objects you are modeling or the design adopted that presumably permits such powerful **action statements**.

Finally, the total effect is the sum of direct effects and indirect effects. For ξ_1 on η_2, we sum two pathways, the first being the direct path of $\xi_1 \rightarrow \eta_2$, the second $\xi_1 \rightarrow \eta_1 \rightarrow \eta_2$ (Bollen, 1989, p. 36). Hence, to get the total effect of ξ_1 on η_2 we add γ_{21} to $(\gamma_{11})(\beta_{21})$. We could obtain other effects in the model in an analogous fashion.

15.10 THEORY OF STATISTICAL MODELING: A DEEPER LOOK INTO COVARIANCE STRUCTURES AND GENERAL MODELING

Having provided a cursory overview of path analysis, confirmatory factor analysis, and structural equation models, we now provide a slightly deeper account of how models are fit in general. The following is a very general account of the principles involved in model-fitting, and serves in part to summarize the modeling process, whether the model be one of a simple *t*-test or that of an elaborate structural equation model. In the end, the model-fitting process is remarkably similar in virtually all cases when considered from a **global** perspective. From a pedagogical standpoint, standing on the "summit" of the structural equation model allows us to review the principles of modeling in its full generality. Our discussion is motivated by leaders in the field such as Bollen (1989), Fox (1997), and Mulaik (2009).

The models we have been considering in this chapter go by different names, for instance, **simultaneous linear equations**, **linear causal analysis**, **path analysis**, **structural equation modeling**, **covariance structure modeling**, to name but a few. Regardless of the model, the task typically boils down to solving a set of equations, usually with imposed constraints. As noted by Bollen (1989), the fundamental hypothesis on which virtually all statistical modeling is based can be given by

$$\sum = \sum(\boldsymbol{\theta}) \tag{15.4}$$

where $\sum$ denotes the population covariance matrix of observed variables, $\boldsymbol{\theta}$ is a vector of model parameters unique to the particular model under test, and $\sum(\boldsymbol{\theta})$ represents the covariance of observed variables written (or "reproduced") as a function of the model parameters contained in $\boldsymbol{\theta}$, usually referred to as the **implied** covariance matrix (sometimes designated $\hat{\sum}(\boldsymbol{\theta})$, since it is that "implied" by the model function). Be sure to note that this implied matrix does not exist on real, observed variables. The matrix $\sum$ exists since it is based on real, empirical variables. $\sum(\boldsymbol{\theta})$, on the other hand, is a reproduction based on how well our given model is specified. **Good models imply covariance matrices that reproduce population covariance matrices**. This is precisely what "fitting a model" ultimately means in most contexts. As noted by Bollen:

> The simplicity of this equation is only surpassed by its generality. It provides a unified way of including many of the most widely used statistical techniques in the social sciences. Regression analysis, simultaneous equation systems, confirmatory factor analysis, canonical correlations, panel data analysis, ANOVA, analysis of covariance, and multiple linear indicator models are special cases of $\sum = \sum(\boldsymbol{\theta})$. (Bollen, 1989, p. 2)

Hence, technically speaking, the job of the statistical modeler and scientist becomes one of equating the population covariance matrix $\sum$ with the covariance matrix implied by his or her theory, that of $\sum(\boldsymbol{\theta})$. Because we do not actually know the population covariances, we estimate them using **S**, the sample covariance matrix. Estimators are then sought which will, in some sense, ensure for us that the match between **S** and **S(θ)** is, on average, as **close as possible**. That is, we want our estimates to be ones that in some sense help minimize the distance between **S** and **S(θ)**, leaving our theory to do the rest in helping **S** be as close as possible to **S(θ)**. This logic applies to even the simplest case of fitting a least-squares line to bivariate data. The least-squares criterion guarantees distance will be minimized, but it is up to our theory to get us as close as possible the rest of the way. To formalize this idea further, we require the concept of discrepancy functions.

A **discrepancy function** is a general name used to describe functions that minimize the degree of misfit between **S** and **S(θ)** (Mulaik, 2009, pp. 312–314). Again referring to the case of fitting to data a line of best fit in OLS regression, in the language of discrepancy functions, we were **minimizing the discrepancy** between observed and fitted values by estimating parameters (e.g., intercept and slope coefficients) in such a way that such minimization is assured. The quality of our data and corresponding fit brought us the rest of the way to how small this minimization be. That is, differential calculus provided us with the solutions, the so-called **normal equations** that guaranteed not that the function would be small necessarily, but that it would be minimized. Hence, least-squares is one of the more common so-called "discrepancy functions."

Path analysis and structural equation modeling likewise use discrepancy functions in minimizing the degree of misfit between $\mathbf{S}$ and $\mathbf{S}(\boldsymbol{\theta})$. Rather than seeking to minimize the sum of squared errors as in OLS regression, much of covariance modeling uses maximum likelihood (ML) to minimize the following very general fitting function (Bollen, 1989, p. 107):

$$F_{ML} = \log \left| \sum \boldsymbol{\theta} \right| + \text{tr}\left(\mathbf{S}\sum{}^{-1}(\boldsymbol{\theta})\right) - \log |\mathbf{S}| - (p + q) \tag{15.5}$$

where $\left| \sum \boldsymbol{\theta} \right|$ is the determinant of the population covariance matrix (implied by the model), $\mathbf{S}$ is the sample covariance matrix, $\sum{}^{-1}(\boldsymbol{\theta})$ is the inverse of the population covariance matrix (implied by the model), and p and q refer to the number of observed variables. Minimizing such a fitting function as F_{ML} subject to particular constraints (see Bollen, 1989, p. 106) results in a consistent estimator of hypothesized free parameters (e.g., variances, covariances). Should $\mathbf{S} = \mathbf{S}(\boldsymbol{\theta})$, then $F_{ML} = 0$. In general then, as $\mathbf{S}(\boldsymbol{\theta})$ gets closer and closer to $\mathbf{S}$, we expect a value for F_{ML} closer and closer to zero.

15.11   THE DISCREPANCY FUNCTION AND CHI-SQUARE

We focus now on the maximum likelihood discrepancy function since, as mentioned, it is the one most commonly used for evaluating fit under a wide variety of situations. As well, $(n - 1)F_{ML}$ is distributed as chi-square χ^2 (i.e., the asymptotic distribution of $(n - 1)F_{ML}$ is a χ^2 distribution (Bollen, 1989, p. 110)) which we can use for testing the null hypothesis in (15.4):

$$H_0 : \sum = \sum(\boldsymbol{\theta}) \tag{15.6}$$

against the statistical alternative hypothesis

$$H_1 : \sum \neq \sum(\boldsymbol{\theta}) \tag{15.7}$$

In addition to the chi-square being a test of the null in (15.6), it can be shown also as a test that all residual covariances of the form $\sum - \sum (\boldsymbol{\theta})$ are equal to zero (Bollen, 1989, p. 263). Hence, if our model fits the data perfectly, then χ^2 should approximate 0. To the extent that the matrices of (15.6) differ, yielding the situation in (15.7), we would increasingly expect the covariances to not "match" those of the implied covariance matrix, and thus residual covariances would be unequal to 0. That is, $\sum - \sum (\boldsymbol{\theta}) \neq 0$.

Recall that in traditional hypothesis testing, we ordinarily seek to reject a null hypothesis in favor of a statistical alternative. For instance, recall that in an independent samples t-test, we ordinarily test the null $H_0 : \mu_1 = \mu_2$ against the alternative $H_1 : \mu_1 \neq \mu_2$ and reject the null should we observe a t-statistic that is large enough in absolute value (two-tailed test) to make it improbable under H_0. In the SEM environment, because of the way the null is stated in (15.6), a rejection of the null hypothesis actually indicates an **unfavorable** result for the researcher, because it implies that the observed covariance matrix does not match that of the implied covariance matrix. Hence, achieving the infamous $p < 0.05$ would actually designate a **disappointment** to the researcher in testing her model. To the contrary, she actually seeks to find **support** for the null hypothesis and not reject it. A "confirmation" of the null hypothesis for a path or structural equation model is thus deemed a **positive** result. **That is, we are "cheering for" a chi-square statistic that is not statistically significant.** As summarized by Stevens (2009, p. 361), "contrary to the general rule in hypothesis testing, the researcher would **not** want to reject the null hypothesis, as finding $\sum \neq \sum (\boldsymbol{\theta})$ would mean that the hypothesized model parameters were unable to reproduce **S**. Thus, smaller rather than larger chi-square values are indicative of a 'good fit'."

As a preliminary indicator of the extent to which the fitted model has made use of the information available, one may compare the size of χ^2 with its degrees of freedom. Once again, Jöreskog explained it best:

> If a value of χ^2 is obtained, which is large compared to the number of degrees of freedom, this is an indication that more information can be extracted from the data. One may then try to relax the model somewhat by introducing more parameters. This can be done by relaxing some restrictions on the common factor space or by introducing additional factors or both. If, on the other hand, a value of χ^2 is obtained which is close to the number of degrees of freedom, this is an indication that the model "fits too well." Such a model is not likely to remain stable in future samples and all parameters may not have real meaning. (Jöreskog, 1969, p. 201)

Jöreskog goes on to say that the final determination of model fit cannot be decided on fit statistics alone, but rather must be evaluated primarily on its usefulness. Still, the ratio of χ^2 to that of degrees of freedom can be regarded as a first useful step toward model evaluation.

15.12 IDENTIFICATION

Identification is a property of a statistical model, and though a feature of all models, nowhere does it come to the forefront more so than in SEM models. In classical ANOVA and regression, the identification of parameters is often implicitly assumed and rarely is an analyst confronted with having **unidentified** parameters. Such models are usually parameterized so that identification is assured. In SEM, however, as a consequence of the modeling flexibility available to the researcher, identification of parameters is just one of the many facets to which the user must devote at least some attention. What does it mean to say a parameter is identified? Formally, **a parameter is identified if unique values of the parameter correspond to unique probability distribution functions** (Casella and Berger, 2002). Less formally, a parameter is identifiable if one can estimate a unique value for that parameter.

A simple example will help clarify the concept of identification. Suppose we wish to solve the following equation:

$$y = f(x)$$
$$4 = 2x$$

The solution to the equation is of course 2, since

$$x = \frac{4}{2} = 2$$

That is, the value of "2" is the **unique** value that satisfies the equation $4 = 2x$. In this case, we say that x, the parameter we are seeking a solution for, is **identifiable** because a unique and distinct value for it exists such that the equation $4 = 2x$ is solvable.

Now, consider the following equation:

$$x + y = 10$$

Do unique values for x and y exist? Since a whole host of possibilities exist for what x and y **could** be that will satisfy the equation, we say the parameters x and y are not identified.

As mentioned, identification is not a property of path or SEM models alone. In the analysis of variance, for instance, the so-called **sigma constraint** or "sum-to-zero constraint" (Fox, 2016, pp. 157–158) is typically used to ensure identifiability of parameter estimates. Recall that in the analysis of variance model parameterized as $y_{ij} = \mu + \alpha_i + \varepsilon_{ij}$, we required estimating parameters μ, α_1, α_2, α_3, $\ldots \alpha_j$. However, there are only j means available, and so under this parameterization, we are trying to estimate more parameters than we have **information**. The sigma-constraint sets $\sum \alpha_i = 0$, which reduces the number of parameters to be estimated to the number of j means, leading to a model that is identified.

In structural equation models, we must ensure that each of the parameters of the given model is identified. If every parameter is identified, then the model is said to be identified. As explained by Mulaik (2009, p. 143), "The **identification problem** concerns whether or not one can determine unique values for the unknown parameters using the observed data and constraints placed on other parameters."

For any model, any one of three conditions can be true regarding identification:

- The model is **underidentified**, meaning that there are more parameters to be estimated than there is available information (e.g., $x + y = 10$).
- The model is **just-identified**, meaning that the number of parameters to be estimated is equivalent to the amount of available information (e.g., $4 = 2x$). This is typically the **saturated** model.
- The model is **overidentified**, meaning that there is more information available than there are parameters to be estimated.

Students new to statistical modeling, and especially SEM, are often taken aback by the issue of identification. After all, the teachings in any science regularly encourage students to place inherent value on empirical observations, that is, to allow data to **speak for themselves**. Surely then, it would seem that "tweaking" a statistical model for the purpose of solving an identification issue should have no place in the repertoire of the serious scientist, would it? After all, **data are data**, right? True, data are data, but identification is not about the data, it is about the **model** we are fitting to data. The process of model-building is one of theorizing a **structure** to data. Data on its own are of little use without a theory or narrative imposed on it to help us understand it, imperfect as the model may be. For this, we must ensure that parameters we wish to estimate as a function of the data are identified. Remember, without a **model**, without a **theory**, Darwin is simply noting observations, accumulating data, and never "explains" them. Even the most hardcore anti-theory scientists must at some point "theorize," and many use mathematical or statistical modeling to help describe such theory in precise terms such that they can more easily communicate such theory to other scientists. This is the precise rationale for building mathematical models. Furthermore, without theory, the plane crash is never "explained," but is rather simply observed, just as without theory, explaining the origins of the universe (or the causes of a toothache) is never even attempted.

15.13 DISTURBANCE VARIABLES

Throughout this chapter, we have incorporated unobserved **disturbance variables** into our models. But what exactly is a disturbance in SEM? At first thought, and drawing on our knowledge of linear models studied thus far, it may be tempting to think of disturbances as typical **residuals** one would obtain in a multiple regression analysis. However, it is generally incorrect to equate a disturbance term in SEM with that of a **residual** in multiple regression. To understand why, we quote Mulaik (2009) at some length:

> Disturbance variables represent extraneous influences such as errors of measurement and random shocks that are combined with the effects of exogenous and/or endogenous variables on a given endogenous variable. Disturbances are analogous to unique factors in common factor analysis or errors of measurement in classical test theory. However, disturbances may contain both systematic and unsystematic error. They are usually assumed to be mutually uncorrelated and uncorrelated also with the exogenous variables. Technically disturbances are also exogenous variables, but whatever is contained in them is not of focal interest in contrast to the explicitly named exogenous variables ... Requiring the disturbances to be uncorrelated with the exogenous variables implies that there are no other hidden relevant causes, not explicitly represented in the model, and permits unbiased estimation of the structural coefficients. In other words, the model represents a conception of external reality, and disturbances and their properties are supposed to hold in reality. When these assumptions are violated, the model may be compromised and yield misleading inferences when seemingly confirmed against data ... Disturbances are not residual variables. Residual variables are formed when one partials from a set of variables what can be predicted in them from other variables. They are the result of a mathematical operation. In linear models, residuals are necessarily uncorrelated with the predictor variables on which the partialled components are based. Disturbances, on the other hand, represent other causes of the variables not explicitly represented in the model otherwise, and subjunctively it is possible in some cases to imagine their being correlated with the explicit causal variables within the system and with each other. **The constraints imposed on disturbances, that they are mutually uncorrelated and uncorrelated with exogenous variables of the system, must be satisfied in the real-world situation represented by the model to achieve a closed system of variables in which causal relations can be inferred and structural coefficients estimated without bias. Residual variables become equivalent to disturbances when these constraints are satisfied** [emphasis added]. But if the constraints are not satisfied in the situation represented by the model, for example, there are hidden relevant causes in the disturbances that are correlated with the exogenous variables, then the residuals are not true disturbances and parameter estimates are likely biased." (p. 122)

We quoted Mulaik at length because of the clarity with which he defines what **is** and what is **not** a disturbance variable. In applying his distinctions to our IQ model in Figure 15.1, $d1$ represents **everything else** (including random "shocks") that could be accounting for the prediction of offspring IQ over and above parental IQ. And as emphasized by Mulaik, it is only in the idealistic situation where disturbance terms are mutually uncorrelated with one another **and** uncorrelated with exogenous variables contained in the system that they can quite possibly be equated to residual terms. For our model of IQ transmission, this would suggest that the disturbance $d1$ is uncorrelated with both parental IQ and the disturbance $d2$ associated with third generational IQ. Of course, such is very unlikely to ever hold in practice, but if we could assume it to be true for the substantive setting under consideration, then the disturbance associated with offspring IQ becomes more analogous to a residual such as one would have in a multiple regression model. For a further discussion of disturbance terms, see Jöreskog (1978).

15.14 MEASURES AND INDICATORS OF MODEL FIT

We survey a few of the more common measures of model fit that have been proposed to evaluate SEM models. Recall that in regression analysis, a measure of model fit such as R^2 was meant to evaluate, in general, how well fitted values "regenerated" observed data on a response variable. The extent to which predicted values correlated with observed values was the extent to which we deemed our regression model to **fit the data**. The general idea of fit statistics is no different in path and SEM models, though because of the complexity and multivariable nature of such models, more options exist for evaluating fit, each attempting to overcome shortcomings of competing statistics. However, the general idea is the same across the range of possibilities, and that is to ultimately evaluate how well one's theoretical model accounts for observed empirical data.

We begin by surveying the **chi-square goodness of fit test**, the **root-mean-square residual**, **standardized root-mean-square residual**, and **root mean-square error of approximation**. These are all generally considered **absolute** or **overall** measures of model fit. Such assess the extent to which the hypothesized model fits the data in a global fashion, measuring the extent to which $\mathbf{S} - \mathbf{S(\theta)} \neq 0$. Overall fit measures, however, cannot be used to evaluate models that are just identified, since for these models recall that $\mathbf{S} = \mathbf{S(\theta)}$, and so evaluating overall fit does not make sense. As well, overall measures do not tell us about the performance of separate individual model equations within the global model.

15.15   OVERALL MEASURES OF MODEL FIT

The χ^2 test, already discussed, is the classic fit statistic historically used to assess the general overall and global fit of path and SEM models. For populations not multivariate normal, χ^2 has been found to be biased and the Satorra–Bentler χ^2 is typically preferred (see Hu and Bentler, 1995). Under the null hypothesis, we expect χ^2 to equal zero, and hence the extent to which $\chi^2 > 0$ is the extent to which the hypothesized model is **less well-fitting**. Recall, however, that one of its major weaknesses is that the statistical significance of χ^2 is largely a function of sample size. Given any discrepancy between observed and expected, one merely has to collect an increasingly large sample size to essentially ensure statistical significance. Hence, even if the model was well-fitting, a statistically significant χ^2 would suggest the model not be retained, since recall statistical significance of χ^2 in the context of SEM works against the hypothesized model rather than in support of it.

Other drawbacks with χ^2, as noted by Bollen (1989, p. 266), include the fact that it can be quite sensitive to kurtosis, it requires the covariance matrix to be analyzed, it requires relatively large samples, and is tested under the assumption that H_0 is exactly true. Other limitations of the chi square test include the fact that χ^2 will generally decrease as model complexity increases. As one adds more parameters to one's model, χ^2 will generally diminish, which could give an illusion that a "better" model has been achieved. Surely, we do not want to judge the "goodness" of our model by simply the number of parameters we are estimating. Indeed, recall from Jöreskog (1969) that better-fitting models are generally those for which the ratio of χ^2 to df is relatively small.

As a result of such problems with χ^2, it is seldom interpreted without a simultaneous consideration of other available criteria for assessing model fit. Indeed, as emphasized by Bollen (1989), though χ^2 should always be reported for any structural equation model, it should nonetheless be supplemented with a number of other indices and indicators. We briefly survey some now.

15.15.1   Root Mean Square Residual and Standardized Root Mean Square Residual

The **root mean square residual** (RMR) is an index of fit proposed by Jöreskog and Sörbom (1981). It is essentially a measure of how well a model does **not** fit, since it is based on the **residuals** of the fitted model. The root mean square residual is given by

$$\mathbf{RMR} = \left[2 \sum_{i=1}^{q} \sum_{j=1}^{i} \frac{\left(s_{ij} - s_{ij}' \right)^2}{q(q+1)} \right]^{1/2}$$

where s_{ij} is a given element of the observed covariance matrix $\mathbf{S}$, s'_{ij} is a given element of the model-implied covariance matrix, $\mathbf{S}(\boldsymbol{\theta})$, and q is the number of observed variables for the given model. A look at the equation for RMR reveals that in general, the greater the sum of differences $s_{ij} - s'_{ij}$ for a constant q, the greater the size of the measure. Hence, the smaller the value of RMR, in general, the better the fit of the model (Stevens, 2009). It is easy to see, however, that the differences $s_{ij} - s'_{ij}$ could be large or small depending on the sizes of the variances and covariances of observed variables and their corresponding scales, somewhat analogous to how the covariance for two variables could be small or large in part due to the variability exhibited on each scale. If the raw observations are highly variable, then RMR will tend to be larger than not, and hence assessing model fit in any "absolute" fashion is very difficult using RMR.

A measure that purports to solve the scale issue problem of RMR is the **standardized root mean square residual** (**SRMR**). As the name suggests, SRMR first standardizes residuals $s_{ij} - s'_{ij}$ by dividing by respective standard deviations, $s_i s_j$, that is, $\left(s_{ij} - s'_{ij} \right) / s_i s_j$. Smaller values of SRMR are preferred over larger ones. According to Hu and Bentler (1999), values of 0.08 or less are indicative of good fit.

15.15.2 Root Mean Square Error of Approximation

A final measure of overall model fit discussed here is the **root mean square error of approximation** (RMSEA) (Steiger and Lind, 1980) given by

$$\text{RMSEA} = \sqrt{\frac{1}{n-1}\left(\frac{\chi^2_m - df_m}{df_m} \right)}$$

The extent to which model $\chi^2_m - df_m$ is large relative to df_m, RMSEA will likewise be larger than not. Conversely, the extent to which $\chi^2_m - df_m$ is small relative to df_m is the extent to which RMSEA will approach zero. General cut-offs in the range of 0.01, 0.05, and 0.08 have been proposed to indicate excellent, good, and relatively poor-fitting models (MacCallum, Browne, and Sugawara, 1996). As we will see, the RMSEA is somewhat similar in spirit to that of the Tucker-Lewis index, in that it essentially penalizes one for having "too complex" of a model by the discrepancy $\chi^2_m - df_m$.

Other measures of overall model fit include the **goodness of fit index** (GFI) and the **adjusted goodness of fit index** (AGFI), both proposed by Jöreskog and Sörbom (1986), though not discussed here.

15.16 MODEL COMPARISON MEASURES: INCREMENTAL FIT INDICES

To reiterate, the measures of fit we have so far discussed, that of χ^2, RMR, SRMR, and RMSEA, are all considered **overall** or **absolute** measures of model fit in that they make no attempt to compare the fit of a given model to that of a competing model. Rather, they simply provide an indication of the extent to which $\mathbf{S} - \mathbf{S}(\boldsymbol{\theta}) \neq 0$. Oftentimes, however, we are more interested in comparing the fit of our hypothesized model relative to a simpler model. Model comparison measures, or **incremental fit indices**, attempt to address this need. In what follows, we survey a few of the more popular of such measures. Our discussion of a few of them should give you an idea of how such incremental fit indices generally work.

The first measure we discuss is that of the **normed-fit index** (NFI) (Bentler and Bonett, 1980) given by

$$\Delta_1 = \frac{\chi_b^2 - \chi_m^2}{\chi_b^2} = \frac{F_b - F_m}{F_b}$$

where χ_b^2 and χ_m^2 are obtained chi-square values for the baseline and hypothesized models, respectively, and since $\chi^2 \sim (n-1)F_{ML}$, F_b and F_m are values of the corresponding fitting functions. The baseline model on which F_b is computed is one that shows more restrictions on it compared with the hypothesized model on which F_m is calculated. Indeed, the baseline model is usually quite restrictive and hence the difference $F_b - F_m$ is indicative of how much the respective fitting function for F_m decreases the baseline value of F_b. The difference $F_b - F_m$ is divided by F_b in order to provide a "context" for evaluating $F_b - F_m$, that is, it provides a maximum for evaluating the distance between competing models. Since F_m must be equal to or less than F_b, it stands that when there is **maximum improvement in model fit**, then (Bollen, 1989, p. 270):

$$\Delta_1 = \frac{F_b - 0}{F_b} = 1$$

Conversely, when the hypothesized model provides no improvement, then we would expect Δ_1 to equal approximately 0, since $F_b - F_m = 0$ and so $0/F_b = 0$. Values of 0.95 and higher are typically indicative of a well-fitting model, not in the "absolute" sense as in the case of such statistics as χ^2, but instead relative to a baseline model. As noted by Bollen (1989), however, a weakness of Δ_1 is that it does not incorporate degrees of freedom into its measure. Furthermore, like χ^2, it is also quite sensitive to sample size. As summarized by Bollen (1989, p. 270):

> A limitation of Δ_1 is that it does not control for degrees of freedom. The value of F_m can be reduced by adding parameters. This is analogous to increasing the R^2 for a regression equation by including more explanatory variables. Though the R^2 may improve, the degrees of freedom decrease, and the model becomes more complex. An adjusted R^2 that corrects for degrees of freedom can reveal that a more parsimonious equation has a superior fit.

Bollen's ρ_1 (1989) is similar to another fit statistic called the **Tucker-Lewis Index** (Tucker and Lewis, 1973), which will be discussed shortly. There is a slight difference between these two statistics, and hence for pedagogical purposes, we keep the two statistics distinct as to more easily study the logic of their formulations. Bollen's ρ_1 is given by (Bollen, 1989, p. 272):

$$\rho_1 = \frac{(F_b/df_b) - (F_m/df_m)}{(F_b/df_b)}$$

$$= \frac{(\chi_b^2/df_b) - (\chi_m^2/df_m)}{(\chi_b^2/df_b)}$$

where as before, F_b and χ_b^2 are the corresponding fit function and chi-square, respectively, for the baseline model, and F_m and χ_m^2 are the corresponding fit function and chi-square for the hypothesized model.

Let us examine what ρ_1 actually measures. We first note that it is very similar to our previous measure Δ_1, in that it assesses how the hypothesized model improves overall fit relative to a baseline model. There is, however, an important difference in that ρ_1 divides each χ^2 by its respective degrees of freedom. Why are degrees of freedom relevant in this regard? The logic of ρ_1 is that it rewards the fitting of models that "spend" a smaller number of degrees of freedom in order to improve model fit relative to the baseline model. It essentially evaluates the difference between $\chi_b^2 - \chi_m^2$ relative to χ_b^2, but also relative to degrees of freedom under each model. Generally, values of ρ_1 greater than 0.95 are indicative of well-fitting models in relation to the baseline model. Again, Bollen (1989, p. 273) summarizes the concept best:

> Since introducing additional parameters lowers df, it is possible for ρ_1 to stay the same or to decrease for more complex specifications. Maintained models that have a lower fitting function value with relatively few parameters have higher ρ_1 values than models with the same fit value for a more complicated specification.

A measure of fit related to ρ_1 is the **Tucker-Lewis index (NNFI)** given by:

$$\rho_2 = \frac{\left(\chi_b^2/df_b\right) - \left(\chi_m^2/df_m\right)}{\left(\chi_b^2/df_b\right) - 1}$$

The distinction between ρ_1 and ρ_2 is such that for ρ_1, the best case scenario for evidence of model improvement occurs when $\left(\chi_b^2/df_b\right) - \left(\chi_m^2/df_m\right)$ is equal to $\left(\chi_b^2/df_b\right)$ in the numerator driving the value of ρ_1 toward 1.0. Again, note that in this regard, the improvement in model fit as evidenced by $\left(\chi_b^2/df_b\right) - \left(\chi_m^2/df_m\right)$ is considered relative to $\left(\chi_b^2/df_b\right)$. For ρ_2, $\left(\chi_b^2/df_b\right) - \left(\chi_m^2/df_m\right)$ is not compared relative to $\left(\chi_b^2/df_b\right)$ but rather to $\left(\chi_b^2/df_b\right) - 1$. In this denominator, we have a contrast between the baseline fit and a "best fit" as indicated by "1." That is, ρ_2 puts $\left(\chi_b^2/df_b\right)$ in some context of a best-fit model, something that is not done in ρ_1. Bollen, again, summarizes it best:

> For ρ_2, the best fit is defined as the **expected value** of $\left(\chi_m^2/df_m\right)$. This equals one when the assumptions underlying the chi-square approximation are satisfied for the maintained model, since the expected value of a chi-square variate is its df ... When $\left(\chi_m^2/df_m\right)$ is one, ρ_2 is one, and this is an ideal fit. (Bollen, 1989, p. 273)

Values of ρ_2 greater than 0.90–0.95 are generally indicative of well-fitting models. For further details on ρ_2, see Bollen (1989, pp. 273–274) and Mulaik (2009, pp. 330–333).

A final measure of incremental fit discussed here is the **comparative fit index (CFI)**, given by

$$\text{CFI} = 1 - \frac{\max\left(\chi_m^2 - df_m, 0\right)}{\max\left(\chi_b^2 - df_b, \chi_m^2 - df_m, 0\right)}$$

We can see that the logic of the CFI is similar in spirit to that of ρ_1, only now we are subtracting df from χ^2 (i.e., $\chi_b^2 - df_b$) instead of taking ratios as was done for ρ_1. A value of CFI close to zero suggests that the additional estimated parameters (i.e., increased model complexity) used in generating χ_m^2 are hardly worthwhile. Values of CFI in the range of 0.90 to 0.95 or higher are typically indicative of good fit.

15.17 WHICH INDICATOR OF MODEL FIT IS BEST?

Our purpose in surveying a select sample of fit measures and indices was merely to give you a hint of how models are assessed in the SEM literature. Numerous measures of model fit have been proposed and there exists a whole literature of simulation studies and the like meant to evaluate their performance under conditions of violated assumptions, small to large sample sizes, and so on. A complete evaluation of fit measures is well beyond the scope of this chapter. Hence, deciding on which measure of fit is "best" under a variety of contexts is a decision tree we will not build here. Still, we can offer some guidelines.

According to Hu and Bentler (1998), who have extensively evaluated a variety of fit indices, the SRMR and RMSEA, supplemented with such indices as the TLI or CFI, assuming adequate sample size, are generally recommended in reporting the results of most SEM models. The SRMR and RMSEA, in addition to reporting χ^2, provide an overall assessment of model fit, and when coupled with incremental fit indices such as TLI or CFI, should give an overall adequate account of one's model. The reader is encouraged to consult Jackson, Gillaspy, and Purc-Stephenson (2009) for a useful overview of reporting practices in CFA models and SEM models more generally, which also includes a relevant discussion of fit indices.

What to do if model fit is unsatisfactory? Poor-fitting models should generally either be abandoned or improved. When attempting to improve on model fit, one may conduct **specification searches**, in which tests are performed on model parameters with the goal of estimating how the model would be improved given the fixing, constraining, or freeing of relevant parameters. Specification searches are often guided by computing so-called **modification indices**, which are numerical estimates of how much a model's fit would improve by adjusting parameters of the model. Of course, one can envision how such searches could potentially be misused. Indeed, a well-fitting model that is such because it has undergone a series of specification searches, though perhaps well-fitting **statistically**, may nonetheless be quite meaningless **scientifically**. If you "tweak" a model enough, fit **will** improve, but if the "tweaking" was not based on **your ideas** but rather on an optimization criterion alone, then well-fitting as the final model may be, it will nonetheless be of minimal value from a theory validation point of view. At minimum, extensive cross-validation will be required. If one is to engage in specification searches, then the number and nature of them should be guided primarily by theory. And if one is to extrapolate on one's theoretical predictions based on the results of such a search, he should be upfront about this to an audience when reporting how the well-fitting model **came to be**. We discuss specification searches no further here.

15.18 STRUCTURAL EQUATION MODEL IN R

As a simple demonstration of an SEM model in R, we fit a three-factor CFA model to the Holzinger and Swineford data, using only tests $x1$–$x9$ (recall from Chapter 11, we named the data hs). Hypothesized factors are visual, textual, and speed:

```
> hs.model <- ' visual   =~ x1 + x2 + x3
+ textual =~ x4 + x5 + x6
+ speed = ~ x7 + x8 + x9 '
```

We proceed to fit the model, displaying only partial results below:

```
> fit <- lavaan(hs.model, data = HolzingerSwineford1939, auto.var =
TRUE, auto.fix.first = TRUE, auto.cov.lv.x= TRUE)
> summary(fit, fit.measures = TRUE)

  Number of observations                              301

  Estimator                                            ML
```

```
Minimum Function Test Statistic                    85.306
Degrees of freedom                                     24
P-value (Chi-square)                                0.000
```

Model test baseline model:

```
Minimum Function Test Statistic                   918.852
Degrees of freedom                                     36
P-value                                             0.000
```

We can see that the model was fit using maximum likelihood yielding a statistically significant χ^2 on 24 degrees of freedom.

User model versus baseline model:

```
Comparative Fit Index (CFI)                         0.931
Tucker-Lewis Index (TLI)                            0.896
```

Both the CFI and TLI are reported to be 0.931 and 0.896 respectively, indicating somewhat modest fit.

Root Mean Square Error of Approximation:

```
RMSEA                                               0.092
90 Percent Confidence Interval             0.071    0.114
P-value RMSEA <= 0.05                                0.001
```

RMSEA is reported as 0.092, higher than the preferred cut-off of 0.05 (or lower).

Standardized Root Mean Square Residual:

```
SRMR                                                0.065
```

SRMR of 0.065 meets the preferred cut-off of 0.08 or less for a reasonably well-fitting model.

Parameter estimates:

```
Information                                      Expected
Standard Errors                                  Standard

                    Estimate  Std.err  Z-value  P(>|z|)
Latent variables:
  visual =~
    x1              1.000
    x2              0.554     0.100     5.554    0.000
    x3              0.729     0.109     6.685    0.000
  textual =~
    x4              1.000
    x5              1.113     0.065    17.014    0.000
```

x6	0.926	0.055	16.703	0.000
speed =~				
x7	1.000			
x8	1.180	0.165	7.152	0.000
x9	1.082	0.151	7.155	0.000

Parameter estimates for the factor loadings are given above. We had specified the path of the first variable on each factor at 1.0 through `auto.fix.first = TRUE`, which is why the estimates for $x1$, $x4$, and $x7$ are all equal to 1.0. We can see that all other paths are statistically significant yielding very low p-values. Covariances between latent variables appear below along with significance tests. Variances with respective standard errors are also given.

```
Covariances:
  visual ~~
```

textual	0.408	0.074	5.552	0.000
speed	0.262	0.056	4.660	0.000
textual ~~				
speed	0.173	0.049	3.518	0.000

```
Variances:
```

x1	0.549	0.114
x2	1.134	0.102
x3	0.844	0.091
x4	0.371	0.048
x5	0.446	0.058
x6	0.356	0.043
x7	0.799	0.081
x8	0.488	0.074
x9	0.566	0.071
visual	0.809	0.145
textual	0.979	0.112
speed	0.384	0.086

15.19   HOW ALL VARIABLES ARE LATENT: A SUGGESTION FOR RESOLVING THE MANIFEST-LATENT DISTINCTION

Recall that a latent variable in confirmatory factor analysis or structural equation modeling is generally defined as a variable that is **unobserved** or considered directly **unmeasurable**. We must **infer** its existence by "indicating" it through the measurement of so-called **manifest** variables, which are variables considered to be more measurable.

When one attempts a distinction between manifest versus latent variables, however, one more than not finds oneself in philosophical quicksand, and some kind of "merger" of the two concepts is really the only way out. For instance, consider the traditional interpretation of a characteristic such as weight. One would rarely refer to such an attribute as "latent" but would instead consider it to be quite observable. That is, though potentially subject to slight measurement error, its measurement is relatively straightforward. On the other hand, an attribute such as **intelligence** is more times than not considered unmeasurable, and its very existence must be inferred in reference to variables that are measurable.

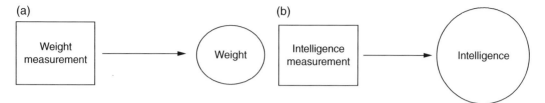

FIGURE 15.7   Weight (a) is less of a latent variable (indicated by smaller circle) than is intelligence (b).

Indeed, one might collect data on quantitative, verbal, and analytical skills, and use these as indicators of intelligence.

The distinction between what is **manifest** versus what is **latent,** however, is, in actuality, quite trivial. When one considers the finer points of such a distinction, one necessarily designates all variables, no matter however seemingly measurable some might be, as **latent variables**. Some latent variables are simply more measurable than others, but this does not remove from the fact that all variables, generally speaking, are being used to indicate some construct. This idea is summarized in Figure 15.7, in which in both cases weight and intelligence are regarded as latent variables indicating their respective constructs. In this sense, all variables (even physical ones measured quite precisely) may be considered latent, some less so than others.

15.20   THE STRUCTURAL EQUATION MODEL AS A GENERAL MODEL: SOME CONCLUDING THOUGHTS ON STATISTICS AND SCIENCE

In our discussion of canonical correlation analysis in Chapter 12, it was remarked that models such as multiple regression can be considered subsumed as "special cases" of the canonical correlation model. Recall we had also concluded that MANOVA (and MANCOVA) could likewise be considered special cases of the wider multivariate multiple regression model.

We now extend this idea to note that virtually all of these models, and more, can in general be considered special cases of the wider structural equation modeling framework. In the spirit of what Bollen (1989) remarked about virtually all models being subsumed under the fundamental identity $\sum = \sum (\mathbf{\theta})$, it becomes clear then that the process of modeling, considered in **totality**, simplifies very much to that of **the fixing and freeing of parameters within a given theory-driven structure in which one has a reasonable statistical theory for estimating unknown parameters**. Most of the remaining details concern themselves with how the given model is parameterized and the research context to which it is applied. As emphasized throughout this book, there is nothing inherently "experimental" about ANOVA models no more than there is anything "correlational" about regression models. Likewise, there is nothing "causal" about structural equation models any more than there is anything "uncausal" about a *t*-test. There is, on the other hand, something inherently experimental about **experimental studies**, just as there is something inherently correlational about **correlational studies**, as there is something inherently causal about **causal studies**. For the scientist, research design always begets statistics (Haddad, 2014, personal communication). Statistics alone can never really tell us **what happened** in the research, which is usually the purpose of the investigation in the first place. Statistical analyses can help in many cases, although if used without care and critical reflection, they can just as equally mislead. When applying a statistical model to data, a researcher should always ask themselves whether it clarifies or confuses the empirical finding, if indeed there is one.

In the end then, for the **scientist**, it is the quality of measurements obtained, the thought process, ingenuity, care and experience that went into conjuring up hypotheses, and the insistence of a convincing research design that will ultimately determine one's success and future discovery. Indeed, the distinction between what is **statistical** versus what is **scientific** has been emphasized throughout this book. Statistics and modeling serve as tremendous **aids** to discovery for the scientist, but they are not panaceas that could ever replace what makes a good scientist a **good scientist**, which usually boils down quite simply to **doing good science**, of which choosing and interpreting a useful statistical model is but one of **many** responsibilities toward this end.

15.21   CHAPTER SUMMARY AND HIGHLIGHTS

- **Path analysis** and **structural equation modeling** are techniques useful for the simultaneous modeling of a network of variables. In the case of SEM, **latent variables,** in addition to **manifest variables,** may be hypothesized, whereas in path analysis, all variables are considered to be observable.

- **Structural equation models** distinguish between **endogenous** and **exogenous** variables. The former are predicted by at least one explanatory variable, while the latter are variables featured as predictors of endogenous variables that are external to the system.

- The phrase **"causal modeling"** has become associated with path analysis (and by extension, SEM models) in part as a result of the methodological context in which such models arose historically, with Wright's work on modeling color in guinea pigs in which a causal "context" was not such a far-fetched idea.

- For an accurate and precise account of what **path coefficients** can and cannot do, one should read (and **re-read**) Burks' 1928 interpretation.

- The true contribution of Wright was not in developing a causal methodology, but rather one of demonstrating how **path coefficients could be decomposed**.

- **Confirmatory factor analysis** solves the rotational indeterminacy problem of EFA by allowing researchers greater flexibility in fixing and freeing parameters, thereby providing more opportunity for parameters to be identified.

- Substantively, the distinction between what constitutes **exploratory** versus **confirmatory** modeling is fuzzy.

- The **structural equation model** $\eta = \mathbf{B}\eta + \mathbf{\Gamma}\xi + \zeta$ can be likened somewhat to the regression model $\mathbf{Y} = \mathbf{XB} + \mathbf{E}$ in some respects, the chief difference being that the former allows one to incorporate **latent structures**, whereas the latter does not, at least not explicitly so.

- A **direct effect** is the prediction of one variable on another, unmediated by any other model variables. An **indirect effect** is the prediction of one variable on another, but mediated by at least one other intervening variable. A **total effect** is the sum of direct and indirect effects.

- As noted by Bollen (1989), much of **statistical modeling** can be reduced to solving the equality $\sum = \sum (\theta)$. Many statistical methods can be conceptualized as special cases of this wider framework.

- A **discrepancy function** is a general name used to describe functions that minimize the degree of misfit between $\mathbf{S}$ and $\mathbf{S}(\theta)$. Ordinary least-squares, maximum likelihood, unweighted, and generalized least-squares are all examples of such discrepancy functions.

- **Identification** is a property of a model. A model may be underidentified, just-identified, or overidentified. Typically, researchers seek to make their models overidentified.

- **Disturbance variables** are conceptualized as influences outside of the model, typically considered unmeasurable, that have an influence on a given endogenous variable. The distinction between what constitutes a disturbance variable and that of a residual is important.

- Several measures and indicators of model fit for SEM have been proposed. They are generally classified into **absolute** (or overall) measures and incremental fit indices. The more popular ones and those generally recommended for minimal inclusion into virtually every SEM model include χ^2, SRMR, RMSEA, TLI, and CFI.
- Poorly-fitting models should be either discarded or improved. **Specification searches**, which generate modification indices, can be conducted and reported so long as one is upfront about this process when reporting findings to an audience.
- In resolving the **manifest** versus **latent** distinction, it is suggested to consider all variables as latent, some less so than others.
- **Structural equation models** can be fit using R, among several other software programs such as AMOS.
- **Statistics** and **modeling** serve as tremendous aids to **discovery**, but choosing and interpreting the correct statistical model is but one of the many responsibilities of the successful scientist.

REVIEW EXERCISES

15.1. Give a definition of **path analysis**, emphasizing its similarities and differences from **regression analysis**.

15.2. Distinguish between **observed** or manifest variables and **latent** variables. Why and how is this distinction somewhat blurry and fuzzy? How can this fuzziness be potentially resolved?

15.3. Discuss the historical influence of how path analysis (and by extension, structural equation modeling), came to be known as "**causal modeling.**"

15.4. Critically evaluate path analysis and structural equation modeling as "**causal**" methodologies. Make an argument for why path and SEM models are or are not more causal than regression models.

15.5. Why is it somewhat reasonable that in Burks' early use of path analysis, it was safe to assume that parental intelligence and environment, at least to some extent, **caused** child's IQ?

15.6. Define a **path coefficient** as used in path analysis.

15.7. Interpret and discuss Burks' quote:

The method [of path coefficients] is limited by the rarity with which we have actual knowledge of causal relations; but it provides a toll of the nicest precision in such situations as do offer an adequate basis for postulating causation. It cannot, itself, uncover what is cause and what is effect, though in the absence of definite knowledge regarding causal relationships between variables, the method '**can be used to find out the logical consequences of any particular hypothesis in regard to them.**' Conservatively stated, in any situation in which we feel justified in drawing conclusions regarding the effects of certain phenomena upon others, the Wright method provides a numerical expression of such conclusions. (Burks, 1928, p. 299)

15.8. What does it mean to **decompose path coefficients** or effects? How was this a contribution of Sewall Wright?

15.9. What does it mean to have a **saturated model**? Why does a saturated model always generate a **perfect** (and hence, typically nonuseful) model fit?

15.10. Explain why having **positive degrees of freedom** for a path analysis model (or structural equation model) is necessary in order to meaningfully evaluate parameter estimates.

15.11. Explain what it means to **fix a parameter** in path analysis. Furthermore, explain why a researcher might want to do this.

15.12. Consider the statement—**the distinction between exploratory and confirmatory models is one more of flavor that it is one of absolute difference.** Comment on what this might mean. Then, counter this argument by discussing how CFA solves the **rotational indeterminacy problem** of EFA.

15.13. Provide a precise definition of a **structural equation model**.

15.14. Define and discuss each component of the **structural equation model** $\eta = B\eta + \Gamma\xi + \zeta$.

15.15. Distinguish between a **direct**, **indirect**, and **total effect**. How are such effects not necessarily indicative of processes inherent in one's data?

15.16. What, according to Bollen (1989), is the **fundamental relation** upon which virtually all statistical modeling is based? Explain the over-reaching concept implied by the equality.

15.17. Distinguish between an **observed** versus an **implied** covariance matrix.

15.18. Describe what is meant by a **discrepancy function**, and give a few examples of some.

15.19. According to Jöreskog, "**If a value of χ^2 is obtained, which is large compared to the number of degrees of freedom, this is an indication that more information can be extracted from the data.**" Interpret and discuss Jöreskog's statement and why such an indicator is meaningful in a general sense.

15.20. What does it mean, in general, to say that a parameter is identified, and how is the concept of **parameter identification** different or similar to that of model identification?

15.21. In the equation $x + 5 = 10$, is x **identified**? Why or why not?

15.22. Distinguish between **underidentified**, **just-identified**, and **overidentified** models.

15.23. Discuss what is meant by a **disturbance variable**, using Mulaik (2009) as a guide to disentangling this theoretically complex topic.

15.24. Distinguish between **overall** indicators of fit and **incremental** indicators. How are they different?

15.25. Discuss how the **root mean square residual** (RMR) is more a measure of how well a model does not fit rather than does fit.

15.26. What is a difficulty with **RMR** and how does **SRMR** attempt to resolve this difficulty?

15.27. Define the **root mean-square error of approximation** (RMSEA), and explain how it assesses model fit.

15.28. Compare the index, ρ_1 with that of ρ_2. Discuss their structural similarities and differences.

15.29. Define the **comparative fit index**, and compare it with ρ_1. How are they similar? Different?

15.30. Consider once more the Holzinger and Swineford (1939) data. Run a **three-factor confirmatory factor analysis** on variables $x1$ through $x9$ as hypothesized by the following model equations. Compare the fit with the CFA performed in this chapter. Is the fit better or worse?

```
visual =~ x1 + x2
textual =~ x4 + x5
speed =~ x7 + x8
```

15.31. Consider again the data (Holzinger Swineford) featured in Exercise 15.30. Run and interpret another confirmatory factor analysis, this time, hypothesizing a one-factor solution:

```
factor =~ x1 + x2 + x3 + x4 + x5 + x6 + x7 + x8 + x9
```

Compare and contrast the model fit to that in Exercise 15.30 and with the three-factor model of this chapter.

Further Discussion and Activities

15.32. Consider how scientists, both natural and social, define the phenomena they study. Would you claim that a characteristic such as weight is more real or less "real" than an attribute such as intelligence? Why or why not? What makes one more **real** than another? Is it the fact that we are able to better define weight that might, in the end, make it more real? What do you think? For some insight into the problem of how concepts in science have been historically defined and operationalized, read Green (1992).

15.33. As discussed in the chapter, path analysis and structural equation modeling have historically been associated with the phrase "**causal modeling**." Causality, however, is an enormous subject on which philosophers have grappled for centuries. Indeed, identifying when causation is actually occurring is challenging. For example, if a person suffers from a headache and she takes medication to alleviate that headache, is it correct to say that the medication **caused** a lessening of symptomology? What would be the difference had we said the medication is **associated** with a lessening of symptomology? Discuss how these statements are different from one another and brainstorm some of the issues involved in identifying just what **does** and **does not** connote causation in science.

REFERENCES

Abdi, H. (2003). Factor rotations. In M. Lewis-Beck, A. Bryman, T. Futing (Eds): *Encyclopedia for research methods for the social sciences. Thousand Oaks, CA: Sage. 978–982.*

Abelson, R. P. (1995). *Statistics as principled argument. London: Psychology Press.*

Agresti, A. (2002). *Categorical data analysis. New York: John Wiley & Sons, Inc.*

Aiken, L. & West, S. G. (1991). *Multiple regression: Testing and interpreting interactions. New York: Sage.*

Aleksandrov, A. D., Kolmogorov, A. N., & Lavrent'ev, M. A. (1999). *Mathematics: Its content, methods and meaning. New York; Dover Publications.*

Anderberg, M. R. (1973). *Cluster analysis for applications. New York: Academic Press.*

Anderson, E. (1935). The irises of the Gaspé peninsula. *Bulletin of the American Iris Society, 59,* 2–5.

Anderson, T. W. (2003). *An introduction to multivariate statistical analysis. New York: John Wiley & Sons, Inc.*

Anton, H. & Rorres, C. (2000). *Elementary linear algebra. New York: John Wiley & Sons, Inc.*

Arbuthnot, J. (1710). An argument for divine providence, taken from the constant regularity observed in the births of both sexes. *Philosophical Transactions, 27,* 186–190 (published in 1711).

Bakan, D. (1966). The test of significance in psychological research. *Psychological Bulletin, 66,* 423–437.

Bar-Hillel, M. (1980). The base-rate fallacy in probability judgments. *Acta Psychologica, 44,* 211–233.

Barnett, R. A., Ziegler, M. R., & Byleen, K. E. (2011). *College mathematics for business, economics, life sciences, and social sciences. New York: Prentice Hall.*

Baron, R. M. & Kenny, D. A. (1986). The moderator-mediator variable distinction in social psychological research: Conceptual, strategic, and statistical considerations. *Journal of Personality and Social Psychology, 51,* 1173–1182.

Bartle, R. G. & Sherbert, D. R. (2011). *Introduction to real analysis. New York: John Wiley & Sons, Inc.*

Bartlett, M. S. (1947). Multivariate analysis. *Journal of the Royal Statistical Society, Series B, 9,* 176–197.

Bartlett, M. S. (1950). Tests of significance in factor analysis. *British Journal of Psychology, 3,* 77–85.

Bartlett, M. S. (1954). A note on multiplying factors for various chi-squared approximations. *Journal of Royal Statistical Society, Series B, 16,* 296–298.

Applied Univariate, Bivariate, and Multivariate Statistics: Understanding Statistics for Social and Natural Scientists, With Applications in SPSS and R, Second Edition. Daniel J. Denis.
© 2021 John Wiley & Sons, Inc. Published 2021 by John Wiley & Sons, Inc.
Companion Website: www.wiley.com/go/denis/appliedstatistics2e

Bates, D., Maechler, M., Bolker, B., & Walker, S. (2014). *lme4: Linear mixed-effects models using Eigen and S4_.* R package version 1.1–7.

Bauer, D. J. (2003). Estimating multilevel linear models as structural equation models. *Journal of Educational and Behavioral Statistics, 28,* 135–167.

Bentler, P. M. & Bonett, D. G. (1980). Significance tests and goodness of fit in the analysis of covariance structures. *Psychological Bulletin, 88,* 588–606.

Berger, J. O. (1993). *Statistical decision theory and Bayesian analysis. New York: Springer.*

Berkson, J. (1938). Some difficulties of interpretation encountered in the application of the chi square test. *Journal of the American Statistical Association, 33,* 526–536.

Berry, D. A. & Lindgren, B. W. (1996). *Statistics: Theory and methods. New York: Duxbury Press.*

Bien, J. & Tibshirani, R. (2011). Hierarchical clustering with prototypes via minimax linkage. *Journal of the American Statistical Association, 106,* 1075–1084.

Bilodeau, M. & Brenner, D. (1999). *Theory of multivariate statistics. New York: Springer.*

Bishara, A. J. & Hittner, J. B. (2015). Reducing bias and error in the correlation coefficient due to nonnormality. *Educational and Psychological Measurement, 75*(5), 785–804.

Bollen, K. A. (1989). *Structural equations with latent variables. New York: John Wiley & Sons, Inc.*

Bollen, K. A. & Curran, P. J. (2006). *Latent curve models: A structural equation perspective. New York: John Wiley & Sons, Inc.*

Bolles, R. C. (1962). The difference between statistical hypotheses and scientific hypotheses. *Psychological Reports, 11,* 639–645.

Box, G. E. P. (1949). A general distribution theory for a class of likelihood criteria. *Biometrika, 36,* 317–346.

Box, G. E. P. (1950). Problems in the analysis of growth and linear curves. *Biometrika, 7,* 362–389.

Box, G. E. P. (1954). Some theorems on quadratic forms applied to the study of analysis of variance problems: I. Effect of inequality of variance in the one-way classification. *Annals of Mathematical Statistics, 25,* 290–302.

Box, G. E. P., Hunter, W. G., & Hunter, J. S. (1978). *Statistics for experimenters: An introduction to design, data analysis, and model building. New York: John Wiley & Sons, Inc.*

Boyer, C. B. (1959). *The history of the calculus and its conceptual development. New York: Dover Publications.*

Boyer, C. B. & Merzbach, U. C. (1991). *A history of mathematics. New York: John Wiley & Sons, Inc.*

Bravais, A. (1846). Analyse mathématique sur les probabilités des erreurs de situation d'un point. *Mémoires présentés par divers savants à l'Académie royale des sciences de l'Institut de France, 9,* 255–332.

Browne, M. W. & Cudeck, R. (1993). Alternative ways of assessing model fit. In: K. A. Bollen & J. S. Long, *Testing structural equation models. Beverly Hills, CA: Sage, pp. 136–162.*

Burks, B. S. (1928). The relative influence of nature and nurture upon mental development: A comparative study of foster parent–foster child resemblance and true parent–true child resemblance. *The Twenty-Seventh Yearbook of the National Society for the Study of Education, 27,* 219–316.

Burnham, K. P. & Anderson, D. R. (2011). AIC model selection and multimodel inference in behavioral ecology: Some background, observations, and comparisons. *Behavioral Ecology and Sociobiology, 65,* 23–35.

Byrne, B. M. (2009). *Structural equation modeling with AMOS: Basic concepts, applications, and programming. New York: Routledge.*

Byrne, B. M. (2016). *Structural equation modeling with AMOS: Basic concepts, applications, and programming. New York: Routledge.*

Campbell, D. T. & Stanley, J. C. (1963). *Experimental and quasi-experimental designs for research. New York: Houghton Mifflin Company.*

Carroll, J. D. & Green, P. E. (1997). *Mathematical tools for applied multivariate analysis. New York: Academic Press.*

Carver, R. (1993). The case against statistical significance testing, revisited. *Journal of Experimental Education, 61,* 287–292.

Casella, G. (2008). *Statistical design. New York: Springer.*

Casella, G. & Berger, R. L. (2002). *Statistical inference. Pacific Grove, CA: Duxbury Press.*

Cattell, R. B. (1966). The scree test for the number of factors. *Multivariate Behavioral Research, 1*, 245–276.

Champely, S. (2020). *pwr: Basic functions for power analysis.* R package version 1.1.1.

Chatfield, C. (2019). *The analysis of time series. New York: Routledge.*

Chatfield, C. & Collins, A. (1981). *Introduction to multivariate analysis. New York: Chapman & Hall.*

Chen, H., Cohen, P. & Chen, S. (2010a). How big is a big odds ratio? Interpreting the magnitudes of odds ratios in epidemiological studies. *Communications in Statistics: Simulation and Computation, 39*, 860–864.

Chen, Y. P., Huang, H. C., & Tu, I. P. (2010b). A new approach for selecting the number of factors. *Computational Statistics and Data Analysis, 54*, 2990–2998.

Clemen, R. T. & Reilly, T. (2004). *Making hard decisions. Pacific Grove, CA: Duxbury Press.*

Cohen, J. (1988). *Statistical power analysis for the behavioral sciences. New York: Routledge.*

Cohen, J. (1990). Things I have learned (so far). *American Psychologist, 45*, 1304–1312.

Cohen, J., Cohen, P., West, S. G., & Aiken, L. S. (2002). *Applied multiple regression/correlation analysis for the behavioral sciences. New Jersey: Lawrence Erlbaum Associates.*

Cohen, J., Cohen, P., West, S. G., & Aiken, L. S. (2003). *Applied multiple regression/correlation analysis for the behavioral sciences. New Jersey: Lawrence Erlbaum Associates.*

Cohen, R. J., Swerdlik, M., & Sturman, E. (2013). *Psychological testing and assessment: An introduction to tests and measurement. New York: McGraw-Hill.*

Cole, D. A., Maxwell, S. E., Arvey, R., & Salas, E. (1994). How the power of MANOVA can both increase and decrease as a function of the intercorrelations among the dependent variables. *Psychological Bulletin, 115*, 465–474.

Cornell, J. E., Young, D. M., Seaman, S. L., & Kirk, R. E. (1992). Power comparisons of eight tests for sphericity in repeated measures designs. *Journal of Educational Statistics, 17*, 233–249.

Cortina, J. M. (1993). What is coefficient alpha? An examination of theory and applications. *Journal of Applied Psychology, 78*, 98–104.

Cortina, J. M. & Nouri, H. (1999). *Effect size for ANOVA designs. Beverly Hills, CA: Sage.*

Courant, R., Robbins, H., & Stewart, I. (1996). *What is mathematics? An elementary approach to ideas and methods. New York: Oxford University Press.*

Cowles, M. (2005). *Statistics in psychology: An historical perspective. London: Psychology Press.*

Crawley, M. J. (2013). *The R book. New York: John Wiley & Sons, Inc.*

da Silva, A. R. (2014). *biotools: Tools for biometry and applied statistics in agricultural science.* R package version 1.2.

Dalal, D. K. & Zickar, M. J. (2011). Some common myths about centering predictor variables in moderated multiple regression and polynomial regression. *Organizational Research Methods, 15*, 339–362.

Dalgaard, P. (2008). *Introductory statistics with R. New York: Springer.*

Daniels, H. E. (1939). The estimation of components of variance. *Journal of the Royal Statistical Society Supplement, 6*, 186–197.

Dawson, B. & Trapp, R. G. (2004). *Basic and clinical biostatistics. New York: McGraw-Hill.*

Dean, A. M. & Voss, D. (1999). *Design and analysis of experiments. New York: Springer.*

Dean, C. B. & Lundy, E. R. (2016). *Overdispersion. Wiley StatsRef: Statistics Reference Online. New York: Wiley.*

DeCarlo, L. T. (1997). On the meaning and use of kurtosis. *Psychological Methods, 2*, 292–307.

Degroot, M. H. & Schervish, M. J. (2002). *Probability and statistics. New York: Addison-Wesley.*

Dehue, T. (2005). History of the control group. *Encyclopedia of Statistics in Behavioral Science, 2*, 829–836.

Demidenko, E. (2004). *Mixed models: Theory and applications. New York: John Wiley & Sons, Inc.*

Denis, D. (2001). The origins of correlation and regression: Francis Galton or Auguste Bravais and the error theorists? *History and Philosophy of Psychology Bulletin, 13*, 36–44.

Denis, D. (2003). Alternatives to null hypothesis significance testing. *Theory & Science, 4*: 1–23, http://theoryandscience.icaap.org/content/vol4.1/02_denis.html.

Denis, D. (2004). The modern hypothesis testing hybrid: R. A. Fisher's fading influence. *Journale de la Société Française de Statistique, 145*, 5–26.

Denis, D. (2019). *SPSS data analysis: For univariate, bivariate, and multivariate statistics.* New York: John Wiley & Sons.

Denis, D. (2020). *Univariate, bivariate, and multivariate statistics using R: Quantitative tools for data analysis and data science.* New York: John Wiley & Sons.

Denis, D. (2020). *Model selection in regression: Statistical and scientific perspectives.* Wiley Online Library.

Denis, D. & Docherty, K. R. (2007). Late nineteenth century Britain: A social, political, and methodological context for the rise of multivariate statistics. *Journale Electronique d'Histoire des Probabilités et de la Statistique, 3*, 1–41.

Denis, D. & Legerski, J. (2006). Causal modeling and the origins of path analysis. *Theory & Science, 17*, 1–21, http://theoryandscience.icaap.org/content/vol7.1/denis.html

Desrosières, A. (1998). *The politics of large numbers.* Massachusetts: Harvard University Press.

Diamond, D. M., Campbell, A. M., Park, C. R., Halonen, J., & Zoladz, P. R. (2007). The temporal dynamics model of emotional memory processing: A synthesis on the neurobiological basis of stress-induced amnesia, flashbulb and traumatic memories, and the Yerkes–Dodson law. *Neural Plasticity, 2007*, 60803.

Dowdy, S., Wearden, S., & Chilko, D. (2004). *Statistics for research.* New York: John Wiley & Sons, Inc.

Draper, N. R. & Smith, H. (1998). *Applied regression analysis.* New York: John Wiley & Sons, Inc.

Dtarazona (1998). *Operant conditioning chamber.* Retrieved from Wikipedia.

Dunham, W. (1994). *The mathematical universe: An alphabetical journey through the great proofs, problems, and personalities.* New York: John Wiley & Sons, Inc.

Dzamonja, M. (2017). Set theory and its place in the foundations of mathematics: A new look at an old question. *Journal of Indian Council of Philosophical Research, 34*, 415–424.

Eckart, C. & Young, G. (1936). The approximation of one matrix by another of lower rank. *Psychometrika, 1*, 211–218.

Edwards, A. L. (1985). *Experimental design in psychological research.* New York: Harper & Row.

Efron, B. (1975). The efficiency of logistic regression compared to normal discriminant analysis. *Journal of the American Statistical Association, 70*, 892–898.

Efron, B. and Tibshirani, R. J. (1993). *An introduction to the bootstrap.* New York: Chapman and Hall.

Einstein, A. J., Jeffery, G. B., & Perrett, W. (trans.) (1922). *Sidelights on relativity. I. Ether and relativity, II. Geometry and experience.* London: Methuen & Co. Ltd.

Eisenhart, C. (1947). The assumptions underlying the analysis of variance. *Biometrics, 3*, 1–21.

Eliopoulos, G. M., Harris, A. D., Lautenbach, E., & Perencevich, E. (2005). A systematic review of quasi-experimental study designs in the fields of infection control and antibiotic resistance. *Clinical Infectious Diseases, 41*, 77–82.

Epskamp, S. (2014). *semPlot: Path diagrams and visual analysis of various SEM packages' output.* R package version 1.0.1.

Estes, W. K. (1997). Significance testing in psychological research: Some persisting issues. *Psychological Science, 8*, 18–20.

Evans, M. J. & Rosenthal, J. S. (2010). *Probability and statistics: The science of uncertainty.* New York: W. H. Freeman and Company.

Everitt, B. S. (2002). *The Cambridge dictionary of statistics.* New York: Cambridge University Press.

Everitt, B. S. (2007). *An R and S-Plus companion to multivariate analysis.* New York: Springer.

Everitt, B. S. & Hothorn, T. (2015). *HSAUR: A handbook of statistical analyses using R.* R package version 1.3–6.

Everitt, B. S., Landau, S., & Leese, M. (2001). *Cluster analysis.* New York: Oxford University Press.

Fancher, R. E. & Rutherford, A. (2011). *Pioneers of psychology: A history.* New York: W.W. Norton & Company.

Faul, F., Erdfelder, E., Buchner, A., & Lang, A. G. (2009). Statistical power analyses using G*Power 3.1: Tests for correlation and regression analyses. *Behavior Research Methods, 41,* 1149–1160.

Faul, F., Erdfelder, E., Lang, A. G., & Buchner, A. (2007). G*Power 3: A flexible statistical power analysis program for the social, behavioral, and biomedical sciences. *Behavior Research Methods, 39,* 175–191.

Federer, W. T. (1955). *Experimental design: Theory and application.* New York: Macmillan Company.

Feller, W. (1968). *An introduction to probability theory and its applications.* New York: John Wiley & Sons, Inc.

Field, A. (2009). *Discovering statistics using SPSS.* London: Sage Publications.

Filzmoser, P. & Gschwandtner, M. (2014). *mvoutlier: Multivariate outlier detection based on robust methods.* R package version 2.0.5.

Finch, H. (2005). Comparison of distance measures in cluster analysis with dichotomous data. *Journal of Data Science, 3,* 85–100.

Fisher, R. A. (1922a). On the mathematical foundations of theoretical statistics. *Philosophical Transactions of the Royal Society of London. Series A, Containing Papers of a Mathematical or Physical Character, 222,* 309–368.

Fisher, R. A. (1922b). On the interpretation of χ^2 from contingency tables, and the calculation of P. *Journal of the Royal Statistical Society, 85,* 87–94.

Fisher, R. A. (1925, 1934). *Statistical methods for research workers.* Edinburgh: Oliver and Boyd.

Fisher, R. A. (1935). *The design of experiments.* Edinburgh: Oliver and Boyd.

Fisher, R. A. (1936). The use of multiple measurements in taxonomic problems. *Annals of Eugenics, 7,* 466–475.

Fisher, R. A. (1947). The analysis of covariance method for the relation between a part and the whole. *Biometrics, 3,* 65–68.

Fletcher, T. D. (2012). *QuantPsyc: Quantitative psychology tools.* R package version 1.5.

Flury, B. (1997). *A first course in multivariate statistics.* New York: Springer.

Flury, B. & Riedwyl, H. (1985). T2 tests, the linear two-group discriminant function, and their computation by linear regression. *American Statistician, 39,* 20–25.

Fox, J. (1997). *Applied regression analysis, linear models, and related methods.* London: Sage Publications.

Fox, J. (2002). *An R and S-plus companion to applied regression.* London: Sage Publications.

Fox, J. (2003). Effect displays in R for generalized linear models. *Journal of Statistical Software, 8,* 1–27.

Fox. J. (2008a). *A mathematical primer for social statistics.* London: Sage Publications.

Fox. J. (2008b). *Applied regression analysis and generalized linear models.* New York: Sage Publications.

Fox, J. (2016). *Applied regression analysis and generalized linear models.* New York: Sage Publications.

Fox, J. & Weisberg, S. (2011). *An R companion to applied regression.* New York: Sage.

Freedman, D. A. (1987). As others see us: A case study in path analysis (with discussion). *Journal of Educational Statistics, 12,* 101–129.

Friendly, M. (1991). *SAS system for statistical graphics.* North Carolina: SAS Institute.

Friendly, M. (2000). *Visualizing categorical data.* North Carolina: SAS Institute.

Friendly, M. (2007). HE plots for multivariate general linear models. *Journal of Computational and Graphical Statistics, 16,* 421–444.

Friendly, M. (2014). *HistData: Data sets from the history of statistics and data visualization.* R package, version 0.7–5.

Friendly, M. (2020). *DataVis.ca.* Retrieved from www.datavis.ca.

Friendly, M., Monette, G., & Fox, J. (2013). Elliptical insights: Understanding statistical methods through elliptical geometry. *Statistical Science, 28,* 1–40.

Furr, R. M. & Bacharach, V. R. (2013). *Psychometrics: An introduction.* London: Sage Publications.

Gabriel, K. R. (1971). The biplot graphic display of matrices with application to principal component analysis. *Biometrika, 58,* 453–467.

Galton, F. (1886). Regression towards mediocrity in hereditary stature. *Journal of the Anthropological Institute*, *15*, 246–263.

Galton, F. (1888). Co-relations and their measurement, chiefly from anthropometric data. *Proceedings of the Royal Society of London*, *45*, 135–145.

Geisser, S. & Greenhouse, S. W. (1958). An extension of Box's result on the use of the *F* distribution in multivariate analysis. *Annals of Mathematical Statistics*, *29*, 885–891.

Gelman, A. & Hill, J. (2007). *Data analysis using regression and multilevel/hierarchical models. New York: Cambridge University Press.*

Gemignani, M. C. (1998). *Calculus and statistics. New York: Dover Publications.*

Gigerenzer, G. (2004). Mindless statistics. *The Journal of Socio-Economics*, *33*, 587–606.

Gigerenzer, G., Swijtink, Z., Porter, T., Datson, Beatty, J., & Kruger, L. (1990). *The empire of chance: How probability changed science and everyday life. New York: Cambridge University Press.*

Gill, J. (2006). *Essential mathematics for political and social research. New York: Cambridge University Press.*

Gill, J. (2014). *Bayesian methods: A social and behavioral sciences approach. New York: Chapman & Hall/CRC.*

Girden, E. R. (1992). *ANOVA: Repeated measures.* Series: Quantitative Applications in the Social Sciences, 07–084, Sage University Paper, Beverly Hills, CA: Sage Publications.

Gnanadesikan, R. (1997). *Methods for statistical data analysis of multivariate observations. New York: John Wiley & Sons, Inc.*

Goodwin, P. & Wright, G. (2004). *Decision analysis for management judgment. New York: John Wiley & Sons, Inc.*

Gosset, W. S. (1908). The probable error of a mean. *Biometrika*, *6*, 1–25.

Gower, J. C. (1988). Classification, geometry and data analysis. In: H. H. Bock (Ed.), *Classification and related methods of data analysis. Amsterdam: North-Holland.*

Grabiner, J. V. (1983). Who gave you the epsilon? Cauchy and the origins of rigorous calculus. *The American Mathematical Monthly*, *90*, 185–194.

Graham, J. M. (2008). The general linear model as structural equation modeling. *Journal of Educational and Behavioral Statistics*, *33*, 485–506.

Green, C. D. (1992). Of immortal mythological beasts: Operationalism in psychology. *Theory & Psychology*, *2*, 287–316.

Green, C. D. (2015). *Classics in the history of psychology.* Retrieved from http://psychclassics.yorku.ca/

Green, S. B. & Yang, Y. (2009). Commentary on coefficient alpha: A cautionary tale. *Psychometrika*, *74*, 121–135.

Guttman, L. (1954). Some necessary conditions for common-factor analysis. *Psychometrika*, *19*, 149–161.

Hacking, I. (1990). *The taming of chance. Edinburgh: Cambridge University Press.*

Hair, J., Black, B., Babin, B., Anderson, R., & Tatham, R. (2006). *Multivariate data analysis. New Jersey: Prentice-Hall.*

Hald, A. (1990). *A history of probability and statistics and their applications before 1750. New York: John Wiley & Sons, Inc.*

Hald, A. (1998). *A history of mathematical statistics: From 1750 to 1930. New York: John Wiley & Sons, Inc.*

Haller, H. & Krauss, S. (2002). Misinterpretations of significance: A problem students share with their teachers? *Methods of Psychological Research Online*, *7*, 1–20.

Hamming, R. W. (1985). *Methods of mathematics: Applied to calculus, probability, and statistics. New York: Dover Publications.*

Hardy, G. H., Wright, E. M., Wiles, A., Heath-Brown, R., & Silverman, J. (2008). *An introduction to the theory of numbers. New York: Oxford University Press.*

Harlow, L., Mulaik, S., & Steiger, J. (2013). *What if there were no significance tests?* Psychology Press: New York.

Harris, R. J. (2001). *A primer of multivariate statistics. New Jersey: Lawrence Erlbaum Associates.*

Hartley, H. O. & Rao, J. N. K. (1967). Maximum likelihood estimation for the mixed analysis of variance model. *Biometrika, 54*, 93–108.

Harville, D. A. (1997). *Matrix algebra from a statistician's perspective. New York: Springer.*

Hastie, T., Tibshirani, R., & Friedman, J. (2009). *The elements of statistical learning: Data mining, inference, and prediction. New York: Springer.*

Hayes, A. F. & Scharkow, M. (2013). The relative trustworthiness of inferential tests of the indirect effect in statistical mediation analysis: Does method really matter? *Psychological Science, 24*, 1918–1927.

Hays, W. L. (1994). *Statistics. Fort Worth, TX: Harcourt College Publishers.*

Healy, M. J. R. (1969). Rao's paradox concerning multivariate tests of significance. *Biometrics, 25*, 411–413.

Hennig, C. (2009). *Mathematical models and reality: A constructivist perspective.* Research Report No. 304, Department of Statistical Science, University College London.

Heston, J. C. (1948). The graduate record examination vs. other measures of aptitude and achievement. *The Journal of Educational Research, 41*, 338–347.

Heumann, C. & Shalabh, M. S. (2016). *Introduction to statistics and data analysis. New York: Springer.*

Hoelter, J. W. (1983). The analysis of covariance structures: Goodness-of-fit indices. *Sociological Methods and Research, 11*, 325–344.

Hoerl, A. E. & Kennard, R. (1970). Ridge regression: Biased estimation for non-orthogonal problems. *Technometrics, 12*, 55–67. Reprinted in *Technometrics, 42* (2000), 80–86.

Hogg, R. V. & Craig, A. T. (1995). *Introduction to mathematical statistics. New Jersey: Prentice Hall.*

Holzinger, K. J. & Swineford, F. A. (1939). A study in factor analysis: The stability of a bi-factor solution. *The Journal of Educational Research, 34*. 697–700.

Hosmer, D. & Lemeshow, S. (2000). *Applied logistic regression. New York: John Wiley & Sons, Inc.*

Hotelling, H. (1931). The generalization of Student's ratio. *The Annals of Mathematical Statistics, 2*, 360–378.

Hotelling, H. (1933). Analysis of a complex of statistical variables into principal components. *Journal of Educational Psychology, 24*, 417–441, 498–520.

Hotelling, H. (1936). Relation between two sets of variates. *Biometrika, 28*, 321–377.

Hotelling, H. (1951). A generalized t^2 test and measure of multivariate dispersion. *Paper presented at the Second Berkeley Symposium on Mathematical Statistics and Probability*, Vol. 1. *University of California Press, pp. 23–41.*

Hothorn, T., Bretz, F., & Westfall, P. (2008). Simultaneous inference in general parametric models. *Biometrical Journal, 50*, 346–363.

Houwelingen, J. C. V. & Cessie, S. L. (1990). Predictive value of statistical models. *Statistics in Medicine, 9*, 1303–1325.

Howell, D. C. (2002). *Statistical methods for psychology. Pacific Grove, CA: Duxbury Press.*

Hsu, J. C. (1996). *Multiple comparisons: Theory and methods. New York: Chapman & Hall/CRC.*

Hu, L. T. & Bentler, P. M. (1995). Evaluating model fit. In: R. H. Hoyle (Ed.), *Structural equation modeling: Concepts, issues and applications. Beverly Hills, CA: Sage.*

Hu, L. T. & Bentler, P. M. (1998). Fit indices in covariance structure modeling: Sensitivity to underparameterized model misspecification. *Psychological Methods, 3*, 424–453.

Hu, L. T. & Bentler, P. M. (1999). Cutoff criteria for fit indexes in covariance structure analysis: Conventional criteria versus new alternatives. *Structural Equation Modeling, 6*, 1–55.

Hunink, M., Glasziou, P., Siegel, J., Weeks, J., Pliskin, J., Elstein, A., & Weinstein, M. (2001). *Decision making in health and medicine: Integrating evidence and values. Edinburgh: Cambridge University Press.*

Hunter, J. E. (1997). Needed: A ban on the significance test. *American Psychological Society, 8*, 3–7.

Huynh, H. & Feldt, L. S. (1970). Conditions under which mean square ratios in repeated measurement designs have exact F distributions. *Journal of the American Statistical Association, 65*, 1582–1589.

Iacobucci, D., Schneider, M. J., Popovich, D. L., & Bakamitsos, G. A. (2016). Mean centering helps alleviate "micro" but not "macro" multicollinearity. *Behavior Research Methods, 48*, 1308–1317.

Izenman, A. J. (2008). *Modern multivariate statistical techniques: Regression, classification, and manifold learning. New York: Springer.*

Jaccard, J. J. & Turrisi, R. (2003). *Interaction effects in multiple regression, Quantitative Applications in the Social Sciences. London: Sage.*

Jackson, D. L., Gillaspy, J. A., & Purc-Stephenson, R. (2009). Reporting practices in confirmatory factor analysis: An overview and some recommendations. *Psychological Methods, 14,* 6–23.

James, G., Witten, D., Hastie, T., & Tibshirani, R. (2013). *An introduction to statistical learning with applications in R. New York: Springer.*

Jarek, S. (2012). *mvnormtest: Normality test for multivariate variables.* R package version 0.1–9.

Johnson, R. A. & Wichern, D. W. (2007). *Applied multivariate statistical analysis. New Jersey: Pearson Prentice Hall.*

Jolliffe, I. T. (2002). *Principal component analysis. New York: Springer.*

Jöreskog, K. G. (1967). Some contributions to maximum likelihood factor analysis. *Psychometrika, 32,* 443–482.

Jöreskog, K. G. (1969). A general approach to confirmatory maximum likelihood factor analysis. *Psychometrika, 34,* 183–202.

Jöreskog, K. G. (1978). Structural analysis of covariance and correlation matrices. *Psychometrika, 43,* 443–477.

Jöreskog, K. G. & Sörbom, D. (1981). *LISREL V: Analysis of linear structural relationships by the method of maximum likelihood. Chicago, IL: National Educational Resources.*

Jöreskog, K. G. & Sörbom, D. (1986). *LISREL VI: Analysis of linear structural relationships by maximum likelihood and least squares methods. Mooresville, IN: Scientific Software.*

Juretig, F. (2013). *eigenprcomp: Computes confidence intervals for principal components.* R package version 1.0.

Kahneman, D. & Tversky, A. (2000). *Choices, values, and frames. New York: Cambridge University Press.*

Kaiser, H. F. (1958). The varimax criterion for analytic rotation in factor analysis. *Psychometrika, 23,* 187–200.

Kaiser, H. F. (1960). The application of electronic computers to factor analysis. *Educational and Psychological Measurement, 20,* 141–151.

Kaiser, H. F. (1961). A note on Guttman's lower bound for the number of common factors. *British Journal of Statistical Psychology, 14,* 1–2.

Kelly, K. T. (2007). Ockham's razor, empirical complexity, and truth-finding efficiency. *Theoretical Computer Science, 383,* 270–289.

Kempthorne, O. (1975). Fixed and mixed models in the analysis of variance. *Biometrics, 31,* 473–486.

Keppel, G. & Wickens, T. D. (2004). *Design and analysis: A researcher's handbook. New Jersey: Prentice Hall.*

Kieseppä, I. A. (2001). Statistical model selection criteria and bayesianism. *Proceedings of the Philosophy of Science Association, 3,* 141–152.

Kim, S. (2012). *ppcor: Partial and semi-partial (part) correlation.* R package version 1.0.

Kirk, R. E. (1995). *Experimental design: Procedures for the behavioral sciences. Pacific Grove, CA: Brooks/Cole Publishing Company.*

Kirk, R. E. (2008). *Statistics: An introduction. Belmont, CA: Thomson Wadsworth.*

Kline, M. (1977). *Calculus: An intuitive and physical approach. New York: Dover Publications.*

Korkmaz, S., Goksuluk, D., & Zararsiz, G. (2014). MVN: An R package for assessing multivariate normality. *The R Journal, 6,* 151–163.

Kruskal, J. B. & Landwehr, J. M. (1983). Icicle plots: Better displays for hierarchical clustering. *The American Statistician, 37,* 162–168.

Kruskal, J. B. & Wish, M. (1978). *Multidimensional scaling. New York: Sage.*

Kuhn, T. S. (2012). *The structure of scientific revolutions: 50th anniversary edition. Chicago, IL: University of Chicago Press.*

Kyburg, H. E. (2009). *Theory and measurement. New York: Cambridge University Press.*

Labarre, A. E. (1961). *Elementary mathematical analysis. London: Addison-Wesley Publishing Company.*

Laird, N. M. & Ware, J. H. (1982). Random-effects models for longitudinal data. *Biometrics, 38*, 963–974.

Lane, D. M. (2016). The assumption of sphericity in repeated-measures designs: What it means and what to do when it is violated. *The Quantitative Methods for Psychology, 12*, 114–122.

Lapane, K. L., Quilliam, B. J., Benson, C., Chow, W., & Kim, M. (2014). One, two, or three? Constructs of the brief pain inventory among patients with non-cancer pain in the outpatient setting. *Journal of Pain Symptom Management, 47*, 325–333.

Larsen, R. J. & Marx, M. L. (2001). *An introduction to mathematical statistics and its applications. New Jersey: Prentice Hall.*

Lattin, J., Carroll, J. D., & Green, P. E. (2003). *Analyzing multivariate data. San Francisco, CA: Brook/Cole, Cengage Learning.*

Lawley, D. N. (1938). A generalization of Fisher's z-test. *Biometrika, 30*, 180–187.

Lawley, D. N. (1958). Estimation in factor analysis under various initial assumptions. *British Journal of Statistical Psychology, 11*, 1–12.

Lawley, D. N. & Maxwell, A. E. (1971). *Factor analysis as a statistical method. London: Butterworths.*

Legendre, A. M. (1805). *Nouvelles méthodes pour la détermination des orbites des comètes. Paris: Courcier.*

Lei, P. & Koehly, L. M. (2003). Linear discriminant analysis versus logistic regression: A comparison of classification errors in the two-group case. *Journal of Experimental Education, 72*, 25–49.

Levin, J. R., Serlin, R. C., & Webne-Behrman, L. (1989). Analysis of variance through simple correlation. *American Statistician, 43*, 32–34.

Ligges, U. & Mächler, M. (2003). Scatterplot3d: An R package for visualizing multivariate data. *Journal of Statistical Software, 8*, 1–20.

Lindley, D. V. (2001). The philosophy of statistics. *Journal of the Royal Statistical Society: Series D (The Statistician), 49*, 293–337.

Lindsey, J. K. (1999). On the use of corrections for overdispersion. *Applied Statistics, 48*, 553–561.

Lix, L. M. & Sajobi, T. T. (2010). Discriminant analysis for repeated measures data: A review. *Frontiers in Quantitative Psychology and Measurement, 1*, 1–9,

Loftus, G. R. (1991). On the tyranny of hypothesis testing in the social sciences. *Contemporary Psychology, 36*, 102–104.

Lukacs, E. (1942). A characterization of the normal distribution. *The Annals of Mathematical Statistics, 13*, 91–93.

Lumley, T. (2020). *Using Fortran code by Alan Miller—Leaps: Regression subset selection.* R package version 2.9.

MacCallum, R. C., Browne, M. W., & Sugawara, H. M. (1996). Power analysis and determination of sample size for covariance structure modeling. *Psychological Methods, 1*, 130–149.

MacCallum, R. C., Widaman, K. F., Zhang, S., & Hong, S. (1999). Sample size in factor analysis. *Psychological Methods, 4*, 84–99.

MacKenzie, D. A. (1981). *Statistics in Britain, 1865–1930: The social construction of scientific knowledge. Edinburgh: Edinburgh University Press.*

MacKinnon, D. P. (2008). *Introduction to statistical mediation analysis. New York: Lawrence Erlbaum Associates.*

MacKinnon, D. P., Krull, J. L., & Lockwood, C. M. (2000). Equivalence of the mediation, confounding and suppression effect. *Prevention Science, 1*, 173–181.

MacKinnon, D. P., Lockwood, C. M., Hoffman, J. M., West, S. G., & Sheets, V. (2002). A comparison of methods to test mediation and other intervening variable effects. *Psychological Methods, 7*, 83–104.

MacQueen, J. (1967). Some methods for classification and analysis of multivariate observations. In: L. LeCam & J. Neyman, *Proceedings of the Fifth Berkeley symposium on mathematical statistics and probability, Vol. 1. Berkeley, CA: University of California Press, pp.* 281–297.

Mahalanobis, P. C. (1936). On the generalized distance in statistics. *Proceedings of the National Institute of Science, 12*, 49–55.

Marais, M. L. & Wecker, W. E. (1994). Correcting for omitted-variables and measurement-error bias in regression with an application to the effect of lead on IQ. *Journal of the American Statistical Association*, *93*, 494–505.

Mardia, K. V. (1970). Measures of multivariate skewness and kurtosis with applications. *Biometrika*, *57*, 519–530.

Mauchly, J. W. (1940). Significance test for sphericity of a normal *n*-variate distribution. *The Annals of Mathematical Statistics*, *11*, 204–209.

Maxwell, S. E. & Delaney, H. D. (2004). *Designing experiments and analyzing data. New Jersey: Lawrence Erlbaum Associates*.

McCullagh, P. & Nelder, J. A. (1989). *Generalized linear models. New York: Chapman & Hall*.

McDonald, R. P. (1999). *Test theory: A unified treatment. New York: Routledge*.

McLeod, A. I. & Xu, C. (2020). *bestglm: Best Subset GLM*. R package version 0.37.3.

Mead, R. (1988). *The design of experiments: Statistical principles for practical application. New York: Cambridge University Press*.

Mecklin, C. J. & Mundfrom, D. J. (2004). An appraisal and bibliography of tests for multivariate normality. *International Statistical Review*, *72*, 123–138.

Meehl, P. (1967). Theory-testing in psychology and physics: A methodological paradox. *Philosophy of Science*, *34*, 103–115.

Meehl, P. (1978). Theoretical risks and tabular asterisks: Sir Karl, Sir Ronald, and the slow progress of soft psychology. *Journal of Consulting and Clinical Psychology*, *46*, 806–834.

Meerschaert, M. M. (2007). *Mathematical modeling. New York: Elsevier*.

Meyer, D., Zeileis, A., & Hornik, K. (2014). *vcd: Visualizing categorical data*. R package version 1.3–2.

Miller, G. A. & Chapman, J. P. (2001). Misunderstanding analysis of covariance. *Journal of Abnormal Psychology*, *110*, 40–48.

Miller, M. B. (1995). Coefficient alpha: A basic introduction from the perspectives of classical test theory and structural equation modeling. *Structural Equation Modeling*, *2*, 255–273.

Miller, R. G. (1981). *Simultaneous statistical inference. New York: Springer*.

Mirisola, A. & Seta, L. (2013). *pequod: Moderated regression package*. R package version 0.0–3.

Montgomery, D. C. (2005). *Design and analysis of experiments. New York: John Wiley & Sons*.

Moore, D., McCabe, G. P., & Craig, B. A. (2014). *Introduction to the practice of statistics. New York: W. H. Freeman*.

Morabia, A. (2005). Epidemiological causality. *History and Philosophy of the Life Sciences*, *27*, 365–379.

Mukhopadhyay, P. (2008). *Multivariate statistical analysis. Singapore: World Scientific Publishing*.

Mulaik, S. A. (1972). *The foundations of factor analysis. New York: McGraw-Hill*.

Mulaik, S. A. (1987). A brief history of the philosophical foundations of exploratory factor analysis. *Multivariate Behavioral Research*, *22*, 267–305.

Mulaik, S. A. (2009). *Linear causal modeling with structural equations. New York: CRC Press*.

Nelder, J. & Wedderburn, R. (1972). Generalized linear models. *Journal of the Royal Statistical Society, Series A*, *135*, 370–384.

Neter, J., Kutner, M. H., Nachtsheim, C. J., & Wasserman, W. (1996). *Applied linear statistical models. New York. McGraw-Hill*.

Neyman, J. & Pearson, E. S. (1928). On the use and interpretation of certain test criteria for purposes of statistical inference: Part I. *Biometrika*, *20A*, 263–294.

Oakes, M. (1986). *Statistical inference: A commentary for the social and behavioral sciences. Chichester: John Wiley & Sons, Inc*.

Olejnik, S. & Algina, J. (2003). Generalized eta and omega squared statistics: Measures of effect size for some common research designs. *Psychological Methods*, *8*, 434–447.

Olkin, I. & Pratt, J. W. (1958). Unbiased estimation of certain correlation coefficients. *Annals of Mathematical Statistics*, *29*, 201–211.

Panik, M. J. (2005). *Advanced statistics from an elementary point of view. New York: Elsevier Academic Press*.

Pearson, K. (1896). Contributions to the mathematical theory of evolution—III, regression, heredity and panmixia. *Philosophical Transactions of the Royal Society of London, 187,* 253–318.

Pearson, K. (1901). On lines and planes of closest fit to systems of points in space. *Philosophical Magazine, 2,* 559–572.

Pearson, K. (1920). Notes on the history of correlation. *Biometrika, 13,* 25–45.

Pedhazur, E. J. (1997). *Multiple regression in behavioral research: Explanation and prediction. Belmont, CA: Wadsworth Publishing.*

Pillai, K. C. S. (1955). Some new test criteria in multivariate analysis. *Annals of Mathematical Statistics, 26,* 117–121.

Pillai, K. C. S. (1956). On the distribution of the largest or the smallest root of a matrix in multivariate analysis. *Biometrika, 43,* 122–127.

Pinheiro, J., Bates, D., DebRoy, S., Sarkar, D., & R Core Team (2014). *nlme: Linear and nonlinear mixed effects models.* R package version 3.1–119.

Pinheiro, J. C. & Bates, D. M. (2000). *Mixed-effects models in S and S-Plus. New York: Springer.*

Pohlert, T. (2014). *The pairwise multiple comparison of mean ranks package (PMCMR).* R package.

Poisson, S. D. (1837). *Recherches sur la probabilité des jugements en matière criminelle et en matière civile, précédés des règles générales du calcul des probabilités. Paris: Bachelier.*

Potthoff, R. F. & Roy, S. N. (1964). A generalized multivariate analysis of variance model useful especially for growth curve problems. *Biometrika, 51,* 313–326.

Preacher, K. J. & Hayes, A. F. (2004). SPSS and SAS procedures for estimating indirect effects in simple mediation models. *Behavior Research Methods, Instruments, & Computers, 36,* 717–731.

Preacher, K. J. & MacCallum, R. C. (2003). Repairing tom swift's electric factor analysis machine. *Understanding Statistics, 2,* 13–43.

Press, S. J. & Wilson, S. (1978). Choosing between logistic regression and discriminant analysis. *Journal of the American Statistical Association, 73,* 699–705.

Proschan, M. A. (2008). The normal approximation to the binomial. *The American Statistician, 62,* 62–63.

Qiu, W. (2013). *powerMediation: Power/sample size calculation for mediation analysis, simple linear regression, logistic regression, or longitudinal study.* R package version 0.1.7.

Raiche, G. (2010). *nFactors: An R package for parallel analysis and non graphical solutions to the Cattell scree test.* R package version 2.3.3.

Raudenbush, S. W. & Bryk, A. S. (2002). *Hierarchical linear models: Applications and data analysis methods. London: Sage Publications.*

Raveh, A. (1985). On the use of the inverse of the correlation matrix in multivariate data analysis. *The American Statistician, 39,* 39–42.

Raykov, T. & Marcoulides, G. A. (2011). *Introduction to psychometric theory. New York: Routledge.*

Rencher, A. C. (1998). *Multivariate statistical inference and applications. New York: John Wiley & Sons, Inc.*

Rencher, A. C. & Christensen, W. F. (2012). *Methods of multivariate analysis. New York: John Wiley & Sons, Inc.*

Revelle, W. (2015) *psych: Procedures for personality and psychological research, Northwestern University, Evanston, IL.*

Rice, J. A. (2006). *Mathematical statistics and data analysis. New York: Cengage Learning.*

Roman, S. (2008). *Advanced linear algebra. New York: Springer.*

Romeu, J. L. & Ozturk, A. (1993). A comparative study of goodness-of-fit tests for multivariate normality. *Journal of Multivariate Analysis, 46,* 309–334.

Rosario-Martinez, H. D. (2013). *phia: Post-hoc interaction analysis.* R package version 0.1–5.

Rosenthal, R., Rosnow, R. L., & Rubin, D. B. (2000). *Contrasts and effect sizes in behavioral research: A correlational approach. Edinburgh: Cambridge University Press.*

Rosseel, Y. (2012). lavaan: An R package for structural equation modeling. *Journal of Statistical Software, 48,* 1–36.

Rouanet, H. & Lépine, D. (1970). Comparison between treatments in a repeated-measurement design: ANOVA and multivariate methods. *British Journal of Mathematical and Statistical Psychology, 23*, 147–163.

Rozeboom, W. W. (1960). The fallacy of the null hypothesis significance test. *Psychological Bulletin, 57*, 416–428.

Russell, B. (1903). *The principles of mathematics. Oregon: Merchant Books.*

Rutherford, A. (2009). *Beyond the box: B.F. Skinner's technology of behaviour from laboratory to life, 1950s–1970s. Toronto: University of Toronto Press.*

Sakamoto, Y., Ishiguro, M., & Kitagawa, G. (1986). *Akaike information criterion statistics. Dordrecht. Reidel Academic Publishing Company.*

Salas, S. L., Hille, E., & Etgen, G. J. (1999). *Calculus: One and several variables. New York: John Wiley & Sons, Inc.*

Salk, R. H., Hyde, J. S., & Abramson, L. Y. (2017). Gender differences in depression in representative national samples: Meta-analyses of diagnoses and symptoms. *Psychological Bulletin, 143*, 783–822.

Salsburg, D. (2002). *The lady tasting tea. How statistics revolutionized science in the twentieth century. New York: Holt Paperbacks.*

Savage, L. J. (1972). *The foundations of statistics. New York: Dover Publications.*

Schäfer, J., Opgen-Rhein, R., Zuber, V. Ahdesmäki, M. A. Silva, P. D., & Strimmer, K. (2014). *corpcor: Efficient estimation of covariance and (partial) correlation.* R package version 1.6.7.

Scheffé, H. (1999). *The analysis of variance. New York: John Wiley & Sons, Inc.*

Schmitt, N. (1996). Uses and abuses of coefficient alpha. *Psychological Assessment, 8(4)*, 350–353.

Schumacker, R. E. & Lomax, R. G. (2010). *A beginner's guide to structural equation modeling. New Jersey: Lawrence Erlbaum Associates.*

Schwarz, G. E. (1978). Estimating the dimension of a model. *Annals of Statistics, 6*, 461–464.

Searle, S. R. (1982). *Matrix algebra useful for statistics. New York: John Wiley & Sons, Inc.*

Searle, S. R., Casella, G., & McCulloch, C. E. (1992). *Variance components. New York: John Wiley & Sons, Inc.*

Selig, J. P. & Preacher, K. J. (2008). *Monte Carlo method for assessing mediation: An interactive tool for creating confidence intervals for indirect effects [computer software].* Retrieved from http://quantpsy.org/

Shao, J. (2003). *Mathematical statistics. New York. Springer.*

Shelby, L. B. & Vaske, J. J. (2008). Understanding meta-analysis: A review of the methodological literature. *Leisure Sciences, 30*, 96–110.

Shipley, B. (2002). *Cause and correlation in biology: A user's guide to path analysis, structural equations and causal inference. New York: Cambridge University Press.*

Shoesmith, E. (1987). The continental controversy over Arbuthnot's argument for divine providence. *Historia Mathematica, 14*, 133–146.

Shrout, P. E. (1997). Should significance tests be banned? Introduction to a special section exploring the pros and cons. *Psychological Science, 8*, 1–2.

Siegel, S. & Castellan, N. J. (1988). *Nonparametric statistics for the behavioral sciences. New York: McGraw-Hill.*

Simeckova, M., Rusch, T., & Simecek, P. (2014). *additivityTests: Additivity tests in the two way Anova with single sub-class numbers.* R package version 1.1–4.

Singer, J. D. & Willett, J. B. (2003). *Applied longitudinal data analysis. New York: Oxford University Press.*

Smith, R. H. (1980). A geometric comparison of principal component analysis with analysis of variance. *Journal of the Royal Statistical Society, Series D, 29*, 91–96.

Snedecor, G. W. (1934). *Calculation and interpretation of analysis of variance and covariance. Ames, IA: Collegiate Press.*

Snedecor, G. W. & Cochran, W. G. (1967). *Statistical methods. Ames, IA: Iowa State University Press.*

Snijders, T. A. & Bosker, R. (1999). *Multilevel analysis: An introduction to basic and advanced multilevel modeling. London: Sage Publications.*

Sobel, M. E. (1982). Asymptotic intervals for indirect effects in structural equations models. In: S. Leinhart (Ed.), *Sociological methodology 1982. San Francisco, CA: Jossey-Bass, pp. 290–312.*

Spearman, C. (1904a). "General intelligence," objectively determined and measured. *The American Journal of Psychology, 15,* 201–293.

Spearman, C. (1904b). The proof and measurement of association between two things. *The American Journal of Psychology, 15,* 72–101.

Steiger, J. H. & Lind, J. C. (1980). *Statistically-based tests for the number of common factors.* Paper presented at the Annual Meeting of the Psychometric Society, Iowa City, IA.

Steinhorst, R. K. (1982). Resolving current controversies in analysis of variance. *The American Statistician, 36,* 138–139.

Stevens, J. P. (2009). *Applied multivariate statistics for the social sciences. New Jersey: Erlbaum.*

Stevens, S. S. (1946). On the theory of scales of measurement. *Science, 103,* 677–680.

Stewart, I. (1995). *Concepts of modern mathematics. New York: Dover Publications.*

Stigler, S. M. (1986). *The history of statistics: The measurement of uncertainty before 1900. London: Belknap Press.*

Strang, G. (1993). *Introduction to linear algebra. Wellesley, MA: Cambridge Press.*

Tabachnick, B. G. & Fidell, L. S. (2007). *Using multivariate statistics. New York: Pearson Education.*

Tang, W., Cui, Y., & Babenko, O. (2014). Internal consistency: Do we really know what it is and how to assess it? *Journal of Psychology and Behavioral Science, 2(2),* 205–220.

Tatsuoka, M. M. (1971). *Multivariate analysis: Techniques for educational and psychological research. New York: John Wiley & Sons, Inc.*

Teetor, P. (2011). *R cookbook. Sebastopol, CA: O'Reilly Media.*

Thouless, R. H. (1935). The tendency to certainty in religious beliefs. *British Journal of Psychology, 26,* 16–31.

Thurstone, L. L. (1947). *Multiple factor analysis. Chicago, IL: University of Chicago Press.*

Timm, N. H. (2002). *Applied multivariate analysis. New York: Springer.*

Tingley, D., Yamamoto, T., Hirose, K., Keele, L., & Imai, K. (2014). mediation: R package for causal mediation analysis. *Journal of Statistical Software, 59,* 1–38.

Tofighi, D. & MacKinnon, D. P. (2011). RMediation: An R package for mediation analysis confidence intervals. *Behavior Research Methods, 43,* 692–700.

Trefethen, L. N. & Bau, D., III (1997). *Numerical linear algebra. Philadelphia, PA: Society for Industrial and Applied Mathematics.*

Tucker, L. R. & Lewis, C. (1973). A reliability coefficient for maximum likelihood factor analysis. *Psychometrika, 38,* 1–10.

Tucker, W. (2009). *The Cattell controversy: race, science, and ideology. Chicago, IL: University of Illinois Press.*

Tukey, J. W. (1949). One degree of freedom for nonadditivity. *Biometrics, 5,* 232–242.

Tukey, J. W. (1977). *Exploratory data analysis. New York: Pearson.*

Upton & Cook (2002). *Oxford dictionary of statistics. New York: Oxford University Press.*

Vaughan, D. (1996). *The challenger launch decision: Risky technology, culture, and deviance at NASA. Chicago, IL: University of Chicago Press.*

Vaughan, G. M. & Corballis, M. C. (1969). Beyond tests of significance: Estimating strength of effects in selected ANOVA designs. *Psychological Bulletin, 72,* 204–213.

Venables, W. N. & Ripley, B. D. (2002). *Modern applied statistics with S. New York: Springer.*

Wackerly, D. D., Mendenhall, W., III, & Scheaffer, R. L. (2002). *Mathematical statistics with applications. Pacific Grove, CA: Duxbury Press.*

Walker, H. M. (1929). *Studies in the history of statistical method. Baltimore, MD: Williams & Wilkins.*

Wall, M. M. & Amemiya, Y. (2000). Estimation for polynomial structural equation models. *Journal of the American Statistical Association, 95,* 929–940.

Ward, J. H. (1963). Hierarchical grouping to optimize an objective function. *Journal of the American Statistical Association, 58*, 236–244.

Warner, R. M. (2013). *Applied statistics: From bivariate through multivariate techniques. London: Sage Publications.*

Welch, B. L. (1947). The generalization of Student's problem when several difference population variances are involved. *Biometrika, 34*, 28–35.

Welch, B. L. (1951). On the comparison of several mean values: An alternative approach. *Biometrika, 38*, 330–336.

Wickham, H. (2009). *ggplot2: Elegant graphics for data analysis. New York: Springer.*

Wilks, S. S. (1946). Sample criteria for testing equality of means, equality of variances and equality of covariances in a normal multivariate distribution. *Annals of Mathematical Statistics, 17*, 257–281.

Wilson, W., Miller, H. L., & Lower, J. S. (1967). Much ado about the null hypothesis. *Psychological Bulletin, 67*, 188–196.

Winer, B. J., Brown, D. R., & Michels, K. M. (1991). *Statistical principles in experimental design. New York: McGraw-Hill.*

Winkler, R. L. (2003). *An introduction to Bayesian inference and decision. Gainesville, FL: Probabilistic Publishing.*

Winter, J. C. F. & Dodou, D. (2012). Factor recovery by principal axis factoring and maximum likelihood factor analysis as a function of factor pattern and sample size. *Journal of Applied Statistics, 39*, 695–710.

Wishart, J. (1934). Statistics in agricultural research. *Journal of the Royal Statistical Society*, Supplement I, *1*: 26–61.

Wolf, H. P. & Bielefeld, U. (2014). *aplpack: Another plot PACKage: stem.leaf, bagplot, faces, spin3R, plotsummary, plothulls, and some slider functions.* R package version 1.3.0.

Woodworth, R. S. (1928). Dynamic psychology. In: C. Murchison (Ed.), *Psychologies of 1925. Worcester, MA: Clark University Press, pp. 111–126.*

Wright, D. B. & London, K. (2009). *Modern regression techniques using R: A practical approach. Beverly Hills, CA: Sage.*

Wright, S. (1920). The relative importance of heredity and environment in determining the piebald pattern of guinea-pigs. *Proceedings of the National Academy of Sciences of the United States of America, 6*, 320–332.

Wright, S. (1934). The method of path coefficients. *Annals of Mathematical Statistics, 5*, 161–215.

Wu, M., Yu, K., & Liu, A. (2009). Estimation of variance components in the mixed-effects models: A comparison between analysis of variance and spectral decomposition. *Journal of Statistical Planning and Inference, 139*, 3962–3973.

Yalcin, I. & Amemiya, Y. (2001). Nonlinear factor analysis as a statistical method. *Statistical Science, 16*, 275–294.

Yeomans, K. A. & Golder, P. A. (1982). The Guttman–Kaiser criterion as a predictor of the number of common factors. *The Statistician, 31*, 221–229.

Yerkes, R. M. & Dodson, J. D. (1908). The relation of strength of stimulus to rapidity of habit formation. *Journal of Comparative Neurology and Psychology, 18*, 459–482.

Yule, G. U. (1896). On the correlation of total pauperism with proportion of out-relief II. *The Economic Journal, 6*, 613–623.

Yule, G. U. (1897). On the theory of correlation. *Journal of the Royal Statistical Society, 60*, 812–854.

Yule, G. U. (1899). An investigation into the causes of changes in pauperism in England, chiefly during the last two intercensal decades (Part I). *Journal of the Royal Statistical Society, 62*, 249–295.

Zabell, S. L. (2008). On student's 1908 article "the probable error of a mean" *Journal of the American Statistical Association, 103*, 1–7.

Zeisel, H. & Kaye, D. (1997). *Prove it with figures: Empirical methods in law and litigation. New York: Springer.*

INDEX

Applied Univariate, Bivariate, and Multivariate Statistics: Understanding Statistics for Social and Natural Scientists, With Applications in SPSS and R, Second Edition. Daniel J. Denis.
© 2021 John Wiley & Sons, Inc. Published 2021 by John Wiley & Sons, Inc.
Companion Website: www.wiley.com/go/denis/appliedstatistics2e

Printed and bound by CPI Group (UK) Ltd, Croydon, CR0 4YY

17/04/2025

14658863-0001